STATISTICAL ANALYSIS

STATISTICAL ANALYSIS

APPLICATIONS TO BUSINESS AND ECONOMICS

CLARK A. HAWKINS
JEAN E. WEBER

University of Arizona

1817

HARPER & ROW, PUBLISHERS, New York

Cambridge, Hagerstown, Philadelphia, San Francisco,
London, Mexico City, São Paulo, Sydney

Sponsoring Editor: Bonnie K. Binkert
Project Editor: Robert Ginsberg
Designer: Michel Craig
Senior Production Manager: Kewal K. Sharma
Compositor: Syntax International Pte. Ltd.
Printer and Binder: Halliday Lithograph Incorporated
Art Studio: Vantage Art, Inc.

STATISTICAL ANALYSIS: Applications to Business and Economics

Library of Congress Cataloging in Publication Data

Hawkins, Clark A
  Statistical analysis.

  Includes bibliographical references and index.
  1. Mathematical statistics.   I. Weber, Jean E.,
joint author.   II. Title.
QA276.H393   519.5   79-20368
ISBN 0-06-042723-X

CONTENTS

PREFACE

The authors consider the material in this book to be appropriate for a two-semester course in inferential statistics taught to advanced undergraduate or beginning graduate students in business, economics, or the social sciences. The text is intended for students who have had no previous course in statistics; a chapter on descriptive statistics provides a basis for the subsequent chapters on inferential statistics. The mathematical prerequisite is a reasonable facility with algebraic manipulations. A few derivations depending on calculus are included in the technical appendices at the end of some chapters, but no material in the body of the text requires calculus. The fact that this book contains more material than most instructors will wish to cover in one or even two semesters should permit some choice of topics depending on individual instructor preference.

Chapter 2 discusses the basic concepts of descriptive statistics, while Chapter 3 discusses time series and index numbers. If students have a background in descriptive statistics, Chapter 2 can be omitted. The index number part of Chapter 3 is descriptive and rather complete, but it is of primary concern only to students of economics. Therefore, this section need not be examined in detail if the instructor wishes only to introduce the subject, or it may be omitted altogether. Time series is deliberately placed in Chapter 3 rather than toward the end of the book, primarily as an explanation of the decomposition of time series. Although this section gives a naive (but frequently useful) forecasting model, the more interesting analytical discussion of time series is contained in Chapters 13 and 14 on regression. Some instructors may decide to omit Chapter 3, and possibly refer to it only for seasonal adjustment in conjunction with regression.

Chapters 4, 5, and 6 concern probability, on which the following chapters heavily depend. Chapter 4 develops the fundamental concepts of probability; this discussion, largely intuitive, is presented without rigorous proof. The appendices of Chapter 4 on factorials and permutations and combinations are given for the benefit of students in need of review.

Chapter 5 is primarily a discussion of some well-known (and some not so well-known) probability distributions that have been found to have useful applications in business and economic analysis. Although few mathematical derivations are given, the discussions of families of distributions are reasonably rigorous. Some instructors, especially in a one-semester course, may wish to omit Chapter 5 or go over it lightly.

Some probability distributions that occur primarily as a result of sampling are known as sampling distributions. Chapter 6 discusses the most common and useful ones. A few of the distributions in Chapters 5 and 6 are somewhat esoteric. They are included so that the student will recognize that there are many types of probability distributions, some of which seem to pop up in strange places by way of application—for example, the Pareto-Levy distribution for stock market prices. The student should also learn that not everything in the world follows a normal distribution, a seemingly ubiquitous misconception at times.

Chapter 7 discusses the usual point and interval estimates for means and proportions, as well as estimates of variance and the determination of sample size. This chapter should not be omitted.

Chapters 8 and 9 are rather detailed in their coverage of hypothesis testing, one of the most important topics of statistical inference. Chapter 8 introduces the null and alternative hypotheses, α and β errors, power of a test, and the impact of sample size on the errors. Chapter 9 discusses various applications of hypothesis testing, which include hypotheses concerning means, differences between means, paired differences, and variances. Hypotheses concerning proportions are taken up in Chapter 10, which also covers contingency tables and goodness of fit, both applications of the chi-square distribution. In a one semester course, Chapter 9 may be omitted, but Chapters 8 and 10 should not be.

Chapter 11 covers some of the more elementary problems in the construction of samples. Chapter 12 discusses analysis of variance in more detail than usual, but not at an advanced mathematical level. Regression (Chapter 13) is preceded by analysis of variance because it is much easier to discuss sums-of-squares for multivariate regression after this subject has been introduced in analysis of variance. Chapter 11 could be omitted, but the first five sections of Chapter 12 are needed for Chapters 13 and 14.

Regression is the subject of Chapters 13 and 14. Chapter 13 concerns bivariate regression and correlation, and especially time series and trend are discussed, since so many regression analyses in economics and business involve the use of time series data. If Chapter 3 was omitted earlier, its first section, which discusses trend and seasonal adjustment, can be covered in conjunction with Chapter 13. In a one-semester course, all or part of Chapter 14 may be omitted. It includes enough material on multivariate linear regression to give the student some idea of the problems and advantages of the approach without discussing all the problems of econometrics. Some instructors may prefer to omit discussions of nonlinear data in Chapters 13 and 14.

Chapters 15 and 16 introduce decision theory and especially the role of Bayesian statistics in this important area. Chapter 15 can be covered lightly or omitted, but we feel the student should be exposed to the Bayesian discussion of Chapter 16.

Chapter 17 concerns nonparametric and distribution-free methods;

some instructors may wish to omit this material. There are two reasons for the discussion of nonparametric statistics. First, the rather obvious one that methods of nonparametric statistics are very useful. Second, students taught to grasp the notion of nonparametric statistics will be better able to focus their understanding of the rest of the book.

To recapitulate, a one-semester course might ideally take up Chapters 2, 4, 6, 7, 8, 10, 12, 13, 16. This would require a somewhat speedier and less in-depth treatment of some of the chapters than would be the case in a two-semester course. A one-semester course should also go briefly into Chapter 5, but could treat Chapters 6 and 12 more lightly than the others. A two-semester sequence would most likely cover the first eight or nine chapters in the first course, and the balance of the book in the second. Some instructors might place more emphasis on Chapter 12 than on 14 or vice versa, or on Chapter 15 than on 17 or vice versa. If some of the later sections of several chapters, for example, Chapters 12, 14, and 17, are discussed only briefly, it should be possible to cover the entire book in two semesters.

Numerous examples, tables, and figures are used throughout this text to illustrate and extend the various distributions, tests, and theories discussed. In trying to cover as many topics as this text attempts, it is difficult to smoothly motivate all initial discussions of new material. However, in order to show the importance of the material and how it might be utilized, we have used examples illustrating concepts and methods from accounting, economics, finance, marketing, production, and real estate. There are also a few examples taken from other areas, such as psychology and medicine. In addition, there are extensive problem sets for each chapter. A solutions manual, providing complete solutions to all problems, is available to instructors.

Although the presentation and examples in this book are original, the theoretical concepts have been developed by many people over a considerable period of time. The authors acknowledge their debt to those individuals—in particular, Hoel and Mood, Cochran and Scheffé, Goldberger and Johnston, and Bradley and Fraser—whose books have influenced our chapters concerning probability and tests of hypotheses, sampling and analysis of variances, regression, and nonparametric inference, respectively. In addition, the authors acknowledge the help received from other references in Appendix A and from their teachers and colleagues. In particular, they wish to thank those who reviewed the manuscript at various stages of its development; they are: C. Darwin Kirksey, Lamar University; Cecil H. Meyers, University of Minnesota, Duluth; James Murtha, Marietta College; Kay Norris, Bellvue Community College; Barry Pasternack, California State University, Fullerton; Tom Ressler, University of Minnesota; and James Vedder, Syracuse University.

Thanks are also due Roberta Hagaman, who methodically checked and rechecked the problem material to insure accuracy. Of course, the authors accept full responsibility for any errors or omissions.

CLARK A. HAWKINS
JEAN E. WEBER

INTRODUCTION

1.1 STATISTICS AND STATISTICAL INFERENCE

Statistics is the art and science of collecting, analyzing, and interpreting data. Statistical methods are of two types, *descriptive* and *inferential*. The purpose of descriptive statistics is to summarize and characterize data by means of relevant measures; the measures are then used for describing the particular data collected. On the other hand, the purpose of inferential statistics is to generalize concerning the characteristics of a larger collection of data on the basis of observations taken on a smaller part of the data. Any scientific investigation, since it seeks general principles, requires inferential statistics.

To clarify the distinction between the types of statistical methods, suppose a stockbroker has a record of the price movements on a particular day of 20 selected stocks on the New York Stock Exchange. For these data he can compute the average price movement of the 20 stocks, the scatter of the individual stock movements about this average, and various other measures. If the intention is to study the market behavior of those particular stocks, the broker considers the measures as descriptive of that behavior on the day involved. The measures summarize or characterize the data, only the descriptive method is used.

However, the broker may be concerned, not only with the price movements of the particular 20 stocks for which there is data, but with the price movements of all the stocks on the New York Stock Exchange. That is, the broker may wish to make generalizations or inferences about all the stocks on the exchange from the data collected for the 20 selected stocks; if so, he must use the inferential method.

It should be noted, as illustrated in the preceding example and discussion, that the distinction between descriptive and inferential statistics is not with respect to the data collected nor the computations made, but with respect to the use of the data. The same data, and the measures computed from them, may be both descriptive of a small group and the basis for inferences about a larger group. The phenomenal increase in the development and use of statistical methods in the past few decades has been characterized by a pronounced shift in emphasis from descriptive statistics to statistical inference. The stereotype of the statistician as a frowning, spectacled person nearly buried under charts and tables is finally fading as the use of statistical inference becomes increasingly common in economics, production, marketing, business research, advertising, politics, medical research, and other areas.

Statistical inference is concerned with making generalizations about a larger group (the population) from observations of a smaller group (the sample). The population is the total group in which the investigator is interested; it may be either finite or infinite. A *finite* population is one whose members can, with sufficient effort, be listed and counted (for example, all the stocks on the New York Stock Exchange). An *infinite* population is one whose members cannot conceivably be counted (for example, all the persons who have lived, are alive, or will ever be alive). Most statistical theory is based on infinite populations; however, it can be used for inferences concerning large finite populations (for instance, the stock market example above) and, with certain modifications, for inferences concerning relatively small populations.

Two general types of problems with which statistical inference is concerned are: estimating population parameters and testing hypotheses about these parameters. A *parameter* is a characteristic or measure of a population; it distinguishes one population from another population. A *statistic* is a characteristic or measure of a sample; it varies from sample to sample of any given population. Generally, inferences involving either estimation or testing of hypotheses are based on a sample characteristic (statistic) which is an estimate of the corresponding population characteristic (parameter).

In the stock market example, the broker might want to estimate the average price movement of all stocks on the New York Exchange or the proportion of stocks that went up or some other characteristic of stocks on the exchange. Note that the parameter being estimated is characteristic of the population of all stocks on the exchange, although it is estimated by a statistic computed from data for a sample of 20 stocks. For somewhat different purposes the broker might be interested in testing the hypothesis that the average price movement of all stocks on the exchange was zero or that the proportion of stocks on the exchange that went up was equal to the proportion that went down or some other hypothesis. Again, note that the hypothesis concerns parameters of the population of all stocks on the exchange, although it is tested using a statistic computed from data for a sample of 20 stocks.

Making generalizations about all stocks on the New York Stock Exchange on the basis of data for a sample of 20 stocks is clearly a risky business and may result occasionally in wrong conclusions. The likelihood that generalizations from a sample to a population are correct depends essentially on two factors: (1) the method of choosing the sample (over which the investigator has some control) and (2) the distribution of the characteristic under study in the population (over which the investigator has no control and of which he may have little knowledge).

Statistical inference can appropriately deal only with random samples. A *random sample* is one whose elements are chosen so that every member of the population has a known chance of being in the sample. Procedures for obtaining random samples are discussed in Chapter 11. For the present discussion, it is sufficient to recognize that a random sample is chosen so that its elements, on the average, are "representative" of the population from which it is drawn.

There is no guarantee that any particular sample, even though randomly drawn, will be "representative" of the population being considered. How well a given random sample will accurately picture the population from which it is drawn depends on how nature chooses to distribute the characteristic under study. Clearly, if the members of a population are homogeneous with respect to the characteristic being studied, a random sample (even a rather small one) may provide a close estimate of the population parameter; on the other hand, a random sample from a heterogenous population might well provide an estimate of the population parameter that would be considerably in error. Investigators can never be certain that a conclusion or generalization arrived at by statistical inference is correct; they can be certain only that it is correct with a known probability under certain circumstances. Thus statistical inference is based on probability theory, and a clear understanding of the fundamentals of probability theory is necessary for any real appreciation of the logic of statistical inference. After a survey of descriptive statistics in Chapters 2 and 3 and an introduction to probability beginning with Chapter 4, the subsequent chapters are primarily concerned with methods of statistical inference and their application in business and economics.

1.2 TO THE STUDENT

Many students view statistics as a required subject in which to learn a few definitions and procedures, as much as possible by memorization—and, having passed the course, they immediately forget everything they learned. Presumably this occurs because they do not appreciate the fact that statistics can be an important and useful tool in business and economic applications and because, after taking one or two statistics courses, they do not use the material in other courses.

However, the situation where statistics stands more or less by itself and is only viewed as a hurdle is rapidly changing. Instructors in the functional areas of business and economics are increasingly incorporating statistical analyses into the teaching of these courses. Not only does the student then appreciate the importance of statistics, but the business and economic subject matter is enriched and can be explored in greater depth. While some of the topics may appear sterile, they are all useful and none is included arbitrarily. Understanding them permits you to understand better the world around you. Being able to apply them puts you in a position to analyze and solve many practical problems. Although this book attempts to emphasize the business and economic applications of the statistical methods discussed, an exhaustive treatment of all applications cannot be made while covering so many topics. Hopefully, students will be encouraged to find additional applications in their own areas of special interest.

DESCRIPTIVE STATISTICS

2.1 INTRODUCTION

The purpose of descriptive statistics is to summarize data in a useful manner. The criterion of usefulness here is taken to be in the requirements of the investigator who sets out to collect and examine the data. If the investigator is interested in certain characteristics of the data collected and if the criterion of usefulness is met, only the analysis of the data by description is required. In many cases, however, summarizing data by one or more descriptive means is the first step in making inferences or generalizations about a larger set of data. In any case it is necessary to have a knowledge of both the terminology and various descriptive methods to be able to understand and analyze statistics.

There are a considerable number of ways of describing a given set of data, this being particularly true of the large and complex sets of data now routinely collected and put into computers. For descriptive analysis the only criterion is that the description be appropriate and useful for the purposes of the investigator. For inference it is also necessary that the particular descriptive measurement have convenient statistical properties (discussed in Chapter 5). It should be noted that many useful descriptive measurements are not used for inference. Descriptive statistics is concerned primarily with four types of summarization: frequency distributions, measures of location or average, measures of dispersion or spread, and measures of symmetry and peakedness. These will be taken up in turn after a brief discussion of summation and product notation.

2.2 SUMMATION AND PRODUCT NOTATION

The Greek capital letter $\sum$, sigma, indicates summation. If a series of numbers are to be added,

$$x_1 + x_2 + x_3 + \cdots + x_n$$

this sum can be written in sigma notation as

$$\sum_{i=1}^{n} x_i = x_1 + x_2 + x_3 + \cdots + x_n$$

The summation expression, $\sum_{i=1}^{n} x_i$, is read "the sum of x sub i where i goes from 1 to n." The values below and above the summation sign, 1 and n, represent the first and last elements being summed and are referred to as the limits of the summation. This does not necessarily mean that

$$x_1 < x_2 < x_3 < \cdots < x_n$$

The subscripts are merely associative numbers identifying individual x's, but they do not indicate relative size of the x's. The only implicit requirement is that the elements must be measured in the same units. For example, if x_1 is the weight of a student in pounds and x_2 is the weight of another student in kilograms, their sum does not have any meaning.

EXAMPLES
1. Sum the numbers 4, 6, 8, 10

$$\sum_{i=1}^{4} x_i = 4 + 6 + 8 + 10 = 28$$

2. A typical rowing team contains 9 members, including the coxwain. If one team's weights are as follows, add them, using summation notation:

$$\sum_{i=1}^{9} x_i = 187 + 174 + 168 + 182 + 189 + 186 + 181 + 179 + 120 = 1566$$

In general, the symbol sigma indicates the sum of the elements represented by the general term over the range given. The general term may be a simple or a relatively complex expression. A few varieties are as follows:

$$\sum_{i=1}^{n} x_i^2 = 1^2 + 2^2 + 3^2 + \cdots + n^2; \qquad \text{where } x_1 = 1, x_2 = 2, \ldots, x_n = n$$

$$\sum_{i=3}^{n} (2x_i - 4) = 2 + 4 + 6 + \cdots + (2n - 4); \qquad \text{where } x_3 = 3, x_4 = 4, \ldots, x_n = n$$

$$\sum_{i=10,20,\ldots}^{100} x_i = \sum_{i=1}^{10} x_{10i} = x_{10} + x_{20} + x_{30} + \cdots + x_{100}$$

The following useful rules for summation are given without proof.

1. If the term after the summation sign is a constant, all elements represented by the general term are constant and equal. The sum of n equal values is equal to n times the constant value. That is,

$$\sum_{1}^{n} 1 = 1 + 1 + 1 + \cdots + 1 = n \qquad \text{where } n \text{ is the number of 1's}$$

$$\sum_{1}^{n} c = c + c + c + \cdots + c = cn \qquad \text{where } c \text{ is a constant}$$

2(a). The sum of a constant times a variable is equal to the constant times the sum of the variable. That is,

$$\sum_{i=1}^{n} cx_i = c \sum_{i=1}^{n} x_i, \qquad \text{where } c \text{ is a constant}$$

The above is sometimes referred to as taking the constant outside the summation sign.

2(b). If the general expression being summed contains a constant term and a variable term, the terms may be separated. That is,

$$\sum_{i=1}^{n} (a + bx_i) = na + b \sum_{i=1}^{n} x_i, \qquad \text{where } a \text{ and } b \text{ are constants}$$

3. The sum of an expression containing two (or more) variable terms added is equal to the total of the two variable terms summed separately. That is,

$$\sum_{i=1}^{n} (x_i + y_i) = \sum_{i=1}^{n} x_i + \sum_{i=1}^{n} y_i$$

4. The square of the sum of terms is *not* equal to the sum of the squared terms. That is,

$$\left(\sum_{i=1}^{n} x_i \right)^2 \neq \sum_{i=1}^{n} x_i^2$$

$$(z + y)^2 \neq z^2 + y^2$$

5. The sum of products is *not* equal to the product of the sums. This is a corollary of rule 4. That is,

$$\sum_{i=1}^{n} (x_i y_i) \neq \left(\sum_{i=1}^{n} x_i \right) \left(\sum_{i=1}^{n} y_i \right)$$

$$x_1 y_1 + x_2 y_2 \neq (x_1 + x_2)(y_1 + y_2)$$

Product notation can be used to indicate the product of a series of terms, just as summation notation can be used to indicate the sum of a series of terms. If n_1 is multiplied by n_2 and then by n_3, and so on to n_k, this product can be written

$$\prod_{i=1}^{k} n_i = n_1 \cdot n_2 \cdot n_3 \cdots n_k$$

The large Greek pi stands for "the product of," just as the large sigma stands for "the sum of" in summation notation.

EXAMPLE
Multiply together the numbers from 1 through 10.

$$\prod_{i=1}^{10} \eta_i = 1 \cdot 2 \cdot 3 \cdot 4 \cdot 5 \cdot 6 \cdot 7 \cdot 8 \cdot 9 \cdot 10 = 3,628,800$$

2.3 FREQUENCY DISTRIBUTIONS

When large amounts of data are collected, interpretation of the data is usually made easier by grouping them into *classes* and by tabulating the number of observations that fall into each class. A table of classes or groups and the corresponding numbers of observations falling into each is called a *frequency distribution*. Such a table shows the distribution of the observations among the classes. Most of the data collected by the U.S. Census Bureau, the Federal Reserve System, the Public Health Service, and various other government agencies are published in the form of frequency distributions. Much of the data published on American corporations, on international trade, and on that obtained from market surveys or research studies are also often presented in this form.

NUMERICAL AND CATEGORICAL DISTRIBUTIONS

If the data are grouped or classified according to numerical size, a frequency distribution is said to be *numerical* or *quantitative*. In 1972 the federal government began distributing to the states some of the tax moneys collected primarily by the federal income tax. This practice, which has continued, is known as "revenue sharing." As an example of a numerical frequency distribution, Table 2.1 shows the amount of revenue given to the various states under this program, in its first year, as a proportion of the states contribution to federal receipts.

If the data are grouped or classified according to some qualitative characteristics, a frequency distribution is said to be *categorical* or *qualitative*. For this type of frequency distribution, observations are not measured, but are only placed into categories and counted. The distribution of stock issues shown in Table 2.2 is an example of a qualitative frequency distribution.

There are three basic steps in constructing any frequency distribution: (1) specifying the classes, (2) sorting the data into classes, (3) tabulating the number of observations in each class. Specification of the classes is essentially arbitrary. The only firm rule is that the classes must be specified so that each observation falls into one and only one class. Once the classes are specified unambiguously, sorting and tabulating are routine mechanical operations. If the data are on punch cards, sorting and tabulating are done in one step by the computer. As a rule of thumb, it is seldom advisable to use fewer than 5 or more than 15 classes. In any particular case, choice of the number of classes is influenced by the range of the data, the number of observations, and the purpose of the investigator.

For qualitative distributions the nature of the data may suggest or even determine the appropriate classes. However, even for qualitative distributions there may be the question of whether and how to combine classes. For example,

TABLE 2.1 1972 SHARED REVENUE BY STATES AS A PROPORTION OF THEIR CONTRIBUTION TO FEDERAL RECEIPTS

Proportion of Federal Receipts	Number of States
0.50–0.75	4
0.75–1.00	17
1.00–1.25	11
1.25–1.50	8
1.50–1.75	5
1.75–2.00	2
2.00–2.25	2
2.25–2.50	0
2.50–2.75	1
	50

Source: U.S. Treasury Department

in the distribution of stocks given, the electric, gas, telephone, and water categories could have collapsed into a utilities category. Unless qualitative data are collected in the classes to be used for the frequency distributions, sorting data into classes can involve arbitrary decisions. This can present a problem, for example, in coding the responses to survey questionnaires.

For quantitative distributions the range of values included in each class must be specified so that each observation falls into one and only one class. The smallest and largest values included in any given class are referred to as the *lower class limit* and the *upper class limit*, respectively. Specification of the class limits depends on the extent to which the observations are rounded. For example, in the distribution of revenue sharing given, data recorded to the nearest hundredth can be sorted into the classes given with no difficulty, but data recorded to the nearest thousandth cannot, unless a rounding rule is

TABLE 2.2 NONCONVERTIBLE PREFERRED STOCK ISSUES BY INDUSTRY CATEGORY 1970–1975

Industry Category	Number of Issues
Electric	363
Gas	143
Telephone	156
Water	83
Industrial	173
Finance	140
Insurance	13
Miscellaneous	55
Total	1126

Source: Securities and Exchange Commission

specified. An observation of .755 could be put into the class .50–.75 or the class .75–1.00, depending on the rounding rule used.

The class limits are chosen to coincide with and include the rounded observations. The actual boundary values of the classes depend on the rounding technique used in recording the observations. When we make observations, the usual practice is to attempt to read one more decimal place than is to be recorded. A standard rounding rule is then used in recording the observations.* The corresponding class boundaries are thus stated at the class limits midway between the upper class limit of one class and the lower class limit of the next class. Since the class boundaries include one decimal place in addition to those used in recording the observations and in specifying the class limits, an observation cannot fall into more than one class.

When we use this procedure, the class boundaries of the revenue sharing distribution previously given are:

Proportion of Federal Receipts
0.505–0.755
0.755–1.005
1.005–1.255
1.255–1.505
1.505–1.755
1.755–2.005
2.005–2.255
2.255–2.505
2.505–2.755

Although the class boundaries are usually midway between successive class limits, as shown above, there are some situations in which a different rounding rule is used. For example, ages are frequently rounded to the last completed year rather than to the nearest year. Boundaries determined in this way are referred to as *directed boundaries*. Note that the class boundaries cannot be determined from the class limits unless the rounding rule is stated.

The *class marks* of a quantitative frequency distribution are the midpoints of the classes; the class mark for a class can be determined by averaging the class limits or the class boundaries. In the revenue sharing example above, the class marks are .625, .875, 1.125, 1.375, 1.625, 1.875, 2.125, 2.375, 2.625. There are no "open classes" in this example, but they are sometimes used if a few observations are much smaller or much larger than the others. In this case, an open class "greater than 2.25" could have been used and simplified the distribution slightly by reducing the number of classes by one. There would be no class mark for this open class. In general, open classes should be avoided unless they can be utilized to reduce the number of classes materially, because their use makes it impossible to determine the approximate range of the data.

---

* The usual rounding rule is the one often used in physical science measurement: if the digit in the additional decimal place is less than 5, round down, and if it is greater than 5, round up. If the additional digit is 5, the most common practice is to round to an even number. Rounding up or down can also be done using a table of random numbers. Use of such a table is discussed in Chapter 11.

A *class interval* is the length of a class and is determined by subtracting the lower class boundary from the upper class boundary. In the revenue sharing example the class intervals are of length .25. Whenever possible, class intervals should be of equal length and should preferably be multiples of 5 or 10 to facilitate reading the distribution table.

PERCENTAGE FREQUENCY DISTRIBUTIONS

In some cases it is more convenient to state class frequencies in terms of percentages of the total number of observations. A frequency distribution given in terms of percentages is referred to as a *relative frequency distribution* or *percentage frequency distribution*. Frequencies are converted to percentages by dividing by the total number of observations and multiplying by 100.

The revenue sharing distribution in Table 2.1 can be given as the following percentage distribution:

Proportion of Federal Receipts	Percentage of States (%)
0.50–0.75	8
0.75–1.00	34
1.00–1.25	22
1.25–1.50	16
1.50–1.75	10
1.75–2.00	4
2.00–2.25	4
2.25–2.50	0
2.50–2.75	2
	100

The frequency distribution of stocks in Table 2.2 can also be given as a percentage distribution:

Industry Category	Percentage of Issues (%)
Electric	32
Gas	13
Telephone	14
Water	7
Industrial	15
Finance	13
Insurance	1
Miscellaneous	5
Total	100

Percentage distributions are particularly useful for presenting distributions of large numbers of observations and for comparing distributions involving different numbers of observations. Percentage distributions can be misleading when small numbers of observations are involved and should not be used in such cases. Whenever data are presented as percentages, the total number of observations should always be given, either on the table or on the graph or however the data is presented.

CUMULATIVE FREQUENCY DISTRIBUTIONS

For some purposes it is convenient to present data in cumulative form so that the distribution gives the frequency of observations less than or more than specified values. For example, the revenue sharing distribution in Table 2.1 could be written as a "less than" cumulative distribution as follows:

Proportion of Federal Receipts	Cumulative Frequency
less than 0.75	4
less than 1.00	21
less than 1.25	32
less than 1.50	40
less than 1.75	45
less than 2.00	47
less than 2.25	49
less than 2.50	49
less than 2.75	50

or as a "more than" cumulative distribution:

Proportion of Federal Receipts	Cumulative Frequency
2.50 or more	1
2.25 or more	1
2.00 or more	3
1.75 or more	5
1.50 or more	10
1.25 or more	18
1.00 or more	29
0.75 or more	46
0.50 or more	50

Cumulative distributions can also be given in terms of percentages rather than frequencies. As with other percentage distributions, cumulative percentage distributions are useful for describing large numbers of observations but may be misleading when the number of observations is small. The number of observations should always be given.

A summary of the frequencies, class marks, cumulative frequencies, and relative frequencies for the expenditures example of Table 2.1 is given below in Table 2.3.

GRAPHICAL REPRESENTATIONS

After data have been tabulated, it is generally useful to represent them in some graphical form. This type of presentation will often readily display facts about and relationships among the data that otherwise might be difficult to observe. In many cases, frequency distributions are much easier to comprehend when they are presented graphically, which is particularly true of distributions involving large numbers of observations. The most common type of graphical representation is a *column diagram*, or more usually referred to as a *histogram*.

TABLE 2.3 SUMMARY OF INFORMATION FROM TABLE 2.1

Number of States	Class Boundaries	Class Marks	Less Than Cumulative Frequency	More Than Cumulative Frequency	Relative or Percentage Frequency (%)
4	0.505–0.755	.625	4	1	8
17	0.755–1.005	.875	21	1	34
11	1.005–1.255	1.125	32	3	22
8	1.255–1.505	1.375	40	5	16
5	1.505–1.755	1.625	45	10	10
2	1.755–2.005	1.875	47	18	4
2	2.005–2.255	2.125	49	29	4
0	2.255–2.505	2.375	49	46	0
1	2.505–2.755	2.625	50	50	2
$\overline{50}$					$\overline{100}$

A histogram is constructed by representing the variable used for determining the classes on the horizontal axis, the class frequencies on the vertical axis, and drawing rectangles whose bases equal the class interval and whose heights equal the corresponding class frequencies. The horizontal axis can be marked in terms of class limits, class boundaries, class marks, or arbitrary coding values. In practice, class limits are usually used to facilitate reading, although the lengths of the rectangles actually correspond to the class boundaries. The revenue sharing distribution of Table 2.1 can be represented by the histogram in Figure 2.1.

Histograms should be used with extreme caution if the class intervals of a frequency distribution are not all equal. When the sizes of rectangles or other figures are compared visually, the comparison tends to be made in terms of areas. This is misleading, since actually the frequencies represented by the heights, not the areas, of the rectangles should be compared. If the class intervals of a distribution cannot be made equal, the histogram should be drawn so that areas representing the class frequencies are comparable.

Usually the most satisfactory solution to this problem is to make all class intervals the same length. However, there are some cases in which equal class intervals are undesirable. For example, if a large proportion of the observations fall in a relatively small part of the total range of the data, it is sometimes appropriate to narrow the intervals in this range in order to provide useful detail. Government reports customarily present frequency distributions of incomes with smaller intervals for incomes in the center of the range, where a large proportion of all incomes fall. In any case, if unequal class intervals are used for a frequency distribution and the corresponding histogram is drawn, the rectangles should be made comparable with respect to area.

Alternatively, a frequency distribution can be represented graphically as a *frequency polygon*. A frequency polygon is drawn by plotting the class frequencies at the class marks and connecting the successive points by straight lines. It also can be drawn by connecting the midpoints of the tops of the rectangles of the histograms. From the distribution of revenue sharing information in Table 2.3 the frequency polygon obtained from the histogram is

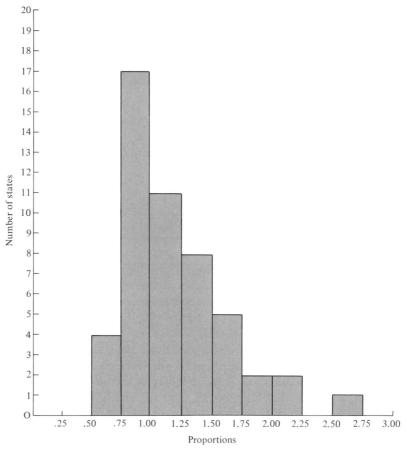

Figure 2.1 Histogram of 1972 shared revenue by states as a proportion of their contribution to federal receipts

Figure 2.2 Frequency polygon superimposed on histogram

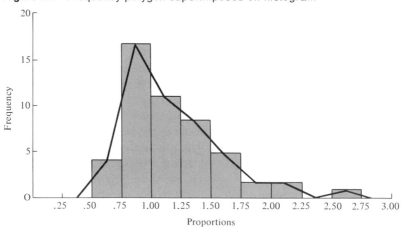

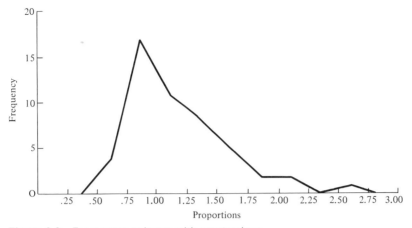

Figure 2.3   Frequency polygon with empty class

given in Figure 2.2. Note that the polygon is completed arbitrarily by drawing a line down to the axis at each end of the distribution, intersecting the axis at the point that would be the midpoint of the next class. Figure 2.3 shows the same polygon drawn without the histogram.

If the method for plotting a frequency polygon is used for a cumulative frequency distribution, the corresponding graphical representation is referred to as an *ogive*. The "less than" ogive for the distribution of revenue sharing taken from Table 2.3 can be represented as in Figure 2.4. Note that the "less

Figure 2.4   "Less than" cumulative frequency distribution

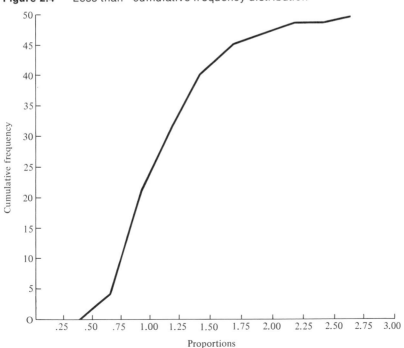

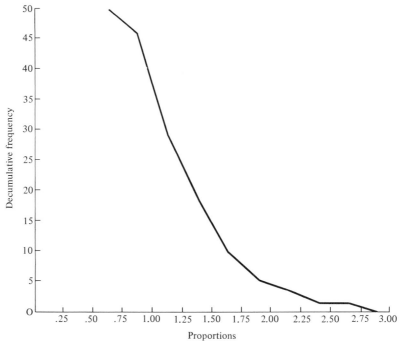

Figure 2.5   "Greater than" decumluative frequency distribution

Figure 2.6   Exxon Corporation Distribution of 1975 revenues (cents per dollar).
(Annual Report)

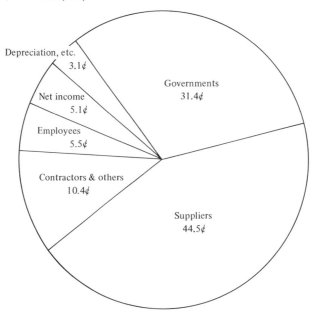

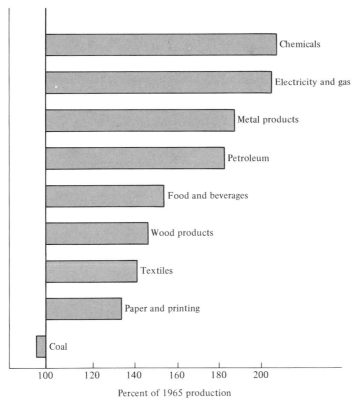

Figure 2.7 World growth of industrial output by industry, 1965–1975. (From *Fortune,* August 1976)

than" cumulative frequencies are plotted at upper class boundaries or, for ease of reading, at upper class limits. Similarly, the cumulative frequencies for a "more than" cumulative distribution are plotted at the lower class boundaries or limits, shown in Figure 2.5 for the information taken from Table 2.3. The "more than" cumulative frequencies are often referred to as "greater than" decumulative frequencies, which are interchangeable terms. The "greater than" decumulative notation is used in Figure 2.5.

In addition to histograms and frequency polygons, there are different ways of presenting data pictorially using various types of charts. Numerous examples of such representations appear in magazines and newspapers; they are often dramatic and effective but have little counterpart or application in inferential statistics. A few illustrations follow.

Three of the most common charts used to show data or to give information are pie charts, bar charts, and line charts. Figure 2.6 shows a pie chart. As in this example, pie charts are often used to illustrate the distribution of revenues. From this chart we can see how a 1975 dollar of revenue for the Exxon Corporation was divided.

Figure 2.7 is a bar chart. It shows the world growth in production for nine major industries from 1965 to 1975. In this case the units of production

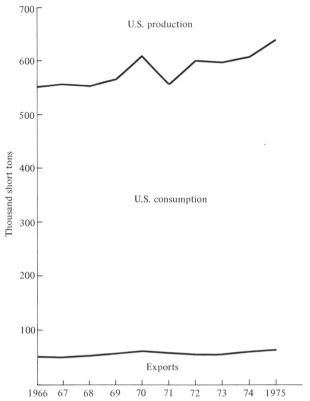

Figure 2.8 Lignite, bituminous, and anthracite coal production 1966–1975. (From Minerals Yearbook and Bureau of Mines News Release)

from one industry to another might be vastly different; therefore the chart illustrates the growth by using 1965 as the base year.

Figure 2.8 shows two line charts: one of U.S. production of coal over the decade 1966 through 1975 and one of our exports of coal over the same period. The upper line is production and the lower line is exports. Notice that the difference between the production and exports would be U.S. consumption.

2.4 MEASURES OF CENTRAL TENDENCY

Frequently, it is desirable to describe a set of observations by a single number that indicates the *central tendency* or location of the data. There are several commonly used measures of central tendency; each of these indicates a central point about which the data tend to cluster, according to a particular criterion. The most appropriate measure of central tendency for any given problem depends partly on the nature of the data and partly on the purpose for which the measure is to be used. The measures of central tendency defined and compared below are: arithmetic mean, median, mode, geometric mean, harmonic mean and weighted mean.

THE ARITHMETIC MEAN

The *arithmetic mean* is defined as the sum of the observations divided by the number of observations. If a set of n observations are denoted by $x_1, x_2, \ldots, x_n$, then their arithmetic mean, denoted by $\bar{x}$, and read "x bar," is given by

$$\bar{x} = \frac{\sum\limits_{i=1}^{n} x_i}{n}$$

The arithmetic mean is sometimes referred to as the *arithmetic average* or, briefly, as the *mean* or the *average*. It is the most commonly used measure of central tendency for purposes of description and is almost always used for purposes of inference.

The numbers listed in Table 2.4 represent a sample of 12 closing prices of stocks on the New York Stock Exchange on a particular day:

TABLE 2.4 CLOSING STOCK PRICES

$63\frac{1}{8}$

$75\frac{1}{4}$

$82\frac{3}{8}$

$80\frac{1}{2}$

$72\frac{1}{8}$

$68\frac{1}{2}$

$80\frac{1}{2}$

$52\frac{3}{8}$

$43\frac{5}{8}$

$70\frac{7}{8}$

$83\frac{1}{8}$

$40\frac{1}{4}$

Their sum is 813 and their mean is

$$\bar{x} = \frac{813}{12} = 67\frac{3}{4}$$

If data are in the form of a frequency distribution, it is not possible to determine exactly their arithmetic mean. However, an approximation can be obtained by assuming that all the observations in a class are evenly distributed or, equivalently, are located at its class mark. The error introduced by this assumption is usually insignificant.

If a frequency distribution has the class marks $x_1, x_2, \ldots, x_k$ and the corresponding class frequencies $f_1, f_2, \ldots, f_k$, then the mean of this distribution

TABLE 2.5   STATE INDEBTEDNESS IN 1975 (IN MILLIONS OF DOLLARS)

Debt	f Number of States	x Class Mark	xf
0–500	19	250	4,750
500–1,000	11	750	8,250
1,000–1,500	6	1,250	7,500
1,500–2,000	5	1,750	8,750
2,000–2,500	1	2,250	2,250
2,500–3,000	3	2,750	8,250
3,000–4,500	2	3,750	7,500
4,500–9,500	2	7,000	14,000
9,500–19,500	1	14,500	14,500
	$\overline{50}$		$\overline{75,750} = \sum xf$

Source: State Government Finances in 1975, Bureau of the Census, U.S. Department of Commerce

is given by*

$$\bar{x} = \frac{\displaystyle\sum_{i=1}^{k} x_i f_i}{\displaystyle\sum_{i=1}^{k} f_i}$$

Table 2.5 lists the 1975 indebtedness of the 50 states in the form of a frequency distribution. Note that not all classes are equal. If this formula is used to calculate the mean state indebtedness, we will obtain

$$\bar{x} = \frac{75750}{50} = \$1515 \text{ million}$$

If the original data were examined and the individual states' actual indebtedness were used, the true mean would be $1442 million. We can see that calculating the mean from a frequency distribution can lead to error. In this instance the error is less than 10 percent, despite the unequal and large intervals at the higher debt levels.

THE MEDIAN

The *median* of a set of observations divides the observations so that half are as large or larger than its value and half are smaller than its value. When the observations are arranged in increasing or decreasing order of magnitude, the median is equal to the value of the middle observation if the number of observations is odd or the mean of the two middle observations if the number of observations is even. That is, if there are n observations, the median is the $((n + 1)/2)$st observation, where the $((n + 1)/2)$st observation is defined as midway between the $(n/2)$st and the $((n + 2)/2)$st observation when n is even.

---

* Note that $\sum f_i$ equals the total number of observations, usually denoted by N; $\sum f_i = N$.

When we determine the median of the stock prices given in Table 2.4, the prices must be ordered say, from smallest to largest:

$40\frac{1}{4}$

$43\frac{5}{8}$

$52\frac{3}{8}$

$63\frac{1}{8}$

$68\frac{1}{2}$

$70\frac{7}{8}$

$72\frac{1}{8}$

$75\frac{1}{4}$

$80\frac{1}{2}$

$80\frac{1}{2}$

$82\frac{1}{4}$

$83\frac{1}{8}$

There are 12 prices, so the median is midway between the sixth and seventh price or

$$\text{Median} = \frac{1}{2}\left(70\frac{7}{8} + 72\frac{1}{8}\right) = 71\frac{1}{2}$$

As is the case for the arithmetic mean, the median of data in the form of a frequency distribution cannot exactly be determined. If the observations are assumed to be evenly distributed in their respective classes, the median of a distribution is the number such that half the area of the corresponding histogram lies to its left and half to its right. In practice, the median of a frequency distribution is determined in two steps: the class in which the median falls is determined and (2) the actual value of the median is calculated. In contrast to the median for ungrouped data that is the $((n + 1)/2)$st observation, the median for grouped data is the number that divides the distribution into two equal parts, each representing a frequency of $n/2$. Thus the median for grouped data is the $(n/2)$st observation.

If L is the lower boundary of the class into which the median falls, f_m is the frequency of this class, c is the class interval, and j is the number of observations needed from this class to total $n/2$, then the median is computed from the formula

$$\text{Median} = L + c\left(\frac{j}{f_m}\right)$$

Note that this formula essentially adds the appropriate proportion of the class interval to its lower boundary to obtain the median. Thus in a sense there are three steps in determining the median of a frequency distribution—(1) the appropriate class is determined, (2) the proportion of that class to be included, j/f_m, is multiplied by the class interval c, and (3) this number is added to the lower boundary.

Table 2.6 lists persons, by age, held in federal prisons as of mid-1974. Since the total is 22,112, then $n/2 = 11,056$. The $(n/2)$th observation will thus fall in the class whose lower boundary is age 26. The class interval is 10, the

TABLE 2.6 SENTENCED
POPULATION CONFINED
IN BUREAU OF PRISON
INSTITUTIONS BY AGE,
JUNE 30, 1974

Age	Number
Under 18	190
18–25	6,577
26–35	8,496
36–50	5,468
Over 50	1,381
Total	22,112

Source: Federal Bureau of
Prisons Statistical Reports.

number of observations needed to total 11,056 from this class is 4287. Using
the above formula for the median, we have

$$\text{Median} = L + c\left(\frac{j}{f_m}\right)$$

$$= 26 + 10\left(\frac{4287}{8496}\right)$$

$$= 26 + 5 = 31$$

Equivalently, the median can be obtained by starting at the upper end
of the distribution and by using the formula

$$\text{Median} = U - c\left(\frac{j'}{f_m}\right)$$

when U is the upper boundary of the class in which the median falls and j' is
the number of observations from this class needed to make $n/2$. In the preceding
illustration

$$\text{Median} = U - c\left(\frac{j'}{f_m}\right)$$

$$= 35 - 10\left(\frac{4207}{8496}\right)$$

$$= 35 - 4 = 31$$

This checks with the answer obtained from the lower boundary.

The median, like the mean, is always defined (except for a distribution
with one or more open classes) and is unique. The most important difference
between the mean and the median from a practical point of view is the effect
of extreme values. The mean is affected by extreme values, the median generally
is not. For this reason, for example, the median is said to be a better description
of income distribution than the mean (see Table 2.7). To show the sensitivity
of the mean to extreme values, using just a few observations, take the distribution

TABLE 2.7   NUMBER OF FAMILIES
BY TOTAL MONEY INCOME IN 1974

Total Money Income	Number of Families
Under $1,000	702,000
1,000–1,999	766,000
2,000–2,999	1,489,000
3,000–3,999	2,040,000
4,000–4,999	2,305,000
5,000–5,999	2,475,000
6,000–6,999	2,478,000
7,000–7,999	2,501,000
8,000–8,999	2,574,000
9,000–9,999	2,629,000
10,000–10,999	2,920,000
11,000–11,999	2,782,000
12,000–12,999	2,859,000
13,000–13,999	2,632,000
14,000–14,999	2,388,000
15,000–19,999	10,032,000
20,000–24,999	5,755,000
25,000–49,999	5,771,000
50,000 and over	614,000
Total	55,712,000

Source: U.S. Department of Commerce Current Population
Reports

of stock prices shown in Table 2.4 and change the last two prices as follows:

$40\frac{1}{4}$

$43\frac{5}{8}$

$52\frac{3}{8}$

$63\frac{1}{8}$

$68\frac{1}{2}$

$70\frac{7}{8}$

$72\frac{1}{8}$

$75\frac{1}{4}$

$80\frac{1}{2}$

$80\frac{1}{2}$

$99\frac{3}{8}$

$152\frac{3}{4}$

The median is unchanged at $71\frac{1}{2}$, but the mean is now $75\frac{5}{8}$ instead of $67\frac{3}{4}$. This difference in sensitivity to extreme values between the mean and the median should be considered in choosing either one for describing a particular set of data. In many cases, both the mean and the median are given to avoid ambiguity.

Table 2.7 gives the 1974 distribution of family income in the United States. Notice that some of the classes are unequal and that the last class is open. The median, using the formula: Median $= L + c(j/f_m)$, is \$12,768. According to the U.S. Department of Commerce the true median family income for 1974 was \$12,836. We can see that the estimated median from the grouped data is very close to the actual. The mean family income for 1974 was \$14,502; this higher figure shows the sensitivity of the mean to the relatively few very high incomes.

In the same sense that the median divides a distribution into two halves, other measures of location can be defined to divide a distribution into any number of parts. Of these measures of location, the most commonly used are *quartiles* which divide a distribution into quarters, *deciles* which divide a distribution into tenths, and *percentiles* which divide a distribution into hundredths. These measures of location are determined by ordering the observations and by counting the appropriate number of observations, as done to determine the median. A quartile, decile, or percentile that falls within a class is determined by a formula analogous to that used for the median. Note that the second quartile, the fifth decile, and the fiftieth percentile are also the median of a distribution. (Figure 2.9 in the next section shows the location of the quartiles for a symmetric distribution.)

THE MODE

The *mode* is the value, class, or category that occurs with the highest frequency. Unlike the mean and the median, the mode is appropriate for qualitative as well as quantitative data. For the distribution of stock issues in Table 2.2 the electric industry is the model category, for closing stock prices in Table 2.5 the mode is $80\frac{1}{2}$, for confined persons in federal prisons in Table 2.6 the modal age group is $26-35$, for family incomes in Table 2.7 the modal class is \$15,000 to \$19,999.

In a sense, the mode is typical of a set of data, since it is the value or class that occurs more often than any other. The mode is easily obtained and requires no calculation. However, it has serious disadvantages and for this reason is used principally for qualitative data where the mean and median are not applicable. The disadvantages of the mode, which make it unusable in any meaningful sense in some cases, are that the mode may not exist or, if it does, it may not be unique. For a particular set of data the mode may fail to exist because no two observations are identical. The mode may not be unique because more than one observation occurs with the maximum frequency.

EXAMPLE

The distributions of invoices in an audit of three retail tire stores are as follows:

INVOICE SIZE CATEGORY	NUMBER OF INVOICES		
	STORE 1	STORE 2	STORE 3
less than \$50	4	3	5
\$50–\$100	6	3	7
\$100–\$150	12	3	15
\$150–\$200	10	3	15
\$200–\$250	8	3	4
over \$250	2	3	3

Notice then, that the mode for Store 1 is the third class, there is no mode for store 2, and store 3 has two modes—the third and fourth classes.

THE GEOMETRIC MEAN

The *geometric mean* of a set of n numbers is the nth root of their product. If a set of n observations are denoted by $x_1, x_2, \ldots, x_n$, then their geometric mean, given as G, is

$$G = \sqrt{x_1 \cdot x_2 \cdot x_3 \cdots x_n}$$

$$G = \left[\prod_{i=1}^{n} x_i \right]^{1/n}$$

The geometric mean is most easily obtained by using logarithms, since

$$\text{Log } G = \frac{\sum\limits_{i=1}^{n} \log x_i}{n}$$

and G can be obtained as the antilogarithm.

From algebra we know that if there were a series of numbers in the following order,

$$4, 8, 12, 16, \ldots$$

they would be said to be in arithmetic progression. If there were a series of numbers in the following order,

$$3, 6, 12, 24, \ldots$$

they would be said to be in geometric progression. If we take the simple geometric progression of three numbers 2, 4, 8, we find their arithmetic average is

$$\frac{2 + 4 + 8}{3} = 4\frac{2}{3}$$

However, their geometric mean is

$$\sqrt[3]{2 \cdot 4 \cdot 8} = 4$$

Intuitively, it would seem that the latter measure giving 4 is a "better" designation of the average of these three numbers than the former.

The geometric mean is thus more appropriate in averaging numbers in geometric progression than the arithmetic mean. Geometric means are most frequently used to average ratios or rates of change. Notice that the geometric mean cannot be used if one of the numbers to be averaged is zero or negative.

From Table 2.8 we can see the census populations of the United States from 1910 to 1970. Except for the decade 1930–1940, the rate of growth of the population between census is nearly the same. During the period 1910–1970, then, we can say that the population was approximately in a geometric progression. The geometric means for the population and for the rate of growth of

TABLE 2.8 POPULATION OF THE UNITED STATES IN MILLIONS

Year	Population	Rate of Growth (Percent)
1910	92.0	—
1920	105.7	15
1930	122.8	16
1940	131.7	7
1950	150.7	14
1960	178.4	18
1970	203.0	14

$$\bar{x}_p = \frac{92.0 + 105.7 + 122.8 + 131.7 + 150.7 + 178.5 + 203.0}{7} = 140.9$$

$$\bar{x}_r = \frac{15\% + 16\% + 7\% + 14\% + 18\% + 14\%}{6} = 14\%$$

$$G_p = \sqrt[7]{92.0 \cdot 105.7 \cdot 122.8 \cdot 131.7 \cdot 150.7 \cdot 178.5 \cdot 203.0} = 135.5$$

$$G_r = \sqrt[6]{15 \cdot 16 \cdot 7 \cdot 14 \cdot 18 \cdot 14} = 13.3\%$$

the population are thus more appropriate as measures of the averages than the arithmetic means.

An example of the use of the geometric mean in business is in the attempt by investment analysts to calculate return on common stocks over time. While the argument for preferring the geometric mean return over other possible

TABLE 2.9 DIVIDENDS, YEAR-END PRICES, AND GEOMETRIC RETURNS

		1965	1966	1967	1968	1969
IBM	Year-end Price	499.000	371.400	627.000	315.000	364.400
	Yearly Dividends	6.000	4.300	4.340	0.000	3.600
GM	Year-end Price	103.400	65.700	82.000	79.100	69.100
	Yearly Dividends	5.250	4.550	0.000	4.300	4.300
CHRIS CRAFT	Year-end Price	22.400	21.700	42.000	39.200	10.500
	Yearly Dividends	0.000	0.240	1.300	0.000	0.293
AMERICAN BRANDS	Year-end Price	38.500	30.700	32.100	37.600	35.600
	Yearly Dividends	1.650	1.800	1.800	0.000	2.000

measures is too lengthy for discussion here, it is difficient to say that many analysts claim it is superior to any others. The formula for the geometric return on a common stock is given by

$$1 + r = \left[\prod_{t=1}^{n} \frac{D_t + P_t}{P_{t-1}} \right]^{1/n}$$

where:

r = geometric mean rate of return
D_t = dividend in year t
P_t = price of the stock at the end of year t
P_{t-1} = price of the stock at the end of year $t - 1$
n = number of years under consideration

Note that no matter which way the price of the stock moves, and even if dividends are zero, the term for each year will always be a positive ratio. The expression is set up as shown and then 1 is subtracted from the expression on the right to obtain r. It is possible to obtain negative geometric rates of return. As an example, Table 2.9 gives dividends and year-end prices for four common stocks listed on the New York Stock Exchange. For the decade 1966–1975, $1 + r$ and r (the geometric rate) are also given. It is interesting to note that of the four, only one had a positive geometric rate of return for the ten years.

THE HARMONIC MEAN

The *harmonic* mean of a set of n numbers is n divided by the sum of the reciprocals of the numbers. If a set of n observations are denoted by

FOR FOUR STOCKS, 1966–1975							
1970	1971	1972	1973	1974	1975	$1 + r$	r
317.600	336.400	402.000	246.600	168.000	224.200		
						.937	− .063
4.800	5.200	5.400	4.480	5.560	6.500		
80.400	80.400	81.100	46.100	30.600	57.500		
						.998	− .002
3.400	3.400	4.450	5.250	3.400	2.400		
8.200	5.400	6.200	2.200	2.400	5.300		
						.872	− .128
0.000	0.000	0.000	0.000	0.000	0.000		
45.100	41.700	42.100	32.200	30.200	38.500		
						1.055	.055
2.100	2.200	2.288	2.379	2.560	2.680		

$x_1, x_2, x_3, \ldots, x_n$, then their harmonic mean, H, is given by

$$H = \frac{n}{\displaystyle\sum_{i=1}^{n} \frac{1}{x_i}}$$

The harmonic mean, although not used very frequently, is useful for problems in which the denominator of a set of comparable observations varies. For example, the harmonic mean is appropriate for averaging price per unit where price is constant and units vary, or where units are constant and price varies. Also, it can be used for averaging speeds where time spent is constant and distances vary, or where distance is constant and speed varies. The use of the harmonic mean in index numbers is discussed in the next chapter.

EXAMPLE
During the four years that Charlie DuBois was boarding his horses, he paid per bale of hay, respectively, $1, \$1.25, \$1.60, \$2.00. What was the average cost of hay per bale over the four years if he spent $1000 per year on hay?

$$\text{Average cost} = \frac{\text{total cost}}{\text{total quantity}} = \frac{\$4000}{1000 + 800 + 625 + 500} = \frac{4000}{2925} = \$1.37$$

or

$$\text{Harmonic mean} = \frac{4}{\dfrac{1}{\$1} + \dfrac{1}{\$1.25} + \dfrac{1}{\$1.60} + \dfrac{1}{\$2.00}} = \$1.37$$

EXAMPLE
A driver spends one hour driving from A to B at an average speed of 30 mph. He then returns from B to A at an average speed of 60 mph. What is the average speed for the trip?
    The distance from A to B must be 30 miles; therefore the time from B to A at 60 mph must have been one-quarter hour.

$$\text{Average speed} = \frac{\text{total distance}}{\text{total time}} = \frac{60}{1\frac{1}{2}} = 40 \text{ mph}$$

or

$$\text{Harmonic mean} = \frac{2}{\frac{1}{30} + \frac{1}{60}} = 40 \text{ mph}$$

If we made the mistake of just averaging the speeds 30 and 60, we would obtain the average speed to be 45 mph, which is incorrect.

THE WEIGHTED MEAN
    The *weighted mean* of a set of numbers is the sum of the n weighted numbers divided by the sum of the n weights. As a special case, the ordinary arithmetic mean discussed in the beginning of this section is a weighted mean, with the weight in each case being 1. However, the term weighted mean usually refers to cases in which at least one of the weights is other than 1.

If a set of n observations are denoted by $x_1, x_2, \ldots, x_n$ and their respective weights are denoted by $w_1, w_2, \ldots, w_n$, then their weighted mean, $\bar{x}_w$, is given by

$$\bar{x}_w = \frac{\sum\limits_{i=1}^{n} w_i n_i}{\sum\limits_{i=1}^{n} n_i}$$

The weights represent the relative importance of the observations—for example, prices are usually weighted by the relative quantities involved. The problem of choosing appropriate weights arises in defining indexes and is discussed in this context in the next chapter.

A special case of the weighted mean is the computation of the overall mean of several sets of data from their individual means and the number of observations in each set. If k sets of data have means $\bar{x}_1, \bar{x}_2, \ldots, \bar{x}_k$ and numbers of observations $n_1, n_2, \ldots, n_k$, respectively, then the overall mean of the observations is

$$\bar{x} = \frac{\sum\limits_{i=1}^{k} n_i \bar{x}_i}{\sum\limits_{i=1}^{k} n_i}$$

where the weights for the means are the respective numbers of observations on which they are based. Note that the formula for obtaining the mean of a frequency distribution is an example of this type of weighted mean.

EXAMPLE
There are three semester examinations in a course, weighted equally, and a final exam weighted twice as much as a semester examination. A student has examination grades of 65, 70, and 85, and a final exam grade of 90. What is his average (mean) for the course?

$$\bar{x} = \frac{(1)65 + (1)70 + (1)85 + (2)90}{1 + 1 + 1 + 2} = \frac{400}{5} = 80$$

2.5 MEASURES OF DISPERSION

A measure of variation or dispersion provides an indication of the scatter of the observations about the measure of their central tendency and is necessary to give a more complete description of a distribution, or to compare properly two or more distributions. For example, the following sets of data (Table 2.10) have the same mean, 50, but vary considerably with regard to spread or dispersion.

There are several commonly used measures of dispersion, each of which indicates some aspect of the spread or variability of the data. For central tendency, the most appropriate measure of dispersion for any given problem depends partly on the nature of the data and partly on the purposes for which the measure is to be used. Measures of dispersion are defined and compared

TABLE 2.10   SETS OF DATA

Set 1	Set 2	Set 3	Set 4
47	60	5	0
53	25	5	250
50	10	95	0
51	60	5	0
49	45	95	50
50	100	95	0

as follows: range, average deviation, standard deviation, and coefficient of variation.

THE RANGE

The *range* of a set of data is defined as the difference between the largest and the smallest observations. The ranges of the sets of data in Table 2.10 are 6, 90, 90, and 250, respectively and can also be written as the interval covered by the data instead of the length of that interval. From Table 2.10 the ranges could thus be written 47–53, 10–100, 5–95, and 0–250, respectively.

The range is easily determined and provides a quick indication of the variability of a set of observations. However, its value depends only on the two extreme observations and gives no indication of the dispersion of the observations between the two extremes. In the preceding example the second and third sets of data have the same range, 90, and even writing the ranges as intervals does not indicate the difference between the dispersion of the two sets of data.

In some problems it is useful to determine the *interquartile range*, which is defined as the third quartile minus the first quartile and thus gives the length of the interval in which the middle 50 percent of the data fall. In a sense, the interquartile range is more informative than the range because it concerns the spread of the middle part of the data and is not affected by one very small or very large value.

Figure 2.9   Quartiles

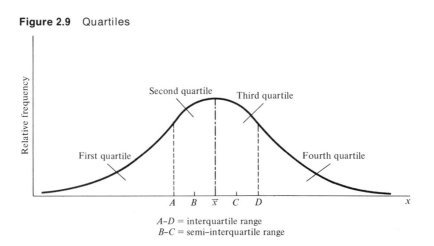

A–D = interquartile range
B–C = semi-interquartile range

In some cases the *semiinterquartile* range, which is defined as one-half the interquartile range, is given. The semiinterquartile range is also referred to as the *quartile deviation*, since it indicates the average amount by which the first and third quartiles deviate from the median.

Figure 2.9 shows the location of the interquartile range and the semi-interquartile range for a symmetric distribution, as well as the location of the quartiles. Note that for this distribution $\bar{x}$ is both the mean and the median.

THE AVERAGE DEVIATION

The dispersion or variation of data is small if the observations are bunched or clustered about their mean and large if the observations are spread out over considerable distance from their mean. Thus, intuitively, the variation of a set of observations can be measured in terms of their deviations about (i.e., differences from) the mean. If a set of n observations are denoted by $x_1, x_2, \ldots, x_n$ and their mean is denoted by $\bar{x}$, then the differences $x_1 - \bar{x}, x_2 - \bar{x}, \ldots, x_n - \bar{x}$ are referred to as *deviations from the mean*. The *average deviation* is defined as the sum of the absolute values of the deviations from the mean divided by the number of observations. The average deviation is given by

$$\frac{\sum_{i=1}^{n} |x_i - \bar{x}|}{n}$$

For example, for the first set of observations in Table 2.10, the deviations from the mean are:

$$47 - 50 = -3$$
$$53 - 50 = 3$$
$$50 - 50 = 0$$
$$51 - 50 = 1$$
$$49 - 50 = -1$$
$$50 - 50 = 0$$

and the average deviation is

$$\frac{3 + 3 + 0 + 1 + 1 + 0}{6} = \frac{4}{3}$$

For the fourth set of observations, the deviations are:

$$0 - 50 = -50$$
$$250 - 50 = 200$$
$$0 - 50 = -50$$
$$0 - 50 = -50$$
$$50 - 50 = 0$$
$$0 - 50 = -50$$

and the average deviation is

$$\frac{50 + 200 + 50 + 50 + 0 + 50}{6} = \frac{200}{3}$$

Recall that each of these sets of observations has mean 50.

When we define the average deviation, the absolute values of the deviations from the mean are used, since the sum of the deviations is always zero:

$$\sum_{i=1}^{n} (x_i - \bar{x}) = \sum_{i=1}^{n} x_i - \sum_{i=1}^{n} \bar{x} = n\bar{x} - n\bar{x} = 0$$

The average deviation is an intuitively appealing measure of variation, because in many cases the magnitudes (absolute values) of the deviations and not their signs are of interest. However, measures involving absolute values are generally inconvenient for statistical purposes and not typically used in inference.

VARIANCE AND STANDARD DEVIATION

The *variance* of a set of n observations is defined as the sum of the squared deviations of the observations from the mean divided by $n - 1$. If a set of n observations are denoted by $x_1, x_2, \ldots, x_n$, then their variance, denoted by s^2, is given by

$$s^2 = \frac{\sum_{i=1}^{n} (x_i - \bar{x})^2}{n - 1}$$

The variance avoids the problem of negative deviations from the mean by squaring them rather than by taking their absolute values. It has convenient statistical properties and is almost always the measure of dispersion used for inference.

The positive square root of the variance is sometimes used, in order to have the measure of dispersion in the same units as the original observations. This measure, denoted by s, is referred to as the *standard deviation:*

$$s = \sqrt{\frac{\sum_{i=1}^{n} (x_i - \bar{x})^2}{n - 1}}$$

EXAMPLE
For the first and fourth sets of observations in Table 2.10 calculate their variances and standard deviation.
First set:

$$s^2 = \frac{9 + 9 + 0 + 1 + 1 + 0}{5} = 4$$

$$s = 2$$

Fourth set:

$$s^2 = \frac{2500 + 40000 + 2500 + 2500 + 0 + 2500}{5} = 10{,}000$$

$$s = 100$$

Variances can sometimes be computed more easily using one of the following formula, which are algebraically equivalent to the formula used above:*

$$s^2 = \frac{n \sum_{i=1}^{n} x_i^2 - \left(\sum_{i=1}^{n} x_i\right)^2}{n(n-1)}$$

or

$$s^2 = \frac{\sum_{i=1}^{n} x_i^2 - \frac{\left(\sum_{i=1}^{n} x_i\right)^2}{n}}{n-1}$$

Notice that the denominator in the variance is $n - 1$ rather than n. When the variance is used only for description, the choice of n or $n - 1$ is arbitrary and n is frequently used. However, for purposes of inference $n - 1$ must be used so that the sample variance is an unbiased estimate of the population variance; that is, the sample variance is equal on the average to the population variance. (This will be discussed further in Chapter 6.)

If data are in the form of a frequency distribution, their approximate standard deviation can be obtained by assuming, as in determining the mean, that all the observations in a class are located at its class mark.

If a frequency distribution of n observations has the class marks x_1, $x_2, \ldots, x_k$ and the corresponding class frequencies $f_1, f_2, \ldots, f_k$, where

$$\sum_{i=1}^{k} f_i = n$$

then the standard deviation of this distribution is given by

$$s = \sqrt{\frac{\sum_{i=1}^{n} (x_i - \bar{x})^2 f_i}{n-1}}$$

---

* $\sum (x_i - \bar{x})^2 = \sum (x_i^2 - 2\bar{x}x_i + \bar{x}^2) = \sum x_i^2 - 2\bar{x}\sum x_i + n\bar{x}^2$

$\qquad = \sum x_i^2 - 2\left(\dfrac{\sum x_i}{n}\right)(\sum x_i) + n\left(\dfrac{\sum x_i^2}{n}\right)$

$\qquad = \sum x_i^2 - \dfrac{(\sum x_i)^2}{n}$

From Table 2.1, revenue shared as a proportion of federal receipts, we have the following:

Class	x (Class Mark)	f (Frequency)	fx	$(x - \bar{x})$	$(x - \bar{x})^2$	$f(x - \bar{x})^2$
0.50–0.75	0.625	4	2.5	−0.565	0.3192	1.2769
0.75–1.00	0.875	17	14.875	−0.315	0.992	1.6868
1.00–1.25	1.125	11	12.375	−0.065	0.0042	0.0465
1.25–1.50	1.375	8	11.000	0.185	0.0342	0.2738
1.50–1.75	1.625	5	8.125	0.437	0.1910	0.9548
1.75–2.00	1.875	2	3.750	0.685	0.4692	0.9385
2.00–2.25	2.125	2	4.250	0.935	0.8742	1.7484
2.25–2.50	2.375	0	0	1.185	1.4042	0.0
2.50–2.75	2.625	1	2.625	1.435	2.0592	2.0592
		50	59.50			8.9849

The mean is

$$\bar{x} = \frac{\sum fx}{n} = \frac{59.5}{50} = 1.19$$

The variance is

$$s^2 = \sum (x - \bar{x})^2 f_i/(n - 1) = \frac{8.9849}{49} = .1795$$

The standard deviation is

$$s = \sqrt{.1795} = .422$$

Note that if a frequency distribution has an open class, its variance cannot be determined for the same reason that its mean cannot be determined—an open class does not have a class mark.

THE COEFFICIENT OF VARIATION
In order to interpret the standard deviation or variance of a set of data, we need some indication of their general magnitude or mean. For example, if a set of observations has a mean of .35 in., a standard deviation of .25 in. is relatively large; however, if the mean is 25.6 in., a standard deviation of .25 in. is relatively small.

Thus it can be seen that a relative measure of dispersion is in some circumstances as important as an absolute measure of dispersion. When we express the variation of a set of data relative to its mean, the most commonly used measure is the *coefficient of variation*, defined as:

$$V = \frac{s}{\bar{x}}$$

If V is multiplied by 100, then the standard deviation is given as a percent of the mean.

An important attribute of the coefficient of variation is that it is independent of units used and therefore is useful in comparing distributions where units are different.

EXAMPLE
Light bulb *A* has a mean life of 1000 hr and a standard deviation of 200 hr. Light bulb *B* has a mean life of 750 hr and a standard deviation of 100 hr. Calculate the relative variation in life for each bulb.

$$V = \frac{s}{x}$$

$$V_A = \frac{200}{1000} = .2$$

$$V_B = \frac{100}{750} = .1333$$

EXAMPLE
The mean exchange rate of the Italian lira to the U.S. dollar in 1976 was *L*820 to $1, with a standard deviation of *L*60 to $1. The mean exchange rate of the German mark to the U.S. dollar in 1976 was *DM*2.54 to $1, with a standard deviation of *DM*.05 to $1. Which of these foreign currencies fluctuated the most against the U.S. dollar?

$$V = \frac{s}{\bar{x}}$$

$$V_L = \frac{60}{860} = .0698$$

$$V_{DM} = \frac{.05}{2.54} = .0197$$

What do these coefficients of variation tell you?

2.6 MEASURES OF SKEWNESS AND PEAKEDNESS

Two distributions that have identical means and standard deviations can differ considerably with regard to shape. Thus measures of central location and variation do not always provide a sufficient description of a distribution. Two additional measures are frequently used: *skewness* an indication of the extent of deviation from symmetry) and *kurtosis* (an indication of peakedness).

SKEWNESS

By definition, a distribution is symmetric about its mean if its right and left halves are mirror images of each other. That is, if a symmetric distribution is folded at its mean, the two halves coincide perfectly (see Figure 2.9). A distribution that is not symmetric is said to be *asymmetric* or *skewed*. If symmetric, the mean and the median coincide; if asymmetric, its skewness can be expressed in terms of the differences between the mean and the median. The *Pearsonian coefficient of skewness*, denoted SK, is defined as

$$SK = \frac{3(\text{mean} - \text{median})}{\text{standard deviation}}$$

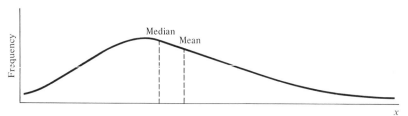

Figure 2.10   Positively skewed distribution

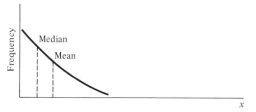

Figure 2.11   Extreme positive skewness

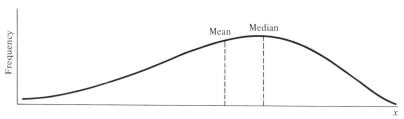

Figure 2.12   Negatively skewed distribution

The Pearsonian coefficient of skewness is positive if the mean exceeds the median; in general, such a positively skewed distribution has a tail to the right, as, for example, in Figure 2.10.

An extreme positive skewness of a distribution would appear as Figure 2.11. Such a case is the second skewness example on page 37.

Similarly, the Pearsonian coefficient of skewness is negative if the median exceeds the mean; in general, such a negatively skewed distribution has a tail to the left, as shown in Figure 2.12.

The most commonly used measure of skewness is denoted α_3 and is defined as the average of the cubed deviations from the mean divided by the cube of the standard deviation. For ungrouped data

$$\alpha_3 = \frac{\frac{1}{n} \sum_{i=1}^{n} (x_i - \bar{x})^3}{s^3}$$

and for grouped data

$$\alpha_3 = \frac{\frac{1}{f_i} \left[\sum_{i=1}^{k} (x_i - \bar{x})^3 f_i \right]}{s^3}$$

Both the Pearsonian coefficient of skewness and α_3 are independent of the units of measurement and are zero for a symmetric distribution.

EXAMPLE

Find the SK and the α_3 measures of skewness for the following distribution of the 1970 sales prices of homes in Palo Verde, California. (Note that this distribution has been used slightly truncated above \$70,000, due to an open topmost class.) Students should obtain this data for 1980 to see tremendous difference.

SALES PRICE OF HOMES	NUMBER OF HOMES (f)	CLASS MARK (x)
0– 5,000	3	2,500
5,000– 9,999	3	7,500
10,000–14,999	7	12,500
15,000–19,999	69	17,500
20,000–24,999	158	22,500
25,000–29,999	347	27,500
30,000–34,999	520	32,500
35,000–39,999	1136	37,500
40,000–44,999	1249	42,500
45,000–49,999	1401	47,500
50,000–54,999	1357	52,500
55,000–59,999	1028	57,500
60,000–64,999	946	62,500
65,000–69,999	781	67,500

$$\bar{x} = \frac{\sum f_x}{\sum f} = \frac{435{,}740{,}000}{9005} = 48{,}389$$

$$\text{Median} = L + c\left(\frac{j}{f_m}\right) = 45{,}000 + 5000\left(\frac{1010.5}{1401}\right) = 48{,}606$$

$$SK = \frac{3(\text{mean} - \text{median})}{s} = \frac{3(48{,}389 - 48{,}606)}{11{,}715} = -0.0556$$

$$\alpha_3 = \frac{\dfrac{1}{f_i}\left[\displaystyle\sum_{i=1}^{n}(x_i - \bar{x})^3 f_i\right]}{s^3} = \frac{\dfrac{1}{9005}(3.5842 \times 10^{15})}{1.6077 \times 10^{12}} = -0.2476$$

We can see from both measures that this distribution is negatively skewed.

EXAMPLE

Table 2.5 lists the distribution of state indebtedness for 1975. Calculate the SK and α_3 measures of skewness for this distribution.

$$\bar{x} = \frac{\sum f_x}{\sum f} = \frac{75{,}750}{50} = 1{,}515$$

$$s = \sqrt{\frac{\sum f(x - \bar{x})^2}{\sum f}} = \sqrt{\frac{2.8143 \times 10^8}{50}} = 2372.5$$

$$\text{Median} = L + c\left(\frac{j}{f_m}\right) = 500 + 500\left(\frac{6}{11}\right) = 772.73$$

$$SK = \frac{3(\text{mean} - \text{median})}{s} = \frac{3(1515 - 772.73)}{2372.5} = 0.939$$

$$\alpha_3 = \frac{\dfrac{1}{f_i}\left[\displaystyle\sum_{i=1}^{n}(x_i - \bar{x})^3 f_i\right]}{s^3} = \frac{\dfrac{1}{50}(2.5479 \times 10^{12})}{1.3354 \times 10^{10}} = 3.8159$$

As indicated, this is an example of an extremely positively skewed distribution, since for a symmetric distribution both of these measures are zero. A plot of this would be approximately as Figure 2.11.

KURTOSIS

Kurtosis is the degree of peakedness of a distribution. If a distribution has a relatively high peak, it is called "leptokurtic," while one with a relatively flat top is called "platykurtic." (See Figure 2.13(a) and (b).) The normal distribution (discussed extensively in Chapter 5), whose peakedness is neither very peaked nor very flat, and is also symmetric, is referred to as "mesokurtic." See Figure 2.13(c).

The most common measure of peakedness or kurtosis is denoted α_4 and is defined as the average of the fourth power of the deviations from the mean divided by the fourth power of the standard deviation. For ungrouped data

$$\alpha_4 = \frac{\frac{1}{n}\sum_{i=1}^{n}(x_i - \bar{x})^4}{s^4}$$

and for grouped data

$$\alpha^4 = \frac{\frac{1}{f_i}\left[\sum_{i=1}^{k}(x_i - \bar{x})^4 f_i\right]}{s^4}$$

Measures of peakedness are more difficult to interpret than measures of skewness and are less frequently used. While kurtosis is an interesting theoretical area in statistics, it has little application. There are other measures of peakedness that have been designed, but they all have two traits in common with α_4: They are rather tedious to calculate and are relative measures of peakedness to the normal distribution. α_4 for the normal distribution is 3, α_4 for leptokurtic distributions would be greater than 3, and for platykurtic distributions would be less than 3.

Figure 2.13   Peakedness

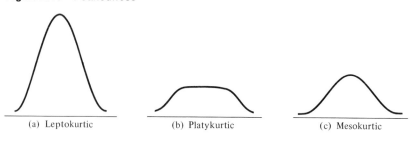

(a) Leptokurtic   (b) Platykurtic   (c) Mesokurtic

EXAMPLE

For the first example under skewness, of the distribution of 1970 sales prices of homes in Palo Verde, California, calculate the α_4 measure of peakedness.

$$\alpha_4 = \frac{\frac{1}{f_i}\left[\sum_{i=1}^{k}(x_i - \bar{x})^4 f_i\right]}{s^4}$$

$$= \frac{\left(\frac{1}{9005}\right)4.3294 \times 10^{20}}{1.8833 \times 10^{16}} = 2.553$$

We can tell then from this computation that in addition to being somewhat negatively skewed, this distribution is slightly less peaked than the normal distribution.

PROBLEMS

2.1 Write the following using summation notation:

(a) $f_1 x_1^2 + f_2 x_2^2 + \cdots + f_n x_n^2$

(b) $x_1 y_1 + x_2 y_2 + \cdots + x_{50} y_{50}$

(c) $1 \cdot 2 + 2 \cdot 3 + 3 \cdot 4 + \cdots + n(n+1)$

(d) $f(5) + f(10) + f(20) + f(40) + f(80)$

(e) $x + (x-1) + x^2 + (x-1)^2 + x^3 + (x-1)^3 + x^4 + (x-1)^4$

(f) $xy + x^2 y^2 + x^3 y^3 + x^4 y^4 + x^5 y^5$

2.2 Write out a number of terms in the sums indicated by the following:

(a) $\sum_{x=1}^{n} 3^x$ (b) $\sum_{x=1}^{5} C_x$ (c) $\sum_{10,20,\ldots}^{100} xf(x)$

(d) $\sum_{i=1}^{40} x_{5i}$ (e) $\sum_{x=0}^{n} f(p + xq)$ (f) $\sum_{i=1}^{5}(100)y_i$

2.3 The marks of a group of students on a statistics exam are grouped into the following classes:

$$10-24$$
$$25-39$$
$$40-54$$
$$55-69$$
$$70-84$$
$$85-99$$

Find the corresponding

(a) Class boundaries

(b) Class marks

(c) Class interval

2.4 Given the following information:

INTERVAL i	WEIGHTS OF SQARKS (lb) (CLASS BOUNDARIES)	FREQUENCY f
1	10.5–20.5	6
2	20.5–30.5	25
3	30.5–40.5	32
4	40.5–50.5	15
5	50.5–60.5	2

Calculate the cumulative frequency based on a "less than" distribution.

2.5 Given the following distribution of weights of students:

CLASS (lb)	f
110–119	2
120–129	6
130–139	11
140–149	14
150–159	9
160–169	5
170–179	3
	50

(a) Determine the class boundaries.
(b) Construct a cumulative frequency table based on a "less than" distribution.
(c) Draw a cumulative frequency curve.
(d) Using the cumulative frequency curve, find the weight above which there are 50 percent of the students.

2.6 A plane flies in the figure of a square. It flies the first side at 100 mph, the second at 200 mph, the third at 300 mph, and the fourth at 400 mph. What is the average speed of the plane?

2.7 Mr. Small has a cottage 120 mi from his office in New York City. One Friday he left his office early and was able to make the trip at an average speed of 60 mph. He overslept Monday morning and so decided not to go to the office. Instead, he drove to Jersey City, 180 mi from the cottage. Since he started late, the traffic was heavy and he averaged only 30 mph on the trip to Jersey City. What was Mr. Small's average speed for the total trip?

2.8 Mrs. Flutterby invests $1000 in each of the following common stocks:

> Great Society Co.—selling at $20/share
> Mississippi Scheme Co.—selling at $50/share
> South Sea Bubble Co.—selling at $100/share

For her total investment of $3000, what is the average price paid by Mrs. Flutterby per share?

2.9 In a class of 30 students there are 17 whose ages in years are 20 but less than 30; 11 whose ages are 30 but less than 40; and 2 whose ages are 40 but less than 50. Obtain the mean age and standard deviation of their ages.

2.10 The heights of students (in inches) in a certain class had the following distribution (figures under column headed "height" are given as directed boundaries):

HEIGHT	f
62–63	2
63–64	6
64–65	14
65–66	16
66–67	8
67–68	3
	50

(a) What is the meaning of directed boundaries.
(b) Calculate the median height of the students (correct to one decimal place).
(c) Calculate the modal height of the students (correct to one decimal place).
(d) Calculate the standard deviation of the heights (correct to one decimal place).

2.11 Why do the median and quartile deviation fall at the same place only in a symmetrical distribution.

2.12 For a distribution with $n = 100$, $Q_1 = 7.5$, and $Q_3 = 9.0$, the range of the distribution is 10.5. Obtain each of the following or explain why it cannot be obtained:
(a) The quartile deviation
(b) The median
(c) The direction of skewness

2.13 The Moonshiners Unofficial Committee on Standards designates a certain type of moonshine as Grade Q. In the Tarpon area Grade Q sells for 5 pints for $1. In the Tampa area it sells at 10 points for $1. In the Tallahassee area it sells at 30 pints for $1. Assume that $1 was spent on Grade Q at each location on a week-end spree. Answer the following concerning the whole week-end:
(a) How many pints of Grade Q were bought?
(b) What was the arithmetic mean price per pint at each location?
(c) What is the price per pint obtained as the simple arithmetic average of the three answers obtained in (b)?
(d) Taking the answer in (c) as the average price per pint, how many pints could be bought for $3?
(e) Why is the answer for (d) different from that for (a)?

2.14 The Gong Telephone Company of Boondard County, Florida started service on January 1, 1971, at which time it had 10,000 telephone poles. (Assume for record-keeping that all of these telephone poles began their lives on this date of initial service.) After some time, these telephone poles deteriorate and must be replaced. Below is a list of the number of original poles still in service as of January 1 of the indicated years:

YEAR	NUMBER OF POLES
1976	1000
1976	2000
1974	5000
1973	7000
1972	9000

Assuming that all poles replaced in a given year were replaced on December 31 of that year, calculate each of the following or state why it cannot be obtained:

(a) Mean age
(b) Median age
(c) Modal age

Note: In the above, age in each case means useful age or useful life.

2.15 Given the following distribution of weekly wages:

INTERVAL NUMBER (i)	DIRECTED BOUNDARIES	FREQUENCY f
1	$ 50–100	15
2	100–150	20
3	150–200	30
4	200–250	15
5	250–300	10
6	300–350	6
7	350–400	4

Calculate the following:
(a) Mean wage
(b) Median wage
(c) Modal wage
(d) Standard deviation of the distribution
(e) Average deviation of the distribution

2.16 The following distribution gives lengths of carbon steel rods to the nearest tenth of an inch.

INTERVAL NUMBER (i)	CLASS LIMITS	FREQUENCY f
1	10.0–11.9	15
2	12.0–13.9	20
3	14.0–15.9	30
4	16.0–17.9	15
5	18.0–19.9	10
6	20.0–21.9	6
7	22.0–23.9	4

Determine:
(a) The class boundaries
(b) The mean
(c) The median
(d) The standard deviation
(e) The mode

2.17 For what type of distribution is it impossible to calculate the Pearson coefficient of skewness?

2.18 If, for a particular distribution

$$\bar{x} = 20$$

$$\text{median} = 22$$

$$s = 10$$

(a) Compute the Pearson Coefficient of Skewness.
(b) Determine the direction in which the distribution is skewed.
(c) Compute the coefficient of variation.

2.19 Professor Theorem has two lab sections for the course he teaches and is curious to know how well the students are attending them. He asks each of the instructors to draw a random sample of six students and to count their absences. The instructors provide the following information:

SAMPLE 1	SAMPLE 2
0	2
2	2
8	4
8	8
8	8
10	12

Compute
(a) Mean
(b) Standard deviation
(c) Pearson Coefficient of Skewness
for each sample. Using this information, summarize the comparison of the samples with respect to location, variation, and shape of the distribution.

2.20 Given the following sample of data:

$$
\begin{array}{r}
11 \\
6 \\
-1 \\
0 \\
10 \\
-2 \\
1 \\
15 \\
-4
\end{array}
$$

Determine the
(a) Mean
(b) Median
(c) Mode
(d) Standard deviation
(e) α_3 measure of skewness
(f) α_4 measure of kurtosis

2.21 Given the following information from the Federal Statistical System of the 1975 value of federal construction jobs in 13 western states (in millions of dollars), calculate,
(a) The Pearson coefficient of skewness
(b) α_3 measure of skewness
(c) α_4 measure of kurtosis

STATE	AMOUNT	$(x - \bar{x})$	$(x - \bar{x})^2$	$(x - \bar{x})^3$	$(x - \bar{x})^4$
1	13.4	−87.7	7,691	−674,526	59,155,942
2	22.4	−78.7	6,194	−487,443	38,361,796
3	25.6	−75.5	5,700	−430,369	32,492,850
4	27.0	−74.1	5,491	−406,869	30,148,994
5	29.3	−71.8	5,155	−370,146	26,576,499
6	56.8	−44.3	1,962	−86,938	3,851,367
7	72.5	−28.6	818	−23,394	669,059
8	102.4	1.3	2	2	3
9	107.9	6.8	46	314	2,138
10	135.6	34.5	1,190	41,064	1,416,695
11	143.6	42.5	1,806	76,766	3,262,539
12	206.9	105.8	11,194	1,184,287	125,297,577
13	370.6	269.5	72,630	19,573,852	5,275,153,207

$$\bar{x} = \frac{1314.0}{13} = 101.1$$

$$s^2 = 9221.5$$

$$\text{median} = 72.5$$

TIME SERIES AND INDEX NUMBERS

3.1 TIME SERIES

The analysis of time series is often taken up as part of regression and will be discussed in that context in Chapter 13. However, time series as well as index numbers are widely used as descriptive measures of the economic environment. Almost daily the news media bombard readers and viewers with some aspect of time series or index numbers, whether it be the inflation rate, the rate of unemployment, the level of automobile production, the seasonal grain harvests, the changes in the wholesale price index, and so forth. Although a detailed study of these concepts belongs in economics courses, students should be aware of the construction and general meaning of either time series or index numbers insofar as they represent a part of statistics. Therefore this chapter discusses the components and decomposition of time series, moving averages, trends, seasonal adjustment, and some varieties and application of index numbers.

A *time series* is a sequence of observations taken on some process that varies over time. There can be many varieties of time series. Some examples might be: the annual Russian wheat harvest, the monthly automobile production of the United States, the weekly Canadian government bond interest rate, the hourly temperature recorded at the Mexico City International airport.

A graph of the annual U.S. production of refined copper from 1956 through 1975 is shown in Figure 3.1. Note that there is a point observation for each year, but connecting these points makes it easier to grasp quickly the variation of the series. Time series may be listed in tabular form, as in Table 3.1 for the information in Figure 3.1, but more often the graphical method is used. The representation of a time series can also be referred to as a line chart (from Section 2.3 in Chapter 2).

Although it may be desirable at times to examine only what occurred in a past sequence of observations, the primary purpose of studying time series in business and economics is to forecast. The most sophisticated and complex forecasting method involves the use of econometric models—an elementary discussion is contained in Chapter 13. This method requires specification of economic relationships and statistically testing them to derive a forecasting

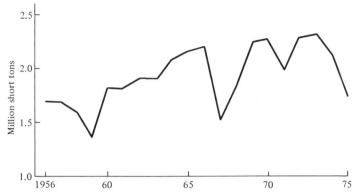

Figure 3.1   Annual U.S. production of refined copper 1956–1975

model. Even with sound economic theories, however, the true process generating economic events is too complicated and contains too many unstable interactions to permit the derivation of an exact explanatory or forecasting equation.

There are two other more descriptive and less mathematical methods of forecasting. These may be described in various ways, but here we call them the *derivative* method and the *symptomatic* method. Under the derivative method, the forecaster attempt to predict certain business conditions or trends by building up the elements that in their parts constitute or contribute to the item or items being forecast. With the symptomatic method, the forecaster attempts to examine those factors that will reveal by their present behavior what economic conditions will be like in the future.

The three preceding methods require the use of time series: In the first to test the econometric models, in the second to build up the forecast from component parts, and in the third to study what are considered to be leading economic indicators.

The analysis of time series in business and economics is to some extent analogous to the study of communication theory in the field of electricity. The electronic engineer examines the signal from the communication medium to isolate the true signal from the noise. Looking at time series, the business

TABLE 3.1 ANNUAL U.S. PRODUCTION OF REFINED COPPER 1956–1975			
1956	1,703,832	1966	2,202,308
1957	1,691,082	1967	1,539,596
1958	1,598,528	1968	1,853,965
1959	1,348,480	1969	2,241,937
1960	1,810,641	1970	2,276,703
1961	1,820,557	1971	1,992,444
1962	1,901,478	1972	2,296,476
1963	1,898,430	1973	2,333,588
1964	2,007,445	1974	2,151,566
1865	2,156,845	1975	1,785,958

Source: Mineral Industry Surveys.

economist or statistician must attempt to isolate the true process generating the time series from the "noise" appearing in the observations. This may not be an easy task for either the engineer or the business analyst.

It is often useful, in general, to divide the series into *trend, cycle*, and *seasonal* components as the first step in time series analysis. After these three are removed from a series, the *random* component is left, analogous to "noise" in a communication system. However, the analogy is not complete. If the random components are plotted (discussed later), the analyst may note "outliers" that are not explained by the other three components. An outlier is an observation that lies noticeably outside the range of the other observations. In marketing if a store ran a special sale or if it ran out of inventory, either of these events could produce observations and thus random plots that would be outliers. If a flood occurred in an area where tourists normally visit, this event could reduce tourist trade dramatically. Although it would appear as random, it can be explained. In examining random, or what appear to be random observations, the analyst should always look for the commonsense explanation of outliers.

The two basic time series models usually utilized are the *additive* and *multiplicative* models. Because it is generally considered to give a better description, we will examine only the multiplicative model:

$$Y = T \cdot S \cdot C \cdot R$$

where:

Y = the observed data

T = the trend component

S = the seasonal component

C = the cyclical component

R = the random component

It should be obvious in the multiplicative model that all the numbers cannot be of the same units of observation. Trend can be considered the most important component and is calculated in the same units as the observed data, while the others are calculated as indexes. This will be explained next in the discussion of the components.

TREND

The trend component of a time series is the general movement of the series in a decreasing, increasing, or flat tendency, or it can be defined as the long-term change in the dependent variable over time. In Figure 3.2 we see the increasing trend for 21 years of the deposits in savings and loan insitutions from 1955 through 1975. However, because a trend continues for some time does not indicate that it will persist forever. Figure 3.3 shows the proved reserves of natural gas in the United States from 1955 through 1975. Although the trend was increasing for most of these years, it has reversed and is now declining. Copper production in Figure 3.1 would seem to indicate no apparent trend.

Other than simply drawing a line through data by eye, there are a number of ways to estimate trend in a time series. The quickest and simplest is the

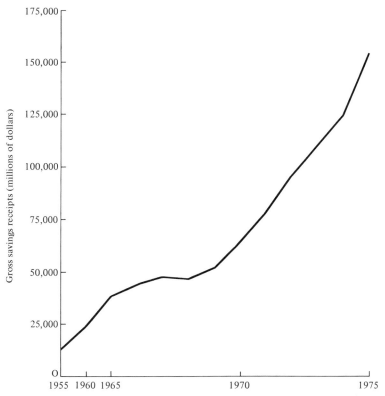

Figure 3.2   Savings flows at savings associations

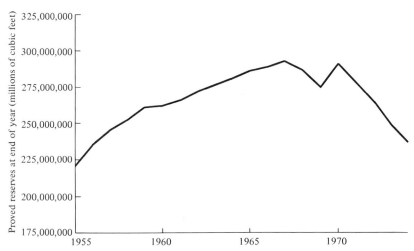

Figure 3.3   Annual estimates of proved natural gas reserves in the United States

"method of averages," discussed below. Note that implicit in the following is the use of "linear" or straight-line trends. Estimating nonlinear trends is much more difficult than estimating linear trends. Nonlinear regression is considered in Chapter 13.

TABLE 3.2   POWER CONSUMPTION IN ARIZONA (1,000,000 kwh)

Month/Year	1970	1971	1972	1973	1974	1975
Jan.	1013	1156	1318	1440	1460	1711
Feb.	950	1090	1247	1353	1378	1558
Mar.	953	1058	1215	1275	1357	1474
Apr.	974	1082	1250	1289	1365	1440
May	1048	1090	1361	1407	1525	1529
June	1292	1268	1550	1760	1882	1815
July	1481	1550	1771	2022	2119	2066
Aug.	1499	1634	1891	2080	2162	2228
Sept.	1402	1615	1800	2055	2142	2169
Oct.	1133	1379	1469	1709	1870	1821
Nov.	971	1162	1258	1451	1581	1495
Dec.	1051	1148	1262	1297	1441	1493

Source: Federal Power Commission

Recall from elementary algebra that the equation for a straight line is $y = a + bx$ where:

y = observations on the vertical axis, the dependent variable

x = observations on the horizontal axis, the independent variable

a = intercept of the line on the vertical axis

b = slope of the line

In the case of a time series, x is time and the letter t will be used following.

The method of averages for fitting a straight time series trend is one in which the algebraic sum of the residuals is zero. (*Residual* is defined as the difference between an observation at a point in time and the value read from the trend line at that point in time.) The method of averages first divides the time series in two. If there are a total of N observations, the t's and y's are summed in two segments as follows:

$$\sum_{1}^{N/2} y = \frac{N}{2} a + b \sum_{1}^{N/2} t$$

$$\sum_{(N/2)+1}^{N} y = \frac{N}{2} a + b \sum_{(N/2)+1}^{N} t$$

Then the parameters a and b are calculated.

Table 3.2 shows electricity consumption by month from 1970 through 1975 in the state of Arizona. Figure 3.4 gives a graph of this data, along with the method of averages trend line, calculated as follows: There are 72 observations, hence $N/2 = 36$, and from the first equation

(1)

$$\sum_{1}^{36} y = 36a + b \sum_{1}^{36} t$$

$$46,391 = 36a + b(666)$$

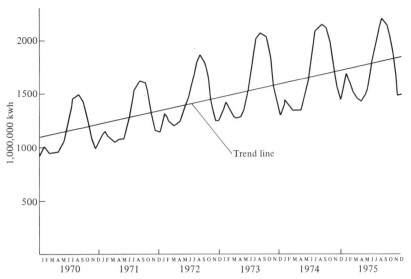

Figure 3.4 Power consumption in Arizona actual data and trend line

From the second equation

(2)
$$\sum_{37}^{72} y = 36a + b\sum_{37}^{72} t$$

$$60{,}219 = 36a + b(1962)$$

Recall from algebra that with two equations, (1) and (2), and two unknowns (in this case a and b) simultaneous equations can be used to solve for the unknowns. Subtracting the first from the second, we have

$$13{,}828 = b(1296)$$

$$b = 10.67, \qquad a = 1091.25$$

Therefore the trend line is

$$y = 1091.25 + 10.67t$$

The actual power consumption figures from Table 3.2 are listed in Table 3.4 under $TSCR$, that is, containing all the time series components. The values of trend calculated from the trend line formula are listed under T in the same table, and the trend line drawn through the data is shown in Figure 3.4.

SEASONAL VARIATION

In addition to trend, the most important estimation component in time series is the seasonal component. An investigator of time series data frequently wishes to determine what the value of the variable would be if seasonality were not present. For example, this is done in determining for the economy seasonally adjusted sales, unemployment, rate of inflation, and so forth. Although the block of time within which the seasonally adjusted values are to be determined may be other than a year, the most frequently analyzed time span is the year, which we will use in this section.

There are a number of ways to isolate the seasonal component of a time series. The simplest and most direct is to adjust each month's values by the ratio of the actual fraction of the year's total that the month's values represent and the proportionate fraction that the month represents of the year as time. Take, for example, the month of May:

$$(\text{actual value for May})\left(\frac{\dfrac{\text{actual value for May}}{\text{total value for year}}}{\dfrac{\text{days in May}}{\text{days in year}}}\right)$$

$$= \text{seasonally adjusted value for May}$$

Instead of using only the figures for the year in question, we can calculate and substitute, an index of a number of years for the above fraction. This procedure would change the fraction to

$$\left(\frac{\dfrac{\text{actual values for May averaged for } n \text{ years}}{\text{total values for year averaged for } n \text{ years}}}{\dfrac{\text{days in May averaged for } n \text{ years}}{\text{days in year averaged for } n \text{ years}}}\right)$$

In either of the above equations if the problem dealt with, say sales days, it would be correct to use days or averaged days as indicated. But if the problem dealt with something such as power consumption that occurs every day, it would be accurate simply to consider a month as one-twelfth of a year.

Although this method is more direct, the one frequently used for seasonal adjustment is known as the "ratio to moving average" method. The essential difference between these two methods consists of substituting a moving average for actual values for May in the fraction. That is, in the first fraction for "actual value for May" a "moving average for May" would be substituted. In the second fraction for "actual values for May averaged for n years" a "moving average values for May averaged for n years" would be substituted. The easiest procedure for computation is to obtain the ratio:

$$\frac{\text{actual value for May}}{\text{moving average value for May}}$$

for May of every year. The average of this ratio over the several years constitutes the seasonal adjustment "index" for May. Then the actual May figure is divided by this index to obtain the May seasonally adjusted value.

A moving average is the type of average that in moving from one period to another drops the first period and adds the next period. For example, a three-month moving average would average values for the months 1, 2, 3; 2, 3, 4; 3, 4, 5; and so forth. However, it should be obvious that for a given month,

TABLE 3.3 SEASONAL INDEX FOR POWER CONSUMPTION IN ARIZONA

RATIO OF ACTUAL TO MOVING AVERAGE

Month/Year	1970	1971	1972	1973	1974	1975	Average or Seasonal Index
Jan.	1.054	1.052	1.057	1.065	1.059	1.090	1.063
Feb.	0.977	0.990	0.990	0.998	0.985	0.985	0.988
Mar.	0.994	0.983	0.982	0.977	0.993	0.989	0.986
Apr.	0.982	1.005	0.980	0.974	0.964	0.972	0.980
May	0.949	0.951	0.981	0.947	0.959	0.959	0.958
Jun.	1.014	0.973	0.993	1.018	1.022	1.006	1.004
Jul.	1.040	1.044	1.019	1.035	1.031	1.015	1.031
Aug.	1.026	1.021	1.039	1.013	1.034	1.010	1.023
Sep.	1.043	1.047	1.047	1.055	1.041	1.046	1.047
Oct.	0.969	0.995	0.973	0.983	1.003	0.996	0.987
Nov.	0.923	0.945	0.946	0.977	0.970	0.933	0.949
Dec.	0.992	0.949	0.956	0.925	0.913	0.968	0.951

say May, the moving average for May would always be April, May, June. The purpose of calculating a moving average is referred to as *smoothing*, that is, to remove some of the fluctuation in a time series and show the underlying components of the series. In calculating a three-month moving average, for example, the investigator is attempting to isolate the seasonal component of the time series. A thirteen-month moving average, on the other hand, would be an attempt to reveal a cycle that may exist in the data that is of longer duration than a year.

To adjust seasonally the power consumption data in Table 3.2, we use a three-month moving average in Table 3.3 to calculate the seasonal index for each month. (To obtain the moving average for the end months of the data shown, we must use consumption for December 1969 and for January 1976.) This index is the average figure for each month in the last column, which when divided into the actual consumption figures yields seasonally adjusted monthly consumption. These figures are given in Table 3.4.

An examination of the time series of power consumption in Arizona and a decomposition of such series might well be the first step in forward planning for the electric power generating companies in that state. The pronounced seasonality in demand would indicate the necessity to make available "peaking" generating equipment to meet the maximum demand in the summer months. Also, the significant rising trend in consumption would require longer range planning for supplying future demand and the investment needed in generating plant, transmission lines, and distribution system. This type of planning would permit a power company to meet customer demand and lessen the chance of brownouts or blackouts.

Recognizing the difference between the component data itself and indexes of components in the decomposition of time series can be confusing. Also, the student should be aware that the first effort is to calculate trend, and the departures from trend are those other components that one tries to estimate in relation to trend. Thus it should be recognized that the seasonally adjusted data given in Table 3.4 only conveys to the investigator approximately what the data would

TABLE 3.4 COMPONENTS OF ELECTRIC
POWER CONSUMPTION IN ARIZONA 1970-1975

Date		$TSCR$	T	S	TS	CR
1970	J	1013	1102	953	1171	0.865
	F	950	1113	962	1100	0.864
	M	953	1123	967	1107	0.861
	A	974	1134	994	1111	0.876
	M	1048	1145	1094	1097	0.955
	J	1292	1155	1287	1160	1.114
	J	1481	1166	1436	1202	1.232
	A	1499	1177	1464	1205	1.244
	S	1402	1187	1339	1243	1.128
	O	1133	1198	1148	1182	0.958
	N	971	1209	1023	1147	0.846
	D	1051	1219	1105	1159	0.907
1971	J	1156	1230	1087	1308	0.884
	F	1090	1241	1103	1226	0.889
	M	1058	1251	1073	1234	0.858
	A	1082	1262	1104	1237	0.875
	M	1090	1273	1138	1220	0.894
	J	1268	1283	1263	1288	0.984
	J	1550	1294	1503	1334	1.162
	A	1634	1305	1596	1336	1.223
	S	1615	1315	1543	1377	1.173
	O	1379	1325	1397	1308	1.054
	N	1162	1337	1224	1269	0.916
	D	1148	1347	1207	1280	0.896
1972	J	1318	1358	1240	1444	0.913
	F	1247	1369	1262	1353	0.922
	M	1215	1379	1232	1360	0.893
	A	1250	1390	1276	1362	0.918
	M	1361	1401	1421	1342	1.014
	J	1550	1411	1544	1417	1.094
	J	1771	1422	1718	1466	1.208
	A	1891	1433	1848	1466	1.290
	S	1800	1443	1719	1511	1.191
	O	1469	1454	1488	1435	1.024
	N	1258	1465	1326	1390	0.905
	D	1262	1475	1327	1403	0.900
1973	J	1440	1485	1355	1579	0.912
	F	1353	1497	1369	1479	0.915
	M	1275	1507	1293	1486	0.858
	A	1289	1518	1315	1488	0.866
	M	1407	1529	1469	1465	0.960
	J	1760	1539	1753	1545	1.139
	J	2022	1550	1961	1598	1.265
	A	2080	1561	2033	1597	1.302
	S	2055	1571	1963	1645	1.249
	O	1709	1582	1732	1561	1.095
	N	1451	1593	1529	1512	0.960
	D	1297	1603	1364	1524	0.851
1974	J	1460	1614	1373	1716	0.851
	F	1378	1625	1395	1606	0.858
	M	1357	1635	1376	1612	0.842
	A	1365	1645	1393	1612	0.847
	M	1525	1657	1592	1587	0.961

TABLE 3.4 COMPONENTS OF ELECTRIC
POWER CONSUMPTION IN ARIZONA 1970–1975 (*continued*)

Date		*TSCR*	*T*	*S*	*TS*	*CR*
	J	1882	1667	1875	1674	1.124
	J	2119	1678	2055	1730	1.225
	A	2162	1689	2113	1728	1.251
	S	2142	1699	2046	1779	1.204
	O	1870	1710	1895	1688	1.108
	N	1581	1721	1666	1633	0.968
	D	1441	1731	1515	1646	0.875
1975	J	1711	1742	1610	1852	0.924
	F	1558	1753	1577	1732	0.900
	M	1474	1763	1495	1738	0.848
	A	1440	1774	1469	1739	0.828
	M	1529	1785	1596	1710	0.894
	J	1815	1795	1808	1802	1.007
	J	2066	1805	2004	1861	1.110
	A	2228	1817	2178	1859	1.198
	S	2169	1827	2072	1913	1.134
	O	1821	1838	1845	1814	1.004
	N	1495	1849	1575	1755	0.852
	D	1493	1859	1570	1768	0.844

be if no seasonal variation were present, and the trend data given only shows approximately what the data would be if no fluctuations from trend existed. Obviously, therefore, it would not be proper to multiply the two data figures together to obtain both trend and seasonal. Instead, they must be combined by multiplying the trend figures by the seasonal index (given in the last column of Table 3.3). Thus we obtain the combination listed under the heading *TS* in Table 3.4. In the formula $Y = TSCR$, then, the *S* multiplied by *T* would not be the *S* in Table 3.4, but the *S* index in Table 3.3 (last column).

CYCLE AND RANDOM

In the decomposition of time series there is little doubt that the cycle component is the most difficult conceptual element. Cycles do seem to be an integral part and natural attribute of human economic endeavor. For example, for many years there has been reference to "business cycles" in overall economic activity, and economists have spent a great deal of time and effort diagnosing "causes" and prescribing "cures" for the elimination of cycles (with less than universal success). In addition, analysts frequently say that certain industries (such as steel and automobiles) are "cyclical," by which is meant that the demand for the product produced rises and falls over time. Despite the existence of cycles, the generality of any analysis of them is open to question, and any results from studying cycles other than describing their historical occurrence is met with skepticism.

The reason for skepticism and the single most important facet of and fundamental difficulty with cycles is *periodicity*. Something being periodic in the analytic sense means that it occurs or appears at regular intervals of time. The phases of the moon are periodic, the return of locusts to certain areas are periodic, the publication of a monthly magazine is periodic, and so forth. If something occurs or appears at regular intervals, then it is predictable, or at least approximately so. If it appears or recurrs only at irregular or nonperiodic intervals, then it is not predictable and cannot be utilized in any straightforward way in forecasting anything influenced by it. What is even worse from the standpoint of the business cycle, it is not possible to say exactly where in a cycle the economy is at a given point in time until afterwards.* Someone once described a business cycle as a bird that flies backwards: the bird can tell where it has been but cannot tell where it is going.

If cycles in business and economic activity are recurrent but not periodic, then estimating a cyclic component from time series data is difficult. We must ask ourselves how to go about isolating it and then, what does it mean? One first step might be to take the position that since seasonality is a type of recurrent periodic cycle, this is the only cycle of less than a year that we will assume the data contains. Having determined that any cycles in the data must be longer than a year, we then ask, "how long?" There are two basic ways to take out the cyclic component, and one of these must be considered somewhat unsatisfactory and definitely *ad hoc*.

The first, and the most defensible, is for the investigator to have some idea of the length of the cycle. Where this comes from, and its validity, represents a less than crystal clear area in time series decomposition. It must derive from some investigation or experience that either the investigator has had or can draw upon, and in any case it faces the problem just discussed: periodicity. For example, Problem 3.1 asks for the cyclical component of annual U.S. automobile production as a three-year cycle, based on historical studies that claim this as the length of the cycle. The three-year cycle in automobile production is put forward only tentatively; however, even taking only the post-World War II experience, there are quite a number of exceptions and departures from a definitive three-year cycle.

The second method of treating the cycle, if one is presumed present but unknown, is to calculate a cyclical index based on smoothing with a three- or five-period (month, quarter, or whatever) moving average in the same manner as constructing the seasonal index. We will not attempt to justify this, but merely state that it is a fairly widespread practice. This method will be used in the example on imported car sales (p. 56).

If a certain month or year moving average is used to remove the cyclical component, then the random component remains and can be estimated as follows: Remove the combination of trend and cycle (TS), previously discussed, from the original data ($TSCR$) by division. What is left is CR. Divide this by C, the cyclic moving average, and the random component remains.

However, there is even a simpler and often more necessary approach to the cyclic and random components. Based on the discussion of the difficulties surrounding interpretation of the cycle, we can argue that the cyclic component can be ignored altogether and that cycle (if any) and random can be viewed jointly as simply random. This would be the case when there was little or no

---

* The classic book on this subject is Arthur F. Burns and Wesley C. Mitchell, *Measuring Business Cycles* (New York: National Bureau of Economic Research, 1946). This book gives not only a description of how the various segments of the business cycle are determined, but also points out the pitfalls in decomposition of time series.

reason to postulate a cycle present in the data, which is the view taken toward electric power consumption in Arizona 1970–1975. There is no reason to suppose that a cycle in the consumption of electric power exists. It would appear reasonable to assume that such consumption consists only of a trend, seasonal variation, and random fluctuations.

Table 3.4 shows the following: original data ($TSCR$), trend (T), seasonally adjusted values (S), combination of trend and seasonal (TS), and the random component (in this case CR). Note that CR is listed as an index; multiplying TS by this index will yield the original data.

The student should recognize that one may talk of the trend figure, the deseasonalized figure, the cyclical figure, but that there is no random figure in the same sense. The random component is simply an unexplained residual, it is the deviation from the combined TSC (or TS if C and R are together) and the observed figure. Hence in Table 3.4 TS together (for January 1972) is 1444; multiplying this by $CR = .913$ will yield 1318, the observed figure. Of course, this is so since .913 was obtained by dividing 1318 by 1444. In other words, the difference between the observed figure and the figure containing both trend and seasonal is the unexplained or random component. For January 1972, $1318 - 1444 = -126$.

In this case, since there is no reason to postulate a cycle in power consumption, C and R were combined as representing the random component. If there is *a priori* reason to assume a cycle of some specified length (as in Problem 3.1), or if the cycle is smoothed (as in the imported car example below), then this cyclical component is identified in the same manner as the seasonal component. Then it is possible to calculate decyclical figures as it is to calculate deseasonal figures. The cyclical index would be calculated similarly to calculating the seasonal index in Table 3.3. Having both indexes, we can calculate TSC from the trend figure. Then the difference between this combined figure and the observed figure would be the random component.

EXAMPLE

Suppose it was desired to develop a short-run sales forecasting model to predict the quarterly sales of imported cars in the United States using known quarterly data for years 1969 through 1976. Use a four-quarter moving average to deseasonalize, and place this average at the third quarter since there is no center to four quarters. Then use the deseasonalized data to calculate the trend line rather than the unadjusted data. (This procedure is sometimes employed, but it is difficult to determine which method is better.) For the following observed data use a three-quarter moving average for the cyclical component index, calculate the random component index as a residual, and show a plot of the random components.

QUARTERLY IMPORT CAR RETAIL SALES IN 1000's
(*AUTOMOBILE FACTS AND FIGURES*)

QUARTER	1969	1970	1971	1972	1973	1974	1975	1976
1	234	272	361	368	465	370	409	329
2	307	341	448	417	512	343	431	381
3	297	326	438	430	426	398	443	421
4	281	340	323	406	377	306	306	369

Now plot the observed data, which follows. Then, calculate the four-quarter moving total, the four-quarter moving average, the seasonal index and the deseasonalized data.

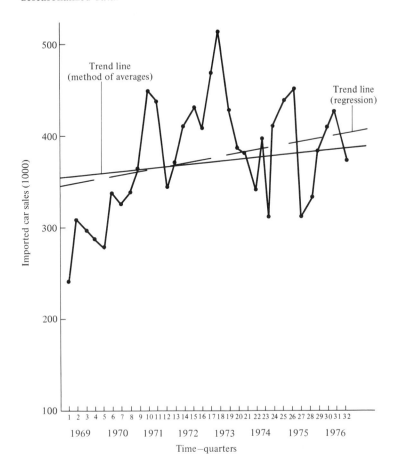

Next, calculate the trend line and trend data from the deasonalized data. Although regression is not considered until Chapter 13, for comparison, trend was determined by regression as well as by the method of averages. Subsequent calculations, however, are made using the method of averages of trend results.

Method of averages: $\bar{X}t_1 = 8.50$ $\bar{X}t_2 = 24.50$

$\bar{X}y_1 = 349.69$ $\bar{X}y_2 = 392.68$

Trend line $= 327.00 + 2.681t$

Trend data:

QUARTER	1969	1970	1971	1972	1973	1974	1975	1976
1	329.69	340.41	351.13	361.85	372.58	383.30	394.02	404.74
2	332.37	343.09	353.81	364.53	375.26	385.98	396.70	407.42
3	335.05	345.77	356.49	367.21	377.94	388.66	399.38	410.10
4	337.73	348.45	359.17	369.89	380.62	391.34	402.06	412.78

QUARTER	1969	1970	1971	1972	1973	1974	1975	1976	SEASONAL INDEX AVERAGE OF SALES DIVIDED BY MOVING AVERAGE	ADJUSTED SEASONAL INDEX
Four-quarter moving total										
1	–	1191	1475	1546	1813	1516	1544	1459		
2	–	1220	1587	1538	1809	1488	1589	1437		
3	1119	1279	1570	1621	1780	1417	1589	1500		
4	1157	1368	1577	1718	1685	1456	1509	–		
Four-quarter moving average										
1	–	297.75	368.75	386.50	453.25	379.00	386.00	364.75		
2	–	305.00	396.75	384.50	452.25	372.00	397.25	359.25		
3	279.75	319.75	392.50	405.25	445.00	354.25	397.25	375.00		
4	289.25	342.00	394.25	429.50	421.25	364.00	377.25	–		
Sales divided by moving average										
1	–	0.9135	0.9790	0.9521	1.0259	0.9763	1.0596	0.9020	0.9720	0.9678
2	–	1.1180	1.1292	1.0845	1.1325	0.9220	1.0850	1.0605	1.0760	1.0719
3	1.0617	1.0195	1.1159	1.0611	0.9573	1.1235	1.1152	1.1227	1.0721	1.0679
4	0.9715	0.9942	0.8193	0.9453	0.8950	0.8407	0.8111	–	0.8967	0.8924
									4.0174	4.0000
De-seasonalized data raw data divided by adjusted seasonalized index										
1	241.79	281.05	373.01	380.24	480.47	382.31	422.61	339.95		
2	286.41	318.13	417.95	389.03	477.66	319.99	402.04	355.44		
3	278.12	305.27	410.15	402.66	398.91	372.69	414.83	394.23		
4	314.88	381.00	361.95	454.95	422.46	342.90	342.90	413.49		

Linear Regression: Trend Line $= 323.12 + 2.916t$

Trend Data:

QUARTER	1969	1970	1971	1972	1973	1974	1975	1976
1	326.03	337.70	349.36	361.03	372.69	384.36	396.02	407.69
2	328.95	340.61	352.28	363.94	375.61	387.27	398.94	410.60
3	331.87	343.53	355.20	366.86	378.53	390.19	401.86	413.52
4	334.78	346.45	358.11	369.78	381.44	393.11	404.77	416.44

The cyclical and random components together can then be taken out as in the case of the electric power consumption by dividing the deseasonalized data (TCR) by the trend data (T), shown in the "CR" column of the following table. If it is desired simply to smooth the (unknown) cycle with, say, a three-period moving average, the results appear in the next column. Taking this moving average as the cyclical component and dividing it into the CR yields the random fluctuation, or random index.

YEAR	CR	THREE-PERIOD MOVING AVERAGE	RANDOM INDEX
1969—1	0.7334	—	—
2	0.8617	0.8084	1.0659
3	0.8301	0.8747	0.9490
4	0.9323	0.8627	1.0807
1970—1	0.8256	0.8950	0.9225
2	0.9272	0.8786	1.0553
3	0.8829	0.9678	0.9123
4	1.0934	1.0129	1.0795
1971—1	1.0623	1.1123	0.9550
2	1.1813	1.1314	1.0441
3	1.1505	1.1132	1.0335
4	1.0077	1.0697	0.9420
1972—1	1.0508	1.0419	1.0085
2	1.0672	1.0715	0.9960
3	1.0965	1.1312	0.9693
4	1.2300	1.2054	1.0204
1973—1	1.2896	1.2642	1.0201
2	1.2729	1.2060	1.0555
3	1.0555	1.1461	0.9208
4	1.1099	1.0543	1.0527
1974—1	0.9974	0.9788	1.0190
2	0.8290	0.9284	0.8929
3	0.9589	0.8880	1.0798
4	0.8762	0.9692	0.9040
1975—1	1.0726	0.9875	1.0862
2	1.0136	1.0428	0.9720
3	1.0387	0.9696	1.0713
4	0.8529	0.9117	0.9355
1976—1	0.8399	0.8563	0.9808
2	0.8724	0.8924	0.9776
3	0.9613	0.9463	1.0159
4	1.0017	—	—

In any problem that is being examined by time series decomposition or by linear regression, the random index should be plotted. (Most computer programs associated with regression analysis will automatically plot the residuals, but this will be further discussed in Chapter 13.) Below is a plot of the random index given in the previous table. If one had a model that was perfect, this value would always be zero. Even so, statistical theory which will be discussed later in the book indicates that the expected value of this random component should be zero. Therefore, in the plot, it is expected that the random observations should lie above and below but not far away from the horizontal $\bar{R} = 1$ line as shown. An outlier, then, is an observation that lies noticeably away from the line and not close to the bulk of the others. In the random index plot for the imported car sales data, there appears to be no distinct outliers.

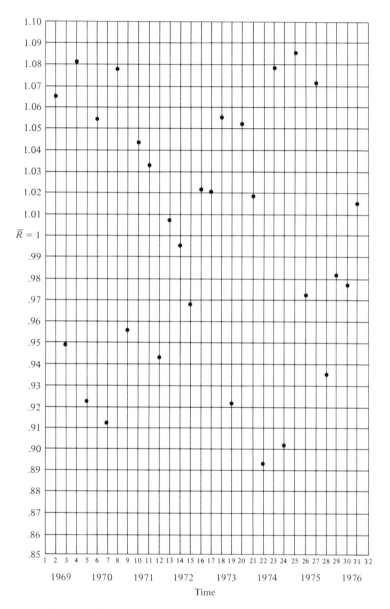

Using the formula for sales estimates of (seasonal index) × (trend data) × (cyclical index) × (random index), we obtain the following values that the model would estimate for the same period as the actual observed data. In addition, a sales forecast is given using the model for the four quarters of 1977. In the forecast, the random index is assumed to be 1. To obtain the cyclical index for this four-quarter forecast, we plot the three-period moving average cyclical index (from the previous table) as shown. There is no clear cyclical pattern; however, we

extend with a dotted line, by eye, what the cyclical index will be expected to look
like in the following four quarters. From the graph, then, these figures are put into
the above formula for the cyclical index.

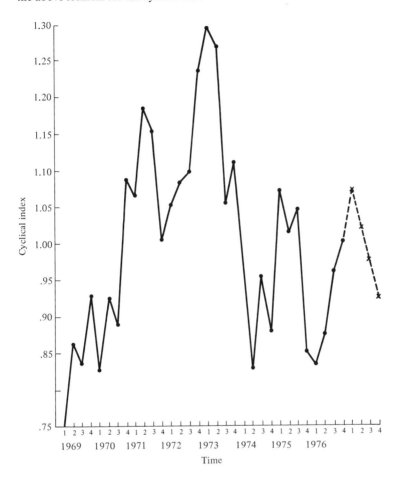

	SALES ESTIMATES								FORECAST
QUARTER	1969	1970	1971	1972	1973	1974	1975	1976	1977
1	—	250.91	344.75	371.12	474.35	377.02	444.27	322.68	432.25
2	327.23	359.84	467.76	415.33	540.43	306.25	418.94	372.45	459.42
3	281.86	297.42	452.62	416.78	392.26	429.75	474.59	427.68	438.17
4	303.66	367.03	304.26	414.29	396.86	276.62	286.28	—	349.60

The student is encouraged to go to the library and obtain the actual quarterly
imported retail car sales in the United States for 1977 from a recent edition of
Automobile Facts and Figures and to compare these actual figures with those
forecast using this model.

3.2 INDEX NUMBERS

An *index number* is a descriptive statistic that is constructed to indicate the
change in a variable over time; it summarizes the variation or relative variation

in a time series variable or in a group of time series variables. The possible indexes that might be considered to measure change over time are virtually limitless. Some examples are: an index to measure the change in crude oil production in the United States, an index of world population growth, an index to measure the melting and building up of glaciers, an index of air pollution in the five largest European cities, and so forth. However, in practice, index numbers are most frequently used in the context of business and economic data, the most important are indexes of changes in price and quantity of commodities and manufactured goods.

In some cases an index simply consists of a periodic listing. For example, the best index of business failures or of housing starts would probably consist of a tabulation of the number of business failures or the number of housing starts, respectively, over the time period of interest. The recording of price movements is usually the best index of price of a single item, say, the price of cocoa beans per pound traded on the Chicago Mercantile Exchange over five years.

The unfortunate aspect of index numbers is that there are problems associated with their construction, interpretation, and use. Perhaps the most difficult problem is in the construction of an index of groups of price movements, for example, an index of price movements of raw materials used in manufacturing in the United States.

Because they are numbers, indexes exude an air of precision which they seldom possess. A common example, and one of which everyone who reads a newspaper or watches television news is aware, is the index that attempts to measure the cost of running an average household. This index, constructed by the Bureau of Labor Statistics of the U.S. Department of Labor, is called the consumer price index. From 1919 to 1945 it was known as the cost of living index, which has a welfare connotation somewhat reduced by the name change. Nevertheless, in a sense this index is thought to indicate the consumer's or household's relative economic situation. Keep in mind, however, that there is no totally appropriate index for the purpose of measuring economic well-being or welfare for individuals or groups.

The consumer price index attempts to measure changes in the cost of a "fixed market basket" of goods and services containing about 400 items. Persons from the Bureau of Labor Statistics check the prices on the items in the basket each month, making some adjustment for quality and comparing the prices with those of the previous months. Some of the problems involved in evaluating the worth of the index as a measure of consumer prices are in the quality adjustments, new goods coming on the market, changes in technolology, different household buying habits, seasonal variations, taxes, and whether an expenditure is of the investment or current expense type.

Two other widely used economic indexes are the wholesale price index and the Federal Reserve Board's index of industrial production, a quantity index. The wholesale price index attempts to measure wholesale prices similarly to the measurement of consumer prices by the consumer price index. The index of industrial production indicates changes in output in the manufacturing sector of the U.S. economy.

In finance, the most frequently used indexes are concerned with changes in stock market prices. Although an individual stock price can be watched to determine movements in relation to its past prices, for "market analysis" some indicator of the movements of prices on the stock market as a whole is required. There is some disagreement (which we will not consider) among financial and investment analysts as to what constitutes a good market index. Three of the most important indexes are the Standard & Poor's 500 stock index, the New York Stock Exchange index, and the Dow Jones industrial, rail, and utility averages. Although the Dow Jones are called averages, they are more properly referred to as indexes. A complete discussion of the economic and financial indexes mentioned can be found in economics and finance texts. However, following a general description of the construction of price and quantity indexes, we will briefly discuss a few stock market and economic indexes.

RELATIVES

As mentioned at the beginning of this section, an index of change in a variable can be constructed by recording the values that the variable takes on at successive times. However, such a listing has limited applicability as an index. In general, examining a single variable as an index requires reduction to a base, so that variation in the variable over time can be studied. Reduction to a base is particularly important when groups of variables are involved in an index. The base of the index is taken as the value of the variable or variables in one of the periods in the time series, often the first time period. The index numbers for the series are then ratios, generally expressed as percentages, of the value in any period with respect to the value in the base period. The index numbers computed in this way are called *relatives;* they indicate variation relative to the base period. The typical quantity or price relative for any period is

$$\text{relative for a given period} = \frac{\text{quantity or price for given period}}{\text{quantity or price for base period}} \times 100$$

The actual price or quantity time series numbers are often called *aggregates.* Table 3.5 shows aggregates and index number construction of relatives for a type of steel production in the United States from 1965 through 1975.

Relatives calculated as in Table 3.5 are sometimes called base relatives, to distinguish them from two others: link relatives and chain relatives. The *link relative* for any period is calculated by dividing the value in that period by the value in the preceding period:

$$\text{link relative for given period} = \frac{\text{quantity or price for given period}}{\text{quantity or price for preceding period}} \times 100$$

The *chain relative* can be calculated as follows: assign 100 as the chain relative for the first period, then calculate the chain relative for any other period by multiplying the link relative for that period by the chain relative for the preceding period, and divide by 100:

chain relative for given period

$$= \frac{\text{link relative for given period} \times \text{chain relative for preceding period}}{100}$$

TABLE 3.5 ELECTRIC PROCESS
PRODUCTION OF STEEL IN U.S.A. 1965–1975*

Year	Aggregates (1000 tons)	Relative to 1965
1965	13,804	100
1966	14,870	107.7
1967	15.089	109.3
1968	16,814	121.8
1969	20,132	145.8
1970	20,162	146.8
1971	20,941	151.7
1972	23,721	171.8
1973	27,759	201.1
1974	28,669	207.7
1975	22,680	164.3

* Aggregate figures in this and subsequent tables are from *Statistical Abstracts of the United States.*

From the definition we see that when a single period is involved, the chain relatives are equal to the base relatives.

UNWEIGHTED INDEX NUMBERS

In constructing an unweighted (or simple) price or quantity index, we give each variable equal weight. That is, the price or quantity of each item in the index is considered of equal importance. In most cases this is a serious disadvantage in interpreting the index. If p_{ij} and q_{ij} represent the price and quantity of the ith commodity in any period, j, and p_{i0} and q_{i0} represent the price and quantity in the base period, then the simple aggregative price index number for the jth period is

$$_0P_j = \frac{\sum_{i=1}^{n} p_{ij}}{\sum_{i=1}^{n} p_{i0}} \times 100 \qquad \text{where there are } n \text{ commodities; } i = 1, 2, \ldots, n$$

$_0P_j$ is the set of price index numbers for $j = 0, 1, \ldots, k$, that is, where j goes from time period 0 (base) to time period k.

Similarly, the simple aggregative quantity index number for the jth period is

$$_0Q_j = \frac{\sum_{i=1}^{n} q_{ij}}{\sum_{i=1}^{n} q_{i0}} \times 100 \qquad \text{where there are } n \text{ commodities; } i = 1, 2, \ldots, n$$

$_0Q_j$ is the set of quantity index numbers for $j = 0, 1, \ldots, k$, that is, where j goes from time period 0 (base) to time period k.

Another frequently used type of unweighted index number is a simple average of the relatives. This index is computed in two steps: (1) the base relatives

TABLE 3.6 VARIOUS GRAINS PRODUCTION
IN MILLIONS OF BUSHEL—U.S.A., 1971–1975

Grain	1971	1972	1973	1974	1975
Corn	5,641	5,573	5,647	4,664	5,767
Oats	881	692	667	614	657
Sorghums	876	809	930	629	758
Soybeans	1,176	1,271	1,547	1,215	1,521
Wheat	1,618	1,545	1,705	1,796	2,134
$\sum\limits_{i=1}^{n} q_i = $ Totals	10,192	9,890	10,496	8,918	10,837
$_0Q_j = \dfrac{\sum\limits_{i=1}^{n} q_{ij}}{\sum\limits_{i=1}^{n} q_{i0}}(100)$	100	97.0	103.0	87.5	106.3

for each item for each period are obtained and (2) the sum of the base relatives for each period is divided by the number of items. The arithmetic mean of price or quantity relatives for the jth period is

$$_AP_j = \frac{\sum\limits_{i=1}^{n} \dfrac{p_{ij}}{p_{i0}}}{n} \times 100$$

$$_AQ_j = \frac{\sum\limits_{i=1}^{n} \dfrac{q_{ij}}{q_{i0}}}{n} \times 100$$

The geometric mean of price or quantity relatives for the jth period is

$$_GP_j = \sqrt{\prod_{i=1}^{n} \frac{p_{ij}}{p_{i0}}(100)}$$

Table 3.6 gives a simple aggregative production index for certain farm crops, where the aggregates are all measured in bushels.

Tables 3.7 and 3.8 show simple aggregative price indexes for various commodities. In Table 3.7 the units of measurement of the commodities are different. In Table 3.8 measurements of the commodities have been converted to the same units. The harmonic mean of price or quantity relatives for the jth period is

$$_HP_j = \frac{n}{\sum\limits_{i=1}^{n} \dfrac{p_{i0}}{p_{ij}}} \times 100$$

$$_HQ_j = \frac{n}{\sum\limits_{i=1}^{n} \dfrac{q_{i0}}{q_{ij}}} \times 100$$

TABLE 3.7 AVERAGE PRICE PER UNIT
FOR VARIOUS COMMODITIES—U.S.A., 1971–1975

Commodity	1971	1972	1973	1974	1975
Cotton ($/lb)	0.282	0.273	0.446	0.429	0.488
Crude Oil ($/bbl)	3.39	3.39	3.89	6.64	8.00
Hay ($/ton)	28.10	31.30	41.71	50.8	51.9
Rice ($/cwt)	5.34	6.73	13.78	11.2	8.74
Wheat ($/bu)	1.34	1.76	3.96	4.09	3.49
$\sum_{i=1}^{n} p_i = $ Totals	38.452	43.453	63.786	73.159	72.618
$_0P_j = \dfrac{\sum_{i=1}^{n} p_{ij}}{\sum_{i=1}^{n} p_{io}}\,n\,(100)$	100	113.0	165.9	190.3	188.9

TABLE 3.8 AVERAGE PRICE* PER UNIT
FOR VARIOUS COMMODITIES—U.S.A., 1971–1975

Commodity	1971	1972	1973	1974	1975
Cotton ($/lb)	0.282	0.273	0.446	0.429	0.488
Crude Oil ($/lb)	0.0110	0.0110	0.0126	0.0216	0.0260
Hay ($/lb)	0.0141	0.0157	0.0209	0.0254	0.0260
Rice ($/lb)	0.0534	0.0673	0.1378	0.1120	0.0874
Wheat ($/lb)	0.0223	0.0293	0.0660	0.0682	0.0582
$\sum_{i=1}^{n} p_i = $ Totals	0.3828	0.3963	0.6833	0.6562	0.6856
$_0P_j = \dfrac{\sum_{i=1}^{n} p_{ij}}{\sum_{i=1}^{n} p_{io}}\,n\,(100)$	100	103.5	178.5	171.4	179.1

* Conversions: cwt = 100 lb, ton = 2000 lb, barrel = 307.68 lb, bu = 60 lb

TABLE 3.9 QUANTITY RELATIVES FOR
AGGREGATES IN TABLE 3.6—BASE YEAR 1971

Grain	1971	1972	1973	1974	1975
Corn	100	98.8	100.1	82.7	102.2
Oats	100	78.5	75.7	69.7	74.6
Sorghums	100	92.4	106.2	71.8	86.5
Soybeans	100	108.1	131.5	103.3	129.3
Wheat	100	95.5	105.5	111.0	131.9
Totals	500	473.3	519.0	438.5	524.5
Arithmetic mean	100	94.7	103.8	87.7	104.9
Geometric mean	100	94.2	102.2	86.2	102.4

$$_GQ_j = \sqrt[n]{\prod_{i=1}^{n} \frac{q_{ij}}{q_{io}}}\,(100)$$

TABLE 3.10 HARMONIC MEAN OF QUANTITY RELATIVES IN TABLE 3.9

Grain	1971	1972	1973	1974	1975
Corn	1.00	1.012	0.999	1.209	0.978
Oats	1.00	1.273	1.321	1.435	1.341
Sorghums	1.00	1.083	0.942	1.393	1.156
Soybeans	1.00	0.925	0.760	0.968	0.773
Wheat	1.00	1.047	0.949	0.901	0.758
$\displaystyle\sum_{i=1}^{n} \frac{q_{i0}}{q_{ij}} = $ Totals	5.00	5.340	4.971	5.906	5.006
Harmonic mean = $\displaystyle {}_HQ_J = \frac{n}{\displaystyle\sum_{i=1}^{n} \frac{q_{i0}}{q_{ij}}} \times 100$	100	93.6	100.6	84.7	99.9

Table 3.9 gives the quantity production relatives for the aggregates in Table 3.6, as well as the arithmetic and geometric means, using 1971 as the base year. The harmonic mean is given in Table 3.10.

WEIGHTED INDEX NUMBERS

Weighted index numbers refer primarily to price indexes. As can be seen in calculating the price index numbers in Table 3.7, an index number makes little sense if the units of measure of the various items are different. When all items are measured in the same units before calculating the price index number, an implicit weighting occurs. Those priced the highest per unit (in this case per pound) influence the index the most, while the items priced the lowest influence the index the least. In order for the price index numbers to reflect more accurately the importance of the different items included, we weight the items by their respective quantities. Although there may be a variety of weightings possible, one of either two methods is most frequently used. The first weights items by the base period quantities, the second by the given period quantities.

The weighted aggregative price index with base period quantities as weights is called the Laspeyres index, given by

$$ {}_LP_j = \frac{\displaystyle\sum_{i=1}^{n} p_{ij}q_{i0}}{\displaystyle\sum_{i=1}^{n} p_{i0}q_{i0}} \times 100 $$

The numerator is obtained by multiplying the price of each of the n items in a given period, j, by the corresponding base period quantity and by summing for the n items. The denominator is obtained by multiplying the price of each of the n items in the base period by its quantity in the base period and by summing for the n items. Essentially the Laspeyres index expresses the relative value of the base period quantities at the prices for any given period.

TABLE 3.11   AVERAGE PRICE IN
DOLLARS PER BUSHEL, 1971–1975

Grain	1971	1972	1973	1974	1975
Corn	1.08	1.57	2.55	3.02	2.49
Oats	0.61	0.73	1.18	1.53	1.46
Sorghums	1.05	1.37	2.14	2.78	2.37
Soybeans	3.03	4.37	5.68	6.64	4.63
Wheat	1.34	1.76	3.95	4.09	3.49

The weighted aggregative price index with given period quantities as weights is called the Paasche index, given by

$$_pP_j = \frac{\sum\limits_{i=1}^{n} p_{ij}q_{ij}}{\sum\limits_{i=1}^{n} p_{i0}q_{ij}} \times 100$$

The numerator here is obtained by multiplying the price of each of r items in a given period, j, by its quantity in the same period and by summing for the r items. The denominator is obtained by multiplying the price of each of the n items in the base period by its quantity in the given period, j, and by summing for the n items. Essentially, the Paasche index expresses the relative value of the given period quantities at base period prices.

Although most weighted index numbers refer to price indexes, Laspeyres and Paasche indexes can also be calculated for quantities. The weights in this case are prices. The Laspeyres quantity index is:

$$_LQ_j = \frac{\sum\limits_{i=1}^{n} p_{i0}q_{ij}}{\sum\limits_{i=1}^{n} p_{i0}q_{i0}} \times 100$$

The Paasche quantity index is:

$$_pQ_j = \frac{\sum\limits_{i=1}^{n} p_{ij}q_{ij}}{\sum\limits_{i=1}^{n} p_{ij}q_{i0}} \times 100$$

Table 3.6 gives the quantities of certain grain production; Table 3.11 (above) gives the prices of these grains per bushel. Tables 3.12 and 3.13 show the calculation of the Laspeyres and Paasche price index, using 1971 as the base.

TABLE 3.12 CALCULATION OF LASPEYRES PRICE INDEX
FOR VARIOUS GRAINS—U.S.A., 1971–1975 (1971 AS BASE)

Grain	(1) Price 1971 p_0	(2) Production 1971 q_0	(3) (1) × (2) p_0q_0	(4) Price 1972	(5) (4) × (2) p_1q_0	(6) Price 1973 p_2	(7) (6) × (2) p_2q_0	(8) Price 1974 p_3	(9) (8) × (2) p_3q_0	(10) Price 1975 p_4	(11) (10) × (2) p_4q_0
Corn	1.08	5641	6,092.3	1.57	8,856.4	2.55	14,384.6	3.02	17,035.8	2.49	14,046.1
Oats	0.61	881	537.4	0.73	643.1	1.18	1,039.6	1.53	1,347.9	1.46	1,286.3
Sorghums	1.05	876	919.8	1.37	1,200.1	2.14	1,874.6	2.78	2,435.3	2.37	2,076.1
Soybeans	3.03	1176	3,563.3	4.37	5,139.1	5.68	6,679.7	6.64	7,808.6	4.63	5,444.9
Wheat	1.34	1618	2,168.1	1.76	2,847.7	3.95	6,391.1	4.09	6,617.6	3.49	5,646.8
Total			13,280.9		18,686.4		30,369.6		35,245.3		28,500.2
$L_j^p = \dfrac{\sum_{i=1}^{n} p_{ij}q_{i0}}{\sum_{i=1}^{n} p_{i0}q_{i0}}(100)$			$\dfrac{(3)}{(3)}$ 100		$\dfrac{(5)}{(3)}$ 140.7		$\dfrac{(7)}{(3)}$ 228.7		265.4		214.6

TABLE 3.13 CALCULATION OF PAASCHE PRICE INDEX
FOR VARIOUS GRAINS—U.S.A., 1971–1975 (1971 AS BASE)

	(1)	(2)	(3)	(4)	(5)	(6)	(7)	(8)	(9)	(10)
	Price 1971	Production 1971	$(1) \times (2)$	Price 1972	Production 1972	$(1) \times (5)$	$(4) \times (5)$	Price 1973	Production 1973	$(1) \times (9)$
Grain	p_0	q_0	$p_0 q_0$	p_1	q_1	$p_0 q_1$	$p_1 q_1$	p_2	q_2	$p_0 q_2$
Corn	1.08	5,641	6,092.3	1.57	5,573	6,018.8	8,749.6	2.55	5,647	6,098.8
Oats	0.61	881	537.4	0.73	692	422.1	505.2	1.18	667	406.9
Sorghums	1.05	876	919.8	1.37	809	849.5	1,108.3	2.14	930	976.5
Soybeans	3.03	1,176	3,563.3	4.37	1,271	3,851.1	5,554.3	5.68	1,547	4,687.4
Wheat	1.34	1,618	2,168.1	1.76	1,545	2,070.3	2,719.2	3.95	1,705	2,284.7
Total			13,280.9			13,211.8	18,636.6			14,454.2

$$P_j^p = \frac{\displaystyle\sum_{i=1}^{n} p_{ij} q_{ij}}{\displaystyle\sum_{i=1}^{n} p_{i0} q_{ij}} \times 100$$

$$\frac{(3)}{(3)} \qquad \frac{(7)}{(6)}$$

$$100 \qquad\qquad 141.1$$

TABLE 3.14 CALCULATION OF PRICE
RELATIVES FROM TABLE 3.11–1971 BASE YEAR

Grain	1971	1972	1973	1974	1975
Corn	100	145.4	236.1	279.6	230.6
Oats	100	119.7	193.4	250.8	239.3
Sorghums	100	130.5	203.8	264.8	225.7
Soybeans	100	144.2	187.5	219.1	152.8
Wheat	100	131.3	294.8	305.2	260.4

Weighted averages of price or quantity relatives may also be calculated as weighted index numbers. The weights in this case are values, which are base year values for the Laspeyres index and given year values for the Paasche index. The Laspeyres weighted average price and quantity relatives are given by

$$_LP_j = \frac{\sum_{i=1}^{n} (p_{i0}q_{i0})\left(\frac{p_{ij}}{p_{i0}}\right)}{\sum_{i=1}^{n} p_{i0}q_{i0}} \times 100$$

$$_LQ_j = \frac{\sum_{i=1}^{n} (p_{i0}q_{i0})\left(\frac{q_{ij}}{q_{i0}}\right)}{\sum_{i=1}^{n} p_{i0}q_{i0}} \times 100$$

The Paasche weighted average price and quantity relatives are given by

$$_PP_j = \frac{\sum_{i=1}^{n} (p_{i0}q_{ij})\left(\frac{p_{ij}}{p_{i0}}\right)}{\sum_{i=1}^{n} p_{i0}q_{i0}} \times 100$$

$$_PQ_j = \frac{\sum_{i=1}^{n} (p_{ij}q_{i0})\left(\frac{q_{ij}}{q_{i0}}\right)}{\sum_{i=1}^{n} p_{ij}q_{i0}} \times 100$$

Table 3.14 gives the price relatives from Table 3.11. Table 3.15 shows the computation of Laspeyres weighted average price relatives based on the information in Tables 3.11 and 3.14.

THREE ECONOMIC INDEXES

As mentioned earlier in Section 3.2, the consumer price index (CPI) is one of the most important, and most watched, economic indicators in the United States. The CPI is a weighted aggregative modified Laspeyres price

TABLE 3.15 CALCULATION OF LASPEYRES WEIGHTED AVERAGE PRICE RELATIVES BASED ON TABLES 3.11 AND 3.14. BASE YEAR VALUES, 1971, AS WEIGHTS. BASE YEAR QUANTITIES FROM TABLES 3.6.

| (1) | (2) | 1972 | | 1973 | | 1974 | | 1975 | |
| | | (3) | (4) | (5) | (6) | (7) | (8) | (9) | (10) |
Grain	Weights $p_{i0}q_{i0}$	Price Relative $\dfrac{p_1}{p_0}$	Weighted Price Relative (2) × (3)	Price Relative $\dfrac{p_2}{p_2}$	Weighted Price Relative (2) × (5)	Price Relative $\dfrac{p_3}{p_3}$	Weighted Price Relative (2) × (7)	Price Relative $\dfrac{p_4}{p_0}$	Weighted Price Relative (2) × (9)
Corn	6,092.3	145.4	885,820	236.1	1,438,392	279.6	1,703,407	230.6	1,404,884
Oats	537.4	119.7	64,327	193.4	103,933	250.8	134,780	239.3	128,600
Sorghums	919.8	130.5	120,034	203.8	187,455	264.8	243,563	225.7	207,599
Soybeans	3,563.3	144.2	513,828	187.5	668,119	219.1	780,719	152.8	544,472
Wheat	2,168.1	131.3	284,672	294.8	639,156	305.3	661,704	260.4	564,573
Totals	13,280.9		1,868,680		3,037,055		3,524,173		2,850,129
$_L P_j$	100		140.7		228.7		265.4		214.6

index, obtained using the formula:

$$\text{CPI}_j = \frac{\sum\limits_{i=1}^{n} p_{ij} q_{ia}}{\sum\limits_{i=1}^{n} p_{i0} q_{ia}}$$

Where q_{ia}, the quantities in a base period, are used as weights. The base period used for the quantities is not the same base period used for the prices, p_{i0}. The p_{ij} are the prices of the commodities in the given period. Essentially, in the CPI the prices in a base period are compared with the prices in a given period, with quantities in a different base period used as weights. The base for the commodity prices is an average of 1977 prices. The base for quantity weights is an average of the quantities purchased by a typical family in the years 1972–1973, as determined by the *Surveys of Consumer Expenditures* for those years. It should be noted that the consumer price index is designed to show changes in the purchasing power of the consumer's dollar, when the consumer is specifically the urban wage and clerical worker.

Computation of the CPI on a month-to-month basis is performed in a chain-linking fashion, in which the index for the previous month is multiplied by the average relative change in price from the previous month to the current month, as follows:

$$\text{CPI}_j = (\text{CPI}_{j-1}) \frac{\sum\limits_{i=1}^{n} p_{i,j-1} q_{ia} \dfrac{p_{ij}}{p_{i,j-1}}}{\sum\limits_{i=1}^{n} p_{i,j-1} q_{ia}}$$

This is equivalent to multiplying the index for the previous month by the relative price change from the previous month to the current month. The CPI differs from the usual chain index, however, since when item substitutions are made, substitute price relatives are used with the weights in the 1972–1973 period for those which they replace.

The construction cost index is a compilation of a number of construction cost indexes for various types of construction which is published monthly by the U.S. Department of Commerce. Some items included hotels, apartments, office buildings, commercial, factory buildings, farm buildings, residential houses, military facilities. The costs of the various types are weighted by their relative importance in construction across the United States. The base year used in 1967, so that the annual composite index represents the ratio between the annual value of new construction in current dollars and the comparable annual total in 1967 dollars.

The actual calculation of the construction cost index is quite complex, and no simple formula can be given. However, it can be understood from the fact that the dollar amounts of 13 components of new construction is divided by the dollar amounts in 1967 dollars deflated using 13 separate indexes, each of which is similar to a consumer price index. Both prices and weights change,

TABLE 3.16 INFLATION—POST WORLD WAR II

| Year | CONSUMER PRICE INDEX (DEPT. OF LABOR) | | CONSTRUCTION COST INDEX (DEPT. OF COMMERCE) | |
	Index	Year change %	Index	Year change %
1947	66.9		54	
1948	72.1	7.8	60	11.1
1949	71.4	−1.0	60	0.0
1950	72.1	1.0	62	3.3
1951	77.8	7.9	68	9.7
1952	79.5	2.2	69	1.5
1953	80.1	0.8	71	2.9
1954	80.5	0.5	71	0.0
1955	80.2	−0.4	73	2.8
1956	81.4	1.5	77	5.5
1957	84.3	3.6	80	3.8
1958	86.6	2.7	81	1.3
1959	87.3	0.8	82	1.2
1960	88.7	1.6	83	1.2
1961	89.6	1.0	84	1.2
1962	90.6	1.1	86	2.4
1963	91.7	1.2	88	2.3
1964	92.9	1.3	90	2.3
1965	94.5	1.7	93	3.3
1966	97.2	2.9	96	3.2
1967	100.0	2.9	100	4.2
1968	104.2	4.2	106	6.0
1969	109.8	5.4	114	7.0
1970	116.3	5.9	121	6.1
1971	121.3	4.3	130	6.9
1972	125.3	3.3	139	6.9
1973	133.1	6.2	148	6.5
1974	147.7	11.0	173	16.9
1975	161.2	9.1	189	9.2
1976	170.5	5.8	198	4.8
1977	181.5	6.5	217	9.9
1978	197.3	9.0	243	12.0

so that the index is an unusual combination of a Paasche and Laspeyres index. The formula may be considered as follows:

$$CCI = \frac{\sum_{i=1}^{13} (VNC_{\text{current } i})}{\sum_{i=1}^{13} (VNC_{67_i})}$$

VNC_{current} = value of new construction in component i in current dollars

VCN_{67} = value of new construction in component i in 1967 dollars, deflated by the component i index

Both the consumer price index and the construction cost index are considered to be measures of inflation. The CPI is generally considered the best overall measure since it includes many items affecting the family budget. But the CCI is regarded as an index of the increase in housing costs, an important component of the family budget. Table 3.16 gives the CPI and the CCI from 1947 through 1977, each with 1967 as the base, as well as the year-to-year changes in percent. A look at these as a measure of inflation indicates that since the beginning of the 1970s there has been much to contend with.

The Federal Reserve Board index of industrial production, which has been published since 1927, is designed to measure changes in the physical volume of output of manufacturing, mining, and utilities in the United States. Currently, the base period for the index is 1967. The index of industrial production is based on monthly data for the three industry sectors and is a weighted arithmetic average of relatives; it is a modified Laspeyres quantity index obtained using the following formula:

$$\text{IIP}_j = \frac{\sum\limits_{i=1}^{n} (p_{ia}q_{i0}) \dfrac{q_{ij}}{q_{i0}}}{\sum\limits_{i=1}^{n} p_{ia}q_{i0}}$$

The p_{ia} are the per unit values added for each of the industry sectors in 1967. Using values added as weights indicates only the contribution of particular industries to output. Although it does not include construction, transportation, agriculture, or the service industries, the index of industrial production is an important barometer of business activity in the nation.

STOCK MARKET INDEXES

The most widely quoted stock index is the Dow Jones industrial average, which was first published in 1897 and has been in its present 30-stock form since 1928. However, there have been substitutions for all of the original stocks except two. The Dow Jones average was originally a simple average of the prices of 30 stocks. Difficulties arise when stocks split; although the stock price might remain the same or increase after the split, the reduced per share price caused by an increased number of shares makes a simple average of the stock prices decrease. This has been progressively corrected by reducing the divisor to reflect splits, until at the beginning of 1977 the divisor was 1.504 with the Dow Jones industrial average in the high 900s, while the average price of the 30 stocks was under $100.

Another frequently used index of stock market prices is the Standard & Poors 500 index. This is a composite of three indexes—the industrial index, composed of 425 stocks; the railroad index of 25 stocks; and the utility index of 50 stocks. Each of the indexes is obtained by first multiplying the price of a stock by the number of shares of that stock outstanding. The market values of each issue so calculated are added and expressed as a percentage of the average total market value of these issues in the years 1941–1943. In other words, an average of 1941–1943 is the base period, and these Standard & Poors indexes are simple weighted aggregative indexes.

In 1966 both the American and New York Stock Exchanges began publishing indexes of stocks traded on their respective exchanges. The American Stock Exchange index is a simple average of the prices of stocks and warrants listed on the exchange. The New York Stock Exchange common stock index is a weighted aggregative index of all common stocks listed on the exchange and is calculated in the same way as the Standard & Poors indexes. In addition, the New York Stock Exchange also publishes four separate indexes: the finance index (75 stocks), the transportion index (76 stocks), the utility index (136 stocks), and the industrial index (1000 stocks).

There are other financial indexes, such as the *New York Times* stock index, the value line average, and Barron's confidence index—all of which are either simple unweighted price indexes or weighted aggregative price indexes.

GENERAL COMMENTS ON INDEX NUMBERS

An index is essentially an average or mean of some type. There is a theorem in algebra that states that if no numbers in a set are zero or negative, then the harmonic mean is less than the geometric mean which is less than the arithmetic mean: $H < G < \bar{X}$. Thus, the arithmetic mean has a relative upward bias and the harmonic mean has a relative downward bias.

Another type of bias possible with index numbers can be examined in the context of the *time reversal test*. This test ascertains whether the index number for period 0 relative to period n is the reciprocal of the index number for period n relative to period 0; that is, whether the index is the same regardless of the direction in time for which it is computed. If direction in time is irrelevant, the product of the two indexes must be 1. The simple geometric mean of relatives satisfies this test, but the simple arithmetic mean and the simple harmonic mean of relatives do not. Thus an index number calculated as a simple geometric mean of relatives is independent of the period taken as a base; an index number of the simple arithmetic or harmonic mean is not independent of the period taken as a base. Bias attributable to the type of mean used in obtaining an index is referred to as *type bias*.

As discussed in the construction of a simple unweighted price index, there is a bias in the direction of the commodity having the highest price; that is, a simple price index has an implicit weight for each price proportional to its magnitude. Every method of weighting is biased from some point of view. Bias attributable to the weighting used in obtaining an index is referred to as *weight bias*. When price levels are gradually rising, weighting by base period quantities or values causes relative downward bias, while weighting by given period quantities or values causes relative upward bias.

The Laspeyres index, since it is an arithmetic average weighted by base period quantities or values, has two biases tending in opposite directions. The Paasche index, since it is a harmonic average weighted by given period quantities or values, has two biases tending in opposite directions. However, it requires more data and more computation, because the quantities in each period serve as weights. For this reason, most published indexes of other than simple averages are of the Laspeyres type.

In interpreting an index, we first determine the type of index that has been computed. Is it a listing, a simple average, a weighted average, or none of these (like the Dow Jones stock averages)? What is the index trying to show and, for what purposes is it to be used?

There is no perfect index for expressing changes in groups of variables over time. Consistency in computation and awareness of the imperfections are probably the most important aspects of constructing and using any index number.

PROBLEMS

3.1 Given below is passenger car production in the United States from 1950 through 1975 (in millions). Assuming a three-year cycle, calculate the trend line, the trend component, the cyclical component, and the random component.

1950	6.636	1963	7.644
1951	5.311	1964	7.745
1952	4.325	1965	9.335
1953	6.132	1966	8.605
1954	5.507	1967	7.413
1955	7.950	1968	8.849
1956	5.807	1969	8.224
1957	6.120	1970	6.550
1958	4.247	1971	8.584
1959	5.599	1972	8.828
1960	6.703	1973	9.667
1961	5.522	1974	7.325
1962	6.943	1975	6.717

3.2 Problem 3.1 gives passenger car production in the United States from 1950 to 1975. Using 1950 as the base, calculate for the remaining years the simple quantity base relatives, the link relatives, and the chain relatives.

3.3 Develop a short-run forecasting model for quarterly shipments of Douglas Fir lumber in the United States using known quarterly data from 1974, 1975, 1976. Use a four-quarter moving average to deseasonalize and center at the third quarter. Use the observed data to calculate the trend line. Use a three-quarter moving average for the cyclical index, and calculate the random index as a residual. Use the model to forecast the shipments in the four quarters of 1977, and compare with actual figures given in the *Survey of Current Business*. The observed data, in millions of board feet, for the three years is as follows:

QUARTER	1974	1975	1976
1	1999	1666	2038
2	2234	1941	2019
3	1829	1917	2046
4	1662	1893	2028

3.4 The average American earned a gross income of $12,000 in 1970 and $17,000 in 1976. In 1970 his or her Social Security tax was $374 and income tax was $1527. In 1976 these taxes were $895 and $2124, respectively. Using the consumer price index (Table 3.16) as the deflator, indicate whether the average American was economically better off in 1970 or 1976.

3.5 Yields to maturity of bonds are an inverse indicator of bond prices, since yields and prices move in opposite directions. The following table gives the average yields in percent for various types of bonds (with the same maturity) for the years 1966 through 1971:

BOND TYPE	1966	1967	1968	1969	1970	1971
Governments	4.75	5.01	6.14	6.53	8.21	7.02
AA						
Industrials	4.82	5.55	6.30	6.75	8.70	7.60
Trust						
certificates	4.85	5.70	6.50	7.10	8.75	7.72
A Utilities	5.00	6.00	7.00	7.30	9.65	8.08

(a) Calculate the simple aggregative yield index numbers using 1966 as a base.
(b) Based on this index, what can be said about bond price movements over this period.
3.6 The table below gives the prices per pound and consumption in millions of pounds for five commodities over four years. Obtain:
(a) The Laspeyre index
(b) The Paasche index
(c) Fisher's ideal index (defined as the square root of the above two indexes multiplied together)

	1974		1975		1976		1977	
COMMODITY	p	q	p	q	p	q	p	q
A	.74	14,200	.80	10,200	.87	16,800	.98	17,400
B	.60	16,000	.68	16,000	.78	17,100	.80	17,200
C	.30	1,200	.32	1,240	.36	1,400	.44	1,380
D	.08	3,400	.09	3,600	.10	3,700	.12	3,900
E	.18	1,720	.20	1,840	.22	2,000	.30	1,820

3.7 Following are monthly domestic whisky sales figure in thousands of tax gallons in the United States for 1971 through 1974. Calculate the trend line and the trend component; also deseasonalized sales, cyclic and random component. Plot the actual data, what is your opinion of the cycle in this time series?

	1971	1972	1973	1974
January	8,213	8,491	9,645	10,827
February	7,677	8,400	8,904	9,606
March	9,852	10,379	11,333	13,063
April	8,531	8,856	10,229	10,869
May	8,291	9,718	11,963	11,667
June	10,151	10,829	10,436	10,942
July	7,576	9,355	8,861	10,184
August	10,677	10,941	10,615	10,454
September	11,735	12,750	11,047	11,392
October	12,345	15,863	16,683	16,002
November	12,194	14,300	14,319	12,320
December	9,594	10,219	9,592	9,692

3.8 The following data show fish caught (million pounds) and value (million dollars) in the United States for four years.

	1970		1973		1974		1975	
	p_0	q_0	p_1	q_1	p_2	q_2	p_3	q_3
Cod	53	6	50	9	59	11	56	13
Flounder	169	23	168	33	156	34	156	43
Haddock	27	6	8	3	8	3	16	5
Halibut	35	9	26	18	18	9	21	14
Herring	79	2	102	6	120	11	119	5
Mackeral	48	2	21	1	22	1	30	1

Use 1970 as base year, calculate Laspeyres and Paasche price indexes for 1973, 1974, and 1975.

What information is needed to compute each of these two indexes?

3.9 From the following table, calculate the simple quantity relatives using 1972 as the base year. What are the geometric average and the geometric rate of growth for each of the categories? What is the meaning of the geometric growth rates in each case?

VARIOUS MEAT SLAUGHTERING IN THE UNITED STATES FROM 1972 TO 1975 (MILLIONS OF HEADS)

	1972	1973	1974	1975
Beef	36.1	34.0	37.3	41.5
Veal	3.2	2.4	3.2	5.4
Lamb & mutton	10.5	9.8	9.1	8.1
Pork	85.7	77.8	83.1	69.8

3.10 Calculate the Laspeyres and Paasche indexes for 1976 and 1972 from the following information on hardbound textbooks.

	1963		1967		1972	
	COPIES SOLD (MIL.)	AVERAGE RECEIPTS PER BOOK ($)	COPIES SOLD (MIL.)	AVERAGE RECEIPTS PER BOOK ($)	COPIES SOLD (MIL.)	AVERAGE RECEIPTS PER BOOK ($)
Elementary	54.9	2.05	85.1	2.12	65.8	2.47
High school	32.9	2.97	36.3	3.38	40.5	3.27
College	24.2	5.44	41.2	5.5	57.1	4.90

3.11 **UNITED STATES EXPORTS OF DOMESTIC AND FOREIGN MERCHANDISE TO CHINA (MILLIONS OF DOLLARS—FROM OVERSEAS BUSINESS REPORTS)**

	1972	1973	1974	1975	1976
Jan.–Mar.	Z*	48	344	70	85
Apr.–June	Z*	103	176	77	34
July–Sept.	4	274	244	52	6
Oct.–Dec.	59	265	58	104	10

* Z—less than $500,000

From the above table calculate the trend line and the trend component. What is the problem with time series decomposition in a case such as this?

3.12 From the following table calculate the simple aggregative price indexes
(a) Using prices and units of measure as given
(b) After converting to a common unit of measure

U.S. PRODUCT		1970
Pig iron	Thousand short ton	91,293
	Avg. val./ton ($)	64.98
Copper	Short ton price	1,719,657
	¢/pd	58.2
Aluminium	Thousand short ton	3,976
	¢/pd	28.7
Bituminous coal	Thousand short ton	596,969
	$/ton FOB mines	
	(sold in open mkt.) ($)	5.89
	Val. FOB mines	6.26
Coking coal	Mil. short ton	66.5
	Avg. mkt. val/ton ($)	27.1

U.S. PRODUCT	1971	1972	1973	1974
Pig iron	81,382	88,876	101,318	95,477
	70.23	77.27	74.43	129.39
Copper	1,522,183	1,664,840	1,717,940	1,597,002
	52.0	51.2	59.5	77.3
Aluminium	3,925	4,122	4,529	4,903
	29.0	26.3	25.3	34.1
Bituminous coal	545,790	584,387	577,574	587,928
	6.66	7.35	8.06	15.16
	7.07	7.66	8.58	15.76
Coking coal	57.4	60.5	64.3	61.6
	30.63	33.21	37.97	73.24

3.13 Monthly unemployment figures (in percent) for the United States from 1972 through 1976 are given in the following table. The figures are the raw or unadjusted percent of the labor force unemployed. The U.S. Department of Labor uses a rather complex method to adjust seasonally the unemployment rates, which is not discussed in this book. Use the ratio to moving average method (with a three-month moving average) described in Section 3.1 to adjust seasonally the unemployment data.

YEAR	JAN.	FEB.	MAR.	APR.	MAY
1972	6.4	6.4	6.1	5.5	5.1
1973	5.5	5.6	5.2	4.8	4.3
1974	5.6	5.7	5.3	4.8	4.6
1975	9.0	9.1	9.1	8.6	8.3
1976	8.8	8.7	8.1	7.4	6.7

YEAR	JUNE	JULY	AUG.	SEPT.	OCT.	NOV.	DEC.
1972	6.2	5.8	5.5	5.4	5.1	4.9	4.7
1973	5.4	5.0	4.7	4.7	4.2	4.5	4.5
1974	5.8	5.6	5.3	5.7	5.5	6.2	6.7
1975	9.1	8.7	8.2	8.1	7.8	7.8	7.8
1976	8.0	7.8	7.6	7.4	7.2	7.4	7.4

AXIOMATIC PROBABILITY

4.1 INTRODUCTION

Many statements or assertions cannot be made with certainty and are said to be uncertain or probabilistic.* *Probability* is a concept intended to express the degree of certainty with which an assertion is made. The use of probability implies both an admission of uncertainty and an attempt to quantify its degree. There is general agreement concerning both the mathematical foundations of probability theory and the calculus of probabilities, that is, the computations performed with probabilities, but much less agreement concerning the interpretation and use of probabilities.

This difference of opinion corresponds essentially to the different types of problems for which probability theory is considered appropriate and reflects the increase in scope of its applications from seventeenth-century considerations of games of chance to twentieth-century considerations of decision making. For the earlier applications of probability theory the *objective* or *relative frequency* interpretation of probabilities is more appropriate. For many of the later applications the *subjective* or *personalistic* interpretation of probabilities is required.

The *objective* or *relative frequency* interpretation of probability is applicable only to experiments or processes of observation that can be repeated indefinitely many times under essentially the same conditions. The probability of a particular outcome, then, is the proportion of the time (or the relative frequency) with which it occurs in the long run. For example, tosses of a coin, rolls of two dice, random samples from a specified population, mass-produced items, and many other occurrences or observations can be considered as resulting from repetitive processes. The probability of a particular outcome (for example, a head, a six on each of the two dice, a particular sample, an item with a specified length) is the proportion of times it would occur in an indefinitely long series of repetitions of the occurrence or observation.

---

* The term *uncertain* here is used simply to imply "not certain." There is a distinction involving probability that is drawn between risk and uncertainty, which is not introduced now to avoid confusion. A discussion of this subject is contained in Chapter 15.

Note that the objective interpretation of probability refers to a repetitive process that generates outcomes that are governed individually by chance and predictable only in terms of relative frequencies. Such an interpretation precludes the consideration of probabilities of unique events—a unique event, by definition, occurs only once and thus a relative frequency interpretation of the probability of its occurrence is not possible. Thus, for example, the probability of rain tomorrow, the probability that Bacon wrote some of the plays attributed to Shakespeare, the probability that the Republicans will win the next presidential election, the probability that the United States will win ten gold medals at the next Olympic Games, and similar propositions are not appropriate for interpretation in terms of objective probabilities.

The *subjective* or *personalistic* interpretation of probability is applicable both to events that can be repeated and to unique events. The subjective or personalistic probability of an event is interpreted as a statement of the strength or degree of personal belief or confidence in its occurrence. In this sense, as a statement of strength of personal belief, the probability of rain tomorrow is as meaningfully interpreted as the probability of obtaining a head when a coin is tossed.

Note that personal probability is subjective, in the sense that it might vary depending on the individual whose degree of belief it represents. Individuals might differ in their degrees of belief or confidence in an outcome, even when considering the same evidence, and thus their (personal) probabilities for the same event would differ.

When used in applications concerning phenomena in the real world, objective probabilities are also subjective in a certain sense. In most cases objective probabilities are assigned to the outcome of experiments or observations that are described by theoretical models; the probabilities of the outcomes are then determined (objectively) by the definitions and assumptions of the models. However, in any particular case a decision must be made with respect to the applicability or appropriateness of the model, which is usually a matter of subjective judgment or belief.

For example, a "fair" coin is a very simple theoretical model and the probabilities of various outcomes on one or more tosses can be calculated in a straightforward manner. However, the appropriateness of these (objective) probabilities for any particular coin depends on whether the coin can be assumed to be fair, a matter of subjective judgment based on the available evidence. The situation is similar for the more complex theoretical models discussed in the following sections. In each application of these models to real phenomena, the relevance of the definitions and assumptions of the model must be considered (subjectively) on the basis of available evidence, although the probabilities of various outcomes are determined (objectively) by these definitions and assumptions.

As mentioned, the mathematical foundations of probability theory and the rules and procedures for computing probabilities are the same regardless of the basis for or the interpretation of the probabilities. Thus these foundations, rules, and procedures are discussed in following sections with only occasional reference to the distinction between probability as a long-run relative frequency and as a measure of confidence or degree of belief. This important distinction in interpretation is noted in some detail in applications later.

The terms *random* and *random variable* are used so often in probability and statistics that an understanding of their meaning is necessary. The definition of a random sample is given in Chapter 1. The term variable is used in Chapter 2 in discussing descriptive statistical measures and summation notation. Statisticians may (and do) quarrel over rigorous definitions of the terms random and random variable; here we will give only reasonably correct definitions that can be understood intuitively.

A *variable* is a symbol or measure that may assume more than one value. If a variable assumes different values, each with a given probability, then it is a *random variable*. The probabilities of the different values may be known or unknown.

Strictly speaking, an event is *random* if its occurrence follows any probability law or distribution. However, statisticians sometimes use the word random to imply that events are equally likely, that is, have equal probabilities of occurrence. For convenience, whenever the expression "at random" or "randomly selected" is used subsequently, it means equally likely events or equal probabilities of occurrence, unless otherwise stated.

4.2 SAMPLE SPACES AND EVENTS

A *set* is a collection of elements; the elements of a set may be people, numbers, possible outcomes of an experiment, items of various types, and so forth. A *subset* is a part of a set defined in some unambiguous way. The *null set*, $\emptyset$ has no members; it is the empty set. The *universe* or *universal set* in any situation is the set to which consideration is limited. The universe under discussion might be, for example, all the common stocks listed on the London Stock Exchange or all the securities in the portfolio of your rich Uncle Henry or all the economists who believe in the labor theory of value. If there are two subsets of a universal set such that every element of the universal set is an element of one and only one of the subsets, then each subset is said to be the *complement* of the other subset. The complement of a set A is denoted by $\overline{A}$.

An experiment is a process of observation that can be repeated under essentially the same conditions; the results of an experiment are *outcomes* and the set of all possible outcomes is the *sample space*, S, of the experiment. A sample space may be finite (if the set of possible outcomes is finite) or infinite (if the set of possible outcomes is infinite). An *event* is a set of outcomes, that is, a subset of the sample space. Note that a subset as a part of a set can refer to the entire set, or, trivially, to the empty set that contains no outcomes. In applications, the probabilities of interest almost invariably refer to the occurrence or nonoccurrence of events.

A Venn diagram is a convenient visual aid used to represent sets and subsets (see Figure 4.1). The sample space or universal set for an experiment is represented by the rectangle, S, and events in the sample space are represented by regions within the rectangle. An element or outcome of an experiment can be thought of as a point in the sample space. Hence, in the diagram, the points of the subset A might represent one type of outcome and the points of the subset $\overline{A}$ might represent all other outcomes. For example, if a coin is tossed, A could represent heads and $\overline{A}$ could represent tails; if a die is thrown, A could represent

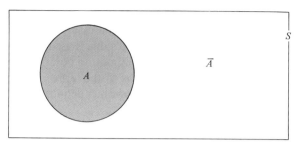

Figure 4.1 Venn diagram

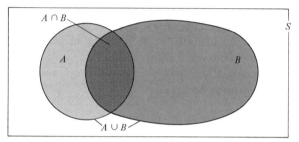

Figure 4.2 Venn diagram—union and intersection

a six and $\bar{A}$ could represent all other outcomes, a one, two, three, four, or five. Since the points in A are not in $\bar{A}$ and both sets together constitute the entire sample space, $\bar{A}$ is the complement of A.

Events that cannot occur in both the same experiment or observation are said to be *mutually exclusive* or *disjoint*. Mutually exclusive events have no outcomes or elements in common, and the occurrence of either event precludes the possibility of occurrence of the other event; n events are mutually exclusive if no two of them have outcomes in common. An event and its complement, for example, A and $\bar{A}$ in Figure 4.1 are mutually exclusive events by definition.

If A and B are two events in the sample space, S, their *union* is denoted by $A \cup B$ and consists of the outcomes contained in A or B, or in both A and B. The *intersection* of A and B is denoted $A \cap B$ and consists of the outcomes contained in both A and B. Union and intersection are illustrated using a Venn diagram in Figure 4.2.

Figure 4.3 Venn diagram—mutually exclusive (disjoint) events

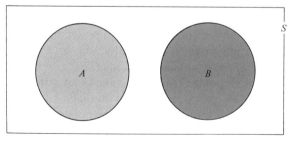

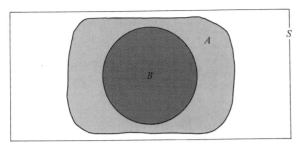

Figure 4.4 Venn diagram: subsets

If A and B are mutually exclusive, then $A \cap B$ contains no outcomes and is the null set. In Figure 4.3, A and B are mutually exclusive (disjoint), and thus $A \cap B = \emptyset$.

If B is a subset of A; that is, if A contains B, this is denoted by $A \supset B$ or, equivalently, by $B \subset A$. For example, suppose the sample space in Figure 4.4 is designated "all the commercial banks in the United States" and A is designated "national banks." In addition, suppose B is designated "national banks with deposits over \$10,000,000." Then it is obvious that A is a subset of S and B is a subset of A.

Sometimes we can understand probability better by thinking of observations or sampling as representing an experiment. The possible outcomes of an experiment—the elements of the sample space—can be represented as a set of points. For example, suppose the experiment is tossing a nickel and a dime simultaneously. The set of possible outcomes, or the sample space for this experiment, can be shown as points, as in Figure 4.5. Although we are not always able to represent possible outcomes as points in rectangular coordinates, in many situations it may be helpful to consider possible outcomes as points in the abstract.

Suppose an experiment consists of tossing a single die. Define event A as the occurrence of an odd number and event B as the occurrence of the numbers 4, 5, 6. These events represent sets of possible outcomes of the experiment (subsets of the sample space), of which, of course, only one outcome can occur (only one side can come up). In addition, there is the possibility that an outcome, for example, a 2, may occur which is neither of these events. The situation can be

Figure 4.5 Sample space for tossing two coins.

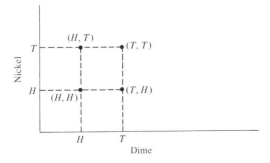

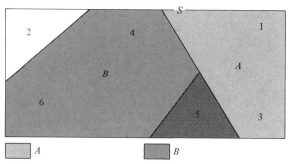

Figure 4.6   Venn diagram—tossing a die

represented by a Venn diagram as in Figure 4.6. Note the following from
Figure 4.6:

$$A = \{1, 3, 5\}$$

$$B = \{4, 5, 6\}$$

$$\bar{A} = \{2, 4, 6\}$$

$$\bar{B} = \{1, 2, 3\}$$

$$A \cup B = \{1, 3, 4, 5, 6\}$$

$$A \cap B = \{5\}$$

Notice one thing in the above problem: if $A \cup B$ is being calculated, the
number 5 is not counted twice even though it occurs in both A and B. Common
sense indicates that this would be foolish, because counting it twice would admit
seven possibilities in throwing a die, when in fact a die has only six sides. If A
and B were written:

$$A = \{1, 3, 5\} \qquad B = \{4, 5, 6\}$$

and then the union of A and B written by just putting these two together:

$$A \cup B = \{1, 3, 5, 4, 5, 6\}$$

the error of counting 5 twice would be made. Looking at the Venn diagram, we
see that this would involve essentially counting the double cross-hatched area
containing 5 twice. To obtain the proper statement of $A \cup B$, we must remove
one of the 5's, which can be done in such situations by eliminating the joint
occurrence; that is, $A \cap B = 5$.

$$A \cup B = \{1, 3, 5\} + \{4, 5, 6\} - \{5\} = \{1, 3, 4, 5, 6\}$$

In general, $A \cup B = A + B - (A \cap B)$.

In this example, addition and subtraction of sets in obtaining $A \cup B$ is
introduced. There are a number of other useful definitions and properties of
sets involving unions, intersections, and complements, which are given below
without proof. If A, B, and C are subsets of S, then the following equalities can be

obtained using Venn diagrams:

1. $A \cup A = A$
2. $A \cap A = A$
3. $A \cup \emptyset = A$
4. $A \cap S = A$
5. $A \cap \emptyset = \emptyset$
6. $A \cup S = S$
7. $A \cup \bar{A} = S$
8. $A \cap \bar{A} = \emptyset$
9. $A \cup B = B \cup A$
10. $A \cap B = B \cap A$
11. $A \cup (B \cup C) = (A \cup B) \cup C$
12. $A \cap (B \cap C) = (A \cap B) \cap C$
13. $A \cup (B \cap C) = (A \cup B) \cap (A \cup C)$
14. $A \cap (B \cup C) = (A \cap B) \cup (A \cap C)$
15. $(\overline{A \cup B}) = \bar{A} \cap \bar{B}$
16. $(\overline{A \cap B}) = \bar{A} \cup \bar{B}$
17. $A \cup B = A + B - (A \cap B)$
18. $A \cup \bar{B} = \bar{B} + (A \cap B)$
19. $A \cap \bar{B} = A - B$
20. $(A - C) \cup C = A \cup C$
21. $A \cap (A - B) = A - B$
22. $(\overline{A - C}) = \bar{A} \cup C$

The following examples illustrate the concepts of sample spaces, unions, intersections, and complements, which are important for understanding probability.

EXAMPLE

A coin is tossed three times successively. The sample space consists of eight possible outcomes:

OUTCOME	FIRST TOSS	SECOND TOSS	THIRD TOSS
1	H	H	H
2	H	H	T
3	H	T	H
4	T	H	H
5	H	T	T
6	T	H	T
7	T	T	H
8	T	T	T

DEFINE THE FOLLOWING EVENTS	SUBJECT OF OUTCOMES
A: 2 heads, 1 tail	A: $\{2, 3, 4\}$
B: 1 head, 2 tails	B: $\{5, 6, 7\}$
C: 3 tosses the same	C: $\{1, 8\}$
D: at least 2 heads	D: $\{1, 2, 3, 4\}$

Then,

$$A \cup B \cup C = S$$
$$A \cap B = A \cap C = B \cap C = \emptyset$$
$$A \cap D = A; \text{ also } D \supset A \text{ or } A \subset D$$
$$A \cup B = \bar{C}$$
$$B \cup C \cup D = S$$
$$\overline{C \cup B} = A$$

EXAMPLE

A subcommittee of three members is chosen from a committee of five members. Denoting the five members by a, b, c, d, and e, the sample space consists of the ten outcomes:

OUTCOME	COMMITTEE
1	*abc*
2	*abd*
3	*abe*
4	*acd*
5	*ace*
6	*ade*
7	*bcd*
8	*bce*
9	*bde*
10	*cde*

DEFINE THE FOLLOWING EVENTS	SUBSET OF OUTCOMES
A: both a and b	A: $\{1, 2, 3\}$
B: d	B: $\{2, 4, 6, 7, 9, 10\}$
C: c and/or e but not d	C: $\{1, 3, 5, 8\}$
D: b and/or c	D: $\{1, 2, 3, 4, 5, 7, 8, 9, 10\}$

Then,

$$B \cup C = S$$
$$A \cap C = \{1, 3\}$$
$$B \cap C = \varnothing$$
$$C \cap D = C; \text{ also, } D \supset C \text{ or } C \subset D$$
$$A \cap D = A; \text{ also, } D \supset A \text{ or } A \subset D$$
$$\bar{B} = C$$
$$B \cup C \cap A = A$$

4.3 ELEMENTARY PRINCIPLES OF PROBABILITY

Probabilities of specific events may be determined in a variety of ways. However, all probabilities, regardless of the manner in which they are obtained, have the following characteristics:

1. A probability lies between zero and one, inclusive
2. The probability of an impossible event is zero
3. The probability of a certain event is 1
4. The sum of the probabilities of all possible mutually exclusive outcomes equals 1

Two principles have been used historically in determining objective probabilities—the principle of insufficient reason and the frequency (or relative frequency) principle.

The principle of insufficient reason states that when there is no basis for assuming that one outcome is more likely to occur than another, the outcomes are assumed to be equally likely to occur. This is based on abstract reasoning and does not depend on experience. Most statisticians reject the principle of insufficient reason on the basis that ignorance concerning the probabilities of alternative outcomes is not justification for assuming the probabilities to be equal.

Unlike the principle of insufficient reason, the relative frequency principle determines probabilities on the basis of experience. Suppose an experiment, for example, tossing a coin, is performed repeatedly under the same conditions. If i is the number of repetitions, and m_i is the number of heads observed after the ith repetition, then the fraction m_i/i calculated at various times during the experiment is the *relative frequency* of heads to the total number of tosses at that point.

If this experiment were actually performed, m_i/i would fluctuate considerably for small i, but as i increased, the amplitude of the fluctuations of heads would become stable. The value toward which the relative frequency tends is defined as the probability of a head occurring on a single toss. The relative frequency principle of probability can be stated as follows: the probability of an event occurring is the relative frequency of its occurrence for a large number of repetitions.

The relative frequency principle is derived from the *law of large numbers*, which can be stated as follows: as the number of repetitions of an experiment or observation gets larger and larger, there is a stronger and stronger tendency for the results to conform, in a ratio sense, to the probability of occurrence of the possible outcomes. The law of large numbers essentially defines probability as the ratio of the number of ways a particular outcome can occur to the total number of possible occurrences.

Suppose a fair coin is tossed over and over and the number of heads and tails are recorded. If the coin is tossed sufficiently many times, the ratio of heads to total tosses can be made arbitrarily close to one half. Note that this does *not* say that if the coin is tossed enough times, the number of heads will equal the number of tails. In fact, the law of large numbers specifies that the absolute number in the numerator of the relative frequency tends to deviate more from the expected number as the repetitions increase, even through the ratio approaches the true probability.

The relative frequency definition of probability does not imply that the number of heads will "even up" with the number of tails if tossing continues long enough. This is an erroneous concept of the law of large numbers, often referred to in everyday speech as the law of averages. One sometimes hears the expression, "the law of averages will work out." This would imply, for instance, that if 5 heads in a row are obtained in tossing a coin, then more tails should be observed in the next sequence of tosses because the number of heads and tails must eventually "even out." But such belief implies that the probability of getting a tail on the next toss is greater than the probability of getting a head, and this cannot be so if the coin is a fair one.

The preceding discussion involves the distinction between dependent and independent events. If a fair coin is tossed repeatedly, the outcome on one toss is not influenced by the outcome of any other toss; the outcomes of the repeated coin tosses are *independent events*. If a card is drawn from a full deck of 52 and then another card is drawn without replacing the first, the outcome of the second draw is influenced by the outcome of the first draw. Hence these are *dependent events*—the second outcome is not independent of the first.

Subjective probability is described in detail in Chapter 16; however, the motivation for subjective probability is apparent when considering the two principles of objective probability. In many cases where it would be desirable to assess a probability, there is neither the (1) belief that the outcomes are equally likely or (2) opportunity to perform repetitive experiments. What does one do? Some persons would say that probability theory cannot help with this problem. Others would say that it is proper to assign as the probability of an event the measure of confidence held as to the occurrence of the event. Such a probability assignment, of course, is subjective.

4.4 THE CALCULATION OF PROBABILITIES

ADDITION RULE

One of the most important rules in the calculation of probability is the addition rule: if two outcomes are not mutually exclusive, the probability of either (or both) occurring is the sum of their separate probabilities of occurrence minus the probability of their joint occurrence:

$$P(A \cup B) = P(A \text{ or } B) = P(A) + P(B) - P(A \cap B)$$

EXAMPLE
The probability of drawing either an ace or a spade from a deck of cards:

$$P(\text{ace or spade}) = P(\text{ace}) + P(\text{spade}) - P(\text{ace and spade})$$

$$= \frac{4}{52} + \frac{13}{52} - \frac{1}{52} = \frac{16}{52}$$

The above statement is the general form of the addition rule that includes mutually exclusive events, because in such a case

$$P(A \cap B) = 0$$

and

$$P(A \cup B) = P(A) + P(B)$$

EXAMPLE
The probability of drawing an ace or a king from a deck of cards:

$$P(\text{ace or king}) = P(\text{ace}) + P(\text{king}) - P(\text{ace and king})$$

$$= \frac{4}{52} + \frac{4}{52} - 0 = \frac{8}{52}$$

CONDITIONAL PROBABILITY

On the basis of both experience and common sense, it is clear that there are many events that are not independent. For example, the probability that a family chosen at random from a city directory will own two cars is less than the probability that a family chosen at random from the country club directory of the same city will own two cars; the probability that a student selected at random from the student directory of a university will make a B average next semester is greater than the probability that a student selected at random from a list of students on probation at the university will make a B average; the probability that a person will develop diabetes is less if he or she is chosen at random from a given group than if chosen at random from the group in which one or both parents had diabetes; the probability that a person has published at least one book is greater if he or she is selected at random from the faculty list of a university than if selected at random from the list of assistant professors at the university; and so forth. Conditional probability is appropriate for studying these kinds of dependent events.

The probability of an event is always with reference (implicitly or explicitly) to a particular sample space; in each of the above examples the probability of an event with reference to a subset of the original sample space (that is, a *reduced sample space*) differs from the probability of that event with reference to the original sample space. The subsets are defined by one or more conditions in addition to those defining the original sample space, and the probabilities associated with events in those subsets are *conditional probabilities.*

Consider the following: of the stocks traded on the New York Stock Exchange on a certain day there were 66 that posted new highs for the year and 6 that posted new lows for the year. Of the new highs, 5 companies could be said to be in the metals industry, of the new lows 3 companies were in metals. Now suppose each of these companies was assigned a number, with the numbers of the lows 1 through 6 and highs 7 through 72; the individual numbers were each written on a small ball, and the balls were placed in a large urn. The balls were then mixed thoroughly so that each one would have an equally likely chance of being selected if one of the balls was drawn from the urn. (As a practical matter this could be done more simply by using a table of random numbers, as discussed in Chapter 11.) Since the experiment has been set up in such a way that an equally likely chance can be assigned to each outcome, the probability assigned to each may be said to be $\frac{1}{72}$ or approximately 0.014. Assume a ball is drawn from the urn and you are told that the ball has a number between 1 and 6. Given this information, what probability can be attached to the event that the ball selected represents a metals company?

The total population is 72 balls, but now there is a restriction or condition placed on the balls to be considered –the subpopulation (or subset) consists of those balls representing firms whose stock prices made new lows. There were 6 new lows, of which 3 were metals companies. Given the subset "new lows," the probability of a metals company being drawn is $\frac{3}{6}$ or $\frac{1}{2}$. Such a probability, based on a subset or reduced sample space, is called a *conditional probability,* written as: P(metals company | new lows). The vertical line represents the word "given," which indicates a conditional probability.

Obtaining a conditional probability involves the calculation of a proportion. The probability that a company is a metals company if (given that) it reached a new low on the day in question is the proportion of companies reaching new lows that are metals companies:

$$P(\text{metals company} \mid \text{new low}) = \frac{p(\text{metals company and new lows})}{P(\text{new lows})}$$

$$= \frac{\frac{3}{72}}{\frac{6}{72}} = \frac{1}{2}$$

The probability in the numerator, P(metals company and new lows), is a joint probability—that of the intersection of metals company and new lows. The probability in the denominator is an unconditional probability (called a marginal probability—that of new lows in the population of new lows and new highs. A conditional probability can be defined as the ratio of a joint probability to a marginal probability. In general, probability refers to unconditional (marginal) probability unless there is an indication that a conditional probability or a joint probability is appropriate. For example, P(metals company "given" new lows) is a conditional probability, while P(metals company "and" new lows) is a joint probability.

The general formula for conditional probability is

$$P(B|A) = \frac{P(A \cap B)}{P(A)}$$

Suppose in the above example that the ball selected from the urn had a number between 7 and 72. In this case the probability that the ball selected represented a metals company is

$$P(\text{metals company} | \text{new highs}) = \frac{P(\text{metals company and new highs})}{P(\text{new highs})}$$

$$= \frac{\frac{5}{72}}{\frac{66}{72}} = \frac{5}{66}$$

If it is given that of the 66 new highs there were four oil companies and of the 6 new lows there were no oil companies, then the calculation of the probability of the ball being an oil company, given that its number was between 7 and 72, would be

$$P(\text{oil company} | \text{new highs}) = \frac{P(\text{oil company and new high})}{P(\text{new highs})}$$

$$= \frac{\frac{6}{72}}{\frac{66}{72}} = \frac{1}{11}$$

The probability of the ball representing an oil company, given a number between 1 and 6, is zero (the probability of an impossible event). This can be obtained formally as follows:

$$P(\text{oil company} | \text{new lows}) = \frac{P(\text{oil company and new lows})}{P(\text{new lows})}$$

$$= \frac{0}{\frac{6}{72}} = 0$$

From this we see that, the joint probability of mutually exclusive events is zero. In this case, oil company and new lows are mutually exclusive; they cannot both occur. The conditional probabilities of mutually exclusive events are also zero; that is, when

$$P(A \cap B) = 0$$

then

$$P(A|B) = 0 = P(B|A)$$

The formula for conditional probability can also be used to calculate the joint probability when a conditional probability and a marginal probability are given.

EXAMPLE

Half of the students in a certain class are male and half are female. Of all students in the class, 30 percent received the grade of A. Of all the students who received A, 40 percent were females.

Question: What is the probability of a randomly selected student from this class being female and having received the grade of A?

Define: Event A = received grade of A

Event F = being female

The following probabilities can be written from the problem:

$$P(A) = .3$$
$$P(F|A) = .4$$

The formula for conditional probability is

$$P(F|A) = \frac{P(A \cap F)}{P(A)}$$

However, $P(A \cap F)$ is desired, that is, the probability of the joint event—received and being female.

$$P(A \cap F) = P(A)P(F|A) = (.3)(.4) = .12$$

Note that the joint probability is calculated from an unconditional (marginal) probability and a conditional probability. This is, in fact, the multiplication rule, which will be stated carefully after a discussion of the potential confusion arising from simultaneous and sequential occurrences. The probabilities in the above example can be represented by a cross-classification table of probabilities, as shown below. The unconditional probabilities of A and not A ($\bar{A}$) and male and female are called marginal probabilities because they appear in the margin of the table. The joint probabilities appear in the cells of the table. The conditional probabilities can be obtained by dividing the appropriate joint probabilities by the corresponding marginal probabilities.

	A	$\bar{A}$	
Female	.12	.38	.50
Male	.18	.32	.50
	.30	.70	

In mentioning mutually exclusive events, we stressed that if events are mutually exclusive, they cannot happen at the same time. $A \cap B = \emptyset$. From this we said that the probability of their joint occurrence is 0. $P(A \cap B) = 0$. The significant point here is that the terminology of joint occurrence or joint events implies *occurrence at one time;* that is, on one experiment. The terminology says nothing about events A and B following one another, being *sequential,* on two experiments. The distinction is important because the mathematics is the same for calculating probabilities of joint events (on one experiment) as for calculating probabilities of sequential events (on repetitive experiments).

It should be noted that this discussion of sequential events concerns only two events taking place in order on two experiments. In the next chapter the occurrence of two events in any order in two or more experiments is discussed.

Until now the notation $(A \cap B)$ has been used to indicate the joint occurrence of A and B simultaneously on the same experiment. This same notation can be extended to sequential events taking place on more than one experiment, but temporarily the notation of $(A \rightarrow B)$ will be used here to indicate sequential events to avoid ambiguity. $(A \rightarrow B)$ means event A followed by event B. It should be noted that in many places both $(A \cap B)$ and $(A \rightarrow B)$ are written (AB). This is a shorthand convenience that will also be used in subsequent sections. Once students are aware of the difference between the two, they can see which is implied from the statement of a problem.

THE MULTIPLICATION RULE

A very important rule in the calculation of probabilities is the multiplication rule, which can be stated as follows: the probability of a joint event or of a sequence of two events is equal to the (marginal) probability of the first event multiplied by the (conditional) probability of the second event given that the first event has occurred.

$$P(A \cap B) \quad \text{or} \quad P(A \rightarrow B) = P(A)P(B)|A)$$

STATISTICAL DEPENDENCE AND INDEPENDENCE

Two events, A and B, are said to be *statistically independent* if the occurrence of one does not affect the probability of occurrence of the other. Thus A and B are statistically independent if

$$P(A|B) = P(A)$$

and

$$P(B|A) = P(B)$$

If A and B are statistically independent, the probability of their joint occurrence is equal to the product of the probabilities of their individual occurrences:

$$P(A \cap B) = P(A)P(B) = P(B)P(A)$$

Statistical independence and dependence are defined for both simultaneous and sequential events. Note that if two events are independent, the probability of their joint occurrence can be written taking the events in either order.

The difference between *dependence* and *independence* can be illustrated using the familiar example of tossing a fair coin. Suppose the following question is asked: If a fair coin is tossed twice, what is the probability of getting two heads?

$$\text{Define}: H_1 = \text{head on first toss}$$

$$H_2 = \text{head on second toss}$$

If the coin is fair, the probability of a head on the first toss is $\frac{1}{2}$:

$$P(H_1) = \frac{1}{2}$$

The probability of a head on the second toss, given that a head occurred on the first toss, is also $\frac{1}{2}$:

$$P(H_2 | H_1) = \frac{1}{2}$$

Hence, by the sequential rule,

$$P(H_1 \rightarrow H_2) = P(H_1)P(H_2 | H_1) = \left(\frac{1}{2}\right)\left(\frac{1}{2}\right) = \frac{1}{4}$$

The events H_1 and H_2 are independent; that is, the one having occurred does not affect the probability of the other occurring. The conditional probability of H_2 is the same as the unconditional probability of H_2:

$$P(H_2 | H_1) = P(H_2)$$

Thus, as mentioned previously, a fair coin has no memory.

When considering statistical independence in one experiment, rather than for sequential events, care must be taken not to confuse independent with mutually exclusive events. Events are mutually exclusive if they cannot occur as outcomes on the same experiment, written in joint probability terms for two events as

$$P(A \cap B) = 0$$

which implies

$$P(A | B) = 0 = P(B | A)$$

since from the multiplication rule,

$$P(A \cap B) = P(A)P(B | A) = P(B)P(A | B)$$

The two types of statistical independence are *independence in time* and *independence of classification*. The first is when the outcomes of successive coin tosses are said to be independent. The second is nonrelated categories of events that can occur on one experiment. Hence independence or dependence of events on one experiment must be distinguished from mutually exclusive or not mutually exclusive events on one experiment.*

EXAMPLE

Suppose from the freshman class entering Arizona University this fall the following characteristics have been determined:

10 percent of the men are left-handed
20 percent of the men have gray eyes
30 percent of the men are 6 feet or more in height

Question: If a male freshman student was selected at random, what would be the probability that he was left-handed, had gray eyes, and was 6 feet or more in height? It seems reasonable to assume that these classifications are independent; that is, being right- or left-handed, color of eyes, and height, should not depend one on another. It is difficult to choose classifications that are known for certain to be independent; there is always the possibility that classifications might be even slightly related. However, assume that these three classifications are truly independent.

Define: L = left-handed male, $P(L) = .10$

G = gray-eyed male, $P(G) = .20$

S = male 6 feet or more in height, $P(S) = .30$

Then the following probabilities can be written:

$$P(L) = P(L|G) = P(L|S) = P(L|G,S) = .10$$

$$P(G) = P(G|L) = P(G|S) = P(G|L,S) = .20$$

$$P(S) = P(S|G) = P(S|L) = P(S|G,L) = .30$$

Therefore

$$P(L \cap G \cap S) = P(L)P(G)P(S) = (.1)(.2)(.3) = .006$$

Thus the probability of a randomly selected male freshman being left-handed, gray-eyed, and 6 feet or more in height is .006. This is the probability of a joint event, with all the events being independent as to classification and the joint event referring to the outcome of a single experiment.

The example previously calculated concerning the probability of a randomly selected student being a female and having received the grade of A is the same type as the above problem, but the classifications there are not independent.

4.5 BAYES' RULE

Bayes' rule, which is an important axiom of conditional probability, is discussed in this section. In Chapter 16 a type of subjective probability analysis, which is based on Bayes' rule and has come to be called Bayesian, is described in more detail.

In the most elementary terms, Bayes' rule provides a formula for obtaining $P(B|A)$ if $P(A|B)$ is known and if the unconditional probabilities $P(A)$ and

---

* Note that mutually exclusive events are dependent, since knowledge of the occurrence of one influences the probability of occurrence of the other and, in fact, makes that probability zero.

$P(B)$ are known or can be calculated.* The conditional probabilities can be written:

$$P(A|B) = \frac{P(AB)}{P(B)}$$

$$P(B|A) = \frac{P(AB)}{P(A)}$$

From these the joint probabilities are:

$$P(AB) = P(B)P(A|B)$$

$$P(AB) = P(A)P(B|A)$$

Things equal to the same thing are equal to each other. Therefore

$$P(B)P(A|B) = P(A)P(B|A)$$

Then Bayes rule is simply:

$$P(B|A) = \frac{P(B)P(A|B)}{P(A)}$$

In Bayesian analysis $P(B|A)$ is referred to as the posterior probability.

EXAMPLE
Consider the example of the students and the grades in Section 4.4. Of the students in a class, 30 percent received the grade of A. Of those that received A, 40 percent are female. Suppose instead of 50 percent, the information is given that 20 percent of the class is female. What is the probability of a student having received an A, given that the student is female?

Define: A = received A, $P(A) = .3$

F = being female, $P(F) = .2$

It is known from the problem that $P(F|A) = .4$. What is $P(A|F)$? The problem can be solved by Bayes rule.

$$P(F|A) = \frac{P(AF)}{P(A)}$$

$$P(AF) = P(A)P(F|A)$$

$$P(A|F) = \frac{P(AF)}{P(F)} = \frac{P(A)P(F|A)}{P(F)} = \frac{(.3)(.4)}{(.2)} = .6$$

In the above example both $P(F)$ and $P(A)$ are given. Suppose $P(A)$ had not been given. What must be given in order to calculate $P(A)$? Note that an A can be given out in the class in only two ways—to a female student or to a male student. Thus the sum of the joint probability of A and female and the joint probability of A and male can be obtained, which is the unconditional probability of A: $P(A) = P(AF) + P(AM)$. In this problem if $P(A)$ were not

* Notice that (AB) is now being used for both $(A \cap B)$ and $(A \to B)$.

given, $P(AF)$ and $P(AM)$ or $P(AF)$ and $P(A|M)$ must be given:

$$P(A) = P(F)P(A|F) + P(M)P(A|M)$$

One of the most well-known applications of Bayes' rule is in solving a variant of what is referred to as "The Three Card Paradox." The statement of the problem runs something like this: You are at a carnival grounds and are being offered to make an even bet on a simple card game at one of the booths. The man at the booth has three cards, one is white on both sides, the second is white on one side and red on the other, and the third is red on both sides. The cards are placed in a small box and thoroughly shuffled. At the bottom of the box is a slit, from which one of the cards is slid out on the table in such a way that the side of the card down on the table can not be seen. The man at the booth wishes to bet you even money that the side of the card down on the table is the same color as the side up. The logic of the argument as to why this is a fair bet is that if the side up is, say red, then the card on the table cannot be the one that is white on both sides. Hence the probability is $\frac{1}{2}$ that it is the card with red on both sides, and the probability is $\frac{1}{2}$ that it is the card with red on one side and white on the other. Presto, it is an eminently even bet. If you are a bit skeptical about this very reasonable argument and ask the booth man to take the other side of the bet, be assured that he will not. The three card paradox is an example of the fact that what seems reasonable logic about probabilities is sometimes wrong. The problem is solved as follows:

THREE CARD PARADOX SOLVED BY BAYES' RULE
cards: #1 #2 #3
cards: WW WR RR
Question: What is the probability of the other side being white if the side up is red, or $P(W|R)$? That is, what is the probability of card #2, given red showing, or $P(2|R)$?

Unconditional Probabilities: $P(1) = \frac{1}{3}$, $P(2) = \frac{1}{3}$, $P(3) = \frac{1}{3}$

Conditionally Probabilities: $P(R|1) = 0$, $P(R|2) = \frac{1}{2}$, $P(R|3) = 1$

$$P(R|2) = \frac{P(R2)}{P(2)}$$

$$P(R2) = P(2)P(R|2)$$

$$P(2|R) = \frac{P(R2)}{P(R)} = \frac{P(2)P(R|2)}{P(R)}$$

To obtain $P(R)$, we get

$$P(R) = P(R1) + P(R2) + P(R3)$$
$$= P(1)P(R|1) + P(2)P(R|2) + P(3)P(R|3)$$
$$= \left(\frac{1}{3}\right)(0) + \left(\frac{1}{3}\right)\left(\frac{1}{2}\right) + \left(\frac{1}{3}\right)(1) = \frac{1}{2}$$

And thus

$$P(2|R) = \frac{\left(\frac{1}{3}\right)\left(\frac{1}{2}\right)}{\left(\frac{1}{2}\right)} = \frac{1}{3}$$

Hence the probability of the other side being white if red is up is

$$P(W\,|\,R) = \frac{1}{3}$$

These calculations of probabilities may be summarized in the following table:

Population	Prior	Conditional	Joint	Posterior
1 WW	$\frac{1}{3}$	0	0	0
2 RW	$\frac{1}{3}$	$\frac{1}{2}$	$\frac{1}{6}$	$\frac{1}{3}$
3 RR	$\frac{1}{3}$	1	$\frac{2}{6}$	$\frac{2}{3}$
			$\frac{3}{6}$	1

4.6 GAMBLING AND ODDS

The investigation of probability theorems was first stimulated by interest in gambling. In the sixteenth century Galileo wrote a treatise on probability after an Italian gambler asked him why, when three dice were thrown, the sum of ten seemed to occur more often than the sum of nine. In the following century the French nobleman and gambler Fermat raised a similar question with the mathematician Pascal. This time the question was why there seemed to be a higher chance of obtaining at least one six in 4 throws of a single die than of obtaining at least one double six in 24 throws of two dice. Fermat thought this was strange, but it appeared to be true from experience. This problem will be solved in the next chapter, and when we compare the probabilities associated with each question, it can only be concluded that the original question was a remarkable example of observed relative frequencies and must have been based on a large number of repetitions.

Dice games are among the most amenable to discussion in terms of probabilities, mainly because if the dice are well made (and not loaded or shaved or fixed in some way), the throws of the dice are independent and probabilities can be calculated exactly. Probabilities for card games are more difficult to calculate, because the trials of passing out cards are dependent—that is, dependent on those cards that have already been passed out. In the United States the most popular dice game is "craps," which is played with two dice. Each die has six symmetrical sides, and on each side are spots, from one to six, respectively. In the following rules for craps, the number used refers to the sum of the spots showing on the up sides of the two dice.

Two dice are rolled:
1. If the number is 7 or 11 on the initial roll, a win
2. If the number is 2, 3, or 12 on the initial roll, a loss
3. If either (1) or (2) occurs, the shooter retains the dice
4. The shooter continues to roll until a point is rolled, that is 4, 5, 6, 8, 9, 10
5. The point rolled is then to be made, and it must be rolled before a 7 to win
6. If a 7 is rolled before the point, this is a loss, and the shooter gives up the dice

The 36 combinations on two dice are:

Numbers	Can Occur	Probability of Occurrence on One Roll
Natural 7	6	$\frac{6}{36} = .1667$
Natural 11	2	$\frac{2}{36} = .0556$
Crap 2	1	$\frac{1}{36} = .0278$
Crap 3	2	$\frac{2}{36} = .0556$
Crap 12	1	$\frac{1}{36} = .0278$
Point 4	3	$\frac{3}{36} = .0333$
Point 5	4	$\frac{4}{36} = .1111$
Point 6	5	$\frac{5}{36} = .1389$
Point 8	5	$\frac{5}{36} = .1389$
Point 9	4	$\frac{4}{36} = .1111$
Point 10	3	$\frac{3}{36} = .0333$
	36	$\frac{36}{36}$ 1.0000

Odds, which are used in betting, are stated differently from probabilities. The probability of throwing a 2 is $\frac{1}{36}$; a 2 can be thrown in only 1 way out of 36. Hence 1 way is favorable and 35 ways are unfavorable out of 36. Odds are the unfavorable chances versus the favorable chances. So the odds against throwing a 2 in one roll are 35 to 1.

The probability of throwing an 11 is $\frac{2}{36}$; 11 can be thrown in 2 ways out of 36, 2 ways are favorable, 34 are unfavorable. So the odds against throwing an 11 in one roll are 17 to 1. The probability of throwing a 7 is $\frac{6}{36}$; a 7 can be thrown in 6 ways out of 36, 6 ways are favorable, 30 are unfavorable. So the odds against throwing a 7 in one roll are 5 to 1, and so forth.

As a problem for this chapter, see if you can calculate the probability of winning in craps when you first pick up the dice to become the shooter. The answer is surprisingly close to .5.

APPENDIX 4.1
FACTORIALS

An exclamation point after a symbol or number means factorial. Thus 5! is read "five factorial," and is defined as

$$5! = 5 \cdot 4 \cdot 3 \cdot 2 \cdot 1 = 120$$

The general expression for a factorial is

$$x! = (x)(x - 1)(x - 2) \cdots (1)$$

Another expression for a factorial that is often used is

$$x! = x(x - 1)! = x(x - 1)(x - 2)! = x(x - 1)(x - 2)(x - 3)!, \ldots$$

In order to make use of factorials, a definition of 1 must be given to zero factorial. That is, $0! = 1$.

APPENDIX 4.2
PERMUTATIONS AND COMBINATIONS

Frequently, it is interesting to determine the number of possible outcomes of an experiment or the number of ways in which an event can occur without enumerating or listing the relevant possibilities. For this purpose of counting we use the rules for determining the number of possible permutations (arrangements) or combinations (subsets) of a set of elements.

A *permutation* of the elements of a set is an arrangement of those elements in a particular order. Note that order is essential in identifying a permutation—a change in order results in a different permutation. A *combination* of the elements of a set is a subset of those elements without regard to order or arrangement. Order is the important factor that distinguishes a permutation from a combination.

Take the letters A, B, C. When taken together there is only 1 combination, but when we consider permutations, they can be arranged in 6 different ways:

$$ABC, BAC, CAB, ACB, BCA, CBA$$

The reasoning by which the 6 ways can be obtained without writing them down is as follows:

1. There are 3 places to be filled in the arrangement.
2. The first place can be filled in 3 ways.
3. Once having filled the first place, we can fill the second in 2 ways. Hence the first 2 places can be filled in $3 \cdot 2 = 6$ ways.
4. Having filled the first 2 places, we can fill the last place only in 1 way. Hence, in total, there are $3 \cdot 2 \cdot 1 = 6$ different ways to arrange the 3 letters.
5. But, $3 \cdot 2 \cdot 1$ is equal to $3!$

The fundamental theorem for permutations can be stated: if one thing can be done n_1 ways, a second n_2 ways, and a third n_3 ways, and so forth, then the number of different ways in which they can be done, when taken altogether is $n_1 \cdot n_2 \cdot n_3 \cdots$. In other words, there are $n!$ permutations of n things when all n things are considered.

Now consider the number of permutations of r things out of n, where $r \le n$. For example, suppose $n = 8$, and $r = 3$. The problem is to determine the number of permutations of 3 things out of 8, or to use the appropriate terminology: the permutations of 8 things taken 3 at a time. The reasoning is: (1) there are 8 ways to fill the first place, (2) there are 7 ways to fill the second place, (3) there are 6 ways to fill the third place. Consequently, there are $8 \cdot 7 \cdot 6 = 336$ permutations of 8 things taken 3 at a time.

The notation for permutations of n things taken r at a time is $_nP_r$, and the general formula can be developed by using the above line of reasoning. If there are r positions to be filled from n things, the first position can be filled in n ways, the second position in $n - 1$ ways, and so forth. Thus the r positions can be filled in $(n)(n - 1)(n - 2) \cdots (n - r + 1)$ ways and

$$_nP_r = (n)(n - 1)(n - 2) \cdots (n - r + 1) = \frac{(n)(n - 1)(n - 2) \cdots (n - r + 1)(n - r)!}{(n - r)!}$$

$$= \frac{n!}{(n - r)!}$$

The formula for combinations can be developed from the formula for permutations. Of course, there is only one combination of n things taken altogether because order does not matter. For combinations of n things taken r at a time, $_nC_r$, there will be more than one, but fewer than the number of permutations. To derive $_nC_r$ from $_nP_r$, consider that each combination or subset of r elements can be arranged or ordered in $r!$ ways. Thus for each combination there are $r!$ permutations, and therefore

$$_nC_r = \frac{_nP_r}{r!} = \frac{n!}{r!(n - r)!}$$

To illustrate the difference between a combination and a permutation, consider this problem: Suppose a committee of 5 persons exists from which a subcommittee of 3 to be selected. How many ways can the subcommittee be selected? Since order or arrangement does not matter, this is a combination problem. That is, the problem can be stated: How many combinations of 3 can be obtained from 5?

$$_5C_3 = \frac{5!}{3!(5 - 3)!} = 10$$

If the members of the subcommittee are considered as ordered or arranged, then permutations are involved. Suppose the choice is to select a president, secretary, treasurer of an organization from a committee of 5 persons. Here the arrangement would matter, that is, the arrangement of who filled the individual offices:

$$_5P_3 = \frac{5!}{2!} = 60$$

One of the interesting properties of combinations is that

$$_nC_r = {}_nC_{n-r}$$

This can be shown by writing the combination formula for both and observing that they are equal:

$$_nC_r = \frac{n!}{r!(n - r)!}$$

$$_nC_{n-r} = \frac{n!}{[(n - r)!n - (n - r)]!} = \frac{n!}{(n - r)!r!}$$

This property is not true for permutations:

$$_nP_r = \frac{n!}{(n-r)!}$$

$$_nP_{n-r} = \frac{n!}{[n-(n-r)]!} = \frac{n!}{r!}$$

One difficult conceptual problem with permutations and combinations often involves what has been previously referred to as the fundamental theorem but is generally called the *multiplication principle*. Although finding the answer may be easy in such a problem, arriving at it in terms of permutations and combinations often presents some difficulty. For example, if a penny, a nickel, and a dime are tossed simultaneously, how many different ways can they fall? The answer can be obtained three ways, all of which are equivalent.

First way: Enumerate the possibilities.

Penny	Nickel	Dime
H	H	H
H	H	T
H	T	H
H	T	T
T	T	T
T	T	H
T	H	T
T	H	H

Second way: Based on the multiplication principle, the first coin can fall 2 ways, the second coin 2 ways, and the third coin 2 ways. Hence $2 \cdot 2 \cdot 2 = 8$ total ways.

Third way: The falling of the first coin represents a combination of 2 things taken 1 at a time. The second and third coins each the same. Therefore the possibilities can be represented by $_2C_1 \cdot _2C_1 \cdot _2C_1 = 8$ ways.

The *multiplication principle* can be stated slightly different from the way in which the fundamental theorem was stated: If an operation consists of k steps or parts, of which the first can be performed in n_1 ways, the second can be performed in n_2 ways, and so forth for k steps until the kth part can be performed in n_k ways, then the entire operation can be performed in $n_1 \cdot n_2 \cdot n_3 \cdots n_k$ ways. Using the product notation, we can write:

$$\prod_{i=1}^{k} n_i$$

The *addition principle* can be stated: If k operations are mutually exclusive (i.e., no two operations can be performed together) and the first operation can be performed in n_1 ways, the second operation in n_2 ways, and so forth for k operations, finally the kth operation can be performed in n_k ways, and then the total number of ways in which the k operations can be performed is:

$$n_1 + n_2 + \cdots + n_k = \sum_{i=1}^{k} n_i.$$

EXAMPLES OF THE ADDITION PRINCIPLE

1. How many numbers can be formed from the integers 1, 2, 5, 6, 8 if an integer is used only once in any number? Noting that a number can have 1, 2, 3, 4, and 5 digits, we must determine the number of possible numbers containing 1, 2, 3, 4, and 5 digits, respectively, and add them together:

$$
\begin{array}{lll}
\text{Number of one-digit numbers} & = {}_5P_1 = & 5 \\
\text{Number of two-digit numbers} & = {}_5P_2 = & 20 \\
\text{Number of three-digit numbers} & = {}_5P_3 = & 60 \\
\text{Number of four-digit numbers} & = {}_5P_4 = & 120 \\
\text{Number of five-digit numbers} & = {}_5P_5 = & 120 \\
\hline
\text{Total possible numbers} & & 325
\end{array}
$$

2. A committee consists of 6 men and 5 women; in how many ways can a subcommittee of 8 be formed to contain at least 3 women? The conditions of the problem are satisfied if the committee contains (a) 5 men and 3 women, or (b) 4 men and 4 women, or (c) 3 men and 5 women. Consider (a) first:

$$\text{The 5 men can be chosen in } {}_6C_5 = \frac{6!}{5!\,1!} = 6 \text{ ways}$$

$$\text{The 3 women can be chosen in } {}_5C_3 = \frac{5!}{3!\,2!} = 10 \text{ ways}$$

Therefore a subcommittee containing 5 men and 3 women can be formed in

$$ {}_6C_5 \cdot {}_5C_3 = (6)(10) = 60 \text{ ways}$$

Considering (b), a subcommittee containing 4 men and 4 women can be formed in

$$ {}_6C_4 \cdot {}_5C_4 = (15)(5) = 75 \text{ ways}$$

Considering (c), a subcommittee containing 3 men and 5 women can be formed in

$$ {}_6C_3 \cdot {}_6C_5 = (20)(1) = 20 \text{ ways}$$

Therefore a subcommittee of 8, consisting of at least 3 women can be formed in a total of

$$60 + 75 + 20 = 155 \text{ ways.}$$

Sometimes restrictions are imposed on the permutations or combinations to be calculated. If a restriction is imposed or if some operation must be performed in a special way, accounting for it first is usually desirable.

EXAMPLE

How many 6-place numbers can be formed from the digits 1, 2, 3, 4, 5, 6, if 3 and 4 are always to occupy the middle two places? The two restricted digits, 3 and 4, can be arranged in 2! ways. The other four digits can be arranged in 4! ways. Hence, in all, $2! \cdot 4! = 48$ numbers are possible.

In the discussion of permutations and the multiplication principle, the elements to be arranged or the operations to be performed are assumed to be distinguishable or distinct. If some of the elements to be arranged or operations to be performed are identical or indistinguishable (or are considered to be equivalent in this sense), then the number of possible (distinguishable or distinct) permutations is reduced. More precisely, if a set of n elements consists of n_i (indistinguishable) elements of one kind, n_2 (indistinguishable) elements of another kind, and so forth for k types of elements of which there are n_k (indistinguishable) elements of the kth kind, then the number of distinguishable permutations of the n elements taken together is:

$$\frac{n!}{n_1!n_2!\cdots n_k}$$

where

$$\sum_{i=1}^{k} n_i = n$$

This formula can be obtained from $_nP_r$ by the same type of argument used for obtaining $_nC_r$ from $_nP_r$. In particular, n distinct elements can be permuted in $n!$ distinct ways. But if n_1 of the n elements are indistinguishable, then only $1/n_1!$ of the $n!$ permutations are distinguishable since every choice of possible positions for the n_1 indistinguishable elements is counted as $n_1!$ permutations instead of 1 permutation. Similarly, if another n_2 of the remaining $n - n_1$ elements are indistinguishable, then only $1/n_2!$ of the $n!/n_1!$ permutations are distinguishable, and so forth for k types of elements of which there are n_k indistinguishable elements of the kth type, from which it follows that the number of distinguishable permutations is

$$\frac{n!}{n_1!n_2!\cdots n_k!}$$

Note that for the special case of a set of n elements consisting of r (indistinguishable) elements of one type and $n - r$ (indistinguishable) elements of another type (i.e., $n_1 = r$, $n_2 = n - r$) the number of (distinguishable) permutations is

$$\frac{n!}{r!(n-r)!} = {}_nC_r$$

That is, since the order of r indistinguishable elements is of no consequence and likewise the order of the $n - r$ indistinguishable elements is of no consequence, the problem is one of combinations –the number of ways in which r (or, equivalently, $n - r$) of the n positions can be chosen.

Similarly, for the more general case of k types of indistinguishable elements, the problem is essentially that of determining the number of ways of partitioning the n positions into k subsets of $n_1, n_2, \ldots, n_k$ positions, respectively.

EXAMPLE
How many arrangements can be made of the letters of Mississippi taken all together?

$$\frac{11!}{1!\,4!\,4!\,2!} = 34{,}650$$

EXAMPLE
In how many ways can 25 cars place in a race by makes if 20 of them are Ferraris
and 5 are Jaguars?

$$\frac{25!}{20!\,5!} = 53{,}130$$

EXAMPLE
In the World Series the American League team and the National League team
play until one team wins 4 games. How many different sequences of winners are
possible?
 Consider first the number of ways in which the American League team can
win; they must win the last game played and 3 other games, but the National
League team may have won 0, 1, 2, or 3 games; this can happen in

$$\frac{3!}{3!\,0!} + \frac{4!}{3!\,1!} + \frac{5!}{3!\,2!} + \frac{6!}{3!\,3!} = 35 \text{ ways}$$

Similarly, the National League team can win in 35 ways; thus the total number
of sequences of winners is 70.

Many problems require the application of various combinations of the
multiplication principle, the addition principle, and the rules for permutations
and combinations. In order to solve such problems, we must consider two
questions:

1. Are the operations or objects considered together or are they mutually
   exclusive? (multiplication principle versus addition principle).
2. Is order important or not? (permutations versus combinations).

Permutations and combinations are frequently useful for determining
probabilities in the case of equally likely events and also in some of the discrete
probability distributions.
 A few simple identities in this area that the student should remember are:

$$_nP_n = n!$$
$$_nC_n = 1$$
$$_nC_r = {_nC_{n-r}}$$
$$_nP_r \neq {_nP_{n-r}}$$

PROBLEMS

4.1 If a die is rolled twice, what is the probability that the first roll yields a 5 or 6 and
 the second roll anything but a 3?
4.2 From a box containing 8 black, 8 white, and 8 red balls, 3 balls are drawn successively
 without replacement. What is the probability that a black ball is not drawn?
4.3 If 3 dice are thrown find the probability that
 (a) All 3 will show 4's
 (b) All 3 will be alike
 (c) 2 will show 4's and the other will not show a 4
 (d) Only 2 will be alike
 (e) All 3 will be different

4.4 An urn contains 5 white balls, 2 blue balls, 4 green balls, and 1 red ball. From the urn 4 balls are drawn at random with replacement.
   (a) What is the probability that 3 white balls and 1 green ball will be drawn?
   (b) What is the probability that 1 white ball, 1 blue ball, 1 green ball, and 1 red ball will be drawn?

4.5 Answer problem 4.4 if the drawing is without replacement.

4.6 The probability it will rain tomorrow is $\frac{1}{3}$. If it rains, the game between teams A and B will be called off with probability $\frac{3}{4}$; if the game is played in the rain team A will win with probability $\frac{3}{4}$. If it does not rain, the game will be played and will be won by team A with probability $\frac{1}{2}$. What is the probability team A will win the game tomorrow?

4.7 The Meadow City Ramblers and the Tough Town Tigers are playing in the Little League World Series. The first team to win 3 games is the winner of the series. For any particular game, the probability that the Ramblers will win is $\frac{1}{4}$ and the probability that the Tigers will win is $\frac{3}{4}$. What is the probability that the Tigers will win the series?

4.8 The king of Outer Moronia has set up the following procedure for selecting his daughter's husband. Each suitor chooses one of three rooms (blue, red, yellow); once inside the room of his choice, he then chooses one of three boxes (bronze, silver, gold). If the box he chooses contains a picture of the princess, he has won her hand in marriage. In the blue room the picture is always in the bronze box, in the red room it is in the silver box half the time and in the gold box half the time (at random); in the yellow room it is always in the gold box. Prince Prob arrives on the scene with a device that assures that he will first choose a room at random (equally likely) and then will choose the bronze box with probability $\frac{1}{2}$, the silver box with probability $\frac{1}{3}$, and the gold box with probability $\frac{1}{6}$. What is the probability that he will win the princess?

4.9 Professor Probability has 20 students in his statistics class. Of these, 10 have taken advanced calculus, 8 have taken matrix algebra, and 5 have taken both advanced calculus and matrix algebra. If the professor chooses a student at random, what is the probability that the student
   (a) Has taken either advanced calculus or matrix algebra, or both
   (b) Has taken advanced calculus but not matrix algebra
   (c) Has taken matrix algebra given that he or she has not taken advanced calculus
   (d) Has taken calculus given that he or she has taken matrix algebra

4.10 Jill, Joan, and Joy walk home from school together every day and each time they choose one of three possible routes A, B, or C. Each girl's father drives home from work every day along route A, B, or C, though not always as early as the girls walk home. If one of their fathers passes the girls, he picks them up. Jill's father drives home early $\frac{1}{3}$ of the time; when he does, he takes route A $\frac{1}{3}$ of the time and route B $\frac{2}{3}$ of the time. Joan's father drives home early $\frac{1}{2}$ of the time; when he does, he takes route A $\frac{1}{4}$ of the time, route B $\frac{1}{4}$ of the time, and route C $\frac{1}{2}$ of the time. Joy's father drives home early $\frac{3}{4}$ of the time; when he does, he always takes route C. If the girls choose each route equally often at random and if the fathers independently choose their early days and their routes at random, what proportion of the time will one of the fathers pick up the girls?

4.11 Three types of fuel can be used for a particular industrial engine and, due to random availability, type 1 is used $\frac{1}{2}$ of the time, type 2 is used $\frac{1}{3}$ of the time, and type 3 is used $\frac{1}{6}$ of the time. Sometimes the engine backfires when it is started; this occurs with probabilities $\frac{1}{4}$, $\frac{1}{8}$, and $\frac{4}{5}$ if the fuel is of type 1, 2, 3, respectively. Its operator observed that the engine backfired when he started it. What is the probability that the fuel was of type 3?

4.12 The Weather Bureau of Centralia City classifies winter days (on the basis of predominant weather conditions) as snowy, rainy, overcast, or sunny. A weather person reports to work one morning and finds that the previous day's classification has been lost; she has been out of town and has no firsthand information on the subject. From past records, she knows that 20 percent of winter days are classified as snowy, 10 percent as rainy, 30 percent as overcast, and 40 percent as sunny. She also knows that the probability that the temperature will go below freezing at night is 70 percent if the day is snowy, 20 percent if rainy, 40 percent if overcast, and 80 percent if sunny.

She observes on the temperature recorder that the temperature the night before did go below freezing.

(a) What is the probability the day before was classified as snowy?

(b) Suppose the precipitation gauge registers a considerable amount for the day before and she is thus sure it was either snowy or rainy; then what is the probability it was snowy?

4.13 A company manufactures light bulbs on four different production lines; the bulbs are sent to a central stockroom to be sampled for defectives. Due to a misunderstanding, the labels identifying the production line from which the bulbs came have been removed. A bulb, obtained by random sampling, is found to be defective. If line 1 produces 50 percent of the bulbs with a defective rate of .1 percent, line 2, 10 percent with a defective rate of .1 percent, line 3, 10 percent with a defective rate of .3 percent, and line 4, 30 percent with a defective rate of .2 percent, what is the probability that the defective bulb did not come from line 1?

4.14 A man owns a house in town and a cabin in the mountains. In any one year the probability of the house being burglarized is .01 and the probability of the cabin being burglarized is .05. For any one year what is the probability that

(a) Both will be burglarized?

(b) One or the other (but not both) will be burglarized?

(c) Neither will be burglarized?

4.15 Half of all the articles bought at an expensive women's accessory shop are later returned. The manager has noticed that if an item is bought by a woman who is accompanied by at least one other woman, it is invariably returned. Items bought under other circumstances are returned 37.5 percent of the time. If an item is returned, what is the probability it was bought by a woman accompanied by at least one other woman?

4.16 A fraternity on campus has 120 members who are registered in several colleges of the university and who maintain average grade points as summarized in the following table:

	GRADE POINT AVERAGE			
	A	B	C	
Arts and science	10	20	10	40
Engineering	20	25	5	50
Agriculture	10	10	10	30
	40	55	25	120

A student is picked at random from the list of fraternity members.

(a) What is the probability that he is registered in agriculture?

(b) What is the probability that he has a B average?

(c) What is the probability that the student is in engineering given that he has a C average?

(d) What is the probability that the student has an A average given that he is in arts and science?

(e) Are maintaining an A average and being registered in arts and science independent?

(f) Are being registered in engineering and maintaining a C average independent?

4.17 The Happy Motoring Company manufactures sedans and sports cars on two production lines as follows:

	SEDANS	SPORTS CARS
Line 1	150	50
Line 2	200	100

From each of these production runs of 500 cars, one car is selected for intensive testing. If the selection is done so that each car from line 1 has probability $\frac{1}{800}$ of being selected and each car from line 2 has probability $\frac{1}{400}$ of being selected,

(a) Determine the probability that the car selected is from line 1 if it is a sedan

(b) Determine the probability that the car selected is a sports car

(c) Are being a sports car and being manufactured on line 2 independent

4.18 Every weekday, Samuel Executive eats lunch either in the coffee shop on the ground floor of the building in which he has his office or in the Tycoon Club of which he is a member. If it is raining and/or he has an early afternoon meeting, Samuel eats lunch in the coffee shop. If it is not raining and he has no early afternoon meeting, Samuel eats in the coffee shop $\frac{1}{6}$ of the time and in the Tycoon Club $\frac{5}{6}$ of the time. If he eats in the coffee shop, Samuel chooses hamburger $\frac{2}{3}$ of the time, if he eats in the Tycoon Club, he chooses hamburger only $\frac{2}{9}$ of the time. If it rains on $\frac{1}{10}$ of the weekdays and Samuel has early afternoon meetings on $\frac{2}{3}$ of the weekdays, what proportion of the weekdays does Samuel eat hamburger for lunch? (Assume that meetings are scheduled in advance and thus independent of the weather conditions.)

4.19 The Peoples party has three candidates (A, B, and C) trying for its presidential nomination. The party planners have the situation analyzed as follows: the probability of winning a key primary to be held shortly is $\frac{2}{3}$ for candidate A, $\frac{1}{6}$ for candidate B, and $\frac{1}{6}$ for candidate C. The candidate who wins this primary has probability $\frac{2}{3}$ of winning the nomination. Each of the losers in the primary has probability $\frac{1}{6}$ of winning the nomination. If nominated, candidates A, B, and C have respective probabilities $\frac{1}{3}$, $\frac{2}{15}$, and $\frac{1}{5}$ of winning the election. Assuming the party planners are correct in their analysis,

(a) What probability has each candidate of winning the election

(b) What probability has the Peoples party of winning the election

4.20 Before installing transistors in its computers, Magnitronics tests them for defects. Experience has established that $\frac{1}{6}$ of the transistors supplied are in fact defective. The test used by Magnitronics detects defective transistors $\frac{9}{10}$ of the time; it also (mistakenly) indicates that nondefective transistors are defective $\frac{1}{2}$ of the time. If this test indicates that a transistor is nondefective, what is the probability that the transistor is in fact nondefective?

4.21 Assume the time is 1877. The Interstellar Company (whose president is Jules Verne) is attempting to fire a projectile at the moon from a place in Florida called Tampa Town. It is called the Columbiad, and its designers have broken the firing down into four phases where malfunctions could occur. The phases are specified in such a way that if a malfunction occurs in any one of them the entire project of trying to get the projectile to the moon will be a failure. The phases are sequential in time and independent of each other; that is, the successful operation of any phase is independent of the successful or unsuccessful operation of any subsequent phase. Of course, if any particular phase malfunctions, there are no subsequent phases because the project fails at that point. The probabilities of a malfunction at each phase are as follows:

Phase 1————.20
Phase 2————.40
Phase 3————.30
Phase 4————.10

The chief engineer of the project is Mr. Marston. He says that if the project is a failure, the probability of losing his job is .7.

(a) What is the probability of the project being a failure?

(b) What is the probability of Mr. Marston losing his job?

4.22 In some gambling casinos there is a game played called Chuck-a-Luck. All bets are made by players against the house on a cloth with squares containing the numbers 1 through 6. The houseman has three dice is a cagelike apparatus, which he turns and the dice fall onto the floor of the cage. If the number you bet shows on one of the dice, the house pays you the amount you bet; if the number you bet shows on two of the dice, the house pays you 2 to 1; and if the number you bet shows on all three dice, the house pays you 3 to 1. Of course, if the number you bet does not show, you lose.

Suppose you place a bet on the number 1, and the dice are turned and let fall. Assuming the dice and apparatus are fair, what is the probability of
(a) Winning even money?
(b) Winning 2 to 1?
(c) Winning 3 to 1?

4.23 A and B are competing for a contract. The probability that A will be awarded the contract is $\frac{1}{2}$, the probability that B will be awarded the contract is $\frac{1}{4}$. If A has the contract, the work will be completed on time with probability $\frac{1}{3}$; if B has the contract it will be completed on time with probability $\frac{1}{2}$. What is the probability that the work will be completed on time?

4.24 The window decorator for LaPetite Boutique has been given 5 black dresses, 2 pink dresses, and 4 red dresses by the manager. From these she is to choose 4 dresses for each of two windows. The decorator decides that there should not be more than 2 dresses of the same color in a window and that pink and red dresses should not appear in the same window. The dresses are of similar styles and she distinguishes them only on the basis of color. If her (completely color blind) assistant chooses two sets of four dresses at random while the decorator is out to lunch, what is the probability that they meet the decorator's criteria?

4.25 The president of Carbor Can Company told his friend, Mr. Bauxite, that the company was going to introduce a new product and would market it for a year as a trial. At the end of the year the product would be judged a success or a failure and continued or discontinued accordingly. The president knew that a competitor was about to introduce one of three products. He told Mr. Bauxite that he thought Carbor's new product had probability .30 of being a success if the competitor introduced product A, .25 if it introduced product B, and .40 if it introduced product C and that he estimated the respective probabilities of introducing A, B, C as $\frac{1}{5}$, $\frac{2}{5}$, $\frac{2}{5}$. Mr. Bauxite was out of the country for a year and a half and noticed on his return that Carbor's product was still on the market. What probabilities, respectively, should he assign to the competitors having introduced product A, B, or C?

4.26 Mrs. Smith is attending a postal auction for her firm, Electronic Suppliers, Inc. The firm has recently lost 3 shipments in the mail due to faulty labels which came off in transit. One shipment carton contained 3 color TVs, 3 black and white TVs, and 6 FM radios worth a total of $2400.00. A second contained 12 black and white TVs and 6 FM radios worth a total of $3000.00. A third contained 6 color TVs and 6 black and white TVs worth a total of $3600.00. A carton, which Mrs. Smith is certain contains one of the firm's lost shipments, comes up for bidding.
(a) If Mrs. Smith considers that the carton could possibly be any of the three shipments but that it is twice as likely to be the least valuable ($2400.00) carton as to be either of the other two, and equally likely between these other two, how much should she be willing to bid for it?
(b) One of the auction officials tells Mrs. Smith that he saw an FM radio caton through a break in the outer carton; assume that each of the inner cartons is equally likely to be visible through a break in an outer carton. Considering also the additional information provided by the auction official, how much should Mrs. Smith be willing to bid for the carton?

4.27 The Board of Directors of Diversified Products, Inc., has decided to introduce a new toy, Tinkerbell Talking Doll. After the doll has been on the market for six months, the Board of Directors will meet to decide whether it has been highly successful, moderately successful, or unsuccessful in terms of sales. At that time they will also decide whether to vary the line of Tinkerbell dolls, abandon the Tinkerbell doll and introduce a line of battery-propelled space ships, or stop manufacturing toys and concentrate on their other products. According to one member of the board, the probability that the doll will be highly successful is $\frac{1}{2}$, the probability that it will be moderately successful is $\frac{1}{3}$, and the probability that it will be unsuccessful is $\frac{1}{6}$. If the Tinkerbell doll is highly successful, the probability that more lines will be introduced is $\frac{2}{3}$, that the spaceship will be introduced is $\frac{1}{6}$, and that no more toys will be manufactured is $\frac{1}{6}$. If the doll is moderately successful, the probability that more lines will be introduced is $\frac{1}{3}$, that the spaceship will be introduced is $\frac{1}{3}$, and that no more toys will be manufactured is $\frac{1}{3}$, If the doll is unsuccessful, the probability that more lines will be introduced is $\frac{1}{6}$, that the spaceship will be introduced is $\frac{1}{2}$, and that no more toys will be introduced is $\frac{1}{3}$, What is the probability that the decision in six months will be to manufacture no more toys?

4.28 In the following probability statements A refers to stock A, B refers to stock B, merger refers to merger of companies A and B, and GNP refers to gross national product. Assume that each of the stocks and gross national product either goes up or down but does not remain the same; that is, for each of these variables the probability of going up plus the probability of going down equals one.

Given:

$$P(B \text{ goes up} | GNP \text{ goes down}) = .20$$
$$P(B \text{ goes up} | GNP \text{ goes up}) = .60$$
$$P(Merger | GNP \text{ goes up}) = .60$$
$$P(Merger, GNP \text{ goes up}) = .30$$
$$P(A \text{ goes up} | B \text{ goes up}) = .90$$
$$P(A \text{ goes up} | B \text{ goes down}) = .40$$

Obtain the following:
(a) $P(GNP \text{ goes up})$
(b) $P(B \text{ goes up})$
(c) $P(A \text{ goes up})$
(d) $P(B \text{ goes up}, GNP \text{ goes down})$
(e) $P(A \text{ goes up}, B \text{ goes down})$

PROBLEMS APPLICABLE TO APPENDIX 4.2

4.29 A company has 7 men qualified to operate a machine that requires 3 operators for each shift.
(a) How many shifts are possible?
(b) In how many shifts will a particular man appear?

4.30 An instructor has written 20 problems. Five of these are so important that she assigns them to every student. In addition, each student is assigned five other problems. In how many ways can she assign problems to a student?

4.31 A kennel owner has a little of 12 puppies for sale, 7 of them are males and the other 5 are females; she has 4 orders for males, 2 orders for females, and 4 orders for either males or females; in how many ways can she fill the orders?

4.32 There are 10 men under consideration for the 5-man Board of Directors of the newly formed Crunch Candy Corporation. Three of these men are known to be sworn mutual enemies and if any one of them is on the board the other two must not be.
(a) In how many ways can the board of Directors be chosen?
(b) After its selection, the Board of Directors will elect from its membership a president and a vice-president; how many possible outcomes are there for such an election?

4.33 Staggering into his club's cloakroom after a hard evening, Mr. Smash saw 5 hats on the shelf. A little later Mr. Crash came into the cloakroom and took a hat that he thought was his. The next day each man discovered that he had someone else's hat. In how many ways could both Mr. Smash and Mr. Crash have gotten the wrong hats? (Hint: This problem is deceptive; best way to work it is "backwards." Note that the number of ways that each man could have gotten any hat is 20 and the number of ways that each man could have gotten the right hat is 1.)

4.34 There are three girls: Alice, Bonnie, and Carol. There are five coins: half dollar, quarter, dime, nickel, penny. How many different possible ways may the coins be divided among the girls, if the stipulation is made that all coins must be passed out in any given division but that all girls need not receive coins in any given division. (For example, giving all five coins to Alice would be a legitimate division.)

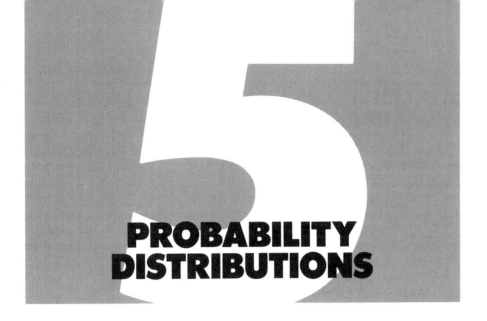

PROBABILITY DISTRIBUTIONS

5.1 INTRODUCTION

In this chapter the definitions and calculus of probabilities developed in Chapter 3 are extended to the definition of the probability distribution of a random variable. Several types of probability distributions and some of their applications are discussed.

RANDOM VARIABLES

For many problems, a particular aspect of the outcome of an experiment or observation is of interest to the investigator. Frequently, a numerical value is naturally associated with this aspect of the outcome—for example, the number of heads, the number of correct answers, the age of the respondent, the price of the product, the value of GNP, the percent unemployed, and so forth. When an experiment results in qualitative outcomes, a numerical code can be assigned for convenience to the outcome categories—for example, 1 for "yes" and 2 for "no"; 1 for red, 2 for blue, and 3 for yellow; 1 for male and 2 for female; 1 for correct and 2 for incorrect, and so forth.

Thus the numerical value associated with a particular aspect of each outcome in the sample space varies and depends on the outcome observed. As discussed in Chapter 3, a variable whose value is determined by the outcome of an experiment or observation is a *random variable*, also referred to as a *stochastic variable* or *stochastic variate*. The values assumed by a random variable are associated with random events in the sample space of a given experiment. A random variable is a real-valued function defined on a sample space. Upper case letters, for example, X, Y, Z, are used to denote random variables; lower case letters, for example, x, y, z, a, b, c, are used to denote particular values of random variables. The expression $P(X = x)$ denotes the probability that the random variable X assumes the value x; $P(X = x)$ is sometimes abbreviated $P(x)$.

There are two general classes of random variables: (1) *discrete*, defined on a discrete sample space and can assume a finite or countably infinite number of values and (2) *continuous*, defined on a continuous sample space and can assume any of the (infinitely many) values within an interval or set of intervals. The observed values of a discrete random variable frequently consist of or are based on the numbers of elements that possess a particular attribute; a continuous random variable results from the measurement on a continuous scale of a characteristic of the elements—length, volume, weight, price, time, temperature, and so forth. In practice, the values a continuous variable can assume are limited by the precision of the instrument used to obtain the measurements; thus the set of possible measures of a continuous variable is finite, due to the limitations of measuring instruments, although the set of its theoretically possible values is infinite. The error involved in assuming continuity is negligible, if the measuring instrument is reasonably precise.

PROBABILITY DISTRIBUTIONS

Many problems involve determining the probabilities with which a random variable assumes specified values in its range of possible values. Thus, just as it is useful to have a list of all the possible outcomes of an observation and their corresponding relative frequencies, it is useful to have a list of all the possible values of a random variable and their corresponding probabilities of occurrence. Such a summary is called a *probability distribution*. A probability distribution provides the same type of information for a random variable that a frequency distribution provides for an empirical variable, with probabilities replacing relative frequencies. For example, if a fair die is thrown, the probability distribution of the outcome is given by

Number on Face of Die	Probability $P(X = x)$
1	$\frac{1}{6}$
2	$\frac{1}{6}$
3	$\frac{1}{6}$
4	$\frac{1}{6}$
5	$\frac{1}{6}$
6	$\frac{1}{6}$
	1

Other sets of possible outcomes of this experiment can be defined and their probability distributions obtained. For example, suppose the possible outcomes are an even number and an odd number. Then the probability distribution is given by

x	$P(X = x)$
Even number	$\frac{1}{2}$
Odd number	$\frac{1}{2}$
	1

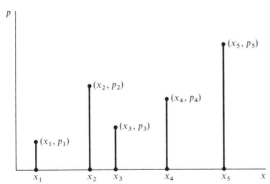

Figure 5.1   Graphical representation of a probability distribution

In general, a probability distribution is any statement of a probability function that has a set of mutually exclusive and exhaustive events for its domain. If a random variable can assume only the values $x_1, x_2, \ldots, x_k$, then

$$P(X = x_i) \geq 0 \qquad i = 1, \ldots, k$$

$$\sum_{i=1}^{k} P(X = x_i) = 1$$

In the preceding discussion of probability distributions we assumed that listing the possible values of the random variable and their corresponding probabilities of occurrence is easy and convenient. If there are relatively few possible events, then this is so. However, if there are a large number of possible events, listing becomes tedious and cumbersome. When a random variable is continuous and thus assumes an infinite number of possible values, its probability distribution cannot be specified by listing. There are two other ways to specify a probability distribution: by a graph and by a mathematical function.

If a random variable X can assume the values $x_1, x_2, \ldots, x_k$ and $p_i = P(X = x_i)$ for $i = 1, 2, \ldots, k$, then the probability distribution can be represented by the set of k pairs of values (x_i, p_i) for $i = 1, 2, \ldots, k$. These pairs can be graphed as shown in Figure 5.1.

A graph of the pairs (x_i, p_i) is sometimes referred to as a graph of a *probability mass function* or *probability function*. Note that the probability of an event can be considered a function because corresponding to each possible value of the random variable, that is, to each possible event, there is one and only one number between zero and 1 inclusive that is its probability.

The probability distribution of the outcome of tossing a fair die, given above by listing, can also be represented graphically as in Figure 5.2.

$$P(X = x) \quad \begin{cases} \frac{1}{6} & \text{if } X = 1, 2, 3, 4, 5, 6 \\ 0 & \text{otherwise} \end{cases}$$

Since $\sum_{x=1}^{6} P(X = x) = 1$, it might seem unnecessary to specify that $P(X = x)$ is zero for other values of x; however, $P(x)$ is a function on the real line and to be mathematically correct it must be defined for all points on the real line.

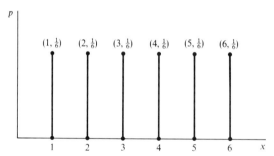

Figure 5.2   Probability distribution for tosses of a fair die

Thus far the discussion of probability distributions has concerned discrete random variables, that is, random variables having a finite number of points in the sample space. For a discrete random variable, the probability distribution is specified in terms of the probabilities with which the random variable assumes various exact values.

The probability distribution for a continuous random variable must be interpreted in a somewhat different manner. Since a continuous random variable can assume an infinite number of values, the probability that it assumes any particular *exact* value is zero. For a continuous random variable, probability is defined in terms of an interval of values; the probability that a continuous random variable assumes any particular value is zero, but the probability that it assumes a value in some specified interval is, in general, not zero and can be specified mathematically if the probability function is known.

The probability distribution for a continuous random variable X is defined in terms of the *probability density* of X at x, denoted by $f(x)$. The probability density function for a continuous random variable is represented by a smooth curve. If $f(x)$ is the probability density function for a continuous random variable X, then $P(a < X < b)$ is represented by the area under the curve $y = f(x)$ between the values a and b, as shown in Figure 5.3. The total area under the curve $y = f(x)$ over the domain of X is 1. The domain is assumed to be the set of all real numbers, although the probability density function may be defined as zero for all real numbers except those in a specified interval or intervals.

Figure 5.3   Graphical representation of a probability density function

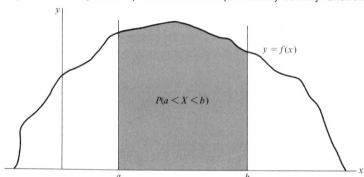

The probability that a random variable assumes a value in a specified interval can always be obtained, whether the random variable is continuous or discrete. The probability that a discrete random variable assumes a value in the interval $a \leq X \leq b$ is the sum of the probabilities of the values in the interval

$$P(a \leq X \leq b) = \sum_{x=a}^{b} P(X = x)$$

The corresponding probability for a continuous random variable can be written as the definite integral* from a to b of $f(x)$:

$$P(a \leq X \leq b) = \int_{a}^{b} f(x)\, dx$$

Since the probability that a continuous random variable X assumes any exact value is zero, $P(X = a) = 0$ and $P(X = b) = 0$. Thus the probability of an interval is the same for a continuous random variable whether or not the endpoints are included. For a discrete random variable, inclusion of the endpoints does make a difference, since $P(X = a) \neq 0$ and $P(X = b) \neq 0$ for a discrete random variable.

CUMULATIVE DISTRIBUTION FUNCTIONS

Representation of the probability distribution of a random variable by a probability mass function for the discrete case and by a probability density function for the continuous case are discussed in the preceding section. Alternatively, the probability distribution for either a discrete or continuous random variable can be represented in terms of cumulative probabilities.

The probability that a random variable X assumes a value less than or equal to a given number a can be written:

$$F(a) = P(X \leq a)$$

The function $F(x)$ is the *cumulative distribution function* of the random variable X. The cumulative distribution function relates the values of X to the corresponding cumulative probabilities. A cumulative distribution function (*cdf*) must satisfy certain mathematical properties that follow from its definition:

1. $0 \leq F(x) \leq 1$
2. If $a < b$, then $F(a) \leq F(b)$
3. $F(\infty) = 1$ and $F(-\infty) = 0$

The cumulative distribution function is uniquely determined by the probability mass function of a discrete random variable or by the probability density function of a continuous random variable. If X is a discrete random

---

* Concepts that are expressed for discrete random variables in terms of sums are expressed for continuous random variables in terms of integrals. The student who is not familiar with integrals and their manipulation, but who reads with this correspondence in mind, should have no difficulty following the discussion. Use of integrals is minimized in the text and integration is not required in solving any of the problems.

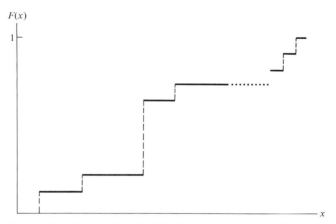

Figure 5.4 Cumulative distribution function for a discrete random variable

variable,

$$F(a) = P(X \le a) = \sum_{x \le a} P(X = a)$$

If X is a continuous random variable,

$$F(a) = P(X \le a) = \int_{-\infty}^{a} f(x)\, dx$$

The cumulative distribution function for a discrete random variable is a step function—a step or discontinuity occurs at each point a for which $P(X = a) > 0$ and the height of the step is equal to $P(X = a)$, as shown in Figure 5.4. Thus if X can assume K values, there are K steps in $F(x)$.

The cumulative distribution function for a continuous random variable is a continuous function, as shown in Figure 5.5. If the *cdf* $F(x)$ of a continuous random variable is known, its density function $f(x)$ can be obtained by the differentiation*

$$f(x) = \frac{d}{dx} F(x) = F'(x)$$

Figure 5.5 Cumulative distribution function for a continuous random variable

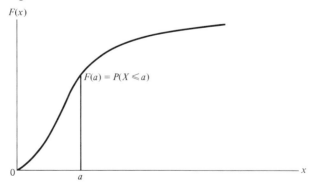

* Differentiation is the reverse operation with respect to integration.

The probability that a continuous random variable assumes a value between a and b can be obtained from its cumulative distribution function

$$P(a \le x \le b) = F(b) - F(a)$$

(for a continuous random variable $\le$ may be replaced by $<$).

All random variables, whether discrete or continuous, have cumulative distribution functions. In most cases, tables of theoretical probability distributions are expressed in terms of the cumulative distribution. Given the cumulative probabilities for various values of a random variable X, the probability of an interval can be obtained by subtraction.

EXPECTATION AND VARIANCE OF A RANDOM VARIABLE

Two general characteristics of any distribution are its central tendency (or location) and dispersion (or variability). Several measures of central tendency and dispersion are defined and discussed in Chapter 2. The expectation and variance of a random variable are generally used as measures of the central tendency and dispersion of its probability distribution. The concepts of expectation and variance of a random variable are important in the theory of probability and statistics.

If the random variable X is discrete, then the *expectation* (or expected value) of X is given by

$$E(X) = \sum_x xP(X = x) = \sum_x xP(x)$$

where the summation is over all values that can be assumed by X and $\sum_x P(x) = 1$.

If the random variable X is continuous, then the *expectation* (or expected value) of X is given by

$$E(X) = \int_{-\infty}^{\infty} xf(x)\,dx$$

where $\int_{-\infty}^{\infty} f(x)\,dx = 1$.

The *variance* of any distribution, denoted by σ^2, is equal to the expectation of the squared deviation of the random variable from its mean:

$$\sigma^2 = E(X - \mu)^2 = \begin{cases} \sum_x (x - \mu)^2 P(x) & \text{if } X \text{ is discrete} \\ \int_{-\infty}^{\infty} (x - \mu)^2 f(x)\,dx & \text{if } X \text{ is continuous} \end{cases}$$

Note that the variance of X is zero if and only if X assumes only one value, that is, if and only if X is constant. The variance increases as the spread of the values of X increases.

The expectation or variance of a random variable may not exist, since the corresponding sum or integral may not converge to a finite value. This mathematical problem does not occur frequently but will be noted for several examples later.

Several probability distributions of particular interest and importance are discussed in the following sections. These are theoretical probability distributions—that is, they are specified by a mathematical function that gives the probability that X will assume each of its possible values. A theoretical probability distribution corresponds to a particular model of the experiment or observation; the probability function is derived mathematically from the assumptions of this model. In practical applications, choice of the appropriate model for the experiment is not always easy. This problem is discussed in later sections.

We will consider examples of discrete probability distributions first. These include the binomial distribution, the Poisson distribution, the hypergeometric distribution, the multinomial distribution, the negative binomial distribution, and the discrete rectangular distribution. Several continuous distributions are then discussed, including the (continuous) rectangular distribution, the exponential distribution, the gamma distribution, the beta distribution, the normal distribution, the log-normal distribution, and the Pareto distribution.

These distributions are discussed briefly in the next sections, primarily for the purpose of illustrating the variety of probability distributions that are possible and thus counteracting the misconception that every random variable has either a binomial or a normal distribution. Several frequently used probability distributions are discussed and illustrated in more detail subsequently.

5.2 EXAMPLES OF DISCRETE PROBABILITY DISTRIBUTIONS

A discrete probability distribution is defined on a discrete sample space, that is, for a finite or countably infinite set of values of a random variable. Of the discrete probability distributions considered here, the binomial, hypergeometric, and discrete uniform distributions are defined for a finite set of values of the random variable; the Poisson and negative binomial distributions are defined for a countably infinite set of values of the random variable.

BERNOULLI PROCESSES

The simplest probability distribution is one having only two mutually exclusive and exhaustive classes. A number of random phenomena have only two possible outcomes. For example, an item from a production line is either good or defective; a tossed coin lands either heads or tails; a person is either married or unmarried, and so forth. Experiments of this type, which consist of a series of independent trials or observations, each of which can have only one of two possible outcomes, are *Bernoulli trials*, named for Jacques Bernoulli (1654–1705). The two event classes and their associated probabilities are a *Bernoulli process*.

It is customary to refer to one of the two possible outcomes of a Bernoulli process as a *success* and to the other as a *failure*. These terms are used only for the purpose of identifying the outcomes and have no connotation of "goodness" of the outcome. The probability of a success is denoted by π; the probability of a failure is denoted by $1 - \pi$. Since a Bernoulli random variable assumes only two possible values, its distribution is sometimes referred to as a *two-point probability distribution.*

THE BINOMIAL DISTRIBUTION

Suppose that a random sample of n trials or observations is obtained from a Bernoulli sample space and that the probability of a success is π for each trial. That is, either the sample space consists of an infinite number of elementary events or sample observations are obtained independently with replacement. The number of successes observed in n trials is a discrete random variable that can assume any integer value between 0 and n inclusive. The distribution of the number of successes in n trials or observations taken from a Bernoulli process is the *binomial distribution.* The binomial distribution can be described not only by listing or graphical methods, but by a mathematical rule or formula.

The binomial probabilities can be obtained as follows: If the probability of occurrence of a success on a trial is π, then the probability of observing, in n trials, first x successes and then $(n - x)$ failures is

$$\overbrace{\pi \cdot \pi \cdots \pi}^{x \text{ terms}} \cdot \overbrace{(1 - \pi)(1 - \pi) \cdots (1 - \pi)}^{(n - x) \text{ terms}} = \pi^x(1 - \pi)^{n - x}$$

But the probability of observing x successes and $(n - x)$ failures in any other order is the same as the probability of observing them in this particular order, since only a rearrangement of the π's and $(1 - \pi)$'s is involved. Thus the probability of observing x successes and $(n - x)$ failures is the product of the number of possible orders and the probability of occurrence in any particular order. The number of possible orders is equal to the number of combinations of n things of which x are of one kind and the remaining $(n - x)$ are of another kind, that is,

$$_nC_x = \frac{n!}{x!(n - x)!}$$

and thus the probability of observing x successes in n trials is given by

$$P(x) = \frac{n!}{x!(n - x)!} \pi^x(1 - \pi)^{n - x} \qquad \text{for } x = 0, 1, \ldots, n$$

Note that the terms in the binomial expansion

$$[(1 - \pi) + \pi]^n = (1 - \pi)^n + n(1 - \pi)^{n-1}\pi + \frac{n(n - 1)}{2}(1 - \pi)^{n-2}\pi^2 + \cdots + \pi^n$$

$$= \sum_{x=0}^{n} \frac{n!}{x!(n - x)!} \pi^x(1 - \pi)^{n - x}$$

correspond to the probabilities of the binomial distribution in the order $x = 0, 1, 2, \ldots, n$. Thus $P(x) \geq 0$ for all x and $\sum_{x=0}^{n} P(x) = [(1 - \pi) + \pi]^n = 1$, as required for a probability distribution.

The binomial is actually a family of distributions that have the same mathematical rule for associating probabilities with values of the random variable. A particular binomial distribution is specified by specifying particular values of the parameters n and π.

A random variable X with probability function given by

$$P(X = x) = \binom{n}{x} \pi^x (1 - \pi)^{n-x} \quad 0 \leq x \leq n$$

is said to have a binomial distribution with parameters n and π. This is sometimes stated as "X has the distribution $B(n, \pi)$" or "X is $B(n, \pi)$."

It can be shown that the mean and variance of the binomial distribution are given by

and

$$\begin{cases} \mu = n\pi \\ \sigma^2 = n\pi(1 - \pi) \end{cases}$$

The binomial distribution has maximum variance (for fixed n) when $\pi = (1 - \pi) = .5$. The binomial probabilities for various values of n and π are given in Table B.1.

The shape of the binomial distribution depends on the values of n and π. The binomial distribution has a single maximum between $x = 0$ and $x = n$; its exact location depends on n and π. When $n\pi$ is an integer, the maximum occurs at $x = n\pi$, the mean of the distribution. The binomial distribution is symmetric if $\pi = \frac{1}{2}$, positively skewed if $\pi < \frac{1}{2}$, and negatively skewed if $\pi > \frac{1}{2}$. For $\pi \neq \frac{1}{2}$ the skewness decreases as n increases. For a given value of π the mean and variance increase as n increases. For a given value of n the mean increases as π increases and the variance increases as π approaches $\frac{1}{2}$. (If $\pi = 0$ or $\pi = 1$, $\mu = n\pi(1 - \pi) = 0$).

EXAMPLE

Of the new products introduced by Galvistron, 75 percent are successes and remain on the market at least five years. What is the probability that exactly 6 of the next 8 products introduced by Galvistron will be successful?

$$P(x) = \frac{n!}{x!(n - x)!} \pi^x (1 - \pi)^{n-x}$$

$$P(6) = \frac{8!}{6!2!} \left(\frac{3}{4}\right)^6 \left(\frac{1}{4}\right)^2$$

$$= \frac{5103}{16384} = .311$$

Note that this example assumes that the probability of success for each new product introduced by Galvistron is .75 and that the success of each new product is independent of the success of any other product.

EXAMPLE

The probability that Mr. Smith, an expert marksman, will hit a target is .80 on each shot. What is the probability that he will hit the target on at least 5 of 6 shots?

$$P(x) = \frac{n!}{x!(n-x)!} \pi^x (1-\pi)^{n-x}$$

$$P(x \geq 5) = f(5) \quad f(6)$$

$$= \frac{6!}{5!1!} \left(\frac{4}{5}\right)^5 \left(\frac{1}{5}\right)^1 + \frac{6!}{6!0!} \left(\frac{4}{5}\right)^6 \left(\frac{1}{5}\right)^0$$

$$= \frac{6144}{15625} + \frac{4096}{15625} = .655$$

EXAMPLE

The management of Premiere Theatre knows from past experience that 20 percent of the complimentary tickets sent to critics are not used. This percentage is apparently independent of the type of play and any other identifiable circumstances. The theatre reserves 10 seats for critics. If 12 complimentary tickets are sent out, what is the probability of accommodating all those critics who actually go to the theatre? (Assume that the probability a critic will not use his or her ticket on any night is .20 and is independent of the behavior of other critics.)

$$P(x) = \frac{n!}{x!(n-x)!} \pi^x (1-\pi)^{n-x}$$

$$P(x \geq 2) = 1 - .7251 = .2749$$

The value of $P(x \geq 2)$ is obtained from Table B.2, which gives values of the cumulative binomial distribution. It could also be obtained from Table B.1, which gives the values of the individual terms of the binomial distribution, as follows:

$$P(x \geq 2) = 1 - P(x < 2) = 1 - [P(0) + P(1)]$$

$$= 1 - [.0687 + .2062]$$

$$= .2749 \quad \text{(same as above)}$$

EXAMPLE

The binomial distribution is appropriate for solving the problem of Pascal and Fermat's that was mentioned in Chapter 3. If a die is tossed 4 times, the probability of obtaining at least one 6 can be obtained as follows:

$$x = \text{number of times a 6 turns up}$$

$$n = 4$$

$$\pi = \frac{1}{6}$$

$$1 - \pi = \frac{5}{6}$$

$$P(x \geq 1) = 1 - P(x = 0)$$

$$P(x = 0) = f(0) = \frac{4!}{0!4!} \left(\frac{1}{6}\right)^0 \left(\frac{5}{6}\right)^4 = \frac{625}{1296} = .482$$

$$P(x \geq 1) = 1 - 0.482 = 0.518$$

If 2 dice are thrown 24 times, the probability of obtaining at least one double 6 is as follows:

$$x = \text{number of times a double 6 turns up}$$

$$n = 24$$

$$\pi = \frac{1}{36}$$

$$(1 - \pi) = \frac{35}{36}$$

$$P(x \geq 1) = 1 - P(x = 0)$$

$$P(x = 0) = f(0) = \frac{24!}{0!24!}\left(\frac{1}{36}\right)^0\left(\frac{35}{36}\right)^{24} = 0.511$$

$$P(x \geq 1) = 1 - 0.511 = .489$$

Thus the probability of throwing one 6 in 4 tosses of a single die is (slightly) greater than the probability of throwing a double 6 in 24 tosses of 2 dice.

THE BINOMIAL DISTRIBUTION OF PROPORTIONS

In many applications an investigator is interested in the proportion of successes rather than in the number of successes. If x is the number of successes in N trials, the proportion of successes is x/n. The proportion of successes is a random variable that assumes decimal values between 0 and 1; the sample proportion is denoted by p, to distinguish it from π, the probability of a success. For given n, the value of p is determined by the value of x; thus

$$P\left(p = \frac{x}{n}\right) = P(X = x)$$

the distribution of the sample proportion therefore is given by the binomial distribution.

The mean and variance of p are given by

and
$$\begin{cases} \mu = \pi \\ \\ \sigma^2 = \dfrac{\pi(1 - \pi)}{n} \end{cases}$$

NOTE

$$E(p) = E\left(\frac{x}{n}\right) = \frac{1}{n}E(x) = \frac{n\pi}{n} = \pi$$

$$\text{var}(p) = \text{var}\left(\frac{x}{n}\right) = \frac{1}{n^2}\text{var}(x) = \frac{n\pi(1 - \pi)}{n^2} = \frac{\pi(1 - \pi)}{n}$$

THE NEGATIVE BINOMIAL (PASCAL) DISTRIBUTION

In developing the binomial distribution, we assume that a fixed number n of Bernoulli trials are observed; the random variable is x—the number of successes. In some problems the sampling procedure is somewhat different—a sequence of independent Bernoulli trials is observed and sampling continues until a fixed number k of successes is observed. Thus the number of observations n is a random variable. The distribution of n can be obtained by the same type of reasoning used to obtain the distribution of x.

The probability of any particular sequence of k successes in n trials is $\pi^k(1 - \pi)^{n-k}$. If sampling continues until the kth success is observed, then the result of the final trial is a success, otherwise observation would have terminated prior to this trial. Since the last trial is a success, $k - 1$ successes must have been observed in the previous $n - 1$ trials. But the number of possible sequences of $k - 1$ successes in $n - 1$ trials is $_{n-1}C_{k-1} = \binom{n-1}{k-1}$. Thus, in sampling from a Bernoulli process with probability of success π, the probability of observing the kth success on the nth trial is

$$P(n) = \binom{n-1}{k-1}\pi^k(1 - \pi)^{n-k} \qquad 0 < k \le n$$

This distribution of the number of trials until the kth success or the trial on which the kth success occurs is the *negative binomial distribution*, sometimes called the *Pascal distribution* after the French mathematician. The terminology negative binomial distribution arises because the probabilities correspond to the terms of the binomial expansion of

$$\left(\frac{1}{\pi} - \frac{1 - \pi}{\pi}\right)^{-k}$$

For the special case $k = 1$, $P(n) = \pi(1 - \pi)^{k-1}$; this probability distribution is referred to as the *geometric distribution* and gives the probability that the first success will occur on the nth trial.

The negative binomial distribution has many applications, particularly in problems involving waiting times and in certain types of industrial sampling, when sampling is continued until a certain number of defectives are observed.

The mean and variance of the negative binomial distribution are given by

and
$$\begin{cases} \mu = \dfrac{k}{\pi} \\[2ex] \sigma^2 = \dfrac{k(1 - \pi)}{\pi^2} \end{cases}$$

The distribution is positively skewed.

The binomial, negative binomial (Pascal), and geometric distributions all involve independent Bernoulli trials with constant probability of success for all trials. The distributions differ with respect to the sampling procedure or stopping rule used: for the binomial distribution, sampling is continued until n trials are observed; for the negative binomial (Pascal) distribution, sampling is continued until k successes are observed; for the geometric distribution, sampling is continued until a success is observed.

Tables B.3 and B.4 are useful for computing probabilities for these distributions. Table B.3 provides the value of $\binom{n}{x}$ for various values of n and x and Table B.4 provides the values of $N!$ and $\log N!$ for various values of N.

EXAMPLE
Market studies have established that 20 percent of the housewives in Oatsville use Happyness detergent. Find the probability that in a random sample of Oatsville housewives, the 25th person interviewed is the 10th user of Happyness detergent.

The negative binomial distribution is appropriate for this problem.

$$P(n) = \binom{n-1}{k-1}\pi^k(1-\pi)^{n-k}$$

$$n = 25, \qquad k = 10, \qquad \pi = .2$$

$$P(25) = \binom{24}{9}(.2)^{10}(.8)^{15}$$

Using Tables B.4 and B.18,

$$\log\binom{24}{9} = 6.1164$$

$$\log(.2)^{10} = 3.0103 - 10$$
$$\log(.8)^{15} = 8.5463 - 10$$

$$\log[P(25)] = \log\binom{24}{9} + \log(.2)^{10} + \log(.8)^{15} = 17.6730 - 20$$

$$P(25) = .0047$$

THE HYPERGEOMETRIC DISTRIBUTION

The binomial distribution is based on the assumption that observations are independent of each other; that is, the probability of a success is constant from trial to trial. When the population is infinite or sampling is with replacement, this assumption is appropriate. However, when the population is finite and sampling is without replacement, successive trials are not identical and independent and the probability of a success changes from trial to trial. The population available for sampling on any trial depends on the number and type of elements sampled (and removed) on previous trials.

Suppose that a random sample of n elements are drawn without replacement from a population of N elements, of which N_1 are of one type and N_2 are of another type and $N_1 + N_2 = N$. The total number of ways in which n elements can be drawn from the set of N elements is

$$_NC_n = \binom{N}{n}$$

The number of ways in which x elements of the first kind can be drawn is $\binom{N_1}{x}$, and the number of ways in which $n - x$ elements of the second kind can be drawn is $\binom{N_2}{n-x}$. Thus the probability that a set of n elements randomly selected from the set of $N = N_1 + N_2$ elements contains x elements of the first type and

$n - x$ elements of the second type is given by

$$P(x) = \frac{\binom{N_1}{x}\binom{N_2}{n-x}}{\binom{N}{n}}$$

The notation can be made more closely comparable to the usual binomial notation by identifying an element of the first type as a success; then $\pi = N_1/N$ and $N_1 = N\pi$; similarly, $1 - \pi = N_2/N$ and $N_2 = N(1 - \pi)$. Thus $P(x)$ can be written:

$$P(x) = \frac{\binom{N\pi}{x}\binom{N-N\pi}{n-x}}{\binom{N}{n}} \qquad x = 0, 1, \ldots, N\pi$$

This probability distribution is the *hypergeometric distribution.*

Note that, from the point of view of probability, sampling repeatedly without replacement is equivalent to taking the same size sample all at once; the hypergeometric distribution is appropriate for both types of problems. Industrial sampling is almost always done without replacement and the hypergeometric distribution is widely used for these applications. The observation of successive identical and independent trials is equivalent to sampling with replacement, and the binomial distribution is appropriate for these problems.*

For sufficiently large finite populations the error arising from using the binomial distribution rather than the hypergeometric distribution is not substantial, as shown in the example on p. 130.

It can be shown that the mean and variance of the hypergeometric distribution are given by

and

$$\begin{cases} \mu = n\pi \\ \\ \sigma^2 = \dfrac{n\pi(1 - \pi)(N - n)}{N - 1} \end{cases}$$

The variance of the hypergeometric distribution differs from that of the binomial distribution by the factor $(N - n)/(N - 1)$, which is referred to as the *finite population correction.* This factor is used in sampling theory to adjust formulas derived assuming infinite populations for use when populations are in fact finite.

Tables B.3 and B.4 are useful for computing probabilities for the hypergeometric distribution. Recall that Table B.3 provides the value of $\binom{n}{x}$ for various values of n and x and Table B.4 provides the values of $N!$ and $\log N!$ for various values of N.

---

* When sampling is done with replacement, the observations are independent; when sampling is done without replacement (or, equivalently, replicates are counted only once), the observations are dependent.

EXAMPLE

In a file of 100 records, 10 have errors. What is the probability of finding at most 2 records having errors if a random sample of 10 records are examined?

For this problem the hypergeometric distribution is appropriate with

$$P(x) = \frac{\binom{N\pi}{x}\binom{N - N\pi}{n - x}}{\binom{N}{n}}$$

$$N = 100, \qquad n = 10, \qquad \pi = .10, \qquad N\pi = 10$$

$$P(x \le 2) = P(0) + P(1) + P(2) = \frac{\binom{10}{0}\binom{90}{10}}{\binom{100}{10}} + \frac{\binom{10}{1}\binom{90}{9}}{\binom{100}{10}} + \frac{\binom{10}{2}\binom{90}{8}}{\binom{100}{10}}$$

$$\log\binom{10}{0} = 0, \qquad \log\binom{90}{10} = 12.7574, \qquad \log\binom{100}{10} = 13.2383; \qquad \text{and}$$

$$\log[P(0)] = \log\binom{10}{0} + \log\binom{90}{10} + \log\binom{100}{10} = 9.5191 - 10$$

$$P(0) = .3305$$

$$\log\binom{10}{1} = 1.0000, \qquad \log\binom{90}{9} = 11.8490, \qquad \log\binom{100}{10} = 13.2383; \qquad \text{and}$$

$$\log[P(1)] = \log\binom{10}{1} + \log\binom{90}{9} - \log\binom{100}{10} = 9.6107 - 10$$

$$P(1) = .4080$$

$$\log\binom{10}{2} = 1.6532, \qquad \log\binom{90}{8} = 10.8894, \qquad \log\binom{100}{10} = 13.2383; \qquad \text{and}$$

$$\log[P(2)] = \log\binom{10}{2} + \log\binom{90}{8} - \log\binom{100}{10} = 9.3040 - 10$$

$$P(2) = .2015$$

$$P(x \le 2) = .3305 + .4080 + .2015 = .9400$$

Using the (inappropriate) binomial distribution, we have

$$P(x) = \binom{n}{x}\pi^x(1 - \pi)^{n - x}$$

$$P(x \le 2) = P(0) + P(1) + P(2)$$

$$= \frac{10!}{0!10!}\left(\frac{1}{10}\right)^0\left(\frac{9}{10}\right)^{10} + \frac{10!}{1!9!}\left(\frac{1}{10}\right)^1\left(\frac{9}{10}\right)^9 + \frac{10!}{2!8!}\left(\frac{1}{10}\right)^2\left(\frac{9}{10}\right)^8$$

$$= (.9)^{10} + (.9)^9 + 45(.01)(.9)^8$$

$$= .34868 + .38742 + .19371 = .92981$$

which does not differ greatly from the answer obtained using the (appropriate) hypergeometric distribution.

There is a generally accepted rule of thumb that the finite population correction is of practical importance only when $n/N \geq .10$, that is, when at least 10 percent of the population is included in the sample. Note that, in this example, n/N is exactly .10.

The difference between the use of the hypergeometric and negative binomial distributions in sampling should be noted carefully. The hypergeometric distribution is appropriate for sampling without replacement, but with the sample size specified. It is appropriate, for example, when the proportion defective is π and the question concerns the proportion of defectives observed in a given size sample. The negative binomial is appropriate for sampling with replacement, but without a fixed sample size; for example, when sampling is to be continued until a certain number of defectives has been observed and the question concerns the trial on which this will occur for a given proportion defective.

THE POISSON DISTRIBUTION

Many processes are observed over time or space. Such processes are characterized by the expected number of successes per unit of time or space, just as the binomial distribution is characterized by the number of successes in n trials. The *Poisson exponential distribution* or *Poisson distribution* is applicable to many processes that are distributed and observed over time or space. This distribution is named for Simeon Poisson, the French mathematician who first described it in 1837.

The assumptions of the Poisson distribution seem to match those of a number of actual processes. This distribution is appropriate, for example, for the number of arrivals at a check-out counter during a specified period, the number of defectives per batch, the number of disintegrating atoms in a fixed time interval from a radioactive substance, the number of telephone calls on a line in a fixed time interval, the number of meteorites on an acre of desert land, the number of connections to a wrong number during a given time period, the number of chromosome interchanges in cells, and so forth.

The Poisson distribution can be derived as a limiting form of the binomial distribution and also, without reference to the binomial distribution, by considering as a process the occurrence of an event over a continuum of time or space. The derivation of the Poisson distribution as a limiting form of the binomial distribution will be discussed first.

Suppose that the random variable to be considered is the number of occurrences of an event in a time period of length t. The time period t can be broken into n equal intervals of length t/n, and these intervals can be considered as n independent Bernoulli trials. The expected number of occurrences is then $n\pi$. However, there is one difficulty with this approach—since the events occur at various points in time, two or more events might occur in a time interval or trial of length t/n. The Bernoulli process considers only occurrence or nonoccurrence of an event; there is no way to consider multiple occurrences. To avoid this problem, the number of intervals must be large enough so that the probability of multiple occurrence of an event in a trial of length t/n is zero for practical purposes. As n increases and the length of the interval decreases, the probability of an occurrence in any interval or trial decreases. Since the division of the time period t into subperiods has no effect on the expected number of occurrences in the period t, $n\pi$ remains constant as n is varied. Thus as n increases, π decreases in such a way that $n\pi$ remains constant. It can be shown that as $n \to \infty$ and $\pi \to 0$ and $n\pi$ remains constant, the binomial distribution

$$P(x) = \binom{n}{x} \pi^x (1 - \pi)^{n-x} \qquad x = 0, \ldots, n$$

approaches the Poisson distribution

$$P(x) = \frac{e^{-n\pi}(n\pi)^x}{x!} \qquad x = 0, \ldots$$

Since $n\pi$ is constant, it is customary to write the Poisson distribution as follows:

$$P(x) = \frac{e^{-\lambda}\lambda^x}{x!} \qquad x = 0, \ldots$$

where $\lambda = n\pi$. Note that in the Poisson distribution, x can assume any non-negative integer value. The mean and variance of the Poisson distribution are both equal to λ; that is

and
$$\begin{cases} \mu = \lambda \\ \sigma^2 = \lambda \end{cases}$$

In general, the Poisson distribution is positively skewed, although it becomes more symmetric as λ increases. If λ is an integer, the Poisson distribution has two modes, located at $x = \lambda - 1$ and $x = \lambda$; if λ is not an integer, the Poisson distribution has a single mode located at the integer value between $\lambda - 1$ and λ.

The Poisson distribution provides a good approximation to the binomial distribution if n is large and π is close to zero. As a rule of thumb, the approximation is reasonably good if $n/\pi > 500$. In these cases the binomial distribution with parameters n and π can be approximated by the Poisson distribution with parameter $\lambda = n\pi$.

The Poisson distribution is given for a number of values of λ in Table B.5; the cumulative Poisson distribution is given in Table B.6.

The Poisson distribution also arises as follows: Consider a sequence of random events occurring in time in such a manner that (1) the probability of an event occurring in a very small time interval t to $t + \Delta t$ is $\alpha\Delta t > 0$, (2) the probability of more than one event occurring in the interval t to $t + \Delta t$ is negligible, and (3) the probability of an event occurring in the interval t to $t + \Delta t$ does not depend on what happened prior to time t. Then it can be shown that the probability of observing x events in a time interval of length t is given by the Poisson distribution with parameter $\lambda = \alpha t$. Thus λ is the expected number of occurrences of the event in a unit interval of time. The parameter λ is sometimes called the *intensity* of the Poisson process.

EXAMPLE
Use the Poisson approximation to compute the probability that (a) two people out of 500 have a birthday on Christmas and (b) at least one person out of 500 has a birthday on Christmas.

For the binomial distribution

$$n = 500$$

$$\pi = \frac{1}{365}$$

Using the Poisson approximation, we get

$$P(x) = \frac{e^{-\lambda}\lambda^x}{x!}$$

$$\lambda = n\pi = 500\left(\frac{1}{365}\right) = 1.37$$

(a)
$$P(x = 2) = P(2)$$

$$= \frac{e^{-1.37}(1.37)^2}{2!}$$

$$= \frac{(.254)(1.8769)}{2}$$

$$= .238$$

(b)
$$P(x \geq 1) = 1 - P(x = 0)$$

$$= 1 - P(0)$$

$$= 1 - \frac{e^{-1.37}(1.37)^0}{0!}$$

$$= 1 - .254$$

$$= .746$$

EXAMPLE
The probability that an office with a vary large switchboard will have an incoming call during a small time interval Δt is .05 Δt, where t is in seconds. Determine the probability that (a) during one minute there are exactly 4 incoming calls and (b) during 20 seconds there are not more than 5 incoming calls.

$$P(x) = \frac{e^{-\lambda}\lambda^x}{x!}$$

(a)
$$P(x = 4) = P(4)$$

$$= \frac{e^{-3}(3)^4}{4!}$$

$$= .16$$

(b)
$$P(x \leq 5) = P(0) + P(1) + P(2) + P(3) + P(4) + P(5)$$

$$= \frac{e^{-1}(1)^0}{0!} + \frac{e^{-1}(1)^1}{1!} + \frac{e^{-1}(1)^2}{2!}$$

$$+ \frac{e^{-1}(1)^3}{3!} + \frac{e^{-1}(1)^4}{4!} + \frac{e^{-1}(1)^5}{5!}$$

$$= .3679 + .3679 + .1839 + .0613 + .0153 + .0031$$

$$= .9994$$

This answer can be obtained more directly from Table B.6, as follows:

$$P(x \le 5) = 1 - P(x \ge 6) = 1 - .0006 = 0.9994$$

THE (DISCRETE) UNIFORM DISTRIBUTION

Suppose that for each trial of an experiment or observation there are k *equally likely* possible outcomes represented by the values $x_1, x_2, \ldots, x_k$ of the random variable X. Then the probability of occurrence of each outcome is $1/k$. The probability distribution of X is given by

$$P(x) = \frac{1}{k} \qquad x = x_1, x_2, \ldots, x_k$$

This is the discrete uniform or rectangular distribution. For the special case $x_i = i$, $i = 1, 2, \ldots, k$, the discrete uniform distribution has mean $(k + 1)/2$. The discrete uniform distribution is useful for many types of problems involving equally likely outcomes.

EXAMPLE
The names on a list have been assigned numbers serially from 1 to 1115. What is the probability that a name chosen at random will have a number between 25 and 50 inclusive?
   The uniform distribution is appropriate for this problem.

$$P(x) = \frac{1}{1115}$$

$$P(25 \le x \le 50) = \frac{26}{1115} = .233$$

5.3 EXAMPLES OF CONTINUOUS PROBABILITY DISTRIBUTIONS

A continuous probability distribution is defined on a continuous sample space for the infinite set of values within an interval or intervals. Of the continuous probability distributions discussed in this section, the normal and Cauchy distributions are defined for all real values of the random variable; the expotential, gamma, and log-normal distributions are defined for all positive values of the random variable; the Pareto distribution is defined for all positive values of the random variable larger than a specified number; the beta distribution is defined for values of the random variable between zero and 1; and the continuous uniform distribution is defined for an interval of a specified length.

THE NORMAL DISTRIBUTION

The normal distribution is by far the most estensively used continuous distribution function in both applied and theoretical statistics. The normal distribution is symmetric, bell-shaped and extends infinitely in both directions along the horizontal axis, as shown in Figure 5.6. Its mean, median, and mode are all equal. The normal distribution is a theoretical distribution specified by a mathematical rule. In practical applications, variables are not exactly normally distributed; however, many random variables have been shown to be approxi-

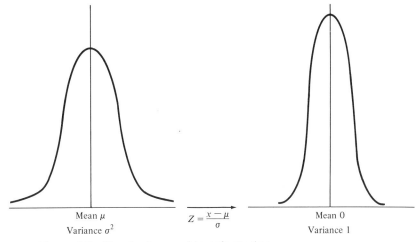

Figure 5.6 Standard normal transformation

mately normally distributed. From the central limit theorem (discussed below), it is known that the mean of a sample obtained from any continuous distribution having finite mean and variance has a limiting normal distribution. Although the central limit theorem does not apply for small samples, it does justify assuming normality for means of very large samples. The normal distribution is given by

$$f(x) = \frac{1}{\beta\sqrt{2\pi}} \, e^{-(1/2)(x-\alpha)^2/\beta^2} \qquad -\infty < x < +\infty$$

where $-\infty < \alpha < +\infty$ and $\beta > 0$.

The normal distribution has mean $\mu = \alpha$ and variance $\sigma^2 = \beta^2$ and thus is frequently written:

$$f(x) = \frac{1}{\sigma\sqrt{2\pi}} \, e^{-(1/2)(x-\mu)^2/\sigma^2} \qquad \infty < x < -\infty$$

By a simple linear transformation, any normal distribution can be transformed so that it has zero mean and unit variance (see Figure 5.6). If X is normally distributed with mean μ and variance σ^2, then

$$Z = \frac{X - \mu}{\sigma}$$

is normally distributed with mean zero and variance 1. This distribution is the *standard normal* or *unit normal distribution*, which is extensively tabled. The fact that any normal distribution can be transformed easily into the standard normal distribution considerably simplifies the computation of areas under a normal curve. Table B.7 gives the areas under the standard normal curve between the mean and the tabled value of z. Since the distribution is symmetric, only one-half of the distribution need be tabled, as shown in Figure 5.7.

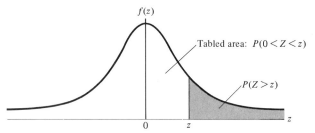

Figure 5.7   Standard normal distribution

EXAMPLE

The length, X, of the iron rods coming off a production line is normally distributed with a mean of 150 cm. and a standard deviation of 2.5 cm. In a very large production run, what proportion of the rods are (a) between 145 centimeters and 155 cm, (b) between 146 cm and 156 cm, and (c) less than 152 cm.

(a)
$$Z = \frac{x - \mu}{\sigma} = \frac{145 - 150}{2.5} = -2$$

$$Z = \frac{x - \mu}{\sigma} = \frac{155 - 150}{2.5} = 2$$

$$P(145 < X < 155) = P(-2 < Z < 2) = .4772 + .4772 = .9544$$

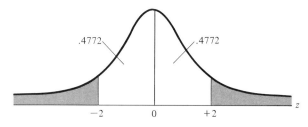

(b)
$$Z = \frac{x - \mu}{\sigma} = \frac{146 - 150}{2.5} = -\frac{4}{2.5} = -1.6$$

$$Z = \frac{x - \mu}{\sigma} = \frac{156 - 150}{2.5} = \frac{6}{2.5} = 2.4$$

$$P(146 < X < 156) = P(-1.6 < Z < 2.4)$$
$$= .4452 + .4918$$
$$= .9370$$

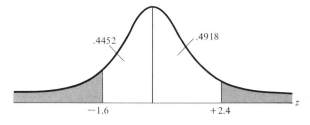

(c)
$$Z = \frac{x - \mu}{\sigma} = \frac{152 - 150}{2.5} = \frac{2}{2.5} = .8$$

$$P(X < 152) = P(Z < .8) = .5 + .2881 = .7881$$

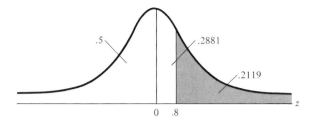

THE EXPONENTIAL DISTRIBUTION

The exponential distribution (see Figure 5.8) is appropriate for many waiting line and servicing problems, in which the probabilities of small values are relatively large and the probabilities of large values are relatively small. In particular, if occurrences of an event follow a Poisson distribution with parameter λ for period of time t, the time until the first occurrence of the event and the time between occurrences of the event have an exponential distribution. The exponential distribution is given by

$$f(x) = \begin{cases} \lambda e^{-\lambda x} & \text{if } x \geq 0 \\ 0 & \text{if } x < 0 \end{cases}$$

where $\lambda > 0$. For the exponential distribution

$$\mu = \lambda \quad \text{and} \quad \sigma^2 = \frac{1}{\lambda^2}$$

The values of e^x for various values of x are given in Table B.8.

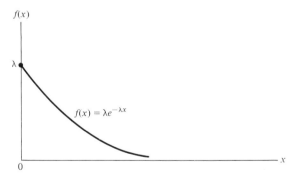

Figure 5.8 Exponential distribution

EXAMPLE

Assuming Poisson distributed arrivals, the time t (in minutes) that a customer spends at a check-out counter has an exponential distribution

$$f(t) = \begin{cases} c e^{-ct} & \text{for } 0 \leq t < \infty \\ 0 & \text{elsewhere} \end{cases}$$

where $c > 0$ is the average number of customers a clerk can serve in one minute. The average time a customer spends in line (the average service time) is $1/c$. The function $f(t)$ is called a service time distribution. The corresponding cumulative distribution function is given by

$$F(t) = 1 - e^{-ct}$$

Thus if, for example, the average service time is 4 minutes, then the probability that a customer will spend less than 6 minutes at the check-out counter is

$$F(6) = 1 - e^{-3/2}$$
$$= 1 - .223 = .777$$

The relationship between the Poisson and exponential distributions is analogous to the relationship between the binomial and geometric distributions. The binomial and Poisson distributions concern the number of occurrences of an event, the geometric distribution concerns the distribution of the number of trials until the first occurrence of an event, and the exponential distribution concerns the distribution of the time until the first occurrence of an event (and between successive occurrences of an event).

THE GAMMA DISTRIBUTION

The gamma distribution (see Figure 5.9) is given by

$$f(x) = \begin{cases} kx^{\alpha-1}e^{-x/\beta} & \text{for } x > 0 \\ 0 & \text{elsewhere} \end{cases}$$

where $\alpha > 0$, $\beta > 0$, and $k = 1/\beta^\alpha\Gamma(\alpha)$. $\Gamma(\alpha)$ is the gamma function that is defined by an integral; it can be shown to satisfy the recursive relationship $\Gamma(\alpha) = (\alpha - 1)\Gamma(\alpha - 1)$. If α is a positive integer, $\Gamma(\alpha) = (\alpha - 1)!$ and k is easily determined; if α is not a positive integer, $\Gamma(\alpha)$ must be obtained from a table of values of the gamma function. In most applications, α is either an integer or a multiple of $\frac{1}{2}$. It can be shown that $\Gamma(\frac{1}{2}) = \sqrt{\pi}$, and thus $\Gamma(\alpha)$ can be evaluated for any multiple of $\frac{1}{2}$ using the recursive relationship $\Gamma(\alpha) = (\alpha - 1)\Gamma(\alpha - 1)$.

For the gamma distribution

$$\mu = \beta(\alpha + 1) \quad \text{and} \quad \sigma^2 = \beta^2(\alpha + 1)$$

Figure 5.9 Gamma distribution

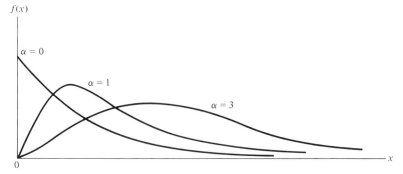

The cumulative gamma distribution, for $x > 0$, is given by

$$F(x) = 1 - \left[1 + \frac{x}{\beta} + \frac{1}{2!}\left(\frac{x}{\beta}\right)^2 + \cdots + \frac{1}{(\alpha - 1)!}\left(\frac{x}{\beta}\right)^{\alpha - 1}\right]e^{-x/\beta}$$

In addition to particular applications, many special cases of the gamma distribution are important in both theoretical and applied statistics. For $\alpha = 1$ the gamma distribution is the exponential distribution of the preceding section. For $\alpha = n/2$ and $\beta > 2$, the gamma distribution is the chi-square distribution, an important distribution discussed in subsequent sections.

EXAMPLE
For a particular city, the daily demand for electric power (in millions of kilowatt hours) has approximately a gamma distribution with $\alpha = 2$ and $\beta = 3$. If the power plant of this city has a daily capacity of 10.5 million kwh, what is the probability that this power supply is inadequate to satisfy demand on any particular day?

$$P(x \geq 10.5) = 1 - F(10.5)$$

$$F(x) = 1 - \left[1 + \frac{x}{\beta} + \frac{1}{2!}\left(\frac{x}{\beta}\right)^2 + \cdots + \frac{1}{(\alpha - 1)!}\left(\frac{x}{\beta}\right)^{\alpha - 1}\right]e^{-x/\beta}$$

$$F(10.5) = 1 - \left[1 + \frac{10.5}{3}\right]e^{-3.5}$$

$$1 - F(10.5) = \left[1 + \frac{10.5}{3}\right]e^{-3.5}$$

$$= 4.5\, e^{-3.5} = .135$$

THE BETA DISTRIBUTION

The Beta distribution (see Figure 5.10) is given by

$$f(x) = \begin{cases} kx^{\alpha - 1}(1 - x)^{\beta - 1} & \text{for } 0 < x < 1 \\ 0 & \text{elsewhere} \end{cases}$$

Figure 5.10   Beta distribution

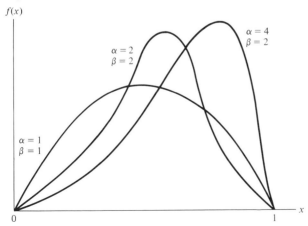

where $\alpha > 0$, $\beta > 0$, and $k = \Gamma(\alpha + \beta)/\Gamma(\alpha)\Gamma(\beta)$. For the special case $\alpha = \beta = 1$, the beta distribution is the uniform distribution discussed later in this section.

Since the range of x for the beta distribution is $0 < x < 1$, this distribution is frequently appropriate for problems involving proportions or probabilities. The beta distribution can be used to represent the distribution of the probability of occurrence of an event from a Bernoulli process when, in a random sample of $n = \alpha + \beta$ observations, α occurrences (successes) and β nonoccurrences (failures) of the event are observed.

The beta distribution has the interesting property that if x has a beta distribution with parameters α and β, then $(1 - x)$ has a beta distribution with the parameters α and β interchanged.* This property is particularly useful if x is the probability of occurrence of an event and $(1 - x)$ is the probability of its nonoccurrence.

The mean and variance of the beta distribution are given by

and

$$
\begin{cases}
\mu = \dfrac{\alpha}{\alpha + \beta} \\[3mm]
\sigma^2 = \dfrac{\alpha\beta}{(\alpha + \beta)^2(\alpha + \beta + 1)}
\end{cases}
$$

The shape of the beta distribution varies widely, depending on the values of the parameters α and β. If $\alpha = \beta$, the distribution is symmetric; if $\alpha < \beta$, the distribution is positively skewed; if $\alpha > \beta$, the distribution is negatively skewed. If $\alpha > 1$ and $\alpha + \beta > 2$, the beta distribution is unimodal with mode equal to $(\alpha - 1)/(\alpha + \beta - 2)$; if $\alpha \le 1$ or $(\alpha + \beta) \le 2$, the distribution is either unimodal with mode at zero or one, U-shaped with nodes at both zero and 1, or uniform on the unit interval.

Unfortunately, the cumulative beta distribution function, sometimes referred to as the incomplete beta function, cannot be written in a simple form and must be obtained from tables.

EXAMPLE
The proportion of "perfect" cars that come from a particular assembly line each week has (approximately) a beta distribution with $\alpha = 15$ and $\beta = 3$. What are the average and standard deviation of the proportion of perfect cars from this assembly line each week?

$$
\mu = \frac{\alpha}{\alpha + \beta} = \frac{15}{18} = .833
$$

$$
\sigma = \sqrt{\frac{\alpha\beta}{(\alpha + \beta)^2(\alpha + \beta + 1)}} = \frac{1}{18}\sqrt{\frac{45}{19}} = .085
$$

Thus the average proportion of "perfect" cars is .833; the standard deviation is .085.

* Let $y = 1 - x$, then $1 - y = x$ and $f(y) = k(1 - y)^{\alpha - 1}y^{\beta - 1}$

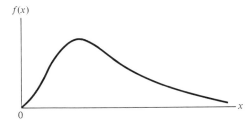

Figure 5.11 Log-normal distribution

THE LOG-NORMAL DISTRIBUTION

If $\ln X$ is normally distributed with mean α and variance β^2, then X has the log-normal distribution given by

$$f(x) = \begin{cases} \dfrac{1}{x\beta\sqrt{2\pi}}\, e^{-(\ln x - \alpha)^2/2\beta^2} & \text{for } x > 0 \\[2mm] 0 & \text{elsewhere} \end{cases}$$

where $\beta > 0$. (See Figure 5.11) For the log-normal distribution

and

$$\begin{cases} \mu = e^{\alpha + (1/2)\beta^2} \\[2mm] \sigma^2 = e^{2\alpha + \beta^2}(e^{\beta^2} - 1) \end{cases}$$

The probability that X is between a and b is given by

$$P(a \le X \le b) = F\left(\frac{\ln b - \alpha}{\beta}\right) - F\left(\frac{\ln a - \alpha}{\beta}\right) \qquad \text{for } 0 < a < b$$

where $F(z)$ is the probability that a standard normal random variable assumes a value less than or equal to z.

The log-normal distribution is appropriate when the value of a random variable can be regarded as representing the joint effect of a large number of independent factors, each of which produces an increase proportional to the value of the random variable at that point in time. The log-normal distribution is extremely positively skewed.

Several biological and economic variables are known to be distributed approximately log-normal. For example, in many cases the size of oil and gas deposits, incomes, and property values are distributed log-normally. Logarithmic transformations are frequently used in analyzing data, as discussed in some detail in Chapter 13; note that if the transformed variable $\ln X$ is normally distributed, the original variable X is log-normally distributed.

EXAMPLE

The size of the oil and gas deposits (in millions of barrels) in a large basin has approximately the log-normal distribution with $\alpha = 1$, $\beta = .4$. Determine (a) the mean and variance of the size of these deposits and (b) the proportion of the deposits in this basin that are between 1 and 3 million barrels.

(a)
$$\mu = e^{\alpha + (1/2)\beta^2} = e^{1 + .08} = e^{1.08} = 2.9447$$
$$\sigma^2 = e^{2\alpha + \beta^2}(e^{\beta^2} - 1) = e^{2 + .16}(e^{.16} - 1)$$
$$= e^{2.16}(1.1735 - 1) = (8.6711)(.1735)$$
$$= 1.5044$$

(b)
$$P(a \leq X \leq b) = F\left(\frac{\ln b - \alpha}{\beta}\right) - F\left(\frac{\ln a - \alpha}{\beta}\right)$$
$$P(1 \leq X \leq 3) = F\left(\frac{\ln 3 - 1}{.4}\right) - F\left(\frac{\ln 1 - 1}{.4}\right)$$
$$= F\left(\frac{1.0981 - 1}{.4}\right) - F\left(\frac{0 - 1}{.4}\right)$$
$$= F(.2465) - F(-2.5)$$
$$= .5972 - .0062$$
$$= .5910$$

THE PARETO DISTRIBUTION

The Pareto distribution is given by

$$f(x) = \begin{cases} \alpha x_0{}^\alpha / x^{\alpha + 1} & \text{for } x > x_0, \text{ where } x_0 \text{ is the minimum value of } y \\ 0 & \text{elsewhere} \end{cases}$$

where $1 < \alpha < 2$.

There are several variations of the Pareto distribution; the distribution shown in Figure 5.12 is sometimes referred to as the *strong Pareto distribution*. For this distribution

and
$$\begin{cases} \mu = \dfrac{\alpha x_0}{\alpha - 1} \\ \\ \sigma^2 \text{ is infinite} \end{cases}$$

The cumulative distribution function for the Pareto distribution is given by

$$F(x) = 1 - \left(\frac{x}{x_0}\right)^{-\alpha} \qquad \text{for } x > x_0$$

Figure 5.12   Pareto distribution

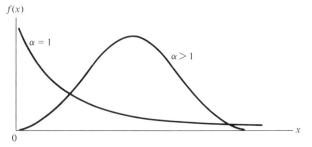

The Pareto distribution has been used to describe the distribution of the size of firms. Another important application is in problems involving the distribution of income. In this context the distribution is frequently written:

$$G(x) = \left(\frac{x}{x_0}\right)^{-\alpha} \qquad \text{for } x > x_0$$

where $G(x)$ is the probability that an income exceeds x; that is, $G(x)$ is the proportion of incomes that exceed x.

EXAMPLE
For a particular population, income in excess of \$4,000 is approximately distributed according to the Pareto distribution with $\alpha = \frac{3}{2}$. Find (a) the average income for this population and (b) the proportion of incomes between \$9,000 and \$16,000.

(a) $$\mu = \frac{\alpha x_0}{\alpha - 1} = \frac{(\frac{3}{2})(4,000)}{\frac{3}{2} - 1} = \frac{6,000}{\frac{1}{2}} = \$12,000$$

(b) $$P(9,000 \leq X \leq 16,000) = F(16,000) - F(9,000)$$

$$= 1 - \left(\frac{4,000}{16,000}\right)^{3/2} - 1 + \left(\frac{4,000}{9,000}\right)^{3/2}$$

$$= \left(\frac{4}{9}\right)^{3/2} - \left(\frac{1}{4}\right)^{3/2}$$

$$= \frac{8}{27} - \frac{1}{8} = \frac{37}{216} = .171$$

Recently there has been considerable discussion of *stable Paretian distributions* in connection with the random walk hypothesis of common stock prices. These distributions have four parameters and may be symmetric, positively skewed, or negatively skewed; their tails are asymptotically distributed according to the Pareto distribution discussed above.

Unfortunately, the stable Paretian family of distributions cannot be written in closed form, except for special cases. The parameter values for these distributions used to describe prices in speculative markets are such that the distributions have finite means but infinite variances. The difference between the stable Paretian hypothesis and the Gaussian (normal) hypothesis of the distribution of prices essentially concerns the specification of the parameter that determines whether the variance is finite or infinite. Although the question is far from settled, there is considerable evidence for the stable Paretian hypothesis; note that this hypothesis implies greater inherent riskiness in the market than is implied by the Gaussian hypothesis.

THE (CONTINUOUS) UNIFORM DISTRIBUTION
The discrete uniform distribution previously discussed is a special case of the continuous uniform or rectangular distribution given by

$$f(x) = \begin{cases} \dfrac{1}{\beta - \alpha} & \text{for } \alpha < x < \beta \\ 0 & \text{elsewhere} \end{cases}$$

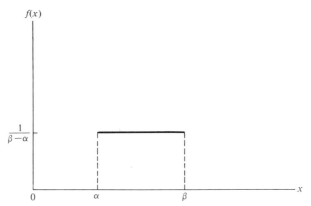

Figure 5.13   Uniform distribution

where α and β are constants. For the continuous uniform distribution

and

$$\begin{cases} \mu = \dfrac{\alpha + \beta}{2} \\[3mm] \sigma^2 = \dfrac{(\beta - \alpha)^2}{12} \end{cases}$$

The continuous uniform distribution (see Figure 5.13) is appropriate for various types of problems involving homogeneity; for example, when measurements are recorded to a specified accuracy, the distribution of rounding errors is uniform. The uniform distribution is also the distribution corresponding to the principle of insufficient reason; that is, it assigns equal probability to all possible outcomes.

EXAMPLE
What is the probability of arriving at a factory during testing if the time of arrival is random and 15 minutes of testing alternates with 25 minutes of assembling? For this problem the uniform distribution is appropriate.

$$f(x) = \frac{1}{40}$$

$$P(\text{assembly}) = \frac{25}{40} = .625$$

testing   assembly

| 0 | 15 | 40 |
| α | | β |

Thus the probability of arriving during testing is .625.

5.4 THE CENTRAL LIMIT THEOREM

The central limit theorem is one of the most important theorems in statistics; it is a very remarkable result from a theoretical point of view and pro-

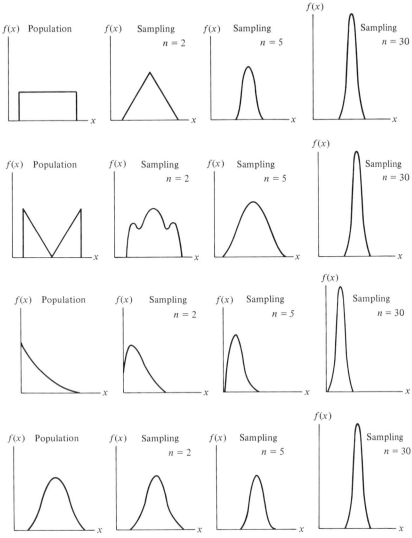

Figure 5.14   Illustration of the central limit theorem (population and sampling distributions)

vides the basis for many applications of normal distribution theory. Essentially, the theorem states that regardless of the form of the original population (provided only that its mean and variance are finite), the distribution of the sample mean is approximately normal as the sample size becomes large. This theorem justifies the preoccupation of statisticians with normal distribution theory, particularly since the tendency toward normality is surprisingly strong even for relatively small samples. Figure 5.14 illustrates the approach to normality of sample means for several population distributions.

The central limit theorem can be stated and proved in various forms and for varying degrees of generality. The form in which it is given below is appropriate for subsequent applications; its proof can be found in any mathematical statistics text (see Appendix A).

Central Limit Theorem   If X is a random variable having finite mean μ and finite variance σ^2, then the probability distribution of the random variable $(\bar{x} - \mu)/(\sigma/\sqrt{n})$ approaches the standard normal distribution as the sample size n becomes infinite.

The central limit theorem justifies using a normal distribution having mean μ and variance σ^2/n as an approximation for the distribution of $\bar{x}$ when n is sufficiently large (rules of thumb for "sufficiently large" vary, but $n \geq 50$ is frequently used). The binomial distribution is probably the distribution most frequently approximated by the normal distribution. The approximation is quite accurate, even for samples that are not very large, as illustrated below.

There are two readily apparent sources of error in using the normal distribution as an approximation for the binomial distribution:

1. The normal distribution is *always* symmetric; the binomial distribution is symmetric only if $\pi = (1 - \pi) = .5$ (regardless of the value of n). Thus the more nearly equal π and $1 - \pi$ are, the better the approximation is for any given n.
2. The normal distribution is continuous, the binomial distribution is discrete. Thus the larger n is, the better the approximation is (for any given values of π and $1 - \pi$) since the binomial distribution produces more possible values for x as n increases. The approximation is improved by a "correction for discontinuity" which is made by considering x to be the interval from $x - \frac{1}{2}$ to $x + \frac{1}{2}$ rather than a discrete value. When we use the correction for discontinuity, the central limit theorem states that if X is binomial with proportion π, then

$$Z = \frac{X - \frac{1}{2} - n\pi}{\sqrt{n\pi(1 - \pi)}}$$

is approximately standard normal.

A good rule of thumb for taking account of the problems of both asymmetry and discreteness is to use the normal approximation for the binomial distribution only if $np > 5$, $n(1 - p) > 5$ and $n > 10$.

EXAMPLE
The proportion of defective transistors produced by a high precision machine is .005. The probability of a defective is constant for each transistor produced—that is, it is independent of the time at which the last defective was produced. In a production run of 10,000 transistors, what is the probability that more than 60 are defective?

The normal approximation to the binomial is appropriate with a correction for continuity.

$$Z = \frac{x - \frac{1}{2} - n\pi}{n\pi(1 - \pi)} = \frac{60 - \frac{1}{2} - 50}{\sqrt{(10,000)(.005)(.995)}}$$

$$= \frac{9.5}{\sqrt{49.75}} = 1.35$$

$$P(X > 60) = P(Z > 1.35) = .0885$$

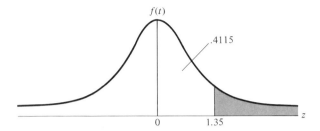

5.5 JOINT PROBABILITY DISTRIBUTIONS

The preceding sections discuss probability distributions of one discrete or continuous random variable. In many problems, more than one aspect of the outcome of an experiment or observation is of interest and therefore more than one random variable is defined. For example, in a particular research situation, the relationship between GNP and unemployment rate, quantity and price, hair color and eye color, height and weight, or sex and answer to a specific question might be of interest to the investigator.

DISCRETE RANDOM VARIABLES

If two discrete random variables X and Y are defined on the sample space of outcomes of an experiment, then the probability that X has the value x and Y has the value y is written $P(x, y)$. The possibilities $P(x, y)$ specified for all possible pairs of values (x, y) in the range of X and Y are the *joint probability distribution* of the random variables X and Y.

If X and Y are discrete random variables having joint probabilities $P(x, y)$, then the probability distribution of X is given by

$$g(x) = \sum_y P(x, y)$$

where the summation is over all values of y in the range of Y for which $P(x, y)$ is defined. Similarly, the probability distribution of Y is given by

$$h(y) = \sum_x P(x, y)$$

where the summation is over all values of x in the range of X for which $P(x, y)$ is defined. When X and Y are jointly distributed, $g(x)$ and $h(y)$ are the *marginal distribution of* X and the *marginal distribution of* Y, respectively.

When X and Y are jointly distributed, the conditional probability that X has the value x given that Y has the value y is

$$g_1(x|y) = \frac{P(x, y)}{h(y)} \qquad \text{for } h(y) \neq 0$$

Note that this definition is analogous to the definition of the conditional probability of the occurrence of an event. The conditional probabilities $g_1(x|y)$

specified for all values of X are the *conditional probability* distribution of X given that Y has the value y. Similarly,

$$h_1(y|x) = \frac{f(x, y)}{g(x)} \qquad \text{for } g(x) \neq 0$$

is the *conditional probability distribution* of Y given that X has the value x.

The random variables X and Y are said to be *independent* if $P(x, y) = g(x)h(y)$ for all x and y in the range of X and Y.

EXAMPLE
The random variable X can assume the values x_1, x_2, x_3, and the random variable Y can assume the values y_1, y_2. The joint probability distribution of X and Y is given by the following table; for example, $P(x_1, y_1) = .10$

		Y		
		y_1	y_2	
	x_1	.10	.15	.25
X	x_2	.35	.10	.45
	x_3	.25	.05	.30
		.70	.30	

The marginal distributions of X and Y are obtained by summing the appropriate rows and columns. Thus

$$g(x_1) = \sum_{y_1, y_2} P(x_1, y) = P(x_1, y_1) + P(x_1, y_2)$$

$$= .1 + .15 = .25$$

Similarly,

$$g(x_2) = .45$$

$$g(x_3) = .30$$

$$h(y_1) = \sum_{x_1, x_2, x_3} P(x, y_1) = p(x_1, y_1) + P(x_2, y_1) + P(x_3, y_1)$$

$$= .1 + .35 + .25 = .70$$

Similarly,

$$h(y_2) = .30$$

Note that

$$\sum_{i=1}^{3} g(x_i) = \sum_{j=1}^{2} h(y_j) = 1$$

The various conditional probabilities can also be obtained from the table. For example,

$$g_1(x|y) = \frac{P(x, y)}{h(y)}$$

$$P(x_1|y_1) = \frac{P(x_1, y_1)}{h(y_1)} = \frac{.1}{.7} = \frac{1}{7}$$

$$P(x_3|y_2) = \frac{P(x_3, y_2)}{h(y_2)} = \frac{.05}{.30} = \frac{1}{6}$$

Similarly,

$$h_2(y|x) = \frac{P(x, y)}{g(x)}$$

$$P(y_2|x_2) = \frac{P(x_2, y_2)}{g(x_2)} = \frac{.1}{.45} = \frac{2}{9}$$

$$P(y_1|x_3) = \frac{P(x_3, y_1)}{g(x_3)} = \frac{.25}{.30} = \frac{5}{6}$$

Note that X and Y are not independent—in fact, for every pair of values x_i and y_j, $P(x_i, y_j) \neq g(x_i)h(y_j)$. For example,

$$P(x_1, y_1) = .1 \quad \text{and} \quad g(x_1)h(y_1) = (.25)(.70) = .175$$

So that

$$P(x_1, y_1) \neq g(x_1)h(y_1)$$

The concept of a joint probability distribution can be generalized to more than two random variables. If the random variables $X_1, X_2, \ldots, X_k$ are defined on a discrete sample space, the probability that X_1 has the value x_1, X_2 has the value $x_2, \ldots,$ and X_k has the value x_k is written $P(x_1, x_2, \ldots, x_k)$. The probabilities $P(x_1, x_2, \ldots, x_k)$ specified for all sets of values in the ranges of these random variables constitute the multivariate probability distribution of X_1, $X_2, \ldots, X_k$. Similarly, marginal distributions, joint marginal distributions, conditional distributions, and joint conditional distributions can be defined.

The independence or dependence of several jointly distributed random variables can be determined using a generalization of the definitions for two random variables. The random variables $X_1, X_2, \ldots, X_k$ are independent if and only if

$$P(x_1, x_2, \ldots x_k) = \prod_{i=1}^{k} g_i(x_i)$$

for all values in the ranges of these random variables for which $P(x_1, x_2, \ldots x_k)$ is defined; $g_i(x_i)$ are the marginal distributions of the X_i.

The multinomial distribution is now discussed as an example of a discrete multivariate probability distribution. Similarly, the hypergeometric distribution can be generalized to apply to problems in which there are more than two possible outcomes of an event.

THE MULTINOMIAL DISTRIBUTION

The binomial distribution is based on the assumption that there are only two possible outcomes for any trial of an experiment or observation. Although the number of possible outcomes can always be reduced to two by combining categories, this procedure frequently is not appropriate in practice because of the resulting loss of information. If the trials or observations of an experiment are identical and independent and there are more than two possible outcomes for any trial, then the multinomial distribution is appropriate.

If each trial of an experiment has k mutually exclusive possible outcomes $A_1, A_2, \ldots, A_k$ with corresponding constant probabilities $\pi_1, \pi_2, \ldots, \pi_k$, then the joint probability that in n trials of the experiment A_1 will occur x_1 times, A_2 will occur x_2 times, and so forth (where $\sum_{i=1}^{k} x_i = n$) is given by

$$P(x_1, x_2, \ldots, x_k) = \frac{n!}{\prod\limits_{i=1}^{k} x_i!} \prod_{i=1}^{k} \pi_i^{x_i}$$

Note that the terms in the multinomial expression

$$(\pi_1 + \pi_2 + \cdots + \pi_k)^n$$

correspond to the probabilities of the multinomial distribution. The binomial expansion and distribution are the special case of the multinomial expansion and distribution in which $k = 2$. The multinomial distribution can be obtained in a manner similar to the binomial distribution as follows: Consider the sequence of events

$$\overbrace{A_1, \ldots, A_1}^{x_1}, \overbrace{A_2, \ldots, A_2}^{x_2}, \ldots, \overbrace{A_k, \ldots, A_k}^{x_k}$$

Since the trials are identical and independent, the probability of observing this particular sequence of events is

$$\pi_1^{x_1} \pi_2^{x_2} \cdots \pi_k^{x_k} = \prod_{i=1}^{k} \pi_i^{x_i}$$

There are $n!/\prod_{i=1}^{k} x_i!$ equally likely possible sequences, giving $P(x_1, x_2, \ldots, x_k)$ as defined above. Unfortunately, computations for the multinomial distribution are tedious, even using tables, unless n is quite small; convenient accurate approximations are not readily available.

It can be shown that the marginal distribution of X_i for $i = 1, 2, \ldots, k$ is a binomial distribution having parameters n and π_i, $i = 1, 2, \ldots k$. Note that k random variables having a multinomial distribution cannot be independent, since the values of any $k - 1$ of the random variables determine the value of the kth random variable by the condition $\sum_{i=1}^{k} x_i = n$. This is the reason that, for $k = 2$, the multinomial distribution reduces to the binomial distribution of one random variable.

EXAMPLE

The Merry-Go-Round Shop has found that of the people who enter their store, $\frac{9}{16}$ make a purchase, $\frac{3}{16}$ return or exchange merchandise, $\frac{1}{16}$ do not make a purchase or return merchandise but return within 24 hours to make a purchase, and $\frac{3}{16}$ neither make a purchase nor exchange merchandise nor return within 24 hours to make a purchase. What is the probability that of a random sample of 10 people entering the store, 4 will make a purchase; 3 will exchange merchandise; 1 will return to make a purchase; and 2 will not make a purchase, nor exchange a purchase, nor return to make a purchase? The multinomial distribution is appropriate for this problem.

$$P(x_1, x_2, \ldots, x_k) = \frac{n!}{\displaystyle\prod_{i=1}^{k} x_i!} \prod_{i=1}^{k} \pi_i^{x_i}$$

$$P(4, 3, 1, 2) = \frac{10!}{4!\,3!\,1!\,2!} \left(\frac{9}{16}\right)^4 \left(\frac{3}{16}\right)^3 \left(\frac{1}{16}\right)^1 \left(\frac{3}{16}\right)^2$$

$$= \frac{(12,600) \cdot 9^4 \cdot 3^5}{16^{10}}$$

$$\log 12,600 = 4.10037$$

$$\log 9^4 = 3.81696$$

$$\log 3^5 = 2.38560$$

$$\log 16^{10} = 12.0412$$

$$\log[P(4, 3, 1, 2)] = 8.26173 - 10$$

$$P(4, 3, 1, 2) = .0183$$

Thus the probability of the event specified is .00428.

CONTINUOUS RANDOM VARIABLES

The definitions of joint distributions given for discrete random variables also apply to continuous random variables, if probability density functions are replaced by probability distributions and integrals are replaced by summations.

The joint density function $f(x, y)$ of two continuous random variables X and Y is a function such that

$$f(x, y) \geq 0 \qquad \text{for all real } x \text{ and } y$$

and

$$\int_{-\infty}^{\infty} \int_{-\infty}^{\infty} f(x, y)\, dx\, dy = 1$$

The probability of occurrence of the joint event $(a \leq X \leq b,\ c \leq Y \leq d)$ is given by

$$\int_{a}^{b} \int_{c}^{d} f(x, y)\, dy\, dx = \int_{c}^{d} \int_{a}^{b} f(x, y)\, dx\, dy$$

If $f(x, y)$ is the joint density function of the continuous random variables X and Y, then the *marginal density function of* X is given by

$$g(x) = \int_{-\infty}^{\infty} f(x, y)\, dy$$

and the *marginal density function of y* is given by

$$h(y) = \int_{-\infty}^{\infty} f(x, y)\, dx$$

When X and Y are jointly distributed, the conditional densities of X and Y can be obtained in a manner similar to that used in obtaining conditional probabilities:

$$f(x \mid Y = y) = \frac{f(x, y)}{h(y)}$$

$$f(y \mid X = x) = \frac{f(x, y)}{g(x)}$$

The conditional density $f(x \mid Y = y)$ is the density of the random variable X, given that the random variable Y assumes the value y. Similarly, the conditional density $f(y \mid X = x)$ is the density of the random variable Y, given that the random variable X assumes the value x.

The random variables X and Y are said to be *independent* if $f(x, y) = g(x)h(y)$. When X and Y are independent,

$$g(x) = f(x \mid Y = y) \qquad \text{for all values of } Y$$

and

$$h(y) = f(y \mid X = x) \qquad \text{for all values of } X$$

which follows from the definitions of independence and conditional densities,

The concept of a joint density function can be generalized to more than two continuous random variables, just as the concept of a joint probability distribution can be generalized to more than two discrete random variables, with integrals replacing summations. If the random variables $X_1, X_2, \ldots, X_k$ are defined on a continuous sample space, their joint density function can be written $f(x_1, \ldots, x_k)$ where

$$f(x_1, \ldots, x_k) \geq 0 \qquad \text{for all real } x_1, \ldots, x_k$$

and

$$\int_{-\infty}^{\infty} \int_{-\infty}^{\infty} \cdots \int_{-\infty}^{\infty} f(x_1, \ldots, x_k)\, dx_1, \ldots, dx_k = 1$$

Similarly, marginal densities, joint marginal densities, conditional densities, and joint conditional densities can be defined.

The random variables $X_1, \ldots, X_k$ are independent if and only if

$$f(x_1, \ldots, x_k) = \prod_{i=1}^{k} g_i(x_i)\, dx_i$$

for all values in the ranges of these random variables for which $f(x_1, \ldots, x_k)$ is defined; $g_i(x_i)$ are the marginal densities of the x_i.

The bivariate normal distribution and a bivariate exponential distribution are discussed below as examples of continuous joint probability distributions.

THE BIVARIATE NORMAL DISTRIBUTION

The bivariate normal density function for independent random variables X_1 and X_2 is given by

$$f(x_1, x_2) = \frac{1}{2\pi\beta_1\beta_2} e^{-(1/2)[(x_1-\alpha_1)^2/\beta_1^2 + (x_2-\alpha_2)^2/\beta_2^2]}$$

$$\text{for} \quad -\infty < x_1 < +\infty \quad \text{and} \quad -\infty < x_2 < +\infty$$

where $-\infty < \alpha_1, \alpha_2 < +\infty$ and $\beta_1, \beta_2 > 0$. Since X_1 and X_2 are independent, the joint cumulative density function $F(x_1, x_2)$ is obtained as the product of the cumulative density functions $F(x_1)$ and $F(x_2)$. That is, $F(x_1, x_2) = F(x_1) \cdot F(x_2)$. As noted previously, $F(x_1)$ and $F(x_2)$ are easily obtained from tables by transforming to the unit normal distribution,

Since X_1 and X_2 are independent, $f(x_1, x_2) = g(x_1) \cdot h(x_2)$ where the marginal densities are given by

$$g(x_1) = \frac{1}{\beta_1\sqrt{2\pi}} e^{-(1/2)[(x_1-\alpha_1)^2/\beta_1^2]}$$

$$h(x_2) = \frac{1}{\beta_2\sqrt{2\pi}} e^{-(1/2)[(x_2-\alpha_2)^2/\beta_2^2]}$$

The bivariate normal distribution and its generalization to k independent variables are extremely important in statistical theory. Almost all standard multivariate statistical inference is based on the assumption of joint normality of the population distributions. The bivariate and multivariate normal distributions and their properties are discussed in detail in subsequent chapters, particularly those concerning analysis of variance and regression. The following example concerns independent random variables that are jointly normally distributed.

EXAMPLE

For a large chemical plant the amount of polythene sheeting produced, which depends on the scheduling of production in other parts of the plant, is normally distributed with mean 200 and standard deviation 10 (measurements are in thousands of pounds). Orders for this sheeting from outside the plant (it is also used by the plant in making other products) are independent of the amount produced and are normally distributed with mean 175 and standard deviation 20 (in thousands of pounds). What is the probability that on a given day the amount produced is less than 175,000 pounds and the amount ordered is greater than 175,000 pounds?

$$f(x_1, x_2) = g(x_1) \cdot h(x_2) = P(x_1 < 175) \cdot P(x_2 > 175)$$

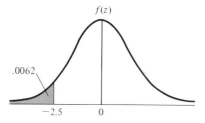
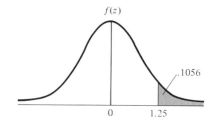

$$Z = \frac{175 - 200}{10} = -2.5 \qquad Z = \frac{175 - 150}{20} = 1.25$$

$$P(Z < -2.5) = .0062 \qquad P(Z > 1.25) = .1056$$

$$P(x_1 < 175) \cdot P(x_2 > 175) = (.0062)(.1056) = .00065$$

Thus the probability of the specified event is .00065.

THE BIVARIATE EXPONENTIAL DISTRIBUTION

The bivariate exponential density is given by

$$f(x_1, x_2) = \begin{cases} c_1 c_2 e^{-(c_1 x_1 + c_2 x_2)} & \text{for } x_1, x_2 > 0 \\ 0 & \text{elsewhere} \end{cases}$$

where $c_1, c_2 > 0$. It can be shown that the corresponding cumulative density is given by

$$F(x_1, x_2) = \begin{cases} (1 - e^{-c_1 x_1})(1 - e^{-c_2 x_2}) & \text{for } x_1, x_2 > 0 \\ 0 & \text{elsewhere} \end{cases}$$

The marginal density of x_1 is

$$g(x_1) = c_1 e^{-c_1 x_1} \qquad \text{for } x_1 > 0$$

and the marginal density of x_2 is

$$h(x_2) = c_2 e^{-c_2 x_2} \qquad \text{for } x_2 > 0$$

Note that $g(x_1) \cdot h(x_2) = f(x_1, x_2)$ so that x_1 and x_2 are independent; thus $g(x_1 | x_2) = g(x_1)$ and $h(x_2 | x_1) = h(x_2)$ for $x_1, x_2 > 0$.

EXAMPLE
A customer makes purchases at a large department store that require his going through two check-out counters. The times (in minutes) spent at the two counters have the bivariate exponential density function.

$$f(t_1, t_2) = \frac{1}{15} e^{-(t_1/3 + t_2/5)}$$

Find the probability that the customer spends no more than 3 minutes in the first line and no more than 6 minutes in the second line.

$$F(t_1, t_2) = (1 - e^{-c_1 t_1})(1 - e^{-c_2 t_2})$$
$$= (1 - e^{-1})(1 - e^{-1.2})$$
$$= (1 - .368)(1 - .301)$$
$$= (.632)(.699)$$
$$= .442$$

Thus the probability of the specified event is .442.

SUMMARY TABLE

The probability distributions and density functions discussed in this chapter are summarized, with some of their properties, in Table 5.1.

TABLE 5.1   SUMMARY OF PROBABILITY DISTRIBUTIONS

Probability Distribution	Probability Function	Mean	Variance
I. Discrete distributions			
Binomial	$P(x) = \dfrac{n!}{x!(n-x)!} \pi^x (1 - \pi)^{n-x}$ $x = 0, 1, \ldots, n$	$n\pi$	$n\pi(1 - \pi)$
Negative binomial	$P(n) = \dbinom{n-1}{k-1} \pi^k (1 - \pi)^{n-k}$ $0 < k \le n$	$\dfrac{k}{\pi}$	$\dfrac{k(1 - \pi)}{\pi^2}$
Hypergeometric	$P(x) = \dfrac{\dbinom{N\pi}{x}\dbinom{N - N\pi}{n - x}}{\dbinom{N}{n}}$ $x = 0, 1, \ldots, N\pi$	$n\pi$	$\dfrac{n\pi(1 - \pi)(N - n)}{N - 1}$
Poisson	$P(x) = \dfrac{e^{-\lambda}\lambda^x}{x!}$ $x = 0, 1, \ldots$	λ	λ
Uniform	$P(x) = \dfrac{1}{k}$ $x = x_1, x_2, \ldots, x_k$		
II. Continuous distributions			
Normal	$f(x) = \dfrac{1}{\beta\sqrt{2\pi}} e^{-(1/2)/[(x-\alpha)^2/\beta^2]}$ $-\infty < x < \infty$ $\beta > 0$	α	β^2

TABLE 5.1 (*continued*)

II. Continuous distributions (continued)

Exponential	$f(x) = \lambda e^{-\lambda x}$	λ	$\dfrac{1}{\lambda^2}$
	$x \geq 0$ $\lambda > 0$		

Gamma	$f(x) = kx^{\alpha-1}e^{-x/\beta}$	$\beta(\alpha + 1)$	$\beta^2(\alpha + 1)$
	$x > 0$		
	$k = \dfrac{1}{\beta^\alpha \Gamma(\alpha)}$		

Beta	$f(x) = kx^{\alpha-1}(1-x)^{\beta-1}$	$\dfrac{\alpha}{\alpha + \beta}$	$\dfrac{\alpha\beta}{(\alpha+\beta)^2(\alpha+\beta+1)}$
	$0 < x < 1$ $\alpha > 0, \beta > 0$		
	$k = \dfrac{\Gamma(\alpha+\beta)}{\Gamma(\alpha)\Gamma(\beta)}$		

Log-normal	$f(x) = \dfrac{1}{x\beta\sqrt{2\pi}} e^{-(\ln x - \alpha)^2/2\beta^2}$	$e^{\alpha + (1/2)\beta^2}$	$e^{2\alpha + \beta^2}(e^{\beta^2} - 1)$
	$x > 0$ $\beta > 0$		

Pareto	$f(x) = \dfrac{\alpha x_0{}^\alpha}{x^{\alpha+1}}$	$\dfrac{\alpha x_0}{\alpha - 1}$	infinite
	$x > x_0$ $1 < \alpha < 2$		

Uniform	$f(x) = \dfrac{1}{\beta - \alpha}$	$\dfrac{\alpha + \beta}{2}$	$\dfrac{(\beta - \alpha)^2}{12}$
	$\alpha < x < \beta$		

PROBLEMS

5.1 Seven dice are rolled. If rolling a 5 or a 6 is a success, find the probability of obtaining
   (a) Exactly 4 successes
   (b) At most 4 successes

5.2 A dealer has 20 apparently identical cars in stock, of which 4 are defective. If a company buys 2 of these (chosen at random), what is the probability that at least 1 is defective? Compare this answer with the answer obtained using the (inappropriate) binomial distribution.

5.3 Use the Poisson approximation to obtain the probability that, in a random sample of 200 items, there will be no more than 3 defectives if the defective rate is 1 percent.

5.4 Of 10 companies, 5 have annual sales exceeding \$100,000. If 3 of the companies are chosen at random for detailed analysis, what is the probability that 2 of these have sales exceeding \$100,000? Compare this answer with the answer obtained using the (inappropriate) binomial distribution.

5.5 An electronic system is set up so that a component that fails is automatically (instantaneously) replaced by an alternative component. The probability of failure in a small interval of time Δt is $.05 \Delta t$, where t is minutes.

(a) What is the probability of one failure occuring in a period of one hour?

(b) For what time period is the probability of at least one failure .99?

5.6 Polythene is inspected in continuous rolls; the probability of observing a defect in a very small time interval Δt is $.15 \Delta t$, where t is minutes. What are the probabilities of observing

(a) 0 defects in 10 minutes

(b) 1 defect in 10 minutes

(c) 2 defects in 10 minutes

(d) 3 defects in 10 minutes

5.7 The manager of a supermarket puts 5 packages of wild rice on a shelf in the gourmet department and another 5 packages on a shelf in the cereal department. If a customer buys wild rice, the probability that she will buy it in the gourmet department is .6. What is the probability that there will be 3 packages of wild rice left in the cereal department when a customer takes the last box from the shelf in the gourmet department?

5.8 Between 1:00 and 1:05 PM the calls arriving at a switchboard follow the Poisson distribution with parameter $\lambda = 4$.

(a) What is the probability that no calls arrive during this period?

(b) What is the probability that, in a period of three days, a total of one call arrives at the switchboard between 1:00 and 1:05 PM?

5.9 If a car drives into a certain parking lot between 7:00 and 7:30 PM on Friday night, the probability is .8 that its occupants are going to a nearby theater. What is the probability that the tenth car to drive into the lot is the third car whose occupants are going to the theater?

5.10 Two-thirds of the employees of a very large favor establishing a coffee bar. If a random sample of 5 employees are asked whether they favor a coffee bar, what are the probabilities that 0, 1, 2, 3, 4, and 5 of the employees give "yes" responses? If 200 random samples of 5 employees are questioned, in how many samples would 0, 1, 2, 3, 4, and 5 "yes" responses be expected?

5.11 Assuming that the birth of a boy and a girl are equally likely, what is the probability that a couple's fifth child is their second boy? What is the probability that a couple's sixth child is their first girl?

5.12 Customers enter a store at the rate of 180 per hour.

(a) What is the probability that in a 2-minute interval no one enters the store?

(b) For what time interval is the probability .5 that no one enters the store?

5.13 The serial numbers on machines from a particular production run are from 000 to 295 inclusive. If the machines are shipped to buyers in random order with respect to serial number, what is the probability that a buyer will have a machine with a serial number between 100 and 199 inclusive?

5.14 Across a short bridge, which is under construction, traffic runs 5 min in one direction and then 5 min in the other direction. If a car's arrival time is random, what is the probability that it will wait at least 3 min to cross the bridge?

5.15 The time t (in minutes) that a program must wait to be run on a computer follows the exponential distribution

$$f(t) = \frac{1}{10} e^{-t/10}$$

(a) What is the probability that a program will wait no longer than 5 minutes to be run?

(b) What is the probability that a program will wait at least 20 minutes to be run?

5.16 The proportion of "good runs" that are printed by a computer each day has approximately a beta distribution with $\alpha = 16$, $\beta = 4$. What are the mean and standard deviation of the proportion of "good runs" each week?

5.17 In a shoe department where customers take numbers and are served in order by the first clerk available, the time t (in minutes) that a customer waits for a clerk has the exponential distribution

$$f(t) = \frac{1}{5} e^{-t/5}$$

(a) What is the average time that a customer waits for a clerk?

(b) What is the probability that a customer waits between 5 and 10 minutes for a clerk?

5.18 In a large department store, the December sales volume of wool gloves (in thousands of pairs) is approximated by a gamma distribution with $\alpha = 4$, $\beta = 1$.

(a) What is the probability that the store will sell at least 5000 pairs?

(b) What is the average number of pairs sold?

5.19 The joint distribution for the servicing times (in minutes) of two computers (run and serviced independently) is

$$f(t_1, t_2) = \frac{1}{3600} e^{-(t_1/30 + t_2/120)}$$

What is the probability that if both computers stop running at the same time, they will both be serviced in no more than one hour?

5.20 The articles stocked and packed in the shipping department of a mail-order company are stored in two different warehouses. Articles are ordered from the warehouses by phone if the stock runs out before it is replenished by the usual inventory control procedures. The joint distribution of the times required to obtain articles from the two warehouses is

$$f(t_1, t_2) = \frac{1}{50} e^{-(t_1/10 + t_2/5)}$$

If an order requires an article to be sent from each warehouse, what is the probability that both articles will arrive within 10 minutes of the time they are ordered?

5.21 One-half of the people in a large community are regular viewers of television. If 100 investigators each interview 10 individuals at random, how many of the investigators would be expected to report that 3 or fewer people were regular viewers? Obtain the answer

(a) Using the exact method

(b) Using the normal approximation

5.22 The proportion of items that are defective from a large production line is .2. What is the probability that in a random sample of 500 items, 15 or more are defective?

5.23 According to records in the Registrar's office, 5 percent of the students who register in a certain course fail the course. What is the probability that of 6 randomly selected students who have registered for the course, fewer than 3 failed?

5.24 The number of defects that occur on sink tops produced by a baked enamel process is known to follow a Poisson distribution. The average number of defects on a sink top is 2. What is the probability that a sink top has:

(a) No defects

(b) One defect

(c) At most, 2 defects

5.25 The number of items of a certain type that are purchased in a store in a week's time follows a Poisson distribution with $\lambda = 4$. How large a weekly stock is required in order that demand can be met with probability .99?

5.26 Mr. Jones leaves home every workday morning promptly at 7 AM. It takes him between 20 and 35 min to arrive at the commuter train station, and this arrive time has a uniform distribution. If commuter trains leave the station at 7:16, 7:29, 7:44, and 7:57, what is the probability that Mr. Jones will wait 3 min or more for a train?

5.27 The probability of observing a meteorite during a very small time interval Δt is $(1/10)\Delta t$ (where t is measured in minutes); the probability of observing more than one meteorite in a time interval Δt is negligible; and the probability of observing a meteorite during any given time interval is independent of the times of past observations. Determine

(a) The probability of observing one meteorite during an observation period of one-half hour

(b) The probability of observing at least one meteorite during an observation period of one-half hour

5.28 After ordering type C208B transistors, the production manager of Electro Company has to wait for the order to arrive from the supplier and to be delivered from the mailroom to the production line. The distribution of the time t_1 (in days) required for delivery from the supplier is given by

$$f(t_1) = \frac{1}{2} e^{-(1/2)t_1}$$

and the distribution of the time t_2 (in days) required for delivery from the mailroom is given by

$$f(t_2) = 3e^{-3t_2}$$

Assuming that the times required for delivery from the supplier and from the mailroom are independent, find the probability that the time required for delivery from the supplier exceeds three days and the time required for delivery from the mailroom does not exceed one-sixth of a day.

5.29 The Tweedy Shoppe has ordered sports jackets from two companies (Masculine Mandates and Fabulous Fashions). They hope to have both orders in stock for their annual sale, which begins in 15 days and for which they want to start rearranging their stock in 10 days. If the joint distribution of the delivery times of orders from Masculine Mandates (corresponding to t_1) and Fabulous Fashions (corresponding to t_2) is given by

$$f(t_1, t_2) = \frac{1}{150} e^{-(t_1/10 + t_2/15)} \qquad \text{for } t_1, t_2 > 0$$

determine the probability that
(a) Both orders will arive within 15 days
(b) The Masculine Mandates order will arive within 10 days and the Fabulous Fashions order will arrive within 15 days
(c) Neither order will arrive within 15 days

5.30 Idiotubes Inc. is a large TV repair company at which different types of repairs are done at different work benches. Of the TV sets repaired at the tube-replacement bench, .6 need α-tubes and .4 need β-tubes. If this work bench has 5 α-tubes and 4 β-tubes on hand, determine the probability that
(a) There are no β-tubes left when the last α-tube is used
(b) There are no α-tubes left when the last β-tube is used

5.31 Hampden Jewelers, Inc., has five gold rings in stock; three of the rings are plain and two of the rings are decoratively engraved. Of the customers who buy gold rings at Hampden's, $\frac{2}{3}$ buy plain rings.
(a) What is the probability that Hampden's will have no decorated rings left when the last plain ring is sold?
(b) If the customers who buy a gold ring of either type are numbered consecutively, what is the expected value of the number of the customer to whom the third plain ring is sold?

5.32 Programs arrive randomly at a large computing center at the rate of 10 per minute. Use a Poisson approximation to obtain
(a) The probability that more than 2 programs arrive in a period of 12 seconds
(b) The probability that more than 2 programs arrive in a period of 24 seconds

5.33 On the Go-Grip test of finger strength, the average score for males is 110; the standard deviation is 10. The average score for females is 80; the standard deviation is 20. Both score distributions are assumed to be normal. In a random sample of 1500 males and 1000 females,
(a) How many males would be expected to score below the mean for females
(b) How many females would be expected to score above the mean for males

5.34 A particular type of electronic component has probability .25 of immediate failure when it is installed in an audio system. When a component fails, another is installed. Joe spends his time at Audio-Phonics, Inc., installing these components.
(a) What is the probability that (on any day) the fifth component he installs is the third component to fail immediately?
(b) What is the expected value of the trial on which the third failure occurs?

6.1 INTRODUCTION

Any quantity that is determined or computed from sample values of a random variable is referred to as a *statistic*. A statistic is a characteristic of a sample. For example, the sample mean, the number or proportion of occurrences of an event in n trials, the sample range, the sample median, and the sample variance are statistics.

The distribution of a random variable corresponding to a statistic is referred to as a *sampling distribution*. Logically the probability distribution of any random variable could be referred to as a sampling distribution; however, the term usually is reserved for the distribution of a random variable corresponding to a statistic, that is, a sample characteristic. The sampling distribution of a statistic is known only if probability sampling is used.

Statistical inference is concerned with making generalizations about a population on the basis of observations of a sample or samples. A *parameter* is a characteristic or measure of a population; it distinguishes one population from another similar population. A statistic is a characteristic or measure of a sample; it varies from sample to sample of the same population in a manner described by its sampling distribution. The sampling distribution of a statistic is the basis for inferences about the corresponding population parameter.

The concept of sampling distributions is very important both logically and computationally. Variability is inherent in the nature of things and must be taken into account in drawing conclusions from empirical research. For this reason, statistical inference is based on the theory of probability. The sampling distribution of a statistic for samples from a specified population gives the probabilities with which the statistic assumes its possible values. On this basis, inferences can be made about the population.

For large samples, the central limit theorem gives the approximate distribution of the sample mean regardless of the population distribution—provided the population has a finite mean and variance. When the distribution of the population is specified, the distributions of the sample mean and other statistics are determined exactly, although they cannot always be written in simple form.

Frequently, a variable is assumed to be normally distributed. The normal distribution is convenient to work with mathematically and the properties and sampling distributions of statistics based on random samples from normal populations are known. The central limit theorem justifies the assumption of normality for means of large samples. In addition, many variables can reasonably be assumed to be approximately normally distributed. In particular, if a distribution is known to be unimodal and approximately symmetric, it is usually assumed to be approximately normal. Studies indicate that, in general, marked departures from normality invalidate inferences based on the assumption of normality; however, minor departures have little effect on the validity of such inferences.

Sampling distributions and statistical inference are based on the assumption of random sampling. In the following section random sampling is defined for finite and for infinite populations. Most statistical theory is based on infinite populations; however, it is also applicable to large finite populations and, with certain modifications, to relatively small finite populations.

Following the discussion of random sampling, the sampling distributions of various statistics based on random samples from normally distributed populations are discussed. Subsequent sections concern inferences based on these sampling distributions.

6.2 RANDOM SAMPLING

When sampling is from a finite population, the sample size is bounded by the size of the population. When sampling is from an infinite population, there is no bound on the size of the sample—it can be as large as the investigator wishes. Note that if sampling from a finite population is with replacement so that each element can be observed indefinitely many times, then the population is in effect infinite and there is no bound on the sample size.

FINITE POPULATIONS

Suppose a sample $x_1, x_2, \ldots, x_n$ is obtained without replacement from a finite population consisting of N elements, where $x_i, i = 1, 2, \ldots, n$ is the ith element selected; then $x_1, x_2, \ldots, x_n$ is a *random sample of size n from the finite population* if

$$f(x_1, x_2, \ldots, x_n) = \frac{1}{N(N-1)\cdots(N-n+1)}$$

$$= \prod_{i=0}^{n-1} \frac{1}{N-i}$$

for every ordered set of n elements from the finite population. That is, a sample of size n from a finite population of N elements is random if each ordered set of n elements has the sample probability $1/\binom{N}{n}$ of being selected in the sample.

Under these assumptions, the probability that any element x_i is selected in the sample is $1/N$

$$P(x_i) = \frac{1}{N} \qquad i = 1, 2, \ldots, N$$

Note that it is not sufficient for a random sample that each element of the population have the same probability of being included in the sample. The sample must also be selected so that each possible set of elements, that is, each possible set of sample values, is equally likely to be selected. Thus various schemes for sampling systematically, such as taking every nth name on a list, do not result in random samples. If the starting point is determined randomly, every name has an equal chance of being included in the sample, but many samples cannot possibly be selected. For example, successive names will not both be included in a sample. Methods of obtaining random samples are discussed in Chapter 11.

In the terminology of sample survey techniques, a random sample as defined above is referred to as a *strictly random sample*. (See Chapter 11.) However, the term random sample is customarily used, except in rather specialized discussions of sampling.

INFINITE POPULATIONS

The observations $x_1, x_2, \ldots, x_n$ constitute a random sample from a population having the frequency function $f(x)$ if

$$f(x_1, x_2, \ldots, x_n) = f(x_1)f(x_2) \cdots f(x_n)$$

$$= \prod_{i=1}^{n} f(x_i)$$

where $X_1, X_2, \ldots, X_n$ are random variables corresponding to the n trials or observations of the sample. Thus a sample from an infinite population is random if each element of the population has an equal and independent chance of being included in the sample.

CHEBYSHEV'S THEOREM

If $X_1, X_2, \ldots, X_n$ are independent and identically distributed random variables having mean μ and variance σ^2, then

and

$$\begin{cases} E(\bar{X}) = \mu_{\bar{x}} = \mu \\[2mm] \text{var}(\bar{X}) = \sigma_{\bar{x}}^2 = \dfrac{\sigma^2}{n} \end{cases}$$

The standard deviation of $\bar{X}$, $\sigma_{\bar{x}} = \sigma/\sqrt{n}$, is the *standard error of the mean*.* The variance of the sample mean approaches zero as n increases, since

---

* If $\bar{X}$ is the mean of a random sample of size n selected from a finite population of size N having mean μ and variance σ^2, then

and

$$E(\bar{X}) = \mu_{\bar{x}} = \mu$$

$$\text{var}(\bar{X}) = \sigma_{\bar{x}}^2 = \frac{\sigma^2}{n} \cdot \frac{N-n}{N-1}$$

This formula for $\sigma_{\bar{x}}^2$, when sampling is from a finite population, differs from the corresponding formula for sampling from an infinite population by the factor $(N-n)/(N-1)$. Note that this is the finite population correction, the same factor by which the variance of the hypergeometric distribution differs from the variance of the binomial distribution. (See Chapter 5.) If N is large relative to n, this factor is negligible. Thus $\sigma_{\bar{x}} = \sigma/\sqrt{n}$ is used as an approximation for the standard deviation of the distribution of $\bar{X}$ for samples obtained without replacement from sufficiently large finite populations.

$\sigma^2/n \to 0$ as $n \to \infty$. This limiting behavior can be stated more precisely using Chebyshev's theorem:

> If X is a random variable whose distribution has mean μ and variance σ^2, then for any positive constant k, greater than or equal to 1, the probability that X assumes a values less than $\mu - k\sigma$ or greater than $\mu + k\sigma$ is less than $1/k^2$. Thus the probability that X assumes a value between $\mu - k\sigma$ and $\mu + k\sigma$ is at least $1 - 1/k^2$.

$$P(|X - \mu| < k\sigma) \geq 1 - \frac{1}{k^2}$$

or, equivalently,

$$P(|X - \mu| > k\sigma) < \frac{1}{k^2}$$

In particular, Chebyshev's theorem states that of the values of a random variable,

75 percent are within 2 standard deviations of the mean
89 percent are within 3 standard deviations of the mean
94 percent are within 4 standard deviations of the mean
96 percent are within 5 standard deviations of the mean

Chebyshev's theorem has the remarkable and useful property that it does not depend on the form of the distribution of X; so long as a random variable has a finite mean and finite variance, the theorem applies. Of course, if the form of a distribution is known, closer bounds on its values can be obtained, as illustrated subsequently for the normal distribution.

For the random variable $\bar{X}$, $E(\bar{X}) = \mu$ and $\sigma_{\bar{x}} = \sigma/\sqrt{n}$. Using Chebyshev's theorem, we have

$$P(|\bar{X} - \mu| > c) < \frac{\sigma^2}{nc^2}$$

where c is an arbitrarily chosen constant. In terms of standard deviations, this inequality can be written:

$$P\left(|\bar{X} - \mu| > \frac{k\sigma}{\sqrt{n}}\right) < \frac{1}{k^2}$$

or, equivalently,

$$P\left(|\bar{X} - \mu| \leq \frac{k\sigma}{\sqrt{n}}\right) \geq 1 - \frac{1}{k^2}$$

This inequality provides what are called "loose bounds"; the interval can be determined more precisely if the form of the distribution of the random variable is known, as shown in the following sections.

6.3 RANDOM SAMPLING FROM NORMAL POPULATIONS

A large part of statistical theory is based on the assumption of random sampling from normal populations. In this section the sampling distributions of several statistics based on random samples from normal distributions are considered. These sampling distributions are used as the basis for the discussions of statistical inference in Chapters 7, 8, and 9.

The following notation is used in subsequent sections: if $x_1, x_2, \ldots, x_n$ are the values of a random sample of n observations, then the sample mean and variance are given by

and

$$
\bar{x} = \frac{\sum\limits_{i=1}^{n} x_i}{n}
$$

$$
s^2 = \frac{\sum\limits_{i=1}^{n} (x_i - \bar{x})^2}{n - 1}
$$

If $x_{11}, x_{12}, \ldots, x_{1n_1}$ are the values of a random sample of n_1 observations from one population and if $x_{21}, x_{22}, \ldots, x_{2n_2}$ are the values of an independent random sample of n_2 observations from another population, then the sample means and variances are given by

$$
\bar{x}_1 = \frac{\sum\limits_{i=1}^{n_1} x_{1i}}{n_1} \qquad\qquad \bar{x}_2 = \frac{\sum\limits_{i=1}^{n_2} x_{2i}}{n_2}
$$

$$
s_1^2 = \frac{\sum\limits_{i=1}^{n_1} (x_{1i} - \bar{x}_1)^2}{n_1 - 1} \qquad s_2^2 = \frac{\sum\limits_{i=1}^{n_2} (x_{2i} - \bar{x}_2)^2}{n_2 - 1}
$$

6.4 SAMPLING DISTRIBUTIONS BASED ON MEANS

Several sampling distributions based on means of random samples from normal populations are discussed in this section. The cases included are: (1) the mean of a random sample from one normal population with known variance, (2) the mean of a random sample from one normal population with unknown variance, (3) the difference between means of random samples from two normal populations with known variances, and (4) the difference between means of random samples from two normal populations with unknown but equal variances.

RANDOM SAMPLING FROM ONE NORMAL POPULATION WITH KNOWN VARIANCE

If a population is normally distributed, then the distribution of the mean of a random sample from the population is normally distributed.

If $\bar{X}$ is the mean of a random sample of size n from a population having a normal distribution with mean μ and variance σ^2, then the sampling distribution of $\bar{X}$ is normal with mean μ and variance σ^2/n. Thus

$$\frac{\bar{x} - \mu}{\sigma/\sqrt{n}} \sim N(0, 1)$$

Note that the central limit theorem states that under general conditions, the distribution of the mean of a random sample from a population of any form approaches normality as the sample size increases. The statement above requires normality of the population distribution; then the corresponding sample means are (precisely) normally distributed for any sample size.

EXAMPLE
The weights of cans of peaches from a production line are normally distributed with a mean of 16.8 oz and a variance of 2.25 oz. From each run of the process, 100 cans are selected randomly and weighted. What is the probability that the average weight of these cans is between 16.5 and 17.1 oz?

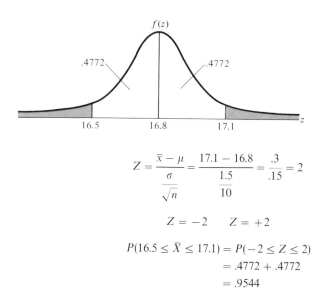

$$Z = \frac{\bar{x} - \mu}{\dfrac{\sigma}{\sqrt{n}}} = \frac{17.1 - 16.8}{\dfrac{1.5}{10}} = \frac{.3}{.15} = 2$$

$$Z = -2 \qquad Z = +2$$

$$P(16.5 \le \bar{X} \le 17.1) = P(-2 \le Z \le 2)$$
$$= .4772 + .4772$$
$$= .9544$$

For this problem the following result can be obtained using Chebyshev's theorem

$$P(16.5 \le \bar{X} \le 17.1) = P(|\bar{X} - \mu| \le .3)$$

$$P\left(|\bar{X} - \mu| \le \frac{k\sigma}{\sqrt{n}}\right) \ge 1 - \frac{1}{k^2}$$

$$\frac{k\sigma}{\sqrt{n}} = .3$$

$$k = \frac{.3\sqrt{n}}{\sigma} = \frac{(.3)(10)}{1.5} = 2$$

The value of k is the same as the value of the unit normal variable Z, since both indicate the number of standard deviations from the mean. Thus, using Chebyshev's theorem, we get

$$P(|\bar{x} - \mu| \le .3) \ge 1 - \tfrac{1}{4}$$
$$P(|\bar{x} - \mu| \le .3) \ge .75$$

Using the normal distribution, we get

$$P(|\bar{x} - \mu| \le .3) = .9544$$

which is considerably more precise. Note that this precision is gained at the expense of generality, since Chebyshev's theorem is distribution-free and allows for all types of distributions—uniform, U-shaped, skewed, and so forth.

RANDOM SAMPLING FROM ONE NORMAL POPULATION WITH UNKNOWN VARIANCE

In practice, the population variance σ^2 is not usually known and must be replaced by an estimate, the sample variance s^2. In this case, inference is based on the sampling distribution of $\dfrac{\bar{x} - \mu}{s/\sqrt{n}}$ rather than on the sampling distribution of $\dfrac{\bar{x} - \mu}{\sigma/\sqrt{n}}$. The exact sampling distribution of $\dfrac{\bar{x} - \mu}{s/\sqrt{n}}$ for random samples from normal populations is the *t distribution* or the *Student-t distribution* with $n - 1$ degrees of freedom. The t distribution is given by

$$f(t) = \frac{\Gamma\left(\dfrac{v + 1}{2}\right)}{\sqrt{\pi v}\,\Gamma\left(\dfrac{v}{2}\right)}\left(1 + \frac{t^2}{v}\right)^{-(v+1)/2} \qquad \text{for } -\infty < t < +\infty$$

where v is the degrees of freedom, a parameter depending on the number of observations in the sample.

Degrees of freedom is sometimes defined as the number of independent variables in the statistic whose distribution is being considered. In this sense the number of independent variables in a mean of n observations is $n - 1$, since if $n - 1$ observations are specified, knowledge of the mean determines the other observation. The t distribution, the distribution of the mean of n observations, has $n - 1$ degrees of freedom. The form of a sampling distribution varies with its degrees of freedom, a parameter or parameters depending on sample size or categorization. Appendix 10.2 discusses degrees of freedom further and summarizes the determination of degrees of freedom for the chi-square, F, and t distributions.

The t distribution was first obtained by W. S. Gosset, who was employed by an Irish brewery that did not permit publication of research by its staff. Gosset chose the pen name "Student" and the distribution is thus sometimes referred to as the Student-t distribution. The t distribution is symmetric and

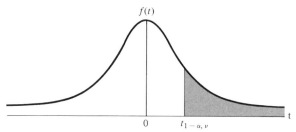

Figure 6.1   *t* distribution

similar in shape to the normal distribution, but has more variability because its variance is estimated from the sample. The tails of the *t* distribution approach the *x*-axis more slowly than the tails of the normal distribution, especially for small degrees of freedom. The *t* distribution has mean 0 and, for $v > 2$, variance $v/(v - 2)$. (See Figure 6.1.) The *t* distribution for various values of *v* is given in Table B.9; the area to the right of the specified value of *t* is tabled, as shown in Figure 6.1.

The *t* distribution usually is not tabled for degrees of freedom greater than 30. The *t* distribution approaches the standard normal distribution as the sample size (or degrees of freedom) increases and for degrees of freedom greater than 30, the standard normal distribution is generally used as an approximation of the *t* distribution.

The *t* distribution has many applications in statistical inference, some of which are discussed in subsequent sections. Its application in sampling from one normally distributed population can be summarized as follows:

If $\bar{x}$ and *s* are the mean and standard deviation of a random sample of size *n* from a normal population having mean μ and variance σ^2, then $\dfrac{\bar{x} - \mu}{s/\sqrt{n}}$ has the *t* distribution with $n - 1$ degrees of freedom.

EXAMPLE
The values of the investment portfolios managed by a large brokerage firm are normally distributed with a mean of \$50,000. A random sample of 25 portfolios has a mean value of \$55,000 and a standard deviation of \$10,000. What proportion of random samples of size 25 have means larger than the mean of this sample?

$$t = \frac{\bar{x} - \mu}{\dfrac{s}{\sqrt{n}}}$$

$$= \frac{55,000 - 50,000}{\dfrac{10,000}{5}}$$

$$= \frac{25,000}{10,000}$$

$$= 2.5$$

$$P(\bar{X} \geq 55,000)$$

$$= P(t \geq 2.5) \approx .01$$

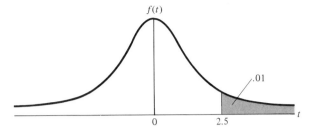

RANDOM SAMPLING FROM TWO INDEPENDENT
NORMAL POPULATIONS WITH KNOWN VARIANCES

In some problems, independent random samples are obtained from two populations and the sampling distribution of a statistic comparing the two samples is required. If two independent populations are normally distributed and their variances are known, then the difference between sample means is normally distributed.

If $\bar{X}_1$ is the mean of a random sample of size n_1 from a normal distribution having mean μ_1 and variance $\sigma_1{}^2$ and $\bar{X}_2$ is the mean of a random sample of size n_2 from an independent normal distribution having mean μ_2 and variance $\sigma_2{}^2$, then

$$\frac{(\bar{x}_1 - \bar{x}_2) - (\mu_1 - \mu_2)}{\sqrt{\dfrac{\sigma_1{}^2}{n_1} + \dfrac{\sigma_2{}^2}{n_2}}}$$

has the standard normal distribution.

EXAMPLE
Two sets of bolts are intended to have different lengths but the same diameter. The diameters of the bolts of type I have a variance of 4 cm; the diameters of the bolts of type II have a variance of 12 cm. Random samples of 10 bolts of type I and 20 bolts of type II are obtained. If the bolts of the two types actually have the same diameter, determine the probability that the sample means will differ by at least 1.5 cm.

$$Z = \frac{(\bar{x}_1 - \bar{x}_2) - (\mu_1 - \mu_2)}{\sqrt{\dfrac{\sigma_1{}^2}{n_1} + \dfrac{\sigma_2{}^2}{n_2}}}$$

$$= \frac{1.5 - 0}{\sqrt{\dfrac{4}{10} + \dfrac{12}{20}}} = 1.5$$

$$
\begin{aligned}
P(|X_1 - X_2| \geq 1.5) &= P(|Z| \geq 1.5) \\
&= .06681 + .06681 \\
&= .13362
\end{aligned}
$$

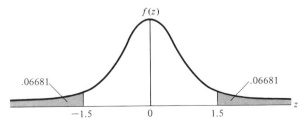

RANDOM SAMPLING FROM TWO
INDEPENDENT NORMAL POPULATIONS
WITH UNKNOWN BUT EQUAL VARIANCES

In practice, the population variances σ_1^2 and σ_2^2 are not usually known and must be replaced by their estimates, the sample variances s_1^2 and s_2^2. If the population variances are equal, the sample variances can be combined or pooled to estimate this common variance; the difference in sample means then has the t distribution. Suppose that there are two normal populations with equal but unknown variances $\sigma_1^2 = \sigma_2^2$ and means μ_1 and μ_2. If $\bar{x}_1$ and s_1^2 are the mean and variance of a random sample of size n_1 from the first population and $\bar{x}_2$ and s_2^2 are the mean and variance of a random sample of size n_2 from the second population, then

$$\frac{(\bar{x}_1 - \bar{x}_2) - (\mu_1 - \mu_2)}{\sqrt{\dfrac{(n_1 - 1)s_1^2 + (n_2 - 1)s_2^2}{n_1 + n_2 - 2}}} = \frac{(\bar{x}_1 - \bar{x}_2) - (\mu_1 - \mu_2)}{\sqrt{\dfrac{\sum (x_{1i} - \bar{x}_1)^2 + \sum (x_{2i} - \bar{x}_2)^2}{(n_1 - 1) + (n_2 - 1)}}}$$

has the t distribution with $n_1 + n_2 - 2$ degrees of freedom.

EXAMPLE

A random sample of 20 stocks from Group A has average annual earnings of $35 with a variance of $16; an independent random sample of 18 stocks from Group B has average annual earnings of $15 with a variance of $12. Assume that the average annual earnings of the two groups of stocks are independently normally distributed with the same variance and means differing by $10. (Mean for Group A > mean for Group B). Find the probability that the sample mean for Group A is at least $20 larger than the sample mean for Group B (as observed).

$$t = \frac{(\bar{x}_1 - \bar{x}_2) - (\mu_1 - \mu_2)}{\sqrt{\dfrac{(n_1 - 1)s_1^2 + (n_2 - 1)s_x^2}{n_1 + n_2 - 2}}}$$

$$= \frac{20 - 10}{\sqrt{\dfrac{(19)(16) + (17)(12)}{20 + 18 - 2}}}$$

$$= \frac{10}{\sqrt{\dfrac{(4)(127)}{36}}} = \frac{30}{\sqrt{127}} = 2.66$$

$$P(\bar{x}_1 - \bar{x}_2 \geq 20) = P(t \geq 2.66) = .0039$$

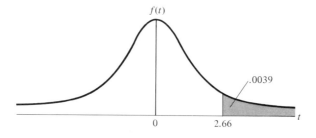

LARGE-SAMPLE APPROXIMATION

For $n_1 > 30$ and $n_2 > 30$, the t distribution can be approximated by a standard normal distribution. That is, if $n_1 > 30$ and $n_2 > 30$, then

$$\frac{(x_1 - x_2) - (\mu_1 - \mu_2)}{\sqrt{\dfrac{s_1^2}{n_1} + \dfrac{s_2^2}{n_2}}} \sim N(0, 1)$$

approximately. This result follows from a two-sample generalization of the central limit theorem.

If two normal populations cannot be assumed to have equal variances and the sample sizes are not large enough for the central limit theorem to apply, then the difference in sample means has the t distribution, but the appropriate degrees of freedom must be computed by a formula—discussed in Section 9.2.

In general, observations can be pooled only if they are obtained from random samples of the same population or from random samples of populations that do not differ with respect to the parameter being estimated. Thus random samples from populations having the same variance can be combined or pooled for the purpose of estimating that variance, although the populations may have different means; however, random samples from populations having unequal variances cannot be pooled for this purposes.

6.5 SAMPLING DISTRIBUTIONS BASED ON VARIANCES

Several sampling distributions based on variances of random samples from normal populations are discussed here. The cases included are: (1) the variance of a random sample from one normal population and (2) the ratio of variances of random samples from two normal populations.

RANDOM SAMPLING FROM
ONE NORMAL POPULATION

If a random variable is normally distributed, the sampling distribution of the variance of samples of size n is based on the chi-square distribution with $n - 1$ degrees of freedom.

> If s^2 is the variance of a random sample of size n from a normal distribution having mean μ and variance σ^2, then $(n - 1)s^2/\sigma^2$ has a chi-square distribution with $n - 1$ degrees of freedom.

As mentioned in Section 5.3, the gamma distribution with parameters $\alpha = v/2$ and $\beta = 2$ is referred to as the chi-square (χ^2) distribution; it is given by

$$f(\chi^2) = \begin{cases} \dfrac{1}{2^{v/2}\Gamma\left(\dfrac{v}{2}\right)} (\chi^2)^{(v-2)/2} e^{-(1/2)\chi^2} & \text{for } x > 0 \\ \\ 0 & \text{elsewhere} \end{cases}$$

where the parameter v is equal to the degrees of freedom. Thus the sampling distribution of s^2 for a sample of n observations is based on the chi-square distribution with $v = n - 1$ degrees of freedom. The chi-square distribution has mean v and variance $2v$; thus the shape of the chi-square distribution depends on its degrees of freedom. The chi-square distribution is given in Table B.10.

LARGE-SAMPLE APPROXIMATION

When $v > 30$, the chi-square distribution can be approximated by a standard normal distribution: If X has a chi-square distribution with v degrees of freedom, then for large v the distribution of $(\sqrt{2\chi^2} - \sqrt{2v})$ is approximately standard normal.

EXAMPLE

The washers used in a particular type of motor are required to be of uniform thickness and are manufactured so that the thickness has a standard deviation of .0001 cm. Determine the probability that the standard deviation of a random sample of 25 washers is at least .00014 cm.

$$\chi^2 = \frac{(n-1)s^2}{\sigma^2}$$

$$= \frac{(24)(.00014)^2}{(.0001)^2}$$

$$= 47.04$$

$$P(s^2 \geq .00014) = P(\chi^2 \geq 47.04) < .01$$

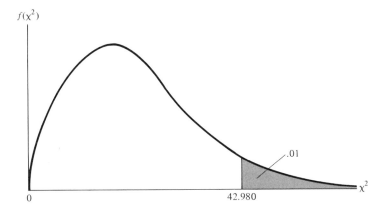

Suppose in this example that the sample size is 50, so that the normal approximation is appropriate.
Then

$$Z = \sqrt{2\chi^2} - \sqrt{2v}$$

$$= \sqrt{\frac{(2)(49)(.00014)^2}{(.0001)^2}} - \sqrt{(2)(49)}$$

$$= \sqrt{192.08} - \sqrt{98}$$

$$= 13.86 - 9.90$$

$$= 3.96$$

$$P(s^2 \geq .00014) = P(Z \geq 3.96) = .00004$$

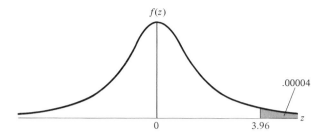

RANDOM SAMPLING FROM
TWO NORMAL POPULATIONS

If two populations are independently normally distributed with variances $\sigma_1{}^2$ and $\sigma_2{}^2$, then the sampling distribution of the ratio of the variances of samples of sizes n_1 and n_2, respectively, is based on the F-distribution with $n_1 - 1$ and $n_2 - 1$ degrees of freedom.

> Suppose $s_1{}^2$ and $s_2{}^2$ are the variances of independent random samples of sizes n_1 and n_2, respectively, from two normal populations having variances $\sigma_1{}^2$ and $\sigma_2{}^2$, respectively. Then, if $\sigma_1{}^2 = \sigma_2{}^2$, the ratio $s_1{}^2/s_2{}^2$ has the F distribution with $n_1 - 1$ and $n_2 - 1$ degrees of freedom.

The F distribution is given by

$$f(F) = \begin{cases} \dfrac{cF^{(v_1/2)-1}}{\left(1 + \dfrac{v_1}{v_2}F\right)^{(v_1+v_2)/2}} & \text{for } x > 0 \\[4mm] 0 & \text{elsewhere} \end{cases}$$

where

$$c = \frac{\Gamma\left(\dfrac{v_1 + v_2}{2}\right)}{\Gamma\left(\dfrac{v_1}{2}\right)\Gamma\left(\dfrac{v_2}{2}\right)}\left(\frac{v_1}{v_2}\right)^{v_1/2}$$

and the parameters v_1 and v_2 are the degrees of freedom. The F distribution is related to the beta distribution; if $X = v_2 Y/v_1(1 - Y)$ where Y has the beta distribution with parameters $\alpha = v_1/2$ and $\beta = v_2/2$, then X has the F distribution with parameters v_1 and v_2. The shape of the F distribution depends on its degrees of freedom; for $v_2 > 2$, it has mean $v_2/(v_2 - 2)$.

The F distribution is given in Table B.11. The F distribution has the interesting property that if X is distributed as F with v_1 and v_2 degrees of freedom, then $1/X$ is distributed as F with v_2 and v_1 degrees of freedom; thus

$$F_{1-\alpha, v_1, v_2} = \frac{1}{F_{\alpha, v_2, v_1}}$$

for example,

$$F_{.05, 6, 8} = \frac{1}{F_{.95, 8, 6}}$$

One very important application of the F distribution concerns the comparison of variances of two normal populations. Many statistical analyses are based on the assumption that population variances are equal, that is, the assumption of homogeneity of variance. The assumption of homogeneity of variance should be tested before an analysis requiring this assumption is performed. One of the most frequently used tests is based on the fact that s_1^2/s_2^2 has the F distribution.

EXAMPLE

The variances of independent random samples of sizes 25 and 20 from two normal populations are, respectively, 79.4 and 25.6. If the population variances are in fact equal, what is the probability of obtaining sample variances that differ by at least this ratio?

$$F = \frac{s_1^2}{s_2^2}$$

$$= \frac{79.4}{25.6}$$

$$= 3.10$$

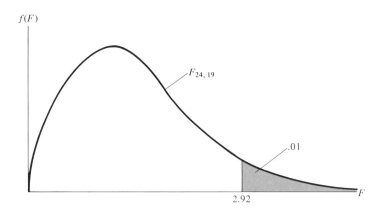

thus

$$P\left(\frac{s_1^2}{s_2^2} \geq 3.10\right) < .01$$

and

$$P\left(\frac{s_1^2}{s_2^2} \geq 3.10 \quad \text{or} \quad \frac{s_2^2}{s_1^2} \geq 3.10\right) < .02$$

Note that if the larger sample variance is the numerator and the smaller sample variance is the denominator, then the probability associated with the corresponding value of F must be doubled to obtain the probability that the variances differ by at least the ratio observed in the samples. This follows from

$$F_{1-\alpha, v_1, v_2} = \frac{1}{F_{\alpha, v_2, v_1}}$$

as mentioned.

The sampling distributions discussed in this chapter are summarized in Table 6.1.

TABLE 6.1   SUMMARY OF SAMPLING DISTRIBUTIONS

Parameter(s) of Normal Distributions	Sampling Distribution of Relevant Statistic
One sample μ	Known variance: $$\frac{\bar{x} - \mu}{\dfrac{\sigma}{\sqrt{n}}} \sim N(0, 1)$$ Unknown variance: $$\frac{\bar{x} - \mu}{\dfrac{s}{\sqrt{n}}} \sim t_{n-1}$$
Two samples $\mu_1 - \mu_2$	Known variances: $$\frac{(\bar{x}_1 - \bar{x}_2) - (\mu_1 - \mu_2)}{\sqrt{\dfrac{\sigma_1^2}{n_1} + \dfrac{\sigma_2^2}{n_2}}} \sim N(0, 1)$$ Equal but unknown variances: $$\frac{(\bar{x}_1 - \bar{x}_2) - (\mu_1 - \mu_2)}{\sqrt{\dfrac{(n_1 - 1)s_1^2 + (n_2 - 1)s_2^2}{n_1 + n_2 - 2}}} \sim t_{n_1 + n_2 - 2}$$
One sample σ^2	$$\frac{(n-1)s^2}{\sigma^2} \sim \chi_{n-1}^2$$
Two samples $\dfrac{\sigma_1^2}{\sigma_2^2}$	$$\frac{s_1^2}{s_2^2} \sim F_{n_1 - 1, n_2 - 1}$$

6.6 RELATIONS AMONG THE NORMAL, CHI-SQUARE, t AND F DISTRIBUTIONS

The sampling distributions discussed in the preceding sections are based on random sampling from normal populations. There are a number of useful mathematical relationships among these sampling distributions, the most important of which are summarized below.

1. If $X_1, X_2, \ldots, X_n$ are independent random variables having the standard normal distribution, then the sum of their squares has the chi-square distribution with n degrees of freedom. That is, if $X_1, X_2, \ldots, X_n \sim$ independently $N(0, 1)$, then

$$\sum_{i=1}^{n} X_i^2 \sim \chi_n^2$$

SPECIAL CASE
If X has the standard normal distribution, then X^2 has the chi-square distribution with one degree of freedom. That is, if $X \sim N(0, 1)$, then $X^2 \sim \chi_1^2$.

2. If X_1 has the standard normal distribution and X_2 has the chi-square distribution with v degrees of freedom, then the ratio X_1 over the square root of X_2 divided by v has the t distribution with v degrees of freedom. That is, if

$$\left.\begin{array}{c} X_1 \sim N(0, 1) \\ \text{and} \\ X_2 \sim \chi_v^2 \end{array}\right\} \quad \text{then} \quad \frac{X_1}{\sqrt{\dfrac{X_2}{v}}} \sim t_v$$

3. If X has the t distribution with v degrees of freedom, then X^2 has the F distribution with degrees of freedom 1 and v. That is, if

$$X \sim t_v \quad \text{then} \quad X^2 \sim F_{1,v}$$

4. If X_1 has the chi-square distribution with v_1 degrees of freedom and X_2 has the chi-square distribution with v_2 degrees of freedom and X_1 and X_2 are independent, then the ratio X_1 divided by v_1 over X_2 divided by v_2 has the F distribution with degrees of freedom v_1 and v_2. That is, if

$$\left.\begin{array}{c} X_1 \sim \chi_{v_1}^2 \\ X_2 \sim \chi_{v_2}^2 \end{array}\right\} \quad \text{then} \quad \frac{\dfrac{X_1}{v_1}}{\dfrac{X_2}{v_2}} \sim F_{v_1,v_2}$$

5. If X has the t distribution with v degrees of freedom, then the distribution of X approaches the standard normal distribution as v approaches infinitity. That is, if

$$X \sim t_v \quad \text{then} \quad X \to N(0, 1) \quad \text{as} \quad v \to \infty$$

6. If X has the F distribution with degrees of freedom v_1 and v_2, then the distribution of $v_1 X$ approaches the chi-square distribution with v_1 degrees of freedom as v_2 approaches infinity. That is, if

$$X \sim F_{v_1, v_2} \quad \text{then} \quad v_1 X \sim X_{v_1}^2 \quad \text{as} \quad v_2 \to \infty$$

PROBLEMS

6.1 The length of the planks cut by a machine has a mean of 6.5 ft and a variance of .25 ft. What is the probability that the average length of a random sample of 25 planks is between 6.0 and 7.0 ft
(a) Using Chebyshev's theorem
(b) Assuming a normal population distribution

6.2 A particular machine has a defective rate of .05. In a random sample of 100, determine the probability of obtaining

   (a) 10 or more defectives using Chebyshev's theorem
   (b) 10 or more defectives using a normal approximation
   (c) 2 or fewer defectives using Chebyshev's theorem
   (d) 2 or fewer defectives using a normal approximation

6.3 The number of transactions per day of a very large group of banks is normally distributed with a mean of 1110.1. If 2.5 percent of the average number of transactions per day for random samples of 36 banks fall above 1129.7, what is the standard deviation of the number of transactions per day for the group of banks?

6.4 The daily number of sales in a large department store has a normal distribution with mean 5541.2 and standard deviation 25.6. On 10 percent of the days, the number of sales is larger than what number?

6.5 The daily wage in a particular industry is normally distributed with a mean of 13.20. For random samples of 25 workers, 9 percent of the sample mean of the daily wages fall below 12.53; what is the standard deviation of the daily wage in this industry?

6.6 In a large savings bank, the average amount per account is $159.32; the standard deviation is $18.00. What is the probability that the average amount in a random sample of 400 accounts is at least $160.00?

6.7 The daily sales volume (in thousands of dollars) for the Bon Ton Department Store has a mean of 515.10 and a standard deviation of 20.

   (a) Assuming that daily sales are normally distributed, 95 percent of the daily sales are no larger than what amount?
   (b) Making no distribution assumptions (other than finite mean and variance), at least 95 percent of the daily sales are no larger than what amount?

6.8 The management of Sudsy Whate Soap Company has decided to change from blue soap flakes to pink soap flakes if at least 60 percent of a random sample of housewives prefer pink soap flakes. Actually, unknown of course to Sudsy White, 50 percent of housewives prefer pink soap flakes. What is the probability that, in the random sample obtained by Sudsy White, at least 60 percent of the housewives say they prefer pink soap flakes if the sample size is

   (a) 100 housewives?
   (b) 400 housewives?

6.9 Suppose that 90 percent of the school-age children in the United States are known to have been vaccinated for smallpox. Each of 250 investigators from the U.S. Public Health Service obtains an independent random sample of 100 school-age children. In how many of these 250 samples should the investigators expect to find that 85 or fewer of the children have been vaccinated?

6.10 In a large university, 20 percent of the students have a grade average of B+ or higher. What is the probability that in a random sample of 900 students, between 150 and 200 (inclusive) have an average of B+ or higher?

6.11 The school board in Hilly Meadows is considering starting a school lunch program. The school board has decided to start the program if at least 75 percent of the parents in a random sample favor it. Actually (unknown to the school board, of course) two-thirds of the parents in Hilly Meadows favor the program. What is the probability that the school lunch program will be started

   (a) If the school board takes a random sample of 400 parents?
   (b) If the school board takes a random sample of 100 parents?

6.12 The quality control engineer of Tubes Glace Corporation obtained a random sample of 225 glass tubes from a production line; he determined that the average weight of this sample of tubes is 20.5 oz with a standard deviation of 1.5 oz. With what probability can he state that the estimated average weight of 20.5 oz differs from the true weight (for a large production run) by at most 0.25 oz?

6.13 LeCuisine plans to introduce a line of gourmet frozen dinners if, in a random sample of customers of suburban supermarkets, at least 45 percent say they would be interested in such products. Actually (unknown to LeCuisine) 40 percent of the customers would be interested. What is the probability that LeCuisine will introduce the gourmet

frozen dinners if the decision is based on a random sample of
(a) 96 customers?
(b) 2400 customers?

6.14 The specification for a particular type of wire states that it must be of uniform diameter with variance not to exceed .002 in. If the wire is as specified, what is the probability that in a sample of size 10 the variance is at least .004 in.?

6.15 The specification for steel rods states that the standard deviation of the length must not exceed .2 in. If this specification is met, what is the probability that in a sample of size 50, the length has a standard deviation as large as .3 in.?

6.16 The variances of independent random samples of sizes 50 and 100 from two normal populations are, respectively, 125 and 115. If the population variances are in fact equal, what is the probability of obtaining sample variances that differ by at least this amount?

6.17 The weights of sacks of flour are supposed to be normally distributed about a mean of 16 oz. A random sample of 9 sacks has a mean weight of 15.8 oz and a standard deviation of 0.2 oz. If the population mean is in fact 16 oz, what is the probability of obtaining a sample mean at least this small?

6.18 In test runs, 10 randomly chosen models of an experimental engine ran for 18, 19, 20, 23, 21, 19, 25, 16, 22, and 17 min, respectively, on a gallon of standard fuel. Calculate the sample mean and standard deviation of the running times. If the actual mean running time on a gallon of standard fuel is 22 min for this engine, what is the probability of obtaining a sample mean at least this large?

6.19 The Ingot Steel Company produces two qualities of steel cable, which is sold in lengths of 50 ft. Quality A has an average test strength of 175 lb with a standard deviation of 15 lb; quality B has an average test strength of 160 lb with a standard deviation of 10 lb.
(a) What proportion of the grade B cables meet the test for grade A cables?
(b) What proportion of the grade A cables meet the test for grade B cables?

6.20 The lifetimes of a certain type of micro transistor are normally distributed about a mean of 156.83 hr; 1.5 percent of the transistors have a life time greater than 167.68 hr.
(a) What is the standard deviation of the distribution of the life times of the transistors?
(b) What is the probability that the average life time of a sample of 25 transistors is at least 155 hr?

6.21 R. A. Statistician has obtained a random sample of size 225 for the purpose of estimating the average weight of sheets of polythene coming from a reactor. He states to the quality control engineer that with probability .98, the sample average of 10.645 oz which he obtained is no more than .233 oz from the true average. What is the largest that the population variance can be, if his statement is true?

6.22 The daily wages in a particular industry are normally distributed with a mean of $13.20. If 9 percent of the average daily wages of samples of 25 workers fall below $12.53, what is the standard deviation of daily wages in this industry?

6.23 The Crunchy Candy Corporation employs a large number of salesmen. The average length of time these salesmen have been employed by Crunchy is 2.6 years; the standard deviation is 1.6 years. The personnel manager obtains the records of a random sample of 15 salesmen from the corporation files. If no assumption is made concerning the form of the distribution of length of employment, what statement can be made concerning the probability that
(a) The average length of time the random sample of 15 salesmen have worked for Crunchy is less than 5 years?
(b) The average length of time the random sample of 15 salesmen have worked for Crunchy is more than one year? (Assume that the mean and variance of the distribution are finite.)

6.24 The weight of bags of crude ore from a particular mining site has a mean of 150 lb and a standard deviation of 15.36 lb. What statement can be made concerning the probability that the average weight of a sample of 9 bags is between 140 lb and 160 lb
(a) Using Chebyshev's theorem
(b) Assuming that the weights are normally distributed

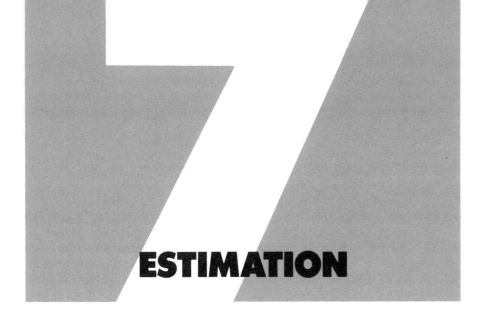

ESTIMATION

7.1 INTRODUCTION

Many problems involve the estimation of an amount or proportion; for example, an investigator may wish to estimate the average income of U.S. residents, the average monthly food budget of families in a particular community, the proportion of defectives produced by a certain process, the share of the market of a particular brand, and so forth. Each of these amounts or proportions is a characteristic or parameter of a particular population distribution, for example, the average income of U.S. residents is a parameter of the distribution of the incomes of these residents.

Parameter estimates are of two general types: point estimates and interval estimates. A *point estimator* of a parameter is a single-valued statistic (that is, a single-valued function of the observed data) intended to give a value "close" to the true parameter value. A particular value of a point estimator is a *point estimate*. A point estimator is, in general, a random variable; for the continuous case, its value *exactly* equals the parameter value with probability zero.

The usefulness of point estimators generally depends in part on knowing how precise they are likely to be. *Interval estimation* is one way of stating the precision or reliability of a point estimate. It provides an interval that is constructed from observed data in such a way that the parameter to be estimated lies within the interval with a specified probability. Such an interval is referred to as a *confidence interval estimate, confidence interval* or *confidence region;* the specified probability is the *confidence coefficient.*

Even when the statement of a problem requires a point estimate, an interval estimate can provide useful supplementary information concerning the precision or reliability of the point estimate. In practice, information concerning reliability of an estimate may be extremely valuable. Interval estimation is closely related to hypothesis testing, which is discussed in Chapter 8.

In the following sections, desirable properties of point estimators, several general methods of obtaining point estimators, and construction of confidence intervals are discussed. Estimation of the proportion π for a binomial distribution and the mean μ and variance σ^2 for a normal distribution are discussed as examples.

7.2 POINT ESTIMATION

The usual procedure in simple point estimation problems is to obtain a random sample from the relevant population, compute the sample statistic corresponding to the population parameter to be estimated, and use this statistic as an estimate of the population parameter. Thus the sample proportion is an estimator of the population proportion, the sample mean is an estimator of the population mean, the sample variance (appropriately defined) is an estimator of the population variance, and so forth.

Although the choice of an estimator seems obvious for many problems, the general problem of choosing the "best" estimator for a population parameter can be quite difficult. Even the most obvious choices should be examined carefully, since some apparent choices are not the "best" estimators, as will be illustrated.

Since point estimators are random variables, the statistical properties of these random variables are the basis for choosing the "best" estimator among the available alternatives. Several properties of estimators generally considered are: unbiasedness, minimum variance (efficiency), minimum expected squared error, sufficiency, and consistency.

Of these properties, unbiasedness, minimum variance, minimum expected squared error, and sufficiency are *small sample properties*, since they are defined for any sample size; consistency is a *large sample* or *asymptotic* property, since it is defined only in the limit as the sample size approaches infinity.

UNBIASED ESTIMATORS

An estimator is an *unbiased* estimator of a parameter if the expected value of the estimator is equal to the value of the parameter; if the expected value of an estimator is not equal to the value of a parameter, then the estimator is a *biased* estimator of the parameter. Denoting the parameter to be estimated by θ and the estimator by $\tilde{\theta}$, we can write:

If $E(\tilde{\theta}) = \theta$, then $\tilde{\theta}$ is an unbiased estimator of θ
If $E(\tilde{\theta}) = \theta + b$, then $\tilde{\theta}$ is a biased estimator of θ and b is its bias

EXAMPLE
The sample mean is an unbiased estimator of the population mean, since $E(\bar{x}) = \mu$; and the sample variance $s^2 = \sum (x_i - \bar{x})^2/(n-1)$ is an unbiased estimator of the population variance, since $E(s^2) = \sigma^2$. However, $E(s) \neq \sigma$, so the sample standard deviation is *not* an unbiased estimator of the population standard deviation. Note that, in general, $[E(y)]^{1/2} \neq E[y^{1/2}]$. (See rule 4, Section 2.2.)

Note that in defining the sample variance, the denominator $n - 1$ must be used instead of n in order to obtain an unbiased estimate of the population variance. This can be shown as follows:

$$E\left[\sum_{i=1}^{n} (x_i - \bar{x})^2\right] = E\left[\sum_{i=1}^{n} \{(x_i - \mu) - (\bar{x} - \mu)\}^2\right]$$

$$= E\left[\sum_{i=1}^{n} (x_i - \mu)^2 - 2\sum_{i=1}^{n} (x_i - \mu)(\bar{x} - \mu) + \sum_{i=1}^{n} (\bar{x} - \mu)^2\right]$$

$$= \left[\sum_{i=1}^{n} E(x_i - \mu)^2 - 2nE(\bar{x} - \mu)^2 + nE(\bar{x} - \mu)^2\right]$$

$$= n\sigma^2 - n\left(\frac{\sigma^2}{n}\right)$$

$$= (n - 1)\sigma^2$$

thus

$$E\left[\frac{\sum_{i=1}^{n} (x_i - \bar{x})^2}{n - 1}\right] = \sigma^2 \quad \text{and} \quad s^2 = \frac{\sum_{i=1}^{n} (x_i - \bar{x})^2}{n - 1}$$

is an unbiased estimate of σ^2.

Unbiasedness is an intuitively appealing property to require of an estimator since, on the average, an unbiased estimator neither underestimates nor overestimates, but is correct. This does not mean that an unbiased estimator is always correct, but only that its errors have an average or expected value of zero.

Although unbiasedness is an intuitively appealing property, there are many circumstances in which unbiasedness of an estimator is less important than other considerations; for example, if an unbiased estimator has relatively large variance, a biased estimator may in some reasonable sense be preferable. (See Figure 7.1.)

MINIMUM VARIANCE ESTIMATORS

The variance of an estimator $\tilde{\theta}$ of a parameter θ is denoted by var($\tilde{\theta}$) and defined by $E[\theta - E(\tilde{\theta})]^2$. If, for every sample size, var($\tilde{\theta}$) exists (is finite) and is at least as small as the variance of any other estimator based on the same size sample, then $\tilde{\theta}$ is a minimum variance estimator of θ.

Without consideration of other properties of an estimator, small or even zero variance is not necessarily desirable. For example, any arbitrary constant, chosen as an estimator of a parameter, has zero variance—even though it has no relation whatever to the parameter being estimated.

As illustrated in Figure 7.1, an unbiased estimator having a large variance is undesirable; an estimator having a small variance but a large bias is also undesirable. However, it seems reasonable (other things being equal) that among estimators having equal bias, the estimator having the smallest variance is preferable; similarly, among estimators having equal variance, the one having the smallest bias is preferable. The unbiased estimator(s) having minimum variance is the *best unbiased estimator* and is said to be *efficient*. This definition

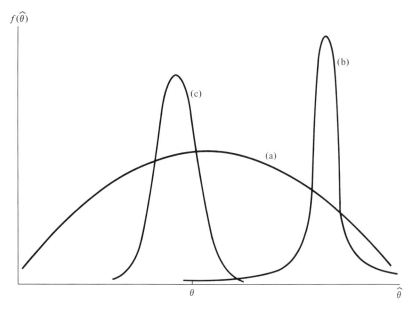

(a) Unbiased estimator, large variance

(b) Extremely biased estimator, small variance

(c) Biased estimator, smaller ESE than (a) or (b)

Figure 7.1   Graphical representation of sampling properties of estimators

of efficiency excludes from consideration all biased estimators and all estimators whose mean or variance is not finite.

MINIMUM EXPECTED SQUARED ERROR

In many cases a useful criterion for choosing among estimators is the minimization of some increasing function of both bias and variance. The function most frequently minimized is the expected squared error (ESE), that is, $E(\theta - \tilde{\theta})^2$.

$$
\begin{aligned}
\text{ESE} = E(\theta - \tilde{\theta})^2 &= E\{[\theta - E(\tilde{\theta})] + [E(\tilde{\theta}) - \tilde{\theta}]\}^2 \\
&= E[\theta - E(\tilde{\theta})]^2 + 2E[\theta - E(\tilde{\theta})][E(\tilde{\theta}) - \tilde{\theta}] + [E(\tilde{\theta}) - \tilde{\theta}]^2 \\
&= (\text{bias } \tilde{\theta})^2 + 0 + \text{var}(\tilde{\theta})
\end{aligned}
$$

In some applications, one can specify a maximum error and consider that errors larger than this maximum are equal to it. Then a truncated squared error is used as the function to be minimized. The concept of a truncated error arises, for example, in the case of a loss function where the loss is total beyond a certain point.

An estimator may have a relatively small expected squared error, even though it is biased and does not have zero variance (see Figure 7.1). Thus, using the minimum expected squared error criterion, there may be biased estimators that are preferable to efficient estimators. Note that minimum expected squared error is defined only for estimators whose mean and variance are finite.

SUFFICIENT ESTIMATORS

An estimator is *sufficient* for estimating a parameter if it uses all the sample information that is relevant to the estimation of the parameter. Knowledge of the individual sample values from which a sufficient estimator is computed provides no additional information that can be used in estimating the parameter. Mathematically, $\tilde{\theta}$ is a sufficient estimator for θ if the conditional joint density of the sample values $x_1, x_2, \ldots, x_n$ given $\tilde{\theta}$ does not depend on the value of θ; then the sample values $x_1, x_2, \ldots, x_n$ provide no information about θ not provided by $\tilde{\theta}$. In this sense a sufficient estimator is one that exhausts all the information in a sample.

EXAMPLE

Suppose two independent random samples of n and m observations, respectively, are drawn from a binomial distribution having parameter π and that there are x successes in the first sample and y successes in the second sample. Then $(x + y)/(n + m)$ is a sufficient estimator of π; equivalently, knowledge of the individual values of x and y provides no information about π in addition to that provided by the sum $x + y$. This can be shown as follows:

$$f(x, y \,|\, x + y) = \frac{f(x, y, x + y)}{g(x + y)}$$

$$= \frac{f(x, y)}{g(x + y)}$$

But

$$f(x, y) = \binom{n}{x}\pi^x(1 - \pi)^{n-x}\binom{m}{y}\pi^y(1 - \pi)^{m-y}$$

$$g(x + y) = \binom{n + m}{x + y}\pi^{x+y}(1 - \pi)^{n+m-x-y}$$

$$f(x, y \,|\, x + y) = \frac{\binom{n}{x}\binom{m}{y}\pi^{x+y}(1 - \pi)^{n-x+m-y}}{\binom{n + m}{x + y}\pi^{x+y}(1 - \pi)^{n+m-x-y}}$$

$$= \frac{\binom{n}{x}\binom{m}{y}}{\binom{n + m}{x + y}}$$

Since $f(x, y \,|\, x + y)$ does not depend on π, $x + y$ is sufficient for the estimation of π.

CONSISTENT ESTIMATORS

An estimator of a parameter is consistent if the probability that the estimator differs from the parameter by more than an arbitrary constant approaches zero as the sample size increases. Mathematically, $\tilde{\theta}$ is a consistent estimator of θ if, for all ε,

$$P\{|\theta - \tilde{\theta}_n| > \varepsilon\} \to 0 \text{ as } n \to \infty$$

where ε is an arbitrarily small constant and $\tilde{\theta}_n$ is the estimator of θ based on a sample of size n. Equivalently, $\tilde{\theta}$ is a consistent estimator of θ if, for all θ, $\lim_{n \to \infty} \tilde{\theta}_n = \theta$.

If an estimator is consistent, the probability that the estimate of a parameter differs from its true value by an arbitrarily small amount tends to zero as the number of observations used in calculating the estimate increases. Consistency is an asymptotic property; if an estimator is consistent, then a sufficiently large sample will almost certainly result in an estimate of specified accuracy. Unfortunately, consistency does not imply that an estimator is useful for samples of small size.

In the limit, a consistent estimator must be unbiased; however, for small samples, a consistent estimator may be biased. A biased estimator may or may not be consistent. It can be shown that an unbiased estimator is consistent if its variance approaches zero as the sample size increases. That is,

$$\text{If} \begin{cases} \tilde{\theta} \text{ is unbiased} \\ \qquad \text{and} \\ \text{var}(\tilde{\theta}) \to 0 \text{ as } n \to \infty \end{cases} \text{then } \tilde{\theta} \text{ is a consistent estimator of } \theta$$

EXAMPLE

As shown above, s^2 is an unbiased estimator of σ^2 and $(n-1)s^2/\sigma^2$ has a chi-square distribution with $(n-1)$ degrees of freedom. Thus the variance of $(n-1)s^2/\sigma^2$ is $2(n-1)$ and the variance of s^2 is

$$2(n-1)\left(\frac{\sigma^2}{n-1}\right)^2 = \frac{2\sigma^4}{n-1}$$

and since $\dfrac{2\sigma^4}{n-1} \to 0$ as $n \to \infty$, s^2 is a consistent estimator of σ^2.

Again, note that consistency is a large sample property and describes the behavior of an estimator as n increases indefinitely. Unbiasedness is a property of estimators that is applicable for small, as well as for large, samples. Consistency implies that errors tend to zero as sample size increases; unbiasedness implies that errors average zero (although they may be large) for any size sample.

7.3 METHODS OF OBTAINING POINT ESTIMATORS

A number of methods have been developed for obtaining point estimators that have many or all of the properties discussed above. Of these, only the method of maximum likelihood is presented here; the method of least squares is discussed later in the context of regression analysis.

METHOD OF MAXIMUM LIKELIHOOD

The *method of maximum likelihood* obtains as a parameter estimate that value for which the probability (or probability density) of obtaining the set of sample values is a maximum. R. A. Fisher, who developed maximum likelihood estimation, referred to this probability (or probability density) as the "likelihood" or "likelihood function". This function is maximized, with respect to the parameter being estimated, to obtain the maximum likelihood estimate of the parameter.

The method of maximum likelihood is widely used and has several advantages. If a sufficient estimator for a parameter exists, the maximum likelihood estimator is sufficient. Maximum likelihood estimators are asymptotically efficient. Maximum likelihood estimators have the invariance property; that is, if $\tilde{\theta}$ is a maximum likelihood estimator of θ and if $g(\theta)$ is continuous, then $g(\tilde{\theta})$ is a maximum likelihood estimator of $g(\theta)$. However, maximum likelihood estimators are not necessarily unbiased.

To obtain a maximum likelihood estimator $\tilde{\theta}$ of the parameter θ based on the observed data $x_1, x_2, \ldots, x_n$, we maximize the likelihood function

$$f(x_1, x_2, \ldots, x_n, \theta) = \sum_{i=1}^{n} f(x_i, \theta)$$

with respect to θ; the parameter θ is thus treated as a variable with respect to which the function is maximized. The form of the distribution must be known in order to write the likelihood function. Maximization is usually most easily accomplished by setting the first derivative with respect to θ of the likelihood function equal to zero and solving for θ; this solution for θ is the maximum likelihood estimator $\hat{\theta}$ of the parameter θ. The maximum likelihood estimator of θ can be defined as follows: Suppose the probability of obtaining the observed sample is obtained for every possible value of the population parameter θ. The value of θ corresponding to the maximum of these probabilities is the maximum likelihood estimate of θ.

The following results concerning maximum likelihood estimators are proved in Appendix 7.1. The maximum likelihood estimator of the parameter π of the binomial distribution is

$$\hat{\pi} = \frac{x}{n}$$

The maximum likelihood estimator of the parameter θ of the exponential distribution is

$$\hat{\theta} = \frac{\sum x_i}{n} = \bar{x}$$

The simultaneous maximum likelihood estimates for the mean and variance of the normal distribution are:

$$\hat{\mu} = \frac{\sum x_i}{n} = \bar{x}$$

$$\hat{\sigma}_2 = \frac{\sum (x_i - \bar{x})^2}{n}$$

And since maximum likelihood estimators have the invariance property, the maximum likelihood estimator of the standard deviation of a normal distribution is

$$\hat{\sigma} = \sqrt{\frac{\sum (x_i - \bar{x})^2}{n}}$$

Note that the maximum likelihood estimator for the variance of a normal distribution is biased.

7.4 INTERVAL ESTIMATION

As noted above, it is often useful to supplement a point estimate with some indication of its reliability or precision. Confidence interval estimation is one method for expressing the precision of a point estimate; this method is widely used since confidence intervals provide an easily interpretable indication of the precision of an estimate.

A confidence interval is an interval constructed from the data in such a way that it contains the true value of the parameter to be estimated with a specified probability. This probability is referred to as the *confidence coefficient.*

The interpretation of the term probability used in this context requires some explanation. Clearly, any particular interval either contains the parameter to be estimated or does not. However, suppose a set of confidence intervals for a particular parameter is constructed in such a way that all the intervals have the same confidence coefficient and one interval is constructed for each of the possible (equally likely) random samples of a specified size. Then the proportion of these intervals that contains the parameter is the confidence coefficient. It is in this sense that a confidence coefficient is a probability.

A confidence interval thus consists of a pair of values between which the parameter to be estimated lies with a specified probability or confidence. A confidence interval may or may not be symmetric about the point estimate of the parameter, depending on whether the sampling distribution of the estimator is or is not symmetric.

A confidence interval for a given parameter and specified confidence coefficient is not unique. The problem of determining which confidence interval estimator for a parameter is shortest, on the average, is closely related to the problem of finding best tests of hypotheses, discussed in Chapters 8 and 9. Confidence interval estimates for means, differences between means, proportions, and variances are presented in the following sections. The intervals discussed are the shortest possible intervals for the problems specified; proofs are not given but can be found in any mathematical statistics text.

7.5 ESTIMATION OF MEANS

If $x_1, x_2, \ldots, x_n$ represent a random sample of size n from a normal distribution having mean μ and variance σ^2, then

$$\bar{x} = \frac{\sum\limits_{i=1}^{n} x_i}{n}$$

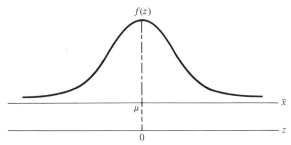

Figure 7.2 Normal distribution—including standard normal axis

is the best unbiased estimator of μ; $\bar{x}$ is consistent and sufficient and is the maximum likelihood estimator of μ.

If X is a normally distributed random variable with mean μ and standard deviation σ, then $Z = (\bar{x} - \mu)/\sigma$ is normally distributed with mean 0 and standard deviation 1. Since $\bar{x}$ is the mean of a random sample of size n from a normal distribution having mean μ and variance σ^2, its distribution is normal with mean μ and variance σ^2/n. Thus $Z = \dfrac{\bar{x} - \mu}{\sigma/\sqrt{n}}$ has the standard normal normal distribution.

The procedure of confidence interval estimation can be illustrated using a rough normal curve drawn with two horizontal scales, as in Figure 7.2. The distribution of $\bar{x}$ is normal if the population is normally distributed and is approximately normal for large samples otherwise (the central limit theorem). The Z scale is for the same distribution transformed so that the mean is zero and the standard deviation is 1. The values of $\bar{x}$ are normally distributed around μ, the mean of the population. The confidence interval for μ is around the sample statistic $\bar{x}$ and is of the form $\bar{x} - C < \mu < \bar{x} + C$.

The confidence interval for μ can be obtained as follows (see Figure 7.3):

$$P\left(z_{\alpha/2} \leq \frac{\bar{x} - \mu}{\sigma/\sqrt{n}} \leq z_{1-\alpha/2}\right) = 1 - \alpha$$

$$P\left(z_{\alpha/2}\frac{\sigma}{\sqrt{n}} \leq \bar{x} - \mu \leq z_{1-\alpha/2}\frac{\sigma}{\sqrt{n}}\right) = 1 - \alpha$$

and the $(1 - \alpha)$ percent confidence interval for μ is given by

$$\bar{x} + z_{\alpha/2}\frac{\sigma}{\sqrt{n}} \leq \mu \leq \bar{x} + z_{1-\alpha/2}\frac{\sigma}{\sqrt{n}}$$

Figure 7.3 Standard normal distribution curve

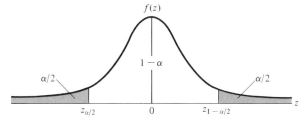

This notation indicates the proportion of the area to the left of Z by its subscript. For symmetric distributions, this notation is unnecessarily cumbersome since, for example, $z_{\alpha/2} = -z_{1-\alpha/2}$. The general notation is used for consistency with other sections.

Note that the confidence interval statement concerning μ is either correct or incorrect for any given interval; however, the interval is constructed in such a way that the probability of its being correct for any particular interval is $1 - \alpha$.

Note that the limits of the confidence interval involve $\bar{x}$ and therefore are subject to sampling variability. Thus, strictly speaking, the probability statement concerns the probability that $\bar{x}$ for a randomly selected sample satisfies both of the inequalities,

$$\bar{x} + z_{\alpha/2}\left(\frac{\sigma}{\sqrt{n}}\right) \le \mu \quad \text{and} \quad \bar{x} + z_{1-\alpha/2}\left(\frac{\sigma}{\sqrt{n}}\right) \ge \mu$$

rather than that μ satisfies them.

EXAMPLE
Suppose the average weight of a random sample of 100 female freshman students at the University of Arizona is 110 lb. If the standard deviation of the weight of freshman girls is 20 lb, obtain a 95 percent confidence interval for the average weight of freshman girls. That is, what interval can be said to contain the true mean, μ, with probability .95?

Let
$$\bar{x} = 110$$
$$\sigma = 20$$
$$n = 100$$

$$\sigma_{\bar{x}} = \frac{\sigma}{\sqrt{n}} = \frac{20}{100} = 2$$

For $\alpha = .05$,
$$z_{\alpha/2} = -z_{1-\alpha/2} = -1.96 \quad \text{(Table B.7)}$$

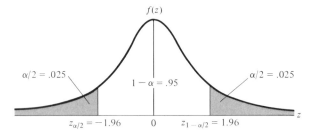

And the 95 percent confidence limits for μ are

$$\bar{x} + z_{\alpha/2}\frac{\sigma}{\sqrt{n}} \le \mu \le \bar{x} + z_{1-\alpha/2}\frac{\sigma}{\sqrt{n}}$$

$$110 - (1.96)(2) \le \mu \le 110 + (1.96)(2)$$
$$110 - 3.92 \le \mu \le 110 + 3.92$$
$$106.08 \le \mu \le 113.92$$

Thus with 95 percent confidence, the population mean μ lies between 106.08 and 113.92. The point estimate of μ is 110.

POPULATION VARIANCE, σ^2 UNKNOWN

The confidence interval estimator for μ given above assumes that the population variance σ^2 is known. If σ^2 is unknown, which is frequently the case, then the t distribution is used rather than the Z distribution. The sample standard error of the mean is $s_{\bar{x}} = s/\sqrt{n}$ and the statistic

$$t = \frac{\bar{x} - \mu}{s_{\bar{x}}}$$

is distributed as t with $n - 1$ degrees of freedom. Thus

$$P\left[t_{\alpha/2,n-1} \le \frac{\bar{x} - \mu}{s_{\bar{x}}} \le t_{1-\alpha/2,n-1}\right] = 1 - \alpha$$

and the $(1 - \alpha)$ percent confidence interval for μ is

$$\bar{x} + t_{\alpha/2,n-1}s_{\bar{x}} \le \mu \le \bar{x} + t_{1-\alpha/2,n-1}s_{\bar{x}}$$

EXAMPLE

A random sample of 25 invoices has been taken from the records of the Kersey Clothing Shop. The average amount of the 25 invoices is $78.50 and the sample standard deviation is $20.00. Obtain a 90 percent confidence interval for the average value of invoices for the shop.

Let

$$\bar{x} = 78.50$$
$$s = 20$$
$$n = 25$$

$$s_{\bar{x}} = \frac{s}{\sqrt{n}} = \frac{20}{5} = 4$$

For $\alpha = .10$,

$$t_{\alpha/2,24} = -t_{1-\alpha/2,24} = -1.711 \quad \text{(Table B.7)}$$

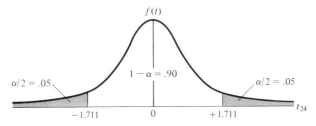

The point estimate of μ is $78.50 and the 90 percent confidence interval for μ is

$$\bar{x} + t_{\alpha/2,n-1}\frac{s}{\sqrt{n}} \le \mu \le \bar{x} + t_{1-\alpha/2,n-1}\frac{s}{\sqrt{n}}$$

$$78.50 - (1.711)(4) \le \mu \le 78.50 + (1.711)(4)$$

$$78.50 - 6.84 \le \mu \le 78.50 + 6.84$$

$$71.66 \le \mu \le 85.34$$

For practical purposes, the normal distribution can be used rather than the t distribution for $n > 30$, even when σ^2 is unknown. Thus the t distribution is used when σ^2 is not known and the sample size is less than 30. If σ^2 is known or the sample size is larger than 30, the normal distribution is used. The finite population correction $\sqrt{(N-n)/(N-1)}$ should be used when the sample size is 10 percent or more of the total population. If the correction is not used in such cases, the estimate of the standard error of the mean is too large and the confidence interval is too wide.

EXAMPLE

From a fleet of 401 similar highway trucks, a random sample of 77 have been weighed empty. The average weight of the trucks in the sample is 5000 lb and the sample standard deviation is 500 lb. Estimate the 95 percent confidence interval for the average empty weight of a truck (a) without using the finite population correction and (b) using the finite population correction.

Let

$$N = 401$$
$$n = 77$$
$$\bar{x} = 5000$$
$$s = 500$$
$$\alpha = .05$$

Since $n > 30$, the normal distribution can be used instead of the t distribution.

(a) $$s_{\bar{x}} = \frac{s}{\sqrt{n}} = \frac{500}{\sqrt{77}} = \frac{500}{8.77} = 57$$

$$\mu = \bar{x} \pm z s_{\bar{x}} = 5000 \pm (1.96)(57) = 5000 \pm 112$$
$$4888 \leq \mu \leq 5112$$

(b) $$s_{\bar{x}} = \frac{s}{\sqrt{n}} \sqrt{\frac{N-n}{N-1}} = \frac{500}{8.77} \sqrt{\frac{401-77}{400}} = \left(\frac{500}{8.77}\right)\left(\frac{18}{20}\right) = 51$$

$$\mu = \bar{x} \pm z s_{\bar{x}} = 5000 \pm (1.96)(51) = 5000 \pm 100$$
$$4900 \leq \mu \leq 5100$$

Note that the confidence interval computed without using the finite population correction is wider.

ONE SIDED ESTIMATORS OF MEANS

In some cases a one-sided rather than a two-sided confidence interval is appropriate. In the one-sided case the probability statement is of the form:

$$P(\mu \geq a) = 1 - \alpha \quad \text{or} \quad P(\mu \leq b) = 1 - \alpha$$

and the confidence coefficient is used to compute a one-sided rather than a two-sided interval estimate for μ. In the one-sided case the area corresponding to the level of significance, α, is in one tail of the distribution of sample means; in the two-sided case $\alpha/2$ of the area is in each tail of the distribution. A similar situation occurs with respect to one-sided and two-sided tests of hypotheses discussed in Chapter 8.

7.6 ESTIMATION OF DIFFERENCES BETWEEN MEANS

The methods used in the preceding section for estimating means can be extended to apply to the problem of estimating the differences between means. There are many problems in which the difference between the means of two populations is of more interest than the individual means. For example, an investigator might be interested in determining the difference between the average yields of two stocks or the difference between the average salaries of salesmen in two types of firms or the difference between the average sales of two brands of a product, and so forth.

If two populations are normally distributed, the differences between means of independent random samples from the populations are normally distributed. As the sample sizes increase, the distribution of the differences between means of independent random samples from two populations approaches the normal distribution, regardless of the form of the population distributions (central limit theorem).

If $\bar{x}_1$ is the mean of a random sample of n_1 observations from a normal distribution having mean μ_1 and variance σ_1^2 and $\bar{x}_2$ is the mean of a random sample of n_2 observations from an independent normal population having mean μ_2 and variance $\sigma_2{}^2$, then $\bar{x}_1 - \bar{x}_2$ is the best estimator of $\mu_1 - \mu_2$; it is consistent and sufficient and therefore is the maximum likelihood estimator.

The distribution of $\bar{x}_1 - \bar{x}_2$ is normal if the population distributions are normal and is approximately normal for large n_1 and n_2, regardless of the form of the population distributions, as already noted. The expected value of the sampling distribution of $\bar{x}_1 - \bar{x}_2$ is $\mu_1 - \mu_2$. The variance of the sampling distribution of $\bar{x}_1 - \bar{x}_2$ depends on whether the population variances $\sigma_1{}^2$ and $\sigma_2{}^2$ are known. The variance of $\bar{x}_1 - \bar{x}_2$ is given by

$$\sigma^2_{\bar{x}_1 - \bar{x}_2} = \sqrt{\frac{\sigma_1{}^2}{n_1} + \frac{\sigma_2{}^2}{n_2}}$$

If $\sigma_1{}^2$ and $\sigma_2{}^2$ are unknown but n_1 and n_2 are large, then the variance of $\bar{x}_1 - \bar{x}_2$ can be estimated by replacing $\sigma_1{}^2$ and $\sigma_2{}^2$ by the sample estimates $s_1{}^2$ and $s_2{}^2$.

$$\sigma^2_{\bar{x}_1 - \bar{x}_2} = \sqrt{\frac{s_1{}^2}{n_1} + \frac{s_2{}^2}{n_2}} \qquad \text{(approximately)}$$

If the population variances $\sigma_1{}^2$ and $\sigma_2{}^2$ are unknown and n_1 and n_2 are not large enough to justify approximation by the sample variances $s_1{}^2$ and $s_2{}^2$, then estimation of the variance of $\bar{x}_1 - \bar{x}_2$ requires the assumption that $\sigma_1{}^2 = \sigma_2{}^2$.* In this case the sampling distribution of $\bar{x}_1 - \bar{x}_2$ is the t distribution with $n_1 + n_2 - 2$ degrees of freedom and the variance of $\bar{x}_1 - \bar{x}_2$ is given by

$$\sigma^2_{\bar{x}_1 - \bar{x}_2} = \sqrt{\frac{(n_1 - 1)s_1{}^2 + (n_2 - 1)s_2{}^2}{n_1 + n_2 - 2}}$$

Thus if the population variances $\sigma_1{}^2$ and $\sigma_2{}^2$ are known, the $1 - \alpha$ percent confidence interval for $\mu_1 - \mu_2$ is given by

$$(\bar{x}_1 - \bar{x}_2) + z_{\alpha/2}\sqrt{\frac{\sigma_1{}^2}{n_1} + \frac{\sigma_2{}^2}{n_2}} \leq \mu_1 - \mu_2 \leq (\bar{x}_1 - \bar{x}_2) + z_{1-\alpha/2}\sqrt{\frac{\sigma_1{}^2}{n_1} + \frac{\sigma_2{}^2}{n_2}}$$

If the population variances $\sigma_1{}^2$ and $\sigma_2{}^2$ are unknown but n_1 and n_2 are large (as a rule of thumb, $n_1 > 30$ and $n_2 > 30$), then the approximate $1 - \alpha$ percent confidence interval for $\mu_1 - \mu_2$ is given by

$$(\bar{x}_1 - \bar{x}_2) + z_{\alpha/2}\sqrt{\frac{s_1{}^2}{n_1} + \frac{s_2{}^2}{n_2}} \leq \mu_1 - \mu_2 \leq (\bar{x}_1 - \bar{x}_2) + z_{1-\alpha/2}\sqrt{\frac{s_1{}^2}{n_1} + \frac{s_2{}^2}{n_2}}$$

If the population variances $\sigma_1{}^2$ and $\sigma_2{}^2$ are unknown but equal, then the $1 - \alpha$ percent confidence interval for $\mu_1 - \mu_2$ is given by

$$(\bar{x}_1 - \bar{x}_2) + t_{\alpha/2}\sqrt{\frac{(n_1 - 1)s_1{}^2 + (n_2 - 1)s_2{}^2}{n_1 + n_2 - 2}}$$

$$\leq \mu_1 - \mu_2 \leq (\bar{x}_1 - \bar{x}_2) + t_{1-\alpha/2}\sqrt{\frac{(n_1 - 1)s_1{}^2 + (n_2 - 1)s_2{}^2}{n_1 + n_2 - 2}}$$

EXAMPLE
Suppose that the average length of a sample of 25 bolts produced by machine 1 is 12.2 in. and the average length of a sample of 20 bolts produced by machine 2 is 10.5 in. If the standard deviation of the lengths of the bolts is 1.5 for machine 1 and 0.5 for machine 2, obtain a 95 percent confidence interval for the difference in the average lengths of the bolts produced by the two machines.
   Since $\sigma_1{}^2 = 2.25$ and $\sigma_2{}^2 = 0.25$ are known, the 95 percent confidence interval is given by

$$(\bar{x}_1 - \bar{x}_2) + z_{\alpha/2}\sqrt{\frac{\sigma_1{}^2}{n_1} + \frac{\sigma_2{}^2}{n_2}} \leq \mu_1 - \mu_2 \leq (\bar{x}_1 - \bar{x}_2) + z_{1-\alpha/2}\sqrt{\frac{\sigma_1{}^2}{n_1} + \frac{\sigma_2{}^2}{n_2}}$$

$$(12.2 - 10.5) - 1.96\sqrt{\frac{2.25}{25} + \frac{.25}{20}} \leq \mu_1 - \mu_2 \leq (12.2 - 10.5) + 1.96\sqrt{\frac{2.25}{25} + \frac{.25}{20}}$$

$$1.7 - 1.96\sqrt{.1025} \leq \mu_1 - \mu_2 \leq 1.7 + 1.96\sqrt{.1025}$$

$$1.07 \leq \mu_1 - \mu_2 \leq 2.33$$

and, with 95 percent confidence, the difference in the average lengths of the bolts produced by the two machines is between 1.07 in. and 2.33 in.

* Actually the assumption that the ratio of $\sigma_1{}^2$ to $\sigma_2{}^2$ is known is sufficient mathematically, but it is customary to simplify computations by considering the case $\sigma_1{}^2 = \sigma_2{}^2$.

EXAMPLE
Now suppose, in the previous example, that the standard deviations of the lengths of the bolts produced by the two machines are not known and that the standard deviations of the bolts in the random samples are $s_1 = 1.2$ and $s_2 = 0.3$, respectively. Since $n_1 = 25$ and $n_2 = 20$, the approximation of σ_1^2 and σ_2^2 by s_1^2 and s_2^2 is not justified. Suppose, that the variances are assumed to be equal; that is, $\sigma_1^2 = \sigma_2^2$. Obtain a 95 percent confidence interval for the difference in the lengths of the bolts produced by the two machines. The 95 percent confidence interval is given by

$$(\bar{x}_1 + \bar{x}_2) + t_{\alpha/2} \sqrt{\frac{(n_1 - 1)s_1^2 + (n_2 - 1)s_2^2}{n_1 + n_2 - 2}}$$

$$\leq \mu_1 - \mu_2 \leq (\bar{x}_1 - \bar{x}_2) + t_{1-\alpha/2} \sqrt{\frac{(n_1 - 1)s_1^2 + (n_2 - 1)s_2^2}{n_1 + n_2 - 2}}$$

where t has degrees of freedom $n_1 + n_2 - 2 = 43$. Note that, for 43 degrees of freedom, t_α can be approximated by z_α. Thus the 95 percent confidence interval is approximately

$$1.7 - 1.96 \sqrt{\frac{(24)(1.44) + (19)(.09)}{43}} \leq \mu_1 - \mu_2 \leq 1.7 + 1.96 \sqrt{\frac{(24)(1.44) + (19)(.09)}{43}}$$

$$1.7 - 1.96 \sqrt{.8435} \leq \mu_1 - \mu_2 \leq 1.7 + 1.96 \sqrt{.8435}$$

$$-0.10 \leq \mu_1 - \mu_2 \leq 3.50$$

Thus, with 95 percent confidence, the difference in the average lengths of the bolts produced by the two machines is between -0.10 in. and 3.50 in. This demonstrates the increase in the length of the confidence interval, that is, the loss in precision, when σ_1^2 and σ_2^2 must be estimated on the basis of relatively small samples.

7.7 ESTIMATION OF PROPORTIONS

Many problems concern estimation of the proportion of a population that possesses a certain characteristic or the proportion of events that are of a specified type. For example, a market research firm might wish to estimate the proportion of families using a certain brand of some product before recommending introduction of a similar product; or a brokerage firm might wish to estimate the proportion of the residents in a city who own stocks before considering whether to establish a branch there. Under the assumptions discussed in Chapter 6, the proportion has a binomial distribution. The sample proportion $p = x/n$ is used as an estimate of the population proportion π.

More precisely, if a random sample of n observations is obtained from a Bernoulli process having parameter π, then the sample proportion $p = x/n$ is the best estimator of π, where x is the number of observations that possess the characteristic of interest in a random sample of size n. This estimator is consistent and sufficient and is the maximum likelihood estimator.

The mean and variance of the sampling distribution of p are:

$$\text{mean} = E(p) = \pi$$

$$\text{variance} = \text{var}(p) = \frac{\pi(1 - \pi)}{n}$$

(see Chapter 5). The standard deviation of p, denoted by σ_p and referred to as the standard error of the proportion is

$$\sigma_p = \sqrt{\frac{\pi(1 - \pi)}{n}}$$

Obtaining an exact confidence interval for a proportion is somewhat complicated. For small samples, confidence intervals can be obtained from tables of the binomial distribution. However, this procedure is tedious unless n is very small. For large samples, the sampling distribution of p is approximately normal. However, the standard error of p depends on the population proportion π, and thus the standard error of p is unknown unless π is known—but π is the parameter estimated by p.

Using the normal approximation, we have

$$P\left(z_{\alpha/2} \le \frac{p - \pi}{\sqrt{\dfrac{\pi(1 - \pi)}{n}}} \le z_{1 - \alpha/2}\right) = 1 - \alpha$$

Rearrangement of the terms gives the $(1 - \alpha)$ percent confidence interval for π, as follows:*

$$p + z_{\alpha/2}\sqrt{\frac{\pi(1 - \pi)}{n}} \le \pi \le p + z_{1 - \alpha/2}\sqrt{\frac{\pi(1 - \pi)}{n}}$$

If n is very large, then this confidence interval can be approximated by replacing the parameter π by its sample estimate p to obtain

$$p + z_{\alpha/2}\sqrt{\frac{p(1 - p)}{n}} \le \pi \le p + z_{1 - \alpha/2}\sqrt{\frac{p(1 - p)}{n}}$$

Note that this problem does not occur in obtaining the confidence interval estimate of the mean μ, because the variance of the estimator $\bar{X}$ does not depend on μ.

EXAMPLE
The Gohigher Brokerage Company is considering opening a branch office in West Blatherskite, an affluent suburb of Megopolis. A random sample of 100 residents was obtained and 80 percent were found to own stocks and bonds. Find a 95 percent confidence interval for the proportion of residents in West B who own stocks and bonds.

Since $(.8)(100) = 80$, and $(.2)(100) = 20$, the normal approximation for the binomial can be used.

$$s_p = \sqrt{\frac{p(1 - p)}{n}} = \sqrt{\frac{(.8)(.2)}{100}} = \sqrt{.0016} = .04$$

---

* The confidence interval for the parameter π can be changed to the confidence interval for the parameter $\mu = np$ by multiplying each term of this equation by n

$$\bar{x} + n z_{\alpha/2}\sqrt{\frac{p(1 - p)}{n}} \le \mu \le \bar{x} + n z_{1 - \alpha/2}\sqrt{\frac{p(1 - p)}{n}}$$

and the 95 percent confidence interval for p is

$$p + z_{\alpha/2}\, s_p \leq \pi \leq p + z_{1-\alpha/2}\, s_p$$
$$.80 - (1.96)(.04) \leq \pi \leq .80 + (1.96)(.04)$$
$$.80 - .078 \leq \pi \leq .80 + .078$$
$$.722 \leq \pi \leq .878$$

The normal approximation provides a symmetric interval estimate around p. The binomial distribution is symmetric only if $\pi = .5$; thus, strictly speaking, this is the only case in which the confidence interval estimate should be symmetric. As noted in Chapter 5, for fixed π the sampling distribution of p becomes more symmetric as n increases. Thus if n is very large and π is not too small, the normal approximation provides a reasonably accurate confidence interval for π. If the normal approximation is not appropriate, an exact confidence interval for π can be obtained using the quadratic formula or confidence interval charts. Note again that an exact confidence interval for π is symmetric only if $\pi = .5$.

The quadratic formula can be used to obtain an exact confidence interval for π, as follows:

$$p + z_{\alpha/2}\sqrt{\frac{\pi(1-\pi)}{n}} \leq \pi \leq p + z_{1-\alpha/2}\sqrt{\frac{\pi(1-\pi)}{n}}$$

$$\pi - p = \pm z_{\alpha/2}\sqrt{\frac{\pi(1-\pi)}{n}}$$

$$\pi^2 - 2\pi p + p^2 = z_{\alpha/2}^2\left(\frac{\pi(1-\pi)}{n}\right)$$

$$\left(1 + \frac{z_{\alpha/2}^2}{n}\right)\pi^2 - \left(2p + \frac{z_{\alpha/2}^2}{n}\right)\pi + p^2 = 0$$

which can be solved using the quadratic formula.

EXAMPLE
For the preceding example, $p = .8$, $n = 100$, and the 95 percent confidence interval for π is

$$p + z_{\alpha/2}\sqrt{\frac{\pi(1-\pi)}{n}} \leq \pi \leq p + z_{1-\alpha/2}\sqrt{\frac{\pi(1-\pi)}{n}}$$

$$\pi^2 - 2\pi p + p^2 = z_{\alpha/2}^2\,\frac{\pi(1-\pi)}{n}$$

$$\pi^2 - 1.6\pi + .64 = 3.84\left(\frac{\pi(1-\pi)}{100}\right)$$

$$1.0384\pi^2 - 1.6384\pi + .64 = 0$$

$$649\pi^2 - 1024\pi + 400 = 0$$

$$\pi = \frac{1024 \pm \sqrt{10176}}{1298} = \frac{1024 \pm 100.88}{1298}$$

Thus the 95 percent confidence interval for π is

$$.711 \leq \pi \leq .867$$

Note that this interval is not symmetric around the point estimate $p = .8$; it is the same length, .156, as the approximate confidence interval given above. This equality does not always occur, but the lengths do not differ appreciably for large n.

A confidence interval estimate of the proportion can be obtained for any sample size using confidence interval charts (see Table B.13). For the example of the Gohigher Brokerage Company, $n = 100$, $p = .8$, and the confidence coefficient is 95 percent; enter the table with $p = .8$ and read approximately .71 and .87 from the bands corresponding to $n = 100$. Thus the 95 percent confidence limits for π read (approximately) from the chart are .71 to .87. Note that the confidence limits obtained from the chart are exact and thus are not symmetric unless $\pi = .5$.

Obtaining confidence intervals for π from charts such as Table B.13 eliminates considerable computation. Unfortunately, a chart is needed for each confidence coefficient desired, and available charts cannot be read with sufficient accuracy for many problems.

The finite population correction must be used in the calculations of estimates of π if n represents 10 percent or more of the population. The finite population correction is used in the calculation of σ_p or s_p, as follows:

$$\sigma_p = \sqrt{\frac{\pi(1 - \pi)}{n}} \sqrt{\frac{N - n}{N - 1}} \qquad s_p = \sqrt{\frac{p(1 - p)}{n}} \sqrt{\frac{N - n}{N - 1}}$$

EXAMPLE
Of a total of 626 dogs in a large kennel, a random sample of 142 were checked for the disease mal-de-pinto. Of the dogs checked, 14 had the disease. Calculate a 98 percent confidence interval for the proportion of dogs in the kennel who have mal-de-pinto.

$$N = 626$$
$$n = 142$$
$$p = .10$$
$$\alpha = .02$$

Since $np > 5$ and $n(1 - p) > 5$, the normal approximation is appropriate and

$$p + z_{\alpha/2}s_p \leq \pi \leq p + z_{1 - \alpha/2}s_p$$

Since n is approximately 10 percent of N, the finite population correction is used in calculating s_p.

$$s_p = \sqrt{\frac{p(1 - p)}{n}} \sqrt{\frac{N - n}{N - 1}} = \sqrt{\frac{(.1)(.9)}{142}} \sqrt{\frac{626 - 142}{625}} = \left(\frac{.3}{12}\right)\left(\frac{22}{25}\right)$$

$$= 0.033$$

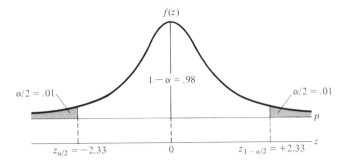

Thus the confidence interval is

$$0.10 - (2.33)(.033) \leq \pi \leq 0.10 + (2.33)(.033)$$

$$0.21 \leq \pi \leq 0.179$$

and the probability is .98 that between 2.1 percent and 17.9 percent of the dogs in the kennel have the disease.

A one-sided confidence interval may also be calculated for a proportion.

7.8 ESTIMATION OF VARIANCES

If $x_1, x_2, \ldots, x_n$ represent a random sample of size n from a normal distribution having mean μ and variance σ^2, then

$$s^2 = \frac{\sum_{i=1}^{n} (x_i - \bar{x})^2}{n - 1}$$

is the best unbiased estimator of σ^2; it is consistent and sufficient. The maximum likelihood estimator of σ^2 is

$$\hat{s}^2 = \frac{\sum_{i=1}^{n} (x_i - \bar{x})^2}{n}$$

which is biased.

Since s^2 is the variance of a random sample of size n from a normal distribution having mean μ and variance σ^2, the statistic $(n - 1)s^2/\sigma^2$ has a chi-square distribution with $n - 1$ degrees of freedom; the chi-square distribution is used to construct an interval estimate for the population variance using the sample variance, as follows:

$$P\left[\chi^2_{\alpha/2,n-1} \leq \frac{(n - 1)s^2}{\sigma^2} \leq \chi^2_{1-\alpha/2,n-1} \right] = 1 - \alpha$$

For ease of notation, denote $\chi^2_{\alpha/2,n-1}$ by $\chi_1{}^2$ and $\chi^2_{1-\alpha/2,n-1}$ by $\chi_2{}^2$. (See Figure 7.4.)

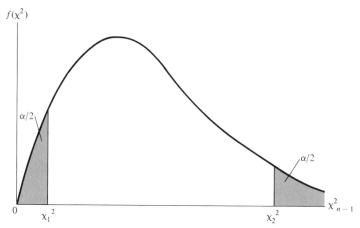

Figure 7.4 Chi-square distribution

By rearranging the terms, we can write the confidence interval for σ^2 as:

$$P\left(\frac{(n-1)s^2}{{\chi_2}^2} \leq \sigma^2 \leq \frac{(n-1)s^2}{{\chi_1}^2}\right) = 1 - \alpha$$

${\chi_1}^2$ and ${\chi_2}^2$ can be obtained for a given confidence coefficient from the tabulated values of the χ^2 distribution in Table B.10. A $(1 - \alpha)$ percent confidence interval for σ can be obtained by taking the square root of the confidence limits for σ^2; however, the result will be a biased confidence interval estimator of σ, since s is a biased estimator of σ.

EXAMPLE

The quality control engineer of the Nutty-Bolt Company is interested in estimating the variance of the lengths of the screws produced by a certain machine. He obtains a random sample of 25 of these screws and finds that the standard deviation of their lengths is 1.1. What 99 percent confidence interval estimate for the variance can the quality control engineer state?

$$P\left(\frac{(n-1)s^2}{{\chi_2}^2} \leq \sigma^2 \leq \frac{(n-1)s^2}{{\chi_1}^2}\right) = 1 - \alpha$$

$$P\left(\frac{(24)(1.21)}{45.559} \leq \sigma^2 \leq \frac{(24)(1.21)}{9.886}\right) = .99$$

$$P(.64 \leq \sigma^2 \leq 2.94) = .99$$

with 99 percent confidence, the variance is between .64 and 2.94.

7.9 ESTIMATING REQUIRED SAMPLE SIZE

In many practical problems it is useful to determine the sample size required to estimate a parameter with specified precision before collecting data — a sample that is too small provides an estimate of insufficient precision; one that is too large is unnecessarily costly. In some cases consideration of the sample size required to obtain an estimate having the minimum precision that would be useful leads to the conclusion that sampling is too costly to be worthwhile. In these cases the required decision should be based on whatever information is available without obtaining additional data by sampling. Determination of the expected worth of additional data (sample observations) is discussed in more detail in Chapter 10.

Estimation of required sample size is closely related to confidence interval estimation; it requires knowledge of the sampling distribution of the relevant statistic and specification of the error to be tolerated and the confidence required. Estimation of the required sample size in estimating means and proportions is described below.

SAMPLE SIZE FOR ESTIMATING MEANS

If the estimate of the mean of a normal population is to be in error by no more than a specified quantity ε, with confidence $(1 - \alpha)$, then

$$\varepsilon = z_{1-\alpha/2}\frac{\sigma}{\sqrt{n}}$$

and the required sample size n is

$$n = \left[\frac{z_{1-\alpha/2}\sigma}{\varepsilon}\right]^2$$

(See Figure 7.5.) This formula for estimating sample size requires knowledge of the population standard deviation. If the population mean is unknown, the population standard deviation is also usually unknown. However, it is sometimes possible to estimate the population standard deviation and to use this estimate in the formula for sample size. For example, for many populations nearly all of the observations can be assumed to be within three standard deviations of the mean. Thus if the range of the data is known, at least approximately, the standard deviation can be estimated as one-sixth of the range. Note that most of the area of a normal distribution is within three standard deviations on either side of the mean, which is also true of most other unimodal distributions. Thus, as a rule of thumb, six standard deviations are assumed to cover the range of a distribution.

Figure 7.5 Sampling distribution of $\bar{X}$

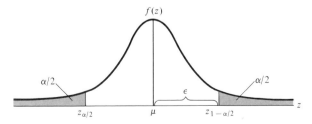

When we use this method, if the range (and thus the standard deviation) is overestimated, the required sample size is also overestimated; similarly, if the range is underestimated, the required sample size is also underestimated. If sampling is costly and/or obtaining the specified precision is important, it may be worthwhile to draw a preliminary sample for the purpose of estimating the standard deviation.

It is customary to use the normal distribution rather than the t distribution in the formula for estimating sample size, even though the standard deviation is estimated rather than known. The sample size involved is usually large enough so that the difference between the two distributions is negligible.

EXAMPLE
An investigator is interested in estimating the average age of the presidents of companies producing electronic computers. She wishes the estimate to be in error by not more than one year with confidence .95. What size sample should she obtain if (a) she is reasonably confident that the ages have a range not exceeding 15 years and (b) she is reasonably confident that the ages have a range not exceeding 20 years?

(a) Estimated $\sigma = \dfrac{15}{6} = 2.5$

$$n = \left[\frac{z_{1-\alpha/2}\sigma}{\varepsilon}\right]^2$$
$$= \left[\frac{(1.96)(2.5)}{1}\right]^2$$
$$= 24.01 \quad \text{or} \quad 24$$

(b) Estimated $\sigma = \dfrac{20}{6} = \dfrac{10}{3}$

$$n = \left[\frac{(1.96)(10/3)}{1}\right]^2$$
$$= 42.64 \quad \text{or} \quad 43$$

SAMPLE SIZE FOR ESTIMATING PROPORTIONS

If the estimate of the proportion for a Bernoulli process is to be in error by no more than a specified quantity ε with confidence $(1 - \alpha)$, then (using the large-sample approximation)

$$\varepsilon = z_{1-\alpha/2}\sqrt{\frac{\pi(1 - \pi)}{n}}$$

and the required sample size n is

$$n = \pi(1 - \pi)\left[\frac{z_{1-\alpha/2}}{\varepsilon}\right]^2$$

This formula for estimating sample size requires knowledge of the population proportion π, which is in fact being estimated. A conservative estimate of the sample size can be obtained by taking $\pi = .5$, since $\pi(1 - \pi)$ is maximized for $0 \leq \pi \leq 1$ if $\pi = .5$. Frequently, a more realistic estimate of required sample size can be obtained by taking π as close to .5 as the investigator thinks is likely to be the case. For example, many events whose probability of occurrence is estimated are known to be relatively rare—perhaps occurring only in a fraction of a percent of the cases observed. Estimating $\pi = .5$ would lead to an unnecessarily large sample in these cases while estimating $\pi = .01$ might actually lead to a conservative estimate of the required sample size.

EXAMPLE
A candidate for national office is interested in the percentage of voters who favor some proposed legislation. His aides questioned a random sample of 2500 voters and found that 35 percent favor the legislation. (a) With what confidence (estimated conservatively) does the candidate know that this percentage is within 2 percent of the true percentage of voters who would say they favor the legislation? (b) What size sample would be needed to know with this same probability that the estimated percentage is within 1 percent of the true percentage?

(a)
$$\varepsilon = z_{1-\alpha/2} \sqrt{\frac{\pi(1-\pi)}{n}}$$

$$.02 = z_{1-\alpha/2} \sqrt{\frac{(.5)(.5)}{2500}}$$

$$z = 2$$

$$P(|Z| \leq 2) = 2(.4772) = .9544$$

(b)
$$n = \pi(1-\pi)\left[\frac{z_{1-\alpha/2}}{\varepsilon}\right]^2$$

$$= (.5)(.5)\left[\frac{2}{.01}\right]^2$$

$$= 10,000$$

SUMMARY TABLE
The estimators discussed in this chapter are summarized in Table 7.1, pp. 202–203.

APPENDIX 7.1
MAXIMUM LIKELIHOOD ESTIMATORS

The procedure for obtaining maximum likelihood estimators is as follows:

1. Obtain the distribution function or likelihood function of the sample, $f(x_1, x_2, \ldots, x_n; \theta)$.
2. Obtain $L = \ln f(x_1, x_2, \ldots, x_n; \theta)$.
3. Determine the value of θ that maximizes L by solving the equation $\partial L/\partial \theta = 0$

This value of θ also maximizes the likelihood. (Maximizing the logarithm of a function also maximizes the function and frequently is less difficult.)

TABLE 7.1 SUMMARY OF ESTIMATORS

Parameter	Point Estimate	Confidence Interval	Estimated Sample Size
μ	$\bar{x} = \dfrac{\sum x_i}{n}$	σ^2 known: $$\bar{x} + z_{\alpha/2}\frac{\sigma}{\sqrt{n}} \leq \mu \leq \bar{x} + z_{1-\alpha/2}\frac{\sigma}{\sqrt{n}}$$ σ^2 unknown: $$\bar{x} + t_{\alpha/2}\frac{s}{\sqrt{n}} \leq \mu \leq \bar{x} + t_{1-\alpha/2}\frac{s}{\sqrt{n}}$$	$n = \left[\dfrac{z_{1-\alpha/2}\,\sigma}{\varepsilon}\right]^2$
$\mu_1 - \mu_2$	$\bar{x}_1 - \bar{x}_2$	σ_1^2 and σ_2^2 known: $$(\bar{x} - \bar{x}_2) + z_{\alpha/2}\sqrt{\frac{\sigma_1^2}{n_1} + \frac{\sigma_2^2}{n_2}} \leq \mu_1 - \mu_2$$ $$\leq (\bar{x}_1 - \bar{x}_2) + z_{1-\alpha/2}\sqrt{\frac{\sigma_1^2}{n_1} + \frac{\sigma_2^2}{n_2}}$$ σ_1^2 and σ_2^2 unknown (large sample): $$(\bar{x}_1 - \bar{x}_2) + z_{\alpha/2}\sqrt{\frac{s_1^2}{n_1} + \frac{s_2^2}{n_2}} \leq \mu_1 - \mu_2$$ $$\leq (\bar{x}_1 - \bar{x}_2) + z_{1-\alpha/2}\sqrt{\frac{s_1^2}{n_1} + \frac{s_2^2}{n_2}}$$	

$\sigma_1^{\,2}$ and $\sigma_2^{\,2}$ unknown but equal:

$$(\bar{x}_1 - \bar{x}_2) + t_{\alpha/2}\sqrt{\frac{(n_1-1)s_1^{\,2} + (n_2-1)s_2^{\,2}}{n_1+n_2-2}} \leq \mu_1 - \mu_2$$

$$\leq (\bar{x}_1 - \bar{x}_2) + t_{1-\alpha/2}\sqrt{\frac{(n_1-1)s_1^{\,2} + (n_2-1)s_2^{\,2}}{n_1+n_2-2}}$$

π

$$p = \frac{x}{n}$$

Small sample:

$$p + z_{\alpha/2}\sqrt{\frac{\pi(1-\pi)}{n}} \leq \pi \leq p + z_{1-\alpha/2}\sqrt{\frac{\pi(1-\pi)}{n}}$$

Large sample:

$$p + z_{\alpha/2}\sqrt{\frac{p(1-p)}{n}} \leq \pi \leq p + z_{1-\alpha/2}\sqrt{\frac{p(1-p)}{n}}$$

$$n = \pi(1-\pi)\left[\frac{z_{1-\alpha/2}}{\varepsilon}\right]^2$$

σ^2

$$s^2 = \frac{\sum(x_i - \bar{x})^2}{n-1} \qquad \frac{(n-1)s^2}{\chi^2_{1-\alpha/2}} \leq \sigma^2 \leq \frac{(n-1)s^2}{\chi^2_{\alpha/2}}$$

For the binomial distribution

$$f(x, \pi) = \binom{n}{x} \pi^x (1 - \pi)^{n-x}$$

$$L = \ln\binom{n}{x} + x \ln \pi + (n - x) \ln(1 - \pi)$$

$$\frac{\partial L}{\partial \pi} = \frac{x}{\pi} - \frac{n - x}{1 - \pi} = 0$$

$$\hat{\pi} = \frac{x}{n}$$

and the maximum likelihood estimator of the parameter θ is the sample proportion $p = x/n$.

For the exponential distribution

$$f(x_1, \ldots, x_n; \theta) = \left(\frac{1}{\theta}\right)^n e^{-(1/\theta)\left(\sum\limits_{i=1}^{n} x_i\right)}$$

$$L = -n \ln \theta - \frac{1}{\theta}\left(\sum_{i=1}^{n} x_i\right)$$

$$\frac{\partial L}{\partial \theta} = -\frac{n}{\theta} + \frac{1}{\theta^2}\left(\sum_{i=1}^{n} x_i\right)$$

$$\hat{\theta} = \frac{\sum\limits_{i=1}^{n} x_i}{n}$$

and the maximum likelihood estimator of the parameter θ is the sample mean

$$\bar{x} = \frac{\sum\limits_{i=1}^{n} x_i}{n}$$

For the normal distribution

$$f(x_1, \ldots, x_n; \mu, \sigma) = \left(\frac{1}{2\pi\sigma^2}\right)^{n/2} e^{-(1/2\sigma^2)\sum\limits_{i=1}^{n} (x_i - \mu)^2}$$

$$L = \frac{n}{2} \ln 2\pi\sigma^2 - \frac{1}{2\sigma^2} \sum_{i=1}^{n} (x_i - \mu)^2$$

$$\frac{\partial L}{\partial \mu} = \frac{1}{\sigma^2} \sum_{i=1}^{n} (x_i - \mu)$$

$$\frac{\partial L}{\partial \sigma^2} = -\frac{n}{2\sigma^2} + \frac{1}{2\sigma^4} \sum_{i=1}^{n} (x_i - \mu)^2$$

The maximum likelihood estimator of μ is $\bar{x} = \sum_{i=1}^{n} x_i/n$ and, substituting $\hat{\mu} = \bar{x}$, the maximum likelihood estimator of σ^2 is

$$\hat{\sigma}^2 = \frac{\sum\limits_{i=1}^{n} (x_i - \bar{x})^2}{n}$$

PROBLEMS

7.1 For a binomial distribution, show that
   (a) $p = x/n$ is an unbiased estimator for the population proportion π and its variance is $\sqrt{\pi(1 - \pi)/n}$
   (b) p is a consistent estimator for π.

7.2 If $\tilde{\theta}$ is an estimator for θ, is minimizing $E|\tilde{\theta} - \theta|$ equivalent to minimizing $|E(\tilde{\theta}) - \theta|$?

7.3 Show that $s^2 = \sum (x_i - \bar{x})^2/(n - 1)$ is an unbiased estimator for σ^2.

7.4 If X_1, X_2, and X_3 are independent random variables having the Poisson distribution with parameter λ, show that the following estimators of λ are unbiased and compare their relative efficiencies:

$$\tilde{\lambda}_1 = \frac{X_1 + X_2 + X_3}{3}$$

$$\tilde{\lambda}_2 = \frac{X_1 + 2X_2 + 3X_3}{6}$$

$$\tilde{\lambda}_3 = \frac{X_1 + 2X_2 + X_3}{4}$$

7.5 For a random sample of 100 students at ABC University, the average monthly allowance is $25.60 with a standard deviation of $3.00.
   (a) Obtain a 95 percent confidence interval for the average monthly allowances of students at ABC University.
   (b) Obtain a 95 percent confidence interval for the standard deviation of the monthly allowances of students at ABC University.

7.6 A random sample of 15 panels has a mean weight of 34.86 lb and a standard deviation of 4.23 lb. Assuming a normal population distribution, find a 99 percent confidence interval for the population mean.

7.7 In a random sample of 25 radio tubes of a particular type, the variance of the number of hours of use before blowing out is 10 hours. Obtain 90 percent confidence limits for σ^2 if the population distribution is normal.

7.8 A new candy vending machine installed in an airport failed 16 times in the first 400 times it was used. Assuming that each use is an independent trial, obtain a .95 confidence interval for the true proportion of failures.

7.9 A large laundry records the number of shirts processed and the number rejected on final inspection due to faulty work. On the basis of a random sample of 300 shirts of which 15 were rejected, the laundry estimated the proportion rejected as .05. What can the management say with 98 percent confidence about the size of the error of this estimate?

7.10 A random sample of 9 sales reports from the files of a very large company has an average of 520.52 words and a standard deviation of 15.3 words. Obtain a 90 percent confidence interval for the average number of words in the company's sales reports.

7.11 An auditor wishes to estimate the proportion of freight bills in a very large file that are in error. If he is to be 95 percent confident that his estimate is not in error by more than 1 percent, how large a sample must he take if
   (a) He has no idea of the true proportion of bills that are in error?
   (b) He can assume from past experience that the true proportion is near .04?

7.12 In a survey to determine attitude toward various design and styling factors, an automobile manufacturer obtained a random sample of 500 owners of its station wagons. If 450 of these owners would prefer to have the third seat in a station wagon facing front, state a .95 confidence interval for the true proportion of owners who have this preference.

7.13 A manufacturing company wishes to determine the proportion of the many items it buys that are discounted from list prices. If the company wants its estimate of this proportion to be in error by not more than 5 percent with confidence .99, how large a sample is required?

7.14 In a random sample of 9 breech blocks for antiaircraft guns, a particular dimension has a mean of .6495 in. and a standard deviation of .0006 in. What can be stated with 99 percent confidence about the error in this estimate of the mean?

7.15 A random sample of 81 orders for a particular type of item has an average total invoice amount of $25.62 with a standard deviation of $2.05. With what confidence can it be stated that the average of the very large number of all such orders does not differ from the sample average by more than $.50?

7.16 A company wishes to estimate with confidence .95 the average impact strength of a particular type of wooden panel with an error not to exceed 25 lb/in.^{2} It is known from other data that the standard deviation is approximately 100 lb/in.^{2} What size sample is needed?

7.17 The quality control engineer of the Explosive Chemical Company obtained the weights of a random sample of 50 packages from a large shipment of dynamite. The standard deviation of the weights of these packages was 20.6 g. Obtain a 99 percent confidence interval for the true standard deviation of the weights of the dynamite packages.

7.18 An insurance company, considering a "safe-driver" plan, wants to estimate the proportion of its large number of policyholders who would qualify for the plan. How large a random sample must be taken from the company's files in order to assert with probability .95 that the difference between the sample proportion and the true proportion does not exceed .02?

7.19 A large university is considering changing to the trimester plan and wants to determine student attitude toward such a change. The college wishes to assert with probability .95 that the estimate of the percent of students favoring trimesters is in error by not more than 4 percent. What size random sample is necessary if
(a) No information about the possible value of the proportion is available?
(b) Experience at other universities indicates that the proportion is about .7?

7.20 A manufacturer wishes to estimate the average lifetime of a particular type of electronic tube. A random sample of 50 tubes is tested and the average lifetime is 17.8 hr; the standard deviation is 0.6 hr. If the investigator is willing to use the sample standard deviation as an adequate estimate of the population standard deviation, what can he state about the size of the error with confidence .98?

7.21 Before bidding on a contract involving the installation of a large computer system, a company wants to be 98 percent confident that its estimate of the time required to install a particular type of component is not in error by more than 5 minutes. The company is reasonably sure that the range of time required does not exceed half an hour. How many (experimental) installations should they time? (Assume the installation times are normally distributed.)

7.22 Under standard test conditions, 16 models of an experimental engine used an average of 2.9 gal of gasoline; the standard deviation was .3 gal. Assuming a normal population distribution, what can be stated with 95 percent confidence about the error involved in estimating the actual average number of gallons of gasoline used by the sample average?

7.23 Suppose that the progeny of certain matings of horses are of the two blood types A and B and that genetic theory predicts that $\frac{5}{8}$ of the progeny have blood type A. Records are kept and, in fact, $\frac{3}{4}$ of the progeny are found to be of blood type A. The geneticist says that she is 95 percent confident that the theory is in error (or does not apply to this case). What is the minimum sample size that would justify this conclusion?

7.24 Practical Polls, Inc., has stated that 64 percent of the American voters favor candidate A for president. Susan Statistician takes a random sample of voters and, of these voters, 56 percent favor candidate A for president. On the basis of his sample, Susan Statistician states that she is 98 percent confident that the Practical Polls estimate is in error.

(a) What is the minimum sample size that would justify this conclusion?

(b) Obtain a 98 percent confidence interval assuming this sample size and based on Susan Statistician's estimate.

7.25 The quality control engineer of the Strong Cable Company wants to estimate the average tensile strength of heavy duty cable manufactured in 15-foot lengths. On the basis of considerations of the possible range of tensile strength, assumed to be normally distributed, he is willing to assume that the standard deviation does not exceed 20.5 lb.

(a) If he takes a random sample of 49 cables, within what error can the quality control engineer be "95 percent certain" that he has estimated the average tensile strength?

(b) If the quality control engineer wants to be "99 percent certain" that his error does not exceed the amount determined in (a), what is the minimum sample size required?

7.26 A random sample of 9 observations has a mean of 169.78 and a standard deviation of 23.4. Assuming a normal population, find 98 percent confidence limits for the population mean.

7.27 The manager of Petrol Products, Inc., wants to estimate the mean and variance of the weights of drums of oil sent from the field station to the refining plant. He obtains a random sample of 25 drums of oil; the average weight is 52.71 lb and the standard deviation is 3.5 lb. Assuming that the weights of the drums of oil are normally distributed, determine 98 percent confidence limits for their mean and variance.

7.28 Assume that $x_1, x_2, x_3,$ and x_4 are independent random observations from a normal distribution having mean μ and variance σ^2.

(a) Determine which of the following estimators of μ are unbiased.

(b) Compare the relative efficiencies of the unbiased estimators:

$$\hat{\mu}_1 = \frac{1}{4}(x_1 + x_2 + x_3 + x_4)$$

$$\hat{\mu}_2 = \frac{1}{4}(x_1 + x_2) + \frac{1}{6}(2x_3 + x_4)$$

$$\hat{\mu}_3 = \frac{x_1}{2} + \frac{x_2}{3} + \frac{x_3}{4} + \frac{x_4}{5}$$

$$\hat{\mu}_4 = \frac{1}{8}(x_1 + x_2) + \frac{1}{4}(x_3 + x_4)$$

$$\hat{\mu}_5 = \frac{x_1}{2} + \frac{x_2}{3} + \frac{x_3}{5} + \frac{x_4}{6}$$

$$\mu_6 = \frac{1}{3}(x_1 + x_2) + \frac{1}{6}(x_3 + x_4)$$

7.29 The quality control engineer of the Nutty-Bolt Company is interested in estimating the variance of the lengths of the screws produced by a certain machine. He obtains a random sample of 25 of these screws; the standard deviation of their lengths is 1.1. What 99 percent confidence interval estimate for the variance of the lengths can the quality control engineer state?

7.30 A candidate for national office is interested in the percentage of voters who favor some proposed legislation. His aides questioned a random sample of 2500 voters and found that 35 percent favor the legislation.

(a) With what probability (estimated conservatively) does the candidate know that this percentage is within 2 percent of the true percentage of voters who would say they favor the legislation?

(b) What size sample would be needed to know with this same probability that the estimated percentage is within 1 percent of the true percentage?

TESTING HYPOTHESES

8.1 INTRODUCTION

Statistical hypothesis testing involves hypotheses that can be examined by statistical analysis; such hypotheses involve parameters of theoretical frequency distributions. This excludes hypotheses such as "the moon is made of green cheese" or "the next world war will kill everyone on the earth." Hence, implicitly it should be noted that throughout this chapter the discussion concerns statistical hypotheses, although the qualifying adjective "statistical" will be dropped hereafter.

First of all, precisely what is a hypothesis? From an English grammar standpoint, a *hypothesis* is a declarative statement. This is important to recognize. Many students get the notion that a hypothesis is an interrogative statement (a question) or an imperative statement (do something). It is neither of these. Another type of statement that students sometimes think is a hypothesis takes the form of a *conditional statement*. Since a conditional statement occurs repeatedly in hypothesis testing, take special note of what it is. A conditional statement is of the *if . . . then* type. That is, *if* something, *then* such and such. Decision rules are conditional statements. The term decision rule is a fancy name for a simple mechanism related to a conditional statement. "If it is cloudy in the morning. I will carry my umbrella to work." The preceding sentence is a decision rule; it tells what to do under certain observable conditions. It is also a conditional statement because it says *if* cloudy, *then* I will carry an umbrella. A hypothesis is not a conditional statement and should not be stated as such. A hypothesis should not be stated as a question nor as an order—it is a simple declarative statement. A hypothesis is often said to be a claim or statement of belief; this is all right because even if it is not *your* belief, it could be some other (hypothetical) person's belief.

The following are examples of statistical hypotheses: (1) The mean weight of freshmen entering the University of Arizona in 1980 is 150 lb. (2) The mean weight of freshmen entering the University of Arizona in 1980 is not 150 lb. (3) The movement of stock market prices approximates a random walk. (4) Forty percent of housewives in Kersey, Pennsylvania, drink Maxwell House Coffee.

(5) The variance in length of 4-in. anchor bolts produced by Ryerson Steel is less than one-sixteenth of an inch.

One of the most difficult conceptual problems in hypothesis testing is stating the appropriate hypothesis to test. There is no simple answer that can be given to this question because "it depends on the situation and what you're after." This is something for which the student must develop an intuitive feel and is absolutely necessary in order to understand hypothesis testing—also, much more important than performing the necessary calculations. The calculations can be easily performed by anyone, but specifying correctly the hypothesis to be tested requires a clear understanding of the problem. This "how to set up the hypothesis" will be referred to throughout the discussion of hypothesis testing, but two obvious points will be made here:

1. The hypothesis to be tested should be set up in such a way that *if* it would be very costly to make the wrong decision when testing the hypothesis, *then* the probability of making the wrong decision is low.
2. The hypothesis to be tested should be set up in such a way that *if* sample results indicate a change in established procedure or accepted opinion, *then* the probability that the sample results lead to the wrong decision is low.

Notice in the above two criteria the appearance of conditional statements and probability. These recur over and over in hypothesis testing. One of the conceptual difficulties with hypothesis testing is the necessity of concentration on making mistakes. Obviously, no researcher wishes to make an incorrect decision. However, by the very nature of the process of observing a sample and making inferences about a population, this possibility exists. Hence, when a conclusion is drawn about a hypothesis, one must make a probability statement. Conditional statements are used because one can never know* whether a hypothesis is true or false. Always, one is thinking *if . . . then*, and making probability statements.

8.2 THE NULL HYPOTHESIS AND ERROR TYPES

In estimation, the problem is that of estimating the value of an unknown population parameter and obtaining confidence limits for the estimate. In testing hypothesis, the problem is that of deciding whether to accept or reject a stated hypothesis concerning a population parameter. For example, the problem might be to decide whether the average diameter of the bolts produced by a machine is $\frac{1}{2}$ in. (as specified) or whether a new orange sorting machine is faster than human sorters or whether a manufacturer's claim of an average lifetime of 40 hr for his electric light bulbs is justified or whether two methods of advertising are equally effective or so forth. In each of these cases the problem must be formulated by stating a hypothesis to be tested in the form of a simple declarative statement.

---

* In this sense "never know" means in terms of the statistical analysis actually being performed. If one did 100 percent sampling one could, of course, find out a true population parameter. Or if the hypothesis is stated in terms of the future, if one can wait for the future, one can then see if the hypothesis was correct or not.

The hypothesis to be tested is called the *null hypothesis*. In order to construct a criterion for testing a given hypothesis, we must also specify an *alternative hypothesis*—which is the hypothesis accepted if the null hypothesis is rejected.* For example, in the problem concerning bolts, the null hypothesis, H_0, and the alternative hypothesis, H_A, might be

$$H_0: \mu = .5$$

$$H_A: \mu \neq .5$$

For the problem concerning sorting oranges, the null and alternative hypotheses might be

$$H_0: \mu_1 = \mu_2$$

$$H_A: \mu_1 > \mu_2$$

where μ_1 represents the machine's average speed and μ_2 represents the average speed of human sorters. For the problem concerning light bulbs, the null and alternative hypotheses might be

$$H_0: \mu = 40$$

$$H_A: \mu < 40$$

For the problem concerning methods of advertising, the null and alternative hypotheses might be

$$H_0: \mu_1 = \mu_2$$

$$H_A: \mu_1 \neq \mu_2 \ (\mu_1 > \mu_2 \ \text{or} \ \mu_1 < \mu_2)$$

A *simple hypothesis* specifies precisely the value of the parameter in question. A *composite hypothesis* specifies a set of values rather than an exact value for the parameter in question. Note that the null hypotheses above are simple and the alternative hypotheses are composite. There is considerable advantage in formulating (or reformulating) a problem so that the null hypothesis is simple, as discussed below. Usually the logic of the situation requires that the alternative hypothesis be composite.

In order to formulate a problem so that the null hypothesis is simple, frequently we need to hypothesize the exact opposite of the statement it is desired to support. Thus if an investigator wishes to demonstrate a difference between groups, processes, methods, effects, and so forth, he or she hypothesizes no difference (and hopes to reject the hypothesis). For this reason the hypothesis is said to be "null"; however, in current usage the term "null hypothesis" refers more generally to any hypothesis whose rejection is considered to be a possible type I error, as now discussed.

---

* There is a third possible decision—to reserve judgement until more data are obtained. Such possibilities are considered in sequential decision theory but are not discussed here.

TYPE I AND TYPE II ERRORS

Whenever a null hypothesis is accepted or rejected on the basis of a sample, there are four possibilities:

1. Accept the hypothesis when it is true
2. Reject the hypothesis when it is true
3. Accept the hypothesis when it is false
4. Reject the hypothesis when it is false

Obviously, (2) and (3) represent wrong decisions. Of course, we always wish to make the correct decision; however, since it is not known which is the correct decision, the potential errors and the probabilities of making them must be considered. It is customary to refer to the first of these errors (rejecting the null hypothesis when it is true) as the *type I error, rejection error*, or α *error*. It is also customary to refer to the second of these errors (accepting the null hypothesis when it is false) as the *type II error, acceptance error*, or β *error*.

A two by two table can be set up to illustrate the possibilities in hypothesis testing.

	H_0 *true*	H_0 *false*
Accept H_0	Correct decision	Type II error (β error)
Reject H_0	Type I error (α error)	Correct decision

NOTE:
Recall (footnote, p. 211) that we assume in this and later discussions that rejecting H_0 and accepting H_A are equivalent. Theoretically, there is another alternative—that of reserving judgment; in most practical applications this possibility is not feasible, unless some method of sequential testing is employed.

Controlling, insofar as possible, the probabilities of these two types of errors is an important aspect of the problem of testing hypotheses. The probability of a type I error is denoted by α; the probability of a type II error is denoted by β. As illustrated below, with a given sample size it is possible to decrease the probability of the other type of error.

Stating the null hypothesis as a simple hypothesis permits an exact specification of α, the probability of committing a type II error. That is, the null hypothesis is set up in such a way that control is maintained over the maximum value that α can assume. For a given n, specifying this maximum probability of the type I error reduces control over the probability of the type II error (exactly why is shown later). Because of this, an important axiom of hypothesis testing states that of the two errors, it is more desirable to avoid the type I error. Hence the null hypothesis should be set up in such a way as to make its rejection when true the more serious of the two potential errors. The approach, then, to specifying α is to keep it relatively low, for example, 0.01, 0.05, 0.10. The actual value of α is specified by researchers; it represents the chance they are willing to take that they will reject the null hypothesis when it is true. Thus the probability of *accepting** the hypothesis when it is true (a correct decision) is $1 - \alpha$. From the above a semantic rule can be stated with reference to setting up the null hypothesis:

> Set up the null hypothesis so that *if* it is true, *then* the probability of accepting and rejecting it are known, and rejection represents the most undesirable consequences.

Notice the *if . . . then* statement in the semantic rule.

THE CRITICAL REGION

Choosing a criterion for testing the null hypothesis, H_0, against an alternative hypothesis, H_A, consists of partitioning the sample space into two sets: a region of acceptance (of H_0) and a region of rejection (of H_0). The rejection region is also called the *critical region*, and its boundary values are called the *critical values*. The probability that a sample point lies in the critical region *if the null hypothesis is true* is the size of the critical region and is equal to α, the probability of committing a type I error. α is also known as the *level of significance* of the test. The value of α as the level of significance in hypothesis testing is analogous to the α level of significance in estimation (Chapter 7). Hypothesis testing and interval estimation are closely related. In this paragraph when referring to the size of the critical region as α, notice the emphasis on *if the null hypothesis is true*. If the null hypothesis is not true, the size of the critical region is unknown; this follows from the previous discussion of the method of stating the null hypothesis. Although an obvious point, reflect for a moment that (1) if the null hypothesis is true, only an α error can be made and (2) if the null hypothesis is false, only a β error can be made. α and β, of course, represent probabilities. Again, notice the conditional and probability statements. Since we do not know whether the null hypothesis is in fact true or false, its acceptance or rejection is implicitly accompanied by probability statements concerning the possible errors.

8.3 ONE-SIDED AND TWO-SIDED TESTS

A test or critical region is said to be *one-sided* (one-tailed) or *two-sided* (two-tailed) according to whether the null hypothesis is rejected for values of the relevant statistic falling into one or both tails of its sampling distribution, respectively. The type of test that is appropriate depends entirely on the nature of the problem.

Suppose the problem is to test the null hypothesis $\theta = \theta_0$. The alternative hypothesis $\theta > \theta_0$ is one-sided, since it specifies that θ is on one side (to the right) of θ_0. Likewise, the alternative $\theta < \theta_0$ is one-sided. However, $\theta \neq \theta_0$ is a two-sided alternative, since it includes values of θ that lie on both sides of θ_0.

---

* The student should note that throughout this chapter the terminology is used of "accepting" a hypothesis. From a decision standpoint this means that when it is accepted, then we will act as if it were true. We can never know if it is true unless we go to 100 percent sampling or, in the case of a future condition, wait for time to pass. Logically speaking, we should say that the evidence does not refute the hypothesis. One can never *know* based on sample evidence. Accepting a hypothesis, then, does not mean that it has been verified, but that it has not been refuted, and we would behave as if it were true.

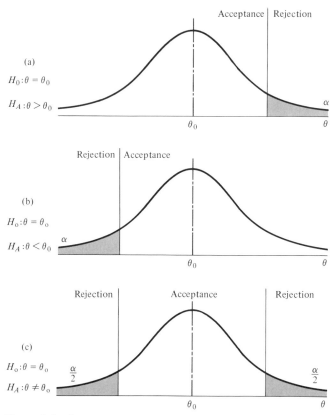

Figure 8.1   Acceptance and rejection regions

Although there are exceptions (discussed later), one-sided tests are usually appropriate for two-sided alternatives. Which tail of a distribution is appropriate for a one-sided test depends on the nature of the alternative hypothesis; the choice of this region should be clear if we keep in mind that rejection of the null hypothesis implies, in this context, acceptance of the alternative hypothesis.

In Figure 8.1 the representation of the sampling distributions and labeling of the rejection regions as α implies that the null hypothesis is true. If the null hypothesis is not true, then there is no α error. However, there is still the possibility of rejecting the null hypothesis, which now represents a correct conclusion; and there is still a possibility of accepting the null hypothesis, which now represents a β error.

The probability of a type II error, β, cannot be specifically designated because, for a given α, β depends on both n and H_A:

1. When n is given, β varies according to the value of H_A
2. When H_A is specified as a particular value (a simple noncomposite hypothesis), β depends on the size of n.

In other words, β has a range of values for a composite alternative hypothesis. In any given situation with the null hypothesis stated, the α specified, and the sample size given, if the null hypothesis is false the probability of accepting it depends on how false it is. Intuitively, if the null hypothesis of a parameter value has been falsely stated as only very slightly different from its correct value, then the probability of accepting it would (and should) be high. On the other hand, if the null hypothesis concerning a parameter value has been falsely stated as greatly different from its correct value, then the probability of accepting it would (and should) be low. To illustrate the range of β's, consider the following problem involving a one-sided test.

In recognition of the growing affluence of households, a large mail-order house is considering the introduction of a new and more expensive line of clothes. Based on research information known to be reliable, the company has con-cluded that the proposed new line will sell successfully if the mean annual income of customer households is $9000 or higher; the new line would not sell successfully if the mean annual household income of customers is below $9000. There are two approaches to the problem, and the choice of approach depends on which potential error is more important to avoid.

(a) In one approach the company is anxious to avoid not introducing the new line when it should be introduced. In this case, the null hypothesis (with some low α risk) would be, $H_0: \mu \geq \$9000$; the average annual household income is $9000 or more.

(b) In the other approach the company is anxious to avoid introducing the new line when it should not be introduced. In this case the null hypothesis (with a correspondingly low α risk) would be, $H_0: \mu < \$9000$; the average annual household income is less then $9000.

Which null hypothesis is appropriate depends, essentially, on the consequences of failure of the new line to sell well. If failure would have serious consequences for the company, such as danger of bankruptcy, then approach (b) should be taken; that is, there should be a low probability of introducing the line when the average annual income is less than $9000. If failure would not be serious and the company was most anxious to introduce the line, if it would prove successful, then approach (a) should be taken; that is, there should be a low probability of not starting the campaign when the average annual income is $9000 or more.

Suppose the situation is such that it is desirable to take approach (a), then

$$H_0: \mu \geq \$9000$$

$$H_A: \mu < \$9000$$

As previously discussed, we can make the null hypothesis a simple one, writing:

$$H_0: \mu = \$9000$$

$$H_A: \mu < \$9000$$

so long as we state the alternative in the proper way. This problem involves a one-sided test because the company would be very pleased if μ were much higher than $9000; it is primarily concerned that μ is at least $9000.

If approach (a) is taken, then the two types of errors are as follows:

Type I error: Rejecting the null hypothesis when the mean is $9000 or more—which would result in the company not introducing the new line when it should.

Type II error: Accepting the null hypothesis when the mean is less than $9000—which would result in the company introducing the new line when it should not.

Suppose the α level is set at .05 or 5 percent, the sample size is $n = 100$, and the known standard deviation of annual incomes is $\sigma = \$500$. The question is "what about the β error?" β is the probability of type II error, the probability of accepting a false null hypothesis. The value of β depends on how false the null hypothesis is, if in fact it is false.

The mean of the sample, $\bar{X}$, is normally distributed, and if $\mu = \$9000$, the distribution of $\bar{X}$ can be represented as follows:

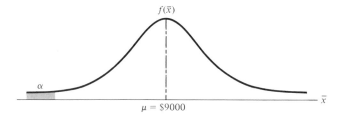

If the mean of the sample falls in the clear area—the region of acceptance—H_0 is accepted. If $\bar{X}$ falls in the shaded area, H_0 is rejected, even though $\mu = \$9000$; this, of course, would be type I error.

If $\mu = \$10,000$, the distribution of $\bar{X}$ can be represented as follows:

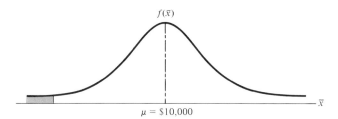

In this case the rejection region is even smaller than when $\mu = \$9000$. Note that $\alpha = 5$ percent exactly when $\mu = \$9000$, but if $\mu = \$10,000$, α is really less than 5 percent. As noted before, α is the maximum value that the probability of type I error can assume.

The rejection region for $\mu = \$8000$ is relatively large, but it does not now constitute α error, because rejection of the null hypothesis is now the correct decision. There is, however, the possibility of type II error, accepting $\mu = \$9000$ when μ is really $8000. This would occur if an $\bar{X}$ sufficiently larger than $8000 was observed. The shaded area is still the region of rejection, and the clear area the region of acceptance. The clear area is now β, the probability of accepting H_0 when it is false. If $\mu = \$7500$, the distribution moves even farther to the

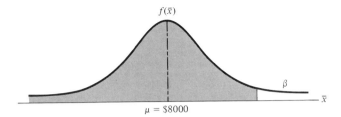

left, and β becomes smaller than for $\mu = \$8000$. A series of distributions for various true values of μ can be drawn, as in Figure 8.2, showing acceptance and rejection regions.

The line dividing acceptance from rejection falls at a particular $\bar{X}$, which is referred to as the *critical value*, designated $\hat{\bar{X}}$. As a good visual aid in working these types of problems, draw a sketch representing the distribution and shade

Figure 8.2 Alpha and beta errors

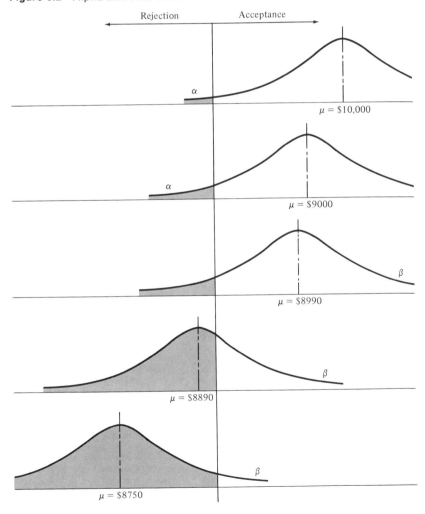

the rejection region while leaving the acceptance region clear. $\overset{+}{X}$ can be found exactly for $\alpha = 5$ percent when $\mu = \$9000$, as follows: for a one-tail test $z = -1.64$

$$z = \frac{\overset{+}{X} - \mu}{\sigma_x} \qquad \sigma_x = \frac{\$500}{\sqrt{100}} = \$50$$

$\overset{+}{X} = \mu - z\sigma_{\bar{x}} = \$9000 - 1.64(50) = \$8928$. This gives the decision rule concerning whether or not to introduce the new line: if $\bar{X} \geq \$8928$, then introduce the new line; if $\bar{X} < \$8928$, then do not introduce the new line.

Suppose now that β is calculated for various μ's. It should be obvious that if $\mu = \$8928$, then $\beta = 0.5$. The β's are calculated from the critical value. If $\mu = \$8990$, then

$$z = \frac{8928 - 8990}{50} = \frac{-62}{50} = -1.24$$

$$\beta = 0.5 + 0.39 = 0.89$$

If $\mu = \$8890$, then

$$z = \frac{8928 - 8890}{50} = \frac{38}{50} = 0.76$$

$$\beta = 0.22$$

If $\mu = \$8750$, then

$$z = \frac{8928 - 8750}{50} = \frac{178}{50} = 3.56$$

$$\beta = 0.001$$

β can be tabulated for various μ's, given α and n. Also, notice that β is defined as the probability of type II error and hence can be tabulated only for $\mu < \$9000$, since for $\mu > \$9000$ there is no β error.

μ	β
$8990	0.89
8928	0.50
8890	0.22
8750	0.001

Repeating again what this tabulation shows: if $\mu =$ the various values in the column, then the probability of erroneously accepting the null hypothesis and of introducing the new line is given by the corresponding values of β. From this example the student can see what is meant in hypothesis testing by stating that α can be specified at a maximum value if the null hypothesis is true but that β can take on a range of values if the null hypothesis is false.

8.4 THE POWER FUNCTION AND THE POWER OF A TEST

As mentioned, it is desirable to formulate the null hypothesis as a simple hypothesis, because this permits the exact specification of α—the (maximum) probability of committing a type I error. In addition, various tests having the same α can then be compared with respect to β. If the alternative hypothesis is composite, then the probability of committing a type II error varies for different values of the alternative, as shown in the preceding example. The function or curve that represents the probability of rejecting the null hypothesis when various alternatives are true is the *power function* of the test. The *power of a test* against a specific alternative is the probability of rejecting the null hypothesis when that specific alternative is true; thus power $= 1 - \beta$, where β is the probability of accepting the null hypothesis when the alternative hypothesis is true (type II error).

If the probability of accepting the null hypothesis is plotted for various alternative hypotheses, then the *operating characteristic function* of the test or (briefly) the *OC-curve* is obtained. Power functions are more frequently used in theoretical discussions; OC-curves are more often used in industrial applications, particularly in quality control problems.

Suppose the problem is to test the null hypothesis

$$H_0: \theta = \theta_0$$

against the alternative hypothesis

$$H_A: \theta \neq \theta_0$$

where θ is a population parameter and θ_0 is a specified constant. Consider tests A, B, and C in Figure 8.3, which are of size α and have the indicated power curves.

Notice that at θ_0 the value of each power function is the probability of committing a type I error; the power functions have the same value α at θ_0. For other values of θ, the values of the power functions are the probabilities of making correct decisions and should be as close to 1 as possible. Thus the test represented by A is preferable to the test represented by B for all $\theta \neq \theta_0$ and A is said to be *uniformly more powerful* than B; B is said to be *inadmissible*. The same clear-cut choice is not possible between A and C or between B and C. A (or B) is preferable if $\theta < \theta_0$ and C is preferable if $\theta > \theta_0$. In this case some additional criterion is needed.* Note that for the alternative hypothesis $\theta < \theta_0$ A or B is each uniformly more powerful than C; for the alternative hypothesis $\theta > \theta_0$, C is uniformly more powerful than either A or B.

In general, when a simple null hypothesis is tested against a composite alternative hypothesis, the probability, α, of committing a type I error is specified.

---

* The criterion of *unbiasedness* might be used in such a case. When testing a simple hypothesis against a composite alternative, a critical region is said to be *unbiased* if the corresponding power function assumes its minimum value at the value of the parameter assumed under the simple null hypothesis. Thus a critical region is unbiased if the probability of rejecting the null hypothesis is least when the null hypothesis is true. Tests A and B are unbiased; test C is biased.

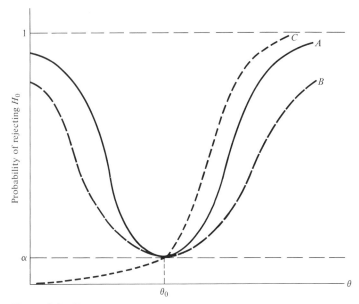

Figure 8.3 Power curves

One critical region of size α is said to be *uniformly more powerful* than another if the probability of *not* committing a type II error with the first is always equal to or greater than the probability of *not* committing a type II error with the second, with inequality holding for at least one value of the parameter in question. If, for a given problem, one critical region of size α is uniformly more powerful than any other critical region of size α, we refer to it as *uniformly most powerful;* unfortunately, uniformly most powerful critical regions seldom exist.

In the previous discussion of one-sided and two-sided tests the probability of rejecting H_0 when H_A is true is represented by the area under the curve representing the sampling distribution of the statistic if H_A is true, which falls in the region of rejection. Thus the appropriate one-sided test will be more powerful against a one-sided alternative than the corresponding two-sided test; as shown by cases A, B, and C which follow. It is assumed throughout that a two-sided test is symmetric—that is, the two parts of the rejection region are equal in area. Usually it is difficult to justify a nonsymmetric two-sided test; however, the discussion is valid for such tests.

CASE 1

$$H_0: \theta = \theta_0$$

$$H_A: \theta > \theta_0$$

In this case for the specification of the alternative hypothesis as $\theta > \theta_0$, the rejection and acceptance regions are as shown in Figure 8.4. These are determined in the usual way from the null hypothesis $\theta = \theta_0$. If the true situation is $\theta = \theta_1$, the power of the test is low. On the other hand, if $\theta = \theta_2$, the probability of falling in the rejection region is substantial, and the power of the test is high. We see from Figure 8.4 that $1 - \beta_2 > 1 - \beta_1$. The power curve for this case is shown in Figure 8.5.

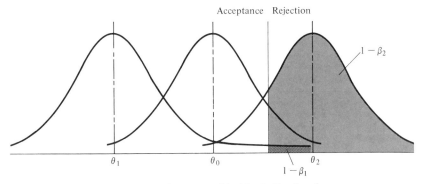

Figure 8.4   Acceptance-rejection, one-sided test, $H_A : \theta > \theta_0$

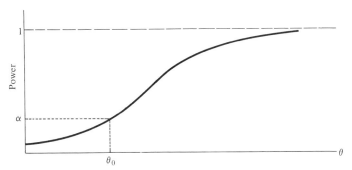

Figure 8.5   Power function, one-sided test, $H_A : \theta > \theta_0$

CASE 2

$$H_0: \theta = \theta_0$$

$$H_A: \theta < \theta_0$$

This case is also a one-sided test, but with the rejection region in the opposite tail from case 1. The rejection and acceptance regions are obtained on the basis of the null hypothesis. The test is relatively powerful for a true situation of $\theta = \theta_1$ in this case and relatively weak for a true situation of $\theta = \theta_2$. As shown in Figure 8.6, $1 - \beta_1 > 1 - \beta_2$. The power curve is shown in Figure 8.7.

Figure 8.6   Acceptance-rejection, one-sided test, $H_A : \theta < \theta_0$

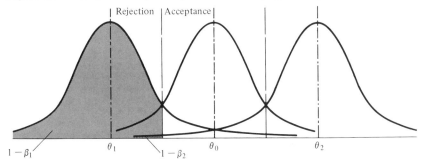

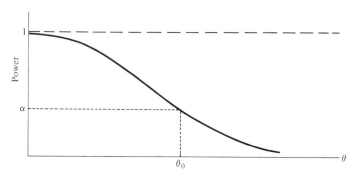

Figure 8.7   Power function, one-sided test, $H_A : \theta < \theta_0$

CASE 3

$$H_0 : \theta = \theta_0$$

$$H_A : \theta \neq \theta_0$$

In this case the test is two sided and the acceptance and rejection regions are shown in Figure 8.8. The farther away the true parameter lies from the null hypothesized parameter (whether larger or smaller), the more powerful the test. The closer the true parameter lies to θ_0, the weaker the test. Notice that $1 - \beta_1 = 1 - \beta_2$ for $|\theta_0 - \theta_1| = |\theta_0 - \theta_2|$. The power curve would appear as in Figure 8.9.

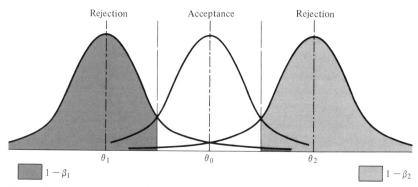

Figure 8.8   Acceptance-rejection, two-sided test

Figure 8.9   Power function, two-sided test

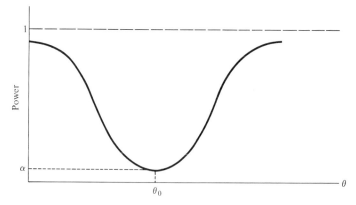

Note one fact for all three cases described. In each case the regions of acceptance and rejection and the power curve are calculated for a given α level. If the α level is raised to, say, $\alpha = 10$ percent from $\alpha = 5$ percent, the region of rejection in each case would get larger and the power curve would shift upward. If α were decreased from 5 to 1 percent, the region of rejection would become smaller and the power curve would shift downward.

8.5 CHANGING THE SAMPLE SIZE

The probability of type II error, β, varies with the distance the falsely hypothesized parameter lies from the true parameter. It also depends on the sample size. These dependences are illustrated in the following examples.

EXAMPLE
Suppose a company accepts a shipment of wires if the average strength is believed to be 200 lb or more and rejects the shipment if the average strength is believed to be less than 200 lb. The company wishes to accept the shipment unless there is strong evidence indicating the contrary. Hence the null and alternative hypotheses are:

$$H_0: \mu = 200$$

$$H_A: \mu < 200$$

Given: $\alpha = 5$ percent, $n = 36$, $\sigma = 20$

$$\sigma_{\bar{x}} = \frac{\sigma}{\sqrt{n}} = \frac{20}{6} = 3.33$$

For a one-tailed test, α of 5 percent, $z = -1.64$

$$-1.64 = \frac{\overset{\star}{X} - 200}{3.33}$$

Critical Value $\overset{\star}{X} = 200 - 5.46 = 194.54$. Calculating β for various possible true values of the alternative hypothesis, we have

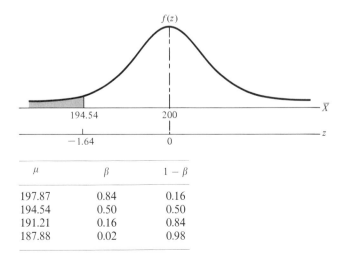

μ	β	$1 - \beta$
197.87	0.84	0.16
194.54	0.50	0.50
191.21	0.16	0.84
187.88	0.02	0.98

Now suppose the sample size is increased to 100. Then

$$\sigma_x = \frac{\sigma}{\sqrt{n}} = \frac{20}{\sqrt{100}} = 2$$

The critical value now becomes $-1.64 = (\check{X} - 200)/2$; $\check{X} = 196.82$. This states
that the shipment should be accepted if $\bar{X} \geq 196.82$. Calculating β for various
possible values of the alternative hypothesis gives lower values of β for the same
μ than before. The $1 - \beta$ values are higher; therefore the second test is more
powerful than the first—due to the increased sample size. This example illustrates
that for a given α and alternative hypothesis, β can be controlled by varying the
sample size.

μ	β	β
197.87	0.70	0.30
194.54	0.10	0.90
191.21	~0	~1.0
187.88	~0	~1.0

EXAMPLE
Consider the mail-order house problem in Section 8.3, but now assume $\sigma = \$1000$,
$\alpha = 2.5$ percent, and β is specified to be 5 percent for the alternative hypothesis of
H_A: $8850.

$$H_0: \mu = \$9000$$

$$H_A: \mu = \$8850$$

The two questions are: (1) What sample size should be used? and (2) What is the
critical value; that is, above what value for $\bar{X}$ should the null hypothesis be
accepted?

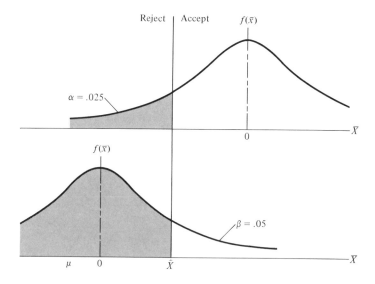

z for one-tailed 5 percent test $= 1.64$
z for one-tailed 2.5 percent test $= 1.96$

$$z = \frac{\hat{\bar{X}} - \mu}{\sigma/\sqrt{n}}$$

$$-1.96 = \frac{\hat{\bar{X}} - 9000}{\frac{1000}{\sqrt{n}}}$$

$$+1.64 = \frac{\hat{\bar{X}} - 8850}{\frac{1000}{\sqrt{n}}}$$

Thus there are two equations in two unknowns. If solved simultaneously, then

$$n = 576$$

$$\hat{\bar{X}} = \$8918$$

Hence take a sample of size 576; if $\bar{X} \geq \$8918$, except H_0; if $\bar{X} < \$8918$, reject H_0.

For a two-tailed test, if β is required not to exceed a certain amount for a specified value of the alternative hypothesis, the sample size must be determined by trial and error, because there are now two critical values to determine instead of one, as well as the sample size, and thus there are three unknowns rather than two. Since there is an upper and a lower limit for the critical values, it is impossible to determine in advance how much of the β error will come from each side of the distribution around the particular value of the alternative hypothesis; the question is how much of the β will be in β_1 and how much in β_2 in the middle sketch of Figure 8.10. Or, stated in another way, it is impossible in

Figure 8.10 Sample size determination for two-tail test

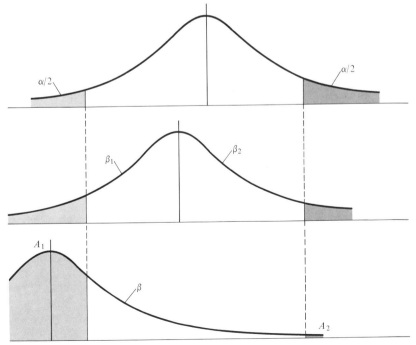

advance to determine how much of the rejection region falls on each side of β; that is, how much of the rejection region will be in A_1 and how much in A_2 in the bottom sketch of Figure 8.10.

EXAMPLE

Suppose a bagging process is filling flour bags for a supermarket with what is supposed to be 5 lb of flour, with a standard deviation of 1.5 lb. The company does not desire to systematically overfill the bags because of the extra cost nor to underfill the bags because of potential legal penalties. The company sets up the null hypothesis, $H_0: \mu = 5.0$ lb, with $\alpha = 10$ percent. It is desired to have β no greater than 15 percent for an alternative hypothesis of $H_A: \mu = 5.4$ lb. Find the sample size and the critical values for accepting and rejecting the null hypothesis. The following sketch illustrates the situation. As mentioned, it is unknown what proportion of the rejection region is represented by A_1 and what proportion by A_2.

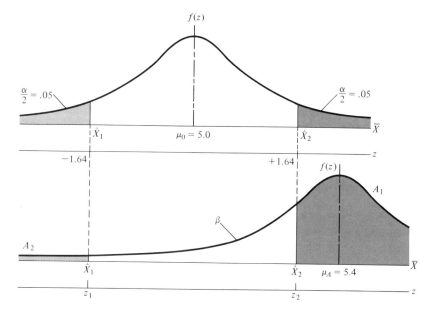

$$z_{\alpha/2} = -1.64 \qquad z_{1-\alpha/2} = +1.64$$

z_1 corresponds to $\breve{X}_1$ and z_2 to $\breve{X}_2$ on the z-axis for the distribution around the alternative hypothesis.

$$z_1 = \frac{\breve{X}_1 - \mu_A}{\sigma_{\bar{x}}} = \frac{\breve{X}_1 - 5.4}{\dfrac{1.5}{\sqrt{n}}}$$

$$z_2 = \frac{\breve{X}_2 - \mu_A}{\sigma_{\bar{x}}} = \frac{\breve{X}_2 - 5.4}{\dfrac{1.5}{\sqrt{n}}}$$

$$z_{\alpha/2} = \frac{\breve{X}_1 - \mu_0}{\sigma_{\bar{x}}} = \frac{\breve{X}_1 - 5.0}{\dfrac{1.5}{\sqrt{n}}} = -1.64$$

$$z_{1-\alpha/2} = \frac{\breve{X}_2 - \mu_0}{\sigma_{\bar{x}}} = \frac{\breve{X}_2 - 5.0}{\dfrac{1.5}{\sqrt{n}}} = +1.64$$

The trial and error calculations are performed on the last two equations by solving, from $z_{\alpha/2}$ and $z_{1-\alpha/2}$ known and n "guessed," for $\bar{X}_1$ and $\bar{X}_2$. These two values are put into the first two equations, along with n, and the equations are solved for z_1 and z_2. Once z_1 and z_2 are obtained, the areas corresponding to A_1 and A_2 can be calculated and the area corresponding to β is obtained to see if it equals the specified β. If it does not, another n must be "guessed" and the computations remade. This process is repeated until the β calculated as a residual is equal to the β designated (in this example $\beta = 0.15$). If the β calculated as a residual is smaller than the β designated, the sample size is too large. If β calculated is larger than β designated, the sample size is too small. Rearranging the above equations for $z_{\alpha/2}$ and $z_{1-\alpha/2}$, we get

$$\bar{X}_1 = 5 - 1.64\left(\frac{1.5}{\sqrt{n}}\right)$$

$$\bar{X}_2 = 5 + 1.64\left(\frac{1.5}{\sqrt{n}}\right)$$

Choose as a start $n = 225$, $\sqrt{n} = 15$. From the immediately preceding two equations,

$$\bar{X}_1 = 4.836 \quad \text{and} \quad \bar{X}_2 = 5.164$$

Using these values of $\bar{X}_1$ and $\bar{X}_2$ and $n = 225$ in the equations for z_1 and z_2,

$$z_1 = -5.64 \quad \text{and} \quad z_2 = -2.36$$

From the normal tables,

$$A_1 + \beta \cong 0.5 \qquad A_2 = 0$$
$$A_1 = 0.49$$

Calculated $\beta = 0.5 - 0.49 = 0.01$

Designated $\beta = 0.15$

Therefore the sample size is too large. Choose as a second trial $n = 81$, $\sqrt{n} = 9$. Calculating $\bar{X}_1$ and $\bar{X}_2$, we have

$$\bar{X}_1 = 4.727 \quad \text{and} \quad \bar{X}_2 = 5.273$$

and then z_1 and z_2, we have

$$z_1 = -4.04 \quad \text{and} \quad z_2 = -0.76$$

Thus
$$A_1 + \beta \cong 0.5 \qquad A_2 = 0$$
$$A_1 = 0.276$$

Calculated $\beta = 0.5 - 0.276 = 0.224$

Designated $\beta = 0.15$

Therefore the sample size is too small. Choose as a third trial $n = 100$, $\sqrt{n} = 10$. Calculating $\bar{X}_1$ and $\bar{X}_2$, we have

$$\bar{X}_1 = 4.754 \quad \text{and} \quad \bar{X}_2 = 5.246$$

and then z_1 and z_2, we have

$$z_1 = -4.30 \quad \text{and} \quad z_2 = -1.02$$

Thus

$$A_1 + \beta \cong 0.5 \qquad A_2 = 0$$

$$A_1 = 0.346$$

Calculated $\beta = 0.5 - 0.346 = 0.154$

Designated $\beta = 0.15$

Therefore the sample size is slightly too large. Choose as a fourth trial $n = (10.1)^2 = 102; \sqrt{n} = 10.1$. Calculating $\overset{*}{\bar{X}}_1$ and $\overset{*}{\bar{X}}_2$ we have

$$\overset{*}{\bar{X}}_1 = 4.756 \quad \text{and} \quad \overset{*}{\bar{X}}_2 = 5.244$$

and then z_1 and z_2, we have

$$z_1 = -4.30 \quad \text{and} \quad z_2 = -1.04$$

Thus

$$A_1 + \beta \cong 0.5 \qquad A_2 = 0$$

$$A_1 = 0.3508$$

Calculated $\beta = 0.5 - 0.351 = 0.149$

Designated $\beta = 0.15$

Therefore the sample size 102 is appropriate for the specified α, β, and H_A, and the corresponding critical values are $\overset{*}{\bar{X}}_1 = 4.756$ and $\overset{*}{\bar{X}}_2 = 5.244$.

EXAMPLE
For the problem of the immediately preceding example, again take $H_0: \mu = 5$, $\alpha = 10$ percent, and $\beta \leq 15$ percent. But consider $H_A: \mu = 5.25$. Since 5.25 is closer to 5 than 5.4, the sample size required for the same α and β should be larger. However, to illustrate the procedure, we then have

$$\overset{*}{\bar{X}}_1 = 4.508$$

$$\overset{*}{\bar{X}}_2 = 5.492$$

$$z_1 = -2.473$$

$$z_2 = 0.807$$

And the calculated $\beta = 0.4933 + 0.2901 = 0.7834$. But the specified $\beta = 0.15$, so n is too small. Take $n = 100$, then

$$\overset{*}{\bar{X}}_1 = 4.754$$

$$\overset{*}{\bar{X}}_2 = 5.246$$

$$z_1 = -3.307$$

$$z_2 = -0.27$$

And the calculated $\beta = 0.5 - 0.0108 - 0.0005 = 0.4887$. But the specified $\beta = 0.15$, so n is too small.

Take $n = 225$, then

$$\tilde{X}_1 = 4.836$$
$$\tilde{X}_1 = 5.164$$
$$z_1 = -4.140$$
$$z_2 = -0.860$$

And the calculated $\beta = 0.5 - 0.3051 = 0.1949$. But the specified $\beta = 0.15$, so n is too small. Take $n = 400$, then

$$\tilde{X}_1 = 4.877$$
$$\tilde{X}_2 = 5.123$$
$$z_1 = -4.973$$
$$z_2 = -1.693$$

And the calculated $\beta = 0.5 - 0.4548 = 0.0452$. But the specified $\beta = 0.15$, so n is too large. Take $n = 256$, then

$$\tilde{X}_1 = 4.846$$
$$\tilde{X}_2 = 5.154$$
$$z_1 = -4.298$$
$$z_2 = -1.021$$

And the calculated $\beta = 0.50 - 0.3463 = 0.1537$. But the specified $\beta = 0.15$, so n is too small. Take $n = (16.1)^2 = 259.21$ or 259, then

$$\tilde{X}_1 = 4.847$$
$$\tilde{X}_2 = 5.153$$
$$z_1 = -4.333$$
$$z_2 = -1.043$$

And the calculated $\beta = 0.5 - 0.3517 = 0.1483$. But the specified $\beta = 0.15$, so n is slightly too small. Take $n = (16.05)^2 = 257.6$ or 258, then

$$\tilde{X}_1 = 4.847$$
$$\tilde{X}_2 = 5.153$$
$$z_1 = -4.333$$
$$z_2 = -1.043$$

And the calculated $\beta = 0.5 - 0.3517 = 0.1483$ (as above). So take $n = 259$ (conservatively, to allow for rounding). Note that this n is considerably larger than $n = 102$ required for $H_A: \mu = 5.4$ with the same α and β.

PROBLEMS: TESTING HYPOTHESES

8.1 It has been claimed that 50 percent of the people in a large community have exactly two colds per year. The interested parties have decided to reject this claim if among 400 people 216 or more say they have two colds per year.
(a) What is the probability of a type I error?
(b) What is the probability of a type II error if, in fact, 40 percent of the people have two colds per year?

8.2 It is hypothesized (by the building committee) that the student population at ABC University is 20 percent female and 80 percent male. A random sample of 400 students is drawn with the intention of accepting the hypothesis if between 60 and 100 of those in the sample are female and rejecting the hypothesis otherwise.
(a) Find the probability of rejecting the hypothesis when it is correct.
(b) Find the probability of accepting the hypothesis when in fact 15 percent of the students are female.
(c) Find the power of the test against the alternative that 50 percent are female.

8.3 A manufacturer wants to produce steel rods with an average length of 150 in. His production process results in batches of 10,000 rods whose lengths have a standard deviation of 30 in. From each batch the quality control engineer measures the lengths of a random sample of 225 rods. If the average length of the 225 rods is between 145 and 155, the batch is accepted; otherwise it is rejected.
(a) What is the probability that a batch with an average length of 150 in. will be rejected?
(b) What is the probability that a batch with an average length of 144 in. will be accepted?
(c) What is the probability that a batch with an average length of 152 in. will be accepted.

8.4 A production process is designed to filled No. $1\frac{1}{4}$ cans with 14.5 oz net weight of sliced pineapple. Although the mean weight varies from time to time, the standard deviation is stable and well established at .64 oz. In order to test incoming lots for weight, a buyer takes a random sample of 30 cans from each incoming lot and determines the average net weight. The buyer does not mind getting too much, but she wishes to protect herself from getting too little.
(a) For what values of $\bar{X}$ should the buyer reject the hypothesis $\mu = 14.5$ if she wants to use a 5 percent level of significance?
(b) Using this criterion, what is the probability she will fail to detect a lot whose actual average weight is only 14.3 oz?

8.5 A mill wants to adjust its sawing machine to cut lumber in 30 ft average lengths. The machine is known to cut lengths having a standard deviation of $\frac{1}{2}$ ft. From each (very large) lot a random sample of 25 pieces of lumber are measured in order to determine whether or not to stop the machine for adjustment. The owners of the mill are concerned about deviations from the 30 ft average in either direction and wish to adjust the machine unnecessarily only 1 percent of the time.
(a) For what values of the sample average should the machine be adjusted?
(b) Using this criterion, what is the probability of failing to stop for adjustment if a lot has a true average length of 30.5 ft?
(c) Using this criterion, what is the probability of stopping for adjustment if a lot has a true average length of 29.4 ft?

8.6 The Strong Steel Company produces cables that are supposed to have an average tensile strength of 2000 lb. The cables are produced in lots of 200,000 and their tensile strength is known to have a standard deviation of 250 lb. The production engineer takes a random sample of 100 cables from each lot and determines their average tensile strength; he is not concerned if this exceeds specification but is concerned if it falls below specification.
(a) For what values of $\bar{X}$ should the production engineer reject a lot if he wishes to reject lots that actually meet specification only 1 percent of the time?
(b) Using the criterion of part (a), with what probability will the production engineer fail to reject a lot having average tensile strength of 1900 lb?

    (c) Using the criterion of part (a), with what probability will he (falsely) reject lots having an average tensile strength of 2100 lb?

8.7 An economist with the Securities and Exchange Commission hypothesized that of the individual investors owning listed preferred stocks 70 percent are female and 30 percent are male. A random sample of 900 individual investors owning listed preferred stocks is obtained. The intention is to accept the hypothesis if between 250 and 290 of the owners are male and to reject it otherwise.

    (a) Find the probability of rejecting the hypothesis when it is correct?

    (b) Find the probability of accepting the hypothesis when in fact 25 percent of the owners are male.

    (c) Find the power of the test against the alternative that 50 percent of the owners are male.

8.8 The Window-Beauty Company wants the traverse curtain rods it produces to have defective mechanisms in not more than one in a thousand (i.e., .001) on the average. The rods are produced in lots of 750,000. The quality control engineer sets up the following test procedure: from each lot a random sample of 10.000 rods is selected and the rods are checked for defective mechanisms. If 16 or more rods are defective, the lot is rejected; otherwise it is accepted.

    (a) What is the probability of rejecting a lot that meets the manufacturer's standard?

    (b) What is the probability of accepting a lot that has .0016 defectives?

8.9 A production process turning out ball-point pens is checked once a day by drawing a random sample of 100 ball-point pens and examining them to see how many are defective and how many are not defective. If too many are defective, the process is stopped and adjustments are made. If the process is thought to be turning out 10 percent or more defectives, the process is stopped; if it is thought to be turning out less than 10 percent defectives, the process is continued. Assumed a significance level of 5 percent, and assume that the error of continuing production when there are 10 percent, or more defectives is of greater seriousness than the error of stopping the process when there are less than 10 percent defectives. Set up a decision rule in terms of number of defective pens in the sample, stating when to stop the process.

8.10 A factory process is placing fertilizer in bags such that the standard deviation of the net contents is 3 lb. If the process is in adjustment, the mean weight placed in a bag is 100 lb. If the process is not in adjustment, it may either overfill or underfill. Every day a sample of 36 bags is taken from the process, and the mean weight of the sample calculated. Since it is very costly to stop the process and adjust the machinery, it has been determined that the probability of doing this when the machinery does not really need adjustment will be permitted to be only .05.

    (a) State the null hypothesis for this test.

    (b) Give the proper decision rule for stopping the machine for adjustment.

    (c) Sketch a curve showing the regions of acceptance and rejection; label the horizontal axis in lb and shade the region of rejection.

    (d) What is the probability of accepting the null hypothesis if the true mean was 96 lb?

8.11 Agatha Anthropologist has arrived on the Alabaster Coast to study the (very large) Yuki tribe. On the basis of preliminary genetic studies, she hypothesizes that $\frac{1}{4}$ of the Yuki tribe have type ABRh+ blood. After some difficulty, Agatha is able to determine the blood types of a random sample of 48 Yukis decides to accept her hypothesis if between 7 and 17 of the sample have blood type ABRh+.

    (a) With what probability will Agatha reject the hypothesis when it is in fact true?

    (b) With what probability will Agatha accept the hypothesis when in fact $\frac{1}{6}$ of the Yukis have type ABRh+ blood?

8.12 A bridge company accepts shipments of tension bars if the average tensile strength of the bars in a shipment is believed to be 400 lb or more. The company is more anxious to avoid sending back acceptable shipments than to avoid keeping understrength shipments. Given the known standard deviation of tension bar strengths to be 40 lb, $n = 25$, and level of significance of 5 percent

    (a) Set up the null hypothesis

    (b) What is the decision rule for accepting shipments?

    (c) What is the β error if the true mean is 397 lb?

    (d) Draw graphs for (b) and (c) and shade the rejection regions.

(e) Redo (b), (c), and (d) assuming the sample standard deviation of a particular sample of 25 is 30 lb.

8.13 A company uses large amounts of wire which is supposed to be diameter 1.000 cm, with a known standard deviation of .150. From large spools of wire, an automatic process winds a type of armature. At random places in the process, pieces of wire from each spool are clipped out and the diameter of each is measured. If it is decided that the diameter of the wire is too small, all armatures wired from that spool must be scrapped. Therefore, the company is most anxious not to scrap armatures that have been wound with correct size wire or larger, and an α level of 1 percent is set. On the other hand, if the wire is too small and it is accepted, the armatures might overheat and burn up the motors in which they are installed. So a β level of 5 percent is specified for an alternative mean diameter of .92.

(a) How should the null hypothesis be stated, and how should the alternative hypothesis be stated?

(b) What sample size should be used?

(c) Above what value for $\bar{X}$ should the null hypothesis be accepted?

8.14 Repeat problem 8.13 above and assume that the wire is desired to be neither small nor larger than 1.000 cm. The α level is set now to be 2 percent, and β is increased to 10 percent for an alternative mean of .92.

(a) What should be the null and alternative hypothesis?

(b) What size sample should be used?

(c) Above what value for $\bar{X}$ should the null hypothesis be accepted?

APPLICATIONS OF TESTING HYPOTHESES

9.1 INTRODUCTION

In this chapter, a number of tests of hypotheses concerning means, differences, and variances are discussed. These tests are more specific than the tests discussed in the previous chapter, though they are very useful and have wide applications in practice. Much of the following material consists of examples illustrating the tests. Recall that a test of hypothesis requires the specification of the sampling statistic and its distribution and the values of the sampling statistic for which the hypothesis is accepted and rejected. The tests of hypotheses are presented in the following format:

1. Null hypothesis
2. Alternative hypothesis
3. Assumptions
4. Sample statistic(s)
5. Sampling distribution
6. Critical region

Tests of hypotheses concerning means, differences, and variances are considered in different sections; in each section, tests for several situations are discussed. These situations differ with respect to the number of populations involved and various assumptions made concerning the populations.

9.2 TESTS CONCERNING MEANS

ONE-SAMPLE TESTS

The hypotheses in this section concern the mean of one normally distributed population. The following cases are considered: (1) known population variance or the central limit theorem applies and (2) unknown population variance.

CASE 1

$H_0: \mu = \mu_0$

$H_A:$ (a) $\mu > \mu_0$
 (b) $\mu < \mu_0$ $\Bigg\}$ possible alternative hypotheses
 (c) $\mu \neq \mu_0$

Assumptions:

A random sample of size n is obtained from a normal population with known variance σ^2 (or the central limit theorem applies and s^2 approximates σ^2).

Sample Statistic:

$$\bar{x} = \frac{\sum_{i=1}^{n} x_i}{n}$$

Sampling Distribution:

$$z = \frac{\bar{x} - \mu_0}{\dfrac{\sigma}{\sqrt{n}}} \text{ standard normal distribution}$$

Critical Region:

(a) $Z \geq z_{1-\alpha}$
(b) $Z \leq z_{\alpha}$
(c) $Z \geq z_{1-\alpha/2}$ and $Z \leq z_{\alpha/2}$

EXAMPLE

A machine is supposed to fill coffee cans with 1 lb of coffee. It is known that the standard deviation of the amount put in the cans is .02 lb. In order to check at a particular time whether the machine is under control (i.e., whether the average amount of coffee per can is 1 lb), a random sample of 25 cans is inspected. The average weight of the cans in the sample is 1.014 lb. Test the hypothesis that the machine is under control at the .01 level of significance.

$$H_0: \mu = 1$$

$$H_A: \mu \neq 1$$

$$z = \frac{\bar{x} - \mu_0}{\dfrac{\sigma}{\sqrt{n}}} = \frac{1.014 - 1}{\dfrac{.02}{\sqrt{25}}} = 3.5$$

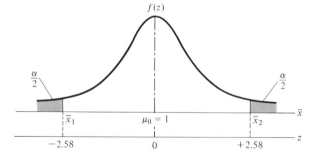

Critical region: $z < -2.58$ and $z > 2.58$
Reject H_0 since $z = 3.5$ falls in critical region (i.e., the machine is not under control is the conclusion).

CASE 2

$H_0: \mu = \mu_0$

$H_A:$ (a) $\mu > \mu_0$
   (b) $\mu < \mu_0$ possible alternative hypotheses
   (c) $\mu \neq \mu_0$

Assumptions:
A random sample of size n is obtained from a normal population with unknown variance. (Small sample so the central limit theorem does not apply.)

Sample Statistics:

$$\bar{x} = \frac{\sum\limits_{i=1}^{n} x_i}{n} \quad \text{and} \quad s^2 = \frac{\sum\limits_{i=1}^{n} (x_i - \bar{x})^2}{n-1}$$

Sampling Distribution:

$$t = \frac{\bar{X} - \mu_0}{\dfrac{s}{\sqrt{n}}} \quad \text{Student's } t \text{ distribution with } n-1 \text{ degrees of freedom}$$

Critical Region:
   (a) $t \geq t_{1-\alpha, n-1}$
   (b) $t \leq t_{\alpha, n-1}$
   (c) $t \geq t_{1-\alpha/2, n-1}$ and $t \leq t_{\alpha/2, n-1}$

EXAMPLE

A study of a random sample of 20 families in a large community determined that their average income is $7150 with a standard deviation of $975. A certain politician claims that the average income in this community is only $6500 (i.e., not more than $6500). On the basis of the sample, is his claim justified? (5 percent level of significance)

$$H_0: \mu = 6500$$

$$H_A: \mu > 6500$$

$$t = \frac{\bar{x} - \mu_0}{\dfrac{s}{\sqrt{n}}} = \frac{7150 - 6500}{\dfrac{975}{\sqrt{20}}} = 1.49$$

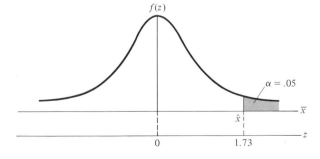

Critical region: $t > 1.73$
Accept H_0 since $t = 1.49$ falls in acceptance region (i.e., accept the claim that the average income is only $6500)

TWO-SAMPLE TESTS

The hypotheses in this section concern the difference between means of two normally distributed populations. The following cases are considered: (1) known population variances or the central limit theorem applies, (2) unknown but equal population variances, and (3) unknown and unequal population variances.

CASE 1

$$H_0: \mu_1 - \mu_2 = \delta$$

$$H_A: \text{(a)} \ \mu_1 - \mu_2 > \delta$$
$$\text{(b)} \ \mu_1 - \mu_2 < \delta \quad\quad \text{possible alternative hypotheses}$$
$$\text{(c)} \ \mu_1 - \mu_2 \neq \delta$$

Assumptions:

Two independent random samples of sizes n_1 and n_2 are obtained from two normal populations having known variances σ_1^2 and σ_2^2 (or the central limit theorem applies and s_1^2 and s_2^2 approximate σ_1^2 and σ_2^2).

Sample Statistics:

$$\bar{x}_1 = \frac{\sum_{i=1}^{n_1} x_{i1}}{n_1} \quad \text{and} \quad \bar{x}_2 = \frac{\sum_{i=1}^{n_2} x_{i2}}{n_2}$$

Sampling Distribution:

$$Z = \frac{\bar{x}_1 - \bar{x}_2 - \delta}{\sqrt{\dfrac{\sigma_1^2}{n_1} + \dfrac{\sigma_2^2}{n_2}}} \quad \text{standard normal distribution}$$

Critical Region:

(a) $Z \geq z_{1-\alpha}$

(b) $Z \leq z_\alpha$

(c) $Z \geq z_{1-\alpha/2}$ and $Z \leq z_{\alpha/2}$

EXAMPLE

The manufacturer of brand 1 cigarettes claims that his cigarettes are no more harmful to health than Brand 2 (filtered) cigarettes. Assuming harmfulness to be associated with nicotine content, the FDA took random samples of 125 cigarettes of brand 1 and 180 cigarettes of brand 2. The average nicotine content in the sample of brand 1 was 24.6 mg with a standard deviation of 1.4 mg; the average nicotine content in the sample of brand 2 was 24.3 mg with a standard deviation of 1.1 mg. At the 5 percent level of significance, is the manufacturer's claim justified?

$$H_0: \mu_1 - \mu_2 = 0$$

$$H_A: \mu_1 - \mu_2 > 0$$

$$z = \frac{\bar{x}_1 - \bar{x}_2 - \delta}{\sqrt{\dfrac{s_1^2}{n_1} + \dfrac{s_2^2}{n_2}}} = \frac{24.6 - 24.3}{\sqrt{\dfrac{(1.4)^2}{125} + \dfrac{(1.1)^2}{180}}} = 2.00$$

(Using the central limit theorem)

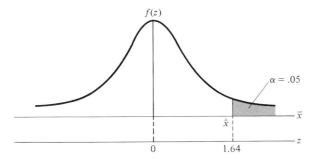

Critical region: $z > 1.645$
Reject H_0 since $z = 2.00$ falls in the critical region (i.e., conclude that brand 1 is more harmful on the basis of nicotine content).

CASE 2

$$H_0: \mu_1 - \mu_2 = \delta$$

$$H_A: \begin{array}{l} \text{(a)} \ \mu_1 - \mu_2 > \delta \\ \text{(b)} \ \mu_1 - \mu_2 < \delta \\ \text{(c)} \ \mu_1 - \mu_2 \neq \delta \end{array} \right\} \quad \text{possible alternative hypotheses}$$

Assumptions:
Two independent random samples of sizes n_1 and n_2 are obtained from two normal populations having unknown but equal variances $\sigma_1^2 = \sigma_2^2 = \sigma^2$ (small samples so the central limit theorem does not apply).

Sample Statistics:

$$\bar{x}_1 = \frac{\sum_{i=1}^{n_1} x_{i1}}{n_1} \quad \text{and} \quad \bar{x}_2 = \frac{\sum_{i=1}^{n_2} x_{i2}}{n_2}$$

$$s_1^2 = \frac{\sum_{i=1}^{n_1} (x_{i1} - \bar{x}_1)^2}{n_1 - 1} \quad \text{and} \quad s^2 = \frac{\sum_{i=1}^{n_2} (x_{i2} - \bar{x}_2)^2}{n_2 - 1}$$

Sampling distribution:

$$t = \frac{\bar{x}_1 - \bar{x}_2 - \delta}{\sqrt{\dfrac{(n_1 - 1)s_1^2 + (n_2 - 1)s_2^2}{n_1 + n_2 - 2} \left(\dfrac{1}{n_1} + \dfrac{1}{n_2} \right)}}$$

Student's-t distribution with $n_1 + n_2 - 2$ degrees of freedom

Critical Region:
    (a) $t \geq t_{1-\alpha, \, n_1 + n_2 - 2}$
    (b) $t \leq t_{\alpha, \, n_1 + n_2 - 2}$
    (c) $t \geq t_{1-\alpha/2, \, n_1 + n_2 - 2}$ and $t \leq t_{\alpha/2, \, n_1 + n_2 - 2}$

EXAMPLE

A manufacturer of power lawn mowers must decide whether to replace the present model with an experimental one. The models seem comparable otherwise, but the manufacturer hopes that the experimental model has a lower oil consumption (a fault of the older model). The manufacturer decides that the replacement would be justified if the experimental model runs at least 15 hr/qt of oil longer. A random sample of 20 experimental model lawn mowers runs an average of 78.2 hr/qt of oil with a standard deviation of 3.2 hr. A random sample of 10 old model lawn mowers runs an average of 50.6 hr/qt. with a standard deviation of 2.5 hr. Should the experimental model replace the old one (1 percent level of significance)?

$$H_0: \mu_1 - \mu_2 = 15$$

$$H_A: \mu_1 - \mu_2 > 15$$

$$t = \frac{78.2 - 50.6 - 15}{\sqrt{\dfrac{(19)(3.2)^2 + (9)(2.5)^2}{20 + 10 - 2}\left(\dfrac{1}{20} + \dfrac{1}{10}\right)}} = 10.8$$

d.f. = 28

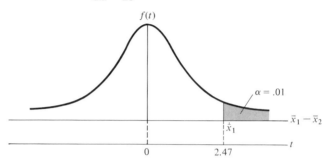

Critical region: $t > 2.47$
Reject H_0 since $t = 10.8$ falls in the critical region (i.e., conclude that the experimental model does use less oil, thus it would be justified to replace the present model with the experimental model)

CASE 3

$$H_0: \mu_1 - \mu_2 = \delta$$

$$H_A: \text{(a)} \ \mu_1 - \mu_2 > \delta$$
$$ \text{(b)} \ \mu_1 - \mu_2 < \delta \Big\} \quad \text{possible alternative hypotheses}$$
$$ \text{(c)} \ \mu_1 - \mu_2 \neq \delta$$

Assumptions:
Two independent random samples of sizes n_1 and n_2 from two normal populations having unknown and unequal variances σ_1^2 and σ_2^2 (small samples so the central limit theorem does not apply).

Sample Statistics:

$$\bar{x}_1 = \frac{\sum_{i=1}^{n_1} x_{i1}}{n_1} \quad \text{and} \quad \bar{x}_2 = \frac{\sum_{i=1}^{n_2} x_{i2}}{n_2}$$

$$s_1^2 = \frac{\sum_{i=1}^{n_1} (x_{i1} - \bar{x}_1)^2}{n_1 - 1} \quad \text{and} \quad s_2^2 = \frac{\sum_{i=1}^{n_2} (x_{i2} - \bar{x}_2)^2}{n_2 - 1}$$

Sampling distribution:
$$t = \frac{\bar{x}_1 - \bar{x}_2 - \delta}{\sqrt{\dfrac{s_1^2}{n_1} + \dfrac{s_2^2}{n_2}}}$$

Student's-t distribution with *approximately* v degrees of freedom

$$v = \frac{\left[\dfrac{s_1{}^2}{n_1} + \dfrac{s_2{}^2}{n_2}\right]^2}{\left[\dfrac{s_1{}^2}{n_1}\right]^2\left[\dfrac{1}{n_1 - 1}\right] + \left[\dfrac{s_2{}^2}{n_2}\right]^2\left[\dfrac{1}{n_2 - 1}\right]} - 2$$

NOTE
v will seldom be an integer, but a good approximation can usually be obtained by using the nearest integer. Since $t_{1-\alpha} > z_{1-\alpha}$ and $t_\alpha < z_\alpha$ for any value of α, if the computed t value falls in the acceptance region using the z distribution, H_0 can be accepted without computing v.

EXAMPLE
Consumer Research Associates, Inc., is interested in comparing average yearly per family consumption of meat in rural and urban areas of a community. In a random sample of 16 rural families, the average yearly consumption of meat is 170.6 lb with a standard deviation of 10.2 lb. In a random sample of 18 urban families, the average yearly consumption of meat is 152.4 lb with a standard deviation of 21.6 lb. Test the hypothesis that the average yearly meat consumption is the same for rural and urban families. (5 percent level of significance).

$$H_0: \mu_1 = \mu_2$$

$$H_A: \mu_1 \neq \mu_2$$

$$t = \frac{\bar{x}_1 - \bar{x}_2 - \delta}{\sqrt{\dfrac{s_1{}^2}{n_1} + \dfrac{s_2{}^2}{n_2}}} = \frac{170.6 - 152.4}{\sqrt{\dfrac{(10.2)^2}{16} + \dfrac{(21.6)^2}{18}}} = 3.20$$

$z_{.975} = 1.96$ so $t = 3.20$ falls in the rejection region for that test and v must be computed.

$$v = \frac{\left[\dfrac{s_1{}^2}{n_1} + \dfrac{s_2{}^2}{n_2}\right]^2}{\left[\dfrac{s_1{}^2}{n_1}\right]^2\left[\dfrac{1}{n_1 - 1}\right] + \left[\dfrac{s_2{}^2}{n_2}\right]^2\left[\dfrac{1}{n_2 - 1}\right]} - 2$$

$$= \frac{\left[\dfrac{(10.2)^2}{16} + \dfrac{(21.6)^2}{18}\right]^2}{\left[\dfrac{(10.2)^2}{16}\right]^2\left[\dfrac{1}{15}\right] + \left[\dfrac{(21.6)^2}{18}\right]^2\left[\dfrac{1}{17}\right]} - 2 = 24.82 - 2 = 22.82$$

to the nearest integer, $v = 23$
Critical region: $t < -2.07$ and $t > 2.07$

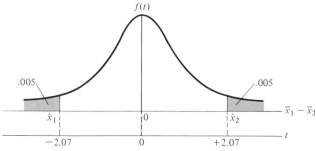

Reject H_0 since $t = 3.20$ falls in the critical region (i.e., reject the hypothesis that average meat consumption is the same for rural and urban families.)

9.3 TESTS INVOLVING DIFFERENCES

In many problems, observations occur in pairs and the hypotheses of interest concern the average differences of these pairs –for example, an investigator might be interested in average sales of a product before and after an advertising campaign (the pairs are before and after observations of the same stores) or in differences in shopping habits of husbands and their wives (the pairs are observations of married couples) or in stated opinions before and after an election (the pairs are before and after observations of the same respondents) or so forth.

In such cases the two samples clearly are not independent and, in effect, instead of having two random samples of observations, the investigator has one random sample of pairs of observations. Hypotheses concerning differences between pairs are tested by considering the differences of pairs to be a random sample from a population of such differences.

Suppose from a random sample of size n, the pairs of observations

$$(x_{11}, x_{12}), (x_{21}, x_{22}), \ldots (x_{n1}, x_{n2})$$

are obtained. From each pair a difference can be computed:

$$d_i = x_{i1} - x_{i2} \text{ for } i = 1, 2, \ldots, n$$

and the null and alternative hypotheses concern $\mu_d = \mu_1 - \mu_2$, the average difference of pairs in the population. The test proceeds as for the one-sample cases previously discussed with the differences $d_i = x_{i1} - x_{i2}$ considered to be a random sample of size n from a population of differences having mean $\mu_d = \delta$ if the null hypothesis is true. The mean of the differences is defined as

$$\mu_d = \frac{\sum d_i}{n} = \frac{\sum x_{i1}}{n} - \frac{\sum x_{i2}}{n} = \mu_1 - \mu_2$$

The assumptions concerning normality and knowledge of population variance must be considered for the population of differences in order to determine which test is appropriate. The following cases are considered: (1) known population variance or the central limit theorem applies and (2) unknown population variance.

$$H_0: \mu_d = \delta$$

$$H_A: \text{(a)} \ \mu_d > \delta$$
$$\text{(b)} \ \mu_d < \delta \Big\} \quad \text{possible alternative hypothesis}$$
$$\text{(c)} \ \mu_d \neq \delta$$

CASE 1 σ_d^2 *KNOWN*

Assumptions:

A random sample of size n is obtained from a normal population of differences with known variance σ_d^2 (or the central limit theorem applies and s_d^2 approximates σ_d^2).

CASE 2 σ_d^2 *UNKNOWN*

Assumptions:

A random sample of size n is obtained from a normal population of differences with unknown variance (small sample so the central limit theorem does not apply).

Sample Statistic:

$$\overline{d} = \frac{\sum\limits_{i=1}^{n} d_i}{n}$$

Sample Statistics:

$$\overline{d} = \frac{\sum\limits_{i=1}^{n} d_i}{n}$$

$$s_d{}^2 = \frac{\sum\limits_{i=1}^{n} (d_i - \overline{d})^2}{n-1}$$

Sampling Distribution:

$$z = \frac{\overline{d} - \delta}{\dfrac{\sigma_d}{\sqrt{n}}}$$

Sampling Distribution:

$$t = \frac{\overline{d} - \delta}{\dfrac{s_d}{\sqrt{n}}},$$

d.f. $n-1$

Critical Region:
(a) $Z \geq z_{1-\alpha}$
(b) $Z \leq z_{\alpha}$
(c) $Z \geq z_{1-\alpha/2}$ and $Z \leq z_{\alpha/2}$

Critical Region:
(a) $t \geq t_{1-\alpha, n-1}$
(b) $t \leq t_{\alpha, n-1}$
(c) $t \geq t_{1-\alpha/2, n-1}$ and
$t \leq t_{\alpha/2, n-1}$

EXAMPLE

According to the advertising of the Low-Cal Delightful Diet Co., faithful use of their product results in an average weight loss of at least 10 lb during the first two weeks of dieting. The FDA suspects false advertising and obtains the following pairs of before and after weights for a random sample of 12 dieters who used the product for two weeks. Can the FDA use these data as a basis for legal action (5 percent level of significance).

DIETER NO.	BEFORE	AFTER	DIFFERENCE (BEFORE MINUS AFTER)
1	126	115	11
2	194	179	15
3	135	124	11
4	179	163	16
5	205	186	19
6	139	137	2
7	142	146	−4
8	172	161	11
9	159	160	−1
10	194	198	−4
11	164	158	6
12	139	125	14

$$\sum_{i=1}^{12} d_i = 96$$

$$\overline{d} = \frac{96}{12} = 8$$

$$H_0: \mu_d = 10$$

$$H_A: \mu_d < 10$$

$$t = \frac{\bar{d} - \delta}{\dfrac{s_d}{\sqrt{n}}} = \frac{8 - 10}{\dfrac{8.01}{\sqrt{12}}} = -.86$$

where $\quad s_d{}^2 = \dfrac{(9 + 49 + 9 + 64 + 121 + 36 + 144 + 9 + 81 + 144 + 4 + 36)}{11}$

$$= \frac{706}{11} = 64.1818$$

$$s_d = 8.01$$

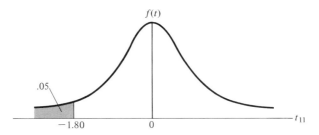

Critical region: $t < -1.80$
Accept H_0 since $t = -.86$ falls in the acceptance region (i.e., conclude that there is no basis for FDA prosecution).

The preceding discussion and examples concern situations in which pairing occurs naturally as a result of repeatedly measuring the same element of the population (e.g., stores or individuals) or as a result of measuring elements that occur in the population as pairs (e.g., husbands and wives). In some cases, even when the observations do not occur naturally as pairs, the investigator may wish to "match" or "pair" them in order to eliminate differences in which he or she is not interested.

For example, suppose a dairy serving a particular community plans to promote a new flavor of ice cream, cranberry crunch, for Thanksgiving and wants to investigate two different methods of promotion: (1) mailing, to known customers of a store, coupons offering 15 cents off the purchase price of a half gallon of cranberry crunch ice cream bought at that store and (2) giving free samples of cranberry crunch ice cream to customers passing by the dairy case where the ice cream is displayed in the store.

One way to investigate the difference between methods of promotion is to assign the two methods to independent random samples of stores in the community and then test the hypothesis that the average sales of the ice cream for stores using the two methods are the same; then $H_0: \mu_1 - \mu_2 = 0$ would be tested against a suitable alternative hypothesis. One obvious drawback, particularly if the samples are small, is the considerable chance that the samples of stores to which the two methods are assigned might be quite different with respect to the amount of ice cream (of any flavor) sold. Thus any observed differences in sales of cranberry crunch ice cream might be attributable largely to this difference between samples of stores rather than to a difference between methods of promotion. In order to avoid such possibilities, the stores in the community could be paired on the basis of average amount of ice cream sold per week (or some other relevant criterion), a random sample of pairs of stores could be chosen, and the two methods could be assigned at random to the members of each pair. The hypothesis to be tested would then be $H_0: \mu_1 = \mu_2 = \mu = 0$. Note that the number of observations is, in effect, cut in half as a result of pairing and, likewise, the number of degrees of freedom is reduced from $2(n-1)$ to $(n-1)$. This is a considerable loss if pairing is not necessary; however, if paired observations are quite similar, the smaller variance resulting from pairing may more than compensate for the loss of degrees of freedom. Pairing is most effective when the units within pairs are similar and there are considerable differences from pair to pair.

The following discussion of (two-sample) tests concerning differences between paired differences is applicable whether pairing occurs naturally or is done by the investigator.

The two-sample test concerning differences of means may also be used when the means themselves are differences of means—that is, the tests are appropriate for comparing paired differences. For example, an investigator might be interested in comparing differences in average sales of a product before and after an advertising campaign in one community with those in another community or in comparing differences in shopping habits of husbands and wives of one income level with those at another income level or in comparing differences in stated opinions before and after an election for registered Democrats with that for registered Republicans or so forth.

Note that the two observations of any pair are not independent; in effect, the investigator has two independent random samples of pairs of observations from two independent populations of pairs.

Suppose the pairs of observations

$$(x_{111}, x_{121}), (x_{112}, x_{122}), \ldots (x_{11n_1}, x_{12n_1})$$

are obtained from a random sample of size n_1 from one population and the pairs of observations

$$(x_{211}, x_{221}), (x_{212}, x_{222}), \ldots, (x_{21n_2}, x_{22n_2})$$

are obtained from a random sample of size n_2 from the other population. Then two sets of differences can be computed:

$$d_{1i} = (x_{11i} - x_{12i}) \qquad \text{for } i = 1, 2, \ldots, n_1$$

and

$$d_{2i} = (x_{21i} - x_{22i}) \qquad \text{for } i = 1, 2, \ldots, n_2$$

The null and alternative hypotheses concern $\mu_{d1} = \mu_{11} - \mu_{12}$ and $\mu_{d2} = \mu_{21} - \mu_{22}$, the average differences of pairs in the two populations. The test proceeds as for the two-sample cases previously discussed. Again, assumptions concerning normality and knowledge of population variances must be considered for the populations of differences in order to determine which test is appropriate. The following cases are considered: (1) known population variances or the central limit theorem applies and (2) unknown population variances.

$$H_0: \mu_{d1} - \mu_{d2} = \delta$$

$$H_A: \begin{array}{ll} \text{(a)} & \mu_{d1} - \mu_{d2} > \delta \\ \text{(b)} & \mu_{d1} - \mu_{d2} < \delta \\ \text{(c)} & \mu_{d1} - \mu_{d2} \neq \delta \end{array} \right\} \quad \text{possible alternative hypotheses}$$

CASE 1   KNOWN VARIANCE

Assumptions:

Two independent random samples of sizes n_1 and n_2 are obtained from two normal populations of differences with known variances σ_{d1}^2 and σ_{d2}^2 (or the central limit theorem applies and s_{d1}^2 and s_{d2}^2 approximate σ_{d1}^2 and σ_{d2}^2).

Sample Statistics:

$$\bar{d}_1 = \frac{\sum\limits_{i=1}^{n_1} d_{1i}}{n_1}$$

$$\bar{d}_2 = \frac{\sum\limits_{i=1}^{n_2} d_{2i}}{n_2}$$

Sampling Distribution:

$$Z = \frac{\bar{d}_1 - \bar{d}_2 - \delta}{\sqrt{\dfrac{\sigma_{d1}^2}{n_1} + \dfrac{\sigma_{d2}^2}{n_2}}}$$

CASE 2   UNKNOWN VARIANCE

Assumptions:

Two independent random samples of sizes n_1 and n_2 are obtained from two normal populations of differences with unknown variances (small samples so the central limit theorem does not apply).

Sample Statistics:

$$\bar{d}_1 = \frac{\sum\limits_{i=1}^{n_1} d_{1i}}{n_1}$$

$$\bar{d}_2 = \frac{\sum\limits_{i=1}^{n_2} d_{2i}}{n_2}$$

$$s_{d1}^2 = \frac{\sum\limits_{i=1}^{n_1} (d_{1i} - \bar{d}_1)^2}{n_1 - 1}$$

$$s_{d2}^2 = \frac{\sum\limits_{i=1}^{n_2} (d_{2i} - \bar{d}_2)^2}{n_2 - 1}$$

Sampling Distribution:
if $\sigma_{d1}^2 = \sigma_{d2}^2$,

$$t = \frac{\bar{d}_1 - \bar{d}_2 - \delta}{\sqrt{\dfrac{(n_1 - 1)s_{d1}^2 + (n_2 - 1)s_{d2}^2}{n_1 + n_2 - 2}\left(\dfrac{1}{n_1} + \dfrac{1}{n_2}\right)}}$$

d.f. $= n_1 + n_2 - 2$

if $\sigma_{d1}^2 \neq \sigma_{d2}^2$

$$t = \frac{\bar{d}_1 - \bar{d}_2 - \delta}{\sqrt{\dfrac{s_{d1}^2}{n_1} + \dfrac{s_{d2}^2}{n_2}}}$$

d.f. (approximate)

$$= \frac{\left(\dfrac{s_{d1}^{2}}{n_1} + \dfrac{s_{d2}^{2}}{n_2}\right)^2}{\left(\dfrac{s_{d1}^{2}}{n_1}\right)^2 \left(\dfrac{1}{n_1 - 1}\right) + \left(\dfrac{s_{d2}^{2}}{n_2}\right)^2 \left(\dfrac{1}{n_2 - 1}\right)}$$

Critical Region:
  (a) $Z \geq z_{1-\alpha}$
  (b) $Z \leq z_{\alpha}$
  (c) $Z \geq z_{1-\alpha/2}$ and $Z \leq z_{\alpha/2}$

Critical Region:
  (a) $t \geq t_{1-\alpha,v}$
  (b) $t \leq t_{\alpha,v}$
  (c) $t \geq t_{1-\alpha/2,v}$ and $t \leq t_{\alpha,v}$

where v is the appropriate degrees of freedom.

EXAMPLE

A county medical society is interested in comparing the changes in cigarette sales in a rural community and in an urban community in which the State Medical School is located after an intensive advertising campaign by the American Cancer Society. The following data concerning total sales of cigarettes (in thousands of cartons) for the month immediately preceding and the month immediately following the campaign were obtained for independent random samples of 10 stores in the rural community and 15 stores in the urban community. Is the average change apparently larger in the urban community? (5 percent level of significance)

STORE NO.	RURAL BEFORE	RURAL AFTER	DIFFERENCE
1	160	145	15
2	111	102	9
3	174	164	10
4	197	186	11
5	127	121	6
6	148	150	-2
7	112	97	15
8	182	178	4
9	97	85	12
10	157	137	20

$$\sum_{i=1}^{10} d_{1i} = 100$$

$$\bar{d}_1 = \frac{100}{10} = 10$$

| | URBAN | | |
STORE NO.	BEFORE	AFTER	DIFFERENCE
1	260	250	10
2	375	376	−1
3	280	275	5
4	304	302	2
5	225	225	0
6	243	235	8
7	265	265	0
8	293	287	6
9	326	320	6
10	359	361	−2
11	238	232	6
12	275	285	−10
13	314	307	7
14	283	280	3
15	349	344	5

$$\sum_{i=1}^{15} d_{2i} = 45$$

$$\bar{d}_2 = \frac{45}{15} = 3$$

$$s_{d1}^2 = \frac{(25 + 1 + 0 + 1 + 16 + 144 + 25 + 36 + 4 + 100)}{9} = \frac{352}{9} = 39.11$$

$$s_{d2}^2 = \frac{(49 + 16 + 4 + 1 + 9 + 25 + 9 + 9 + 9 + 25 + 9 + 169 + 16 + 0 + 4)}{14}$$

$$= \frac{354}{14} = 25.29$$

$$H_0 : \mu_{d1} - \mu_{d2} = 0 \qquad \text{where subscript 1 refers to rural}$$
$$H_A : \mu_{d1} - \mu_{d2} < 0 \qquad \text{and subscript 2 refers to urban}$$

$\sigma_{d1}^2 = \sigma_{d2}^2$ (by test given in next section), so that

$$t = \frac{\bar{d}_1 - \bar{d}_2 - \delta}{\sqrt{\dfrac{(n_1 - 1)s_{d1}^2 + (n_2 - 1)s_{d2}^2}{n_1 + n_2 - 2}\left(\dfrac{1}{n_1} + \dfrac{1}{n_2}\right)}} = \frac{10 - 3}{\sqrt{\dfrac{(9)(39.11) + (14)(25.29)}{10 + 15 - 2}\left(\dfrac{1}{10} + \dfrac{1}{1}\right)}}$$

$$= \frac{7}{\sqrt{\dfrac{706}{23}\left(\dfrac{1}{6}\right)}} = 3.10$$

Critical region: $t < -1.71$

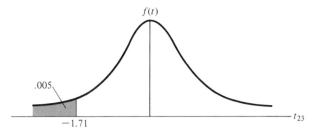

Accept H_0 since $t = 3.10$ falls in the acceptance region (i.e., accept the hypothesis that changes in the two communities were the same).

NOTE
Since $\bar{d}_1 - \bar{d}_2 - \delta > 0$, H_0 would have been accepted without computing the value of t.

There are some practical problems involving differences between differences in which no pairing is possible. For example, an investigator might be interested in comparing the difference in shopping behavior between sexes at one age level with the difference between sexes at another age level or in comparing the difference in average income of urban and rural families in one community with the difference between families in another community or in comparing the difference between average initial salary of letters and science and engineering graduates of one university with the difference between graduates of another university.

In this case the investigator has four independent random samples from four different populations and the null and alternative hypotheses concern $(\mu_1 - \mu_2) - (\mu_3 - \mu_4)$; the methods previously discussed are not appropriate, but the analysis can be performed as follows:

CASE 1

$$H_0: (\mu_1 - \mu_2) - (\mu_3 - \mu_4) = \delta$$

$$H_A: \begin{matrix} \text{(a)} \ (\mu_1 - \mu_2) - (\mu_3 - \mu_4) > \delta \\ \text{(b)} \ (\mu_1 - \mu_2) - (\mu_3 - \mu_4) < \delta \\ \text{(c)} \ (\mu_1 - \mu_2) - (\mu_3 - \mu_4) \neq \delta \end{matrix} \right\} \quad \text{possible alternative hypotheses}$$

Assumptions:
Independent random samples of sizes n_1, n_2, n_3 and n_4 from four normal populations having the same unknown variance σ^2.

Sample Statistics:

$$\bar{x}_1 = \frac{\sum\limits_{i=1}^{n_1} x_{1i}}{n_1} \qquad \bar{x}_2 = \frac{\sum\limits_{i=1}^{n_2} x_{2i}}{n_2} \qquad \bar{x}_3 = \frac{\sum\limits_{i=1}^{n_3} x_{3i}}{n_3} \qquad \bar{x}_4 = \frac{\sum\limits_{i=1}^{n_4} x_{4i}}{n_4}$$

$$s_1{}^2 = \frac{\sum\limits_{i=1}^{n_1} (x_{1i} - \bar{x}_1)^2}{n_1 - 1} \qquad s_2{}^2 = \frac{\sum\limits_{i=1}^{n_2} (x_{2i} - \bar{x}_2)^2}{n_2 - 1}$$

$$s_3{}^2 = \frac{\sum\limits_{i=1}^{n_3} (x_{3i} - \bar{x}_3)^2}{n_3 - 1} \qquad s_4{}^2 = \frac{\sum\limits_{i=1}^{n_4} (x_{4i} - \bar{x}_4)^2}{n_4 - 1}$$

Sampling Distribution:

$$t = \frac{(\bar{x}_1 - \bar{x}_2) - (\bar{x}_3 - \bar{x}_4) - \delta}{\sqrt{s^2 \sum\limits_{i=1}^{4} \frac{1}{n_i}}} \qquad \text{d.f.} = \sum\limits_{i=1}^{4} (n_i - 1) = \left(\sum\limits_{i=1}^{4} n_i \right) - 4$$

where

$$s^2 = \frac{\sum\limits_{i=1}^{4} (n_i - 1)s_i^2}{\left(\sum\limits_{i=1}^{4} n_i\right) - 4}$$

NOTE:
If σ^2 is known, then

$$z = \frac{(\bar{x}_1 - \bar{x}_2) - (\bar{x}_3 - \bar{x}_4) - \delta}{\sqrt{\sigma^2 \sum\limits_{i=1}^{4} \frac{1}{n_i}}}$$

Critical Region:
   (a) $t \geq t_{1-\alpha, v}$
   (b) $t \leq t_{\alpha, v}$
   (c) $t \geq t_{1-\alpha/2, v}$ and $t \leq t_{\alpha/2, v}$

where

$$v = \sum\limits_{i=1}^{4} n_i - 4$$

EXAMPLE
The management of a large department store is interested in comparing the difference between the amounts spent yearly on clothes by girls and boys at age 14 with the difference at age 18. From their files the store has obtained independent random samples and has computed the statistics below. Do the differences between girls and boys seem to be the same at ages 14 and 18 (1 percent level of significance)?

$$\bar{x}_1 = 156.62 \qquad s_1^2 = 25.4 \qquad n_1 = 12 \qquad \text{(girls, age 14)}$$
$$\bar{x}_2 = 100.18 \qquad s_2^2 = 18.2 \qquad n_2 = 16 \qquad \text{(boys, age 14)}$$
$$\bar{x}_3 = 422.43 \qquad s_3^2 = 28.6 \qquad n_3 = 15 \qquad \text{(girls, age 18)}$$
$$\bar{x}_4 = 355.76 \qquad s_4^2 = 27.1 \qquad n_4 = 18 \qquad \text{(boys, age 18)}$$

$$H_0: (\mu_1 - \mu_2) - (\mu_3 - \mu_4) = 0$$
$$H_A: (\mu_1 - \mu_2) - (\mu_3 - \mu_4) \neq 0$$

$$t = \frac{(\bar{x}_1 - \bar{x}_2) - (\bar{x}_3 - \bar{x}_4) - \delta}{\sqrt{s^2 \sum\limits_{i=1}^{4} \frac{1}{n_i}}} = \frac{(156.62 - 100.18) - (422.43 - 355.76)}{\sqrt{\frac{(1413.5)(193)}{(57)(720)}}}$$

$$= -3.97$$

where

$$s^2 = \frac{\sum\limits_{i=1}^{4} (n_i - 1)s_i^2}{\left(\sum\limits_{i=1}^{4} n_i\right) - 4} = \frac{(25.4)(11) + (18.2)(15) + (28.6)(14) + (27.1)(17)}{12 + 16 + 15 + 18 - 4}$$

$$= \frac{1413.5}{57}$$

and

$$\sum_{i=1}^{4} \frac{1}{n_i} = \frac{1}{12} + \frac{1}{16} + \frac{1}{15} + \frac{1}{18} = \frac{193}{720}$$

Critical region: $t > 2.70$ and $t < -2.70$

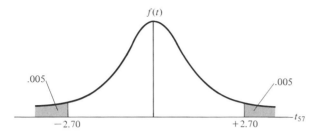

Reject H_0 since $t = -3.97$ falls in the rejection region (i.e., reject they hypothesis that differences between amounts spent by girls and boys is the same for ages 14 and 18).

9.4 TESTS CONCERNING VARIANCES

There are several types of situations in which it is important to test hypotheses concerning population variances, for example, the direct applications of these tests—the variability of a product that must meet rigid specifications must be controlled, the variability of demand must be considered in inventory control, the variabilities of the qualities of the product resulting from two different processes must be considered in deciding between them, and so forth—and the indirect applications of tests concerning variances—such tests are frequently prerequisites for tests concerning other population parameters, such as the t-tests of the preceding sections. In practice, when a test is based on the assumption of equal variances (i.e. homogeneity of variance), the reasonableness of the assumption must be tested in order to determine whether the test is appropriate. The hypotheses in this section concern the variances of one normally distributed population, two normally distributed populations, and more than two populations.

ONE-SAMPLE TESTS

CASE

H_0: $\sigma^2 = \sigma_0^2$

H_A: (a) $\sigma^2 > \sigma_0^2$
      (b) $\sigma^2 < \sigma_0^2$ } possible alternative hypotheses
      (c) $\sigma^2 \neq \sigma_0^2$

Assumptions:

A random sample of size n is obtained from a normal population.

Sample Statistic:

$$(n - 1)s^2 = \sum_{i=1}^{n} (x_i - \bar{x})^2$$

Sampling Distribution:

$$\chi^2 = \frac{(n - 1)s^2}{\sigma_0^2} \qquad \text{chi-square distribution with degrees of freedom } n - 1$$

Critical Region:

(a) $\chi^2 \geq \chi^2_{\alpha, n-1}$

(b) $\chi^2 \leq \chi^2_{1-\alpha, n-1}$

(c) $\chi^2 \geq \chi^2_{\alpha/2, n-1}$ and $\chi^2 \leq \chi^2_{1-\alpha/2, n-1}$

EXAMPLE

An optical company orders glass whose index of refraction must have a standard deviation not exceeding .01. The index of refraction of a random sample of 20 pieces of this glass has a variance of .00023. On this basis, should the glass be accepted or rejected by the optical company at the 1 percent level of significance?

$$H_0: \sigma^2 = .0001$$
$$H_A: \sigma^2 > .0001$$

$$\chi^2 = \frac{(n-1)s^2}{\sigma_0^2} = \frac{(19)(.00023)}{(.01)^2} = 43.7 \qquad \text{d.f.} = 19$$

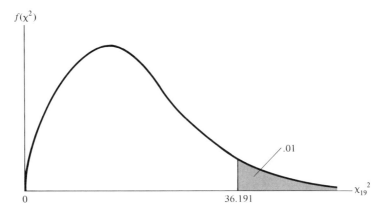

Critical region: $\chi^2 > 36.191$

So Reject H_0—that is, company should reject the glass.

TWO-SAMPLE TESTS

CASE

$$H_0: \sigma_1^2 = \sigma_2^2$$

$H_A:$ (a) $\sigma_1^2 > \sigma_2^2$

(b) $\sigma_1^2 < \sigma_2^2$ } possible alternative hypotheses

(c) $\sigma_1^2 \neq \sigma_2^2$

Assumptions:

Two independent random samples of sizes n_1 and n_2 are obtained from two normal populations having variances σ_1^2 and σ_2^2.

Sample Statistics:

$$s_1^2 = \frac{\sum_{i=1}^{n} (x_{1i} - \bar{x}_1)^2}{n_1 - 1} \quad \text{and} \quad s_2^2 = \frac{\sum_{i=1}^{n} (x_{2i} - \bar{x}_2)^2}{n_2 - 1}$$

Sampling Distribution:

$$F = \frac{s_1^2}{s_2^2} \qquad F \text{ distribution with d.f.} = n_1 - 1, n_2 - 1 \text{ or (equivalently)}$$

$$F = \frac{s_2^2}{s_1^2} \qquad F \text{ distribution with d.f.} = n_2 - 1, n_1 - 1$$

Critical Region:

(a) $\dfrac{s_1{}^2}{s_2{}^2} \geq F_{1-\alpha, n_1-1, n_2-1}$ if $s_1{}^2 > s_2{}^2$

(accept H_0 if $s_1{}^2 < s_2{}^2$)

(b) $\dfrac{s_2{}^2}{s_1{}^2} \geq F_{1-\alpha, n_2-1, n_1-1}$ if $s_2{}^2 > s_1{}^2$

(accept H_0 if $s_2{}^2 > s_1{}^2$)

(c) $\dfrac{s_1{}^2}{s_2{}^2} \geq F_{1-\alpha/2, n_1-1, n_2-1}$ if $s_1{}^2 \geq s_2{}^2$

and

$\dfrac{s_2{}^2}{s_1{}^2} \geq F_{1-\alpha/2, n_2-1, n_1-1}$ if $s_2{}^2 > s_1{}^2$

EXAMPLE

An investigator is interested in performing a *t*-test of the difference between the means of two populations. A random sample from one population has 15 observations and a variance of 22.5. A random sample from the other population has 20 observations and a variance of 4.2. At the 1 percent level of significance, is the assumption of equal variances justified?

$$H_0: \sigma_1{}^2 = \sigma_2{}^2$$

$$H_A: \sigma_1{}^2 \neq \sigma_2{}^2$$

$$F = \frac{s_1{}^2}{s_2{}^2} = \frac{22.5}{4.2} = 5.36 \qquad \text{d.f.} = 14, 19$$

Critical region: $\dfrac{s_1{}^2}{s_2{}^2} \geq 3.19$

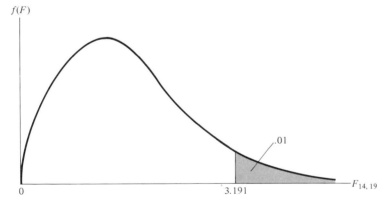

So reject H_0—that is, conclude that the assumption of equal variances is not justified.

TESTS CONCERNING MORE THAN TWO VARIANCES

The tests for equality of variances of more than two populations given in this section are for two cases: (1) samples of the same size and (2) samples of (possibly) different sizes.

CASE 1

$$H_0: \sigma_1{}^2 = \sigma_2{}^2 = \cdots = \sigma_k{}^2$$

$$H_A: \text{at least two variances are different}$$

Assumptions:

Independent random samples of the same size n are obtained from k normal populations having variances.

$$\sigma_1{}^2, \sigma_2{}^2, \ldots \sigma_k{}^2.$$

Sample Statistics:

$$s_i{}^2 = \frac{\sum\limits_{j=1}^{n} (x_{ij} - \bar{x}_i)^2}{n-1} \qquad \text{for } i = 1, 2, \ldots, k$$

Sampling Distribution:

$$F_{\max} = \frac{s_{\max}^2}{s_{\min}^2}$$ Hartley $F_{\max}$ distribution for k variances each having $n-1$ degrees of freedom

Critical Region:

$$\frac{s_{\max}^2}{s_{\min}^2} \geq F_{\max 1 - \alpha, (k, n)}$$

EXAMPLE

An investigator is interested in pooling the data obtained from 5 independent random samples from 5 different presumably normal distributions but is concerned that the population variances might differ. Each sample consists of 16 observations and the respective variances are 56.2, 84.9, 78.3, 37.6 and 98.9. At the 5 percent level of significance, is the assumption of equal variances justified?

$$H_0: \sigma_1{}^2 = \sigma_2{}^2 = \sigma_3{}^2 = \sigma_4{}^2 = \sigma_5{}^2$$

H_A: at least two variances are different

$$F_{\max} = \frac{s_{\max}^2}{s_{\min}^2} = \frac{98.9}{37.6} = 2.63$$

Critical region: $\dfrac{s_{\max}^2}{s_{\min}^2} \geq 4.37$ (see Table B.12 for Hartley $F_{\max}$ statistic)

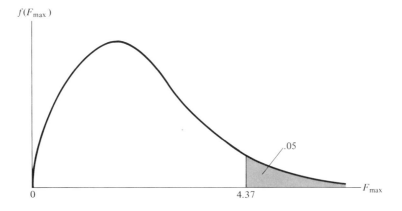

So accept H_0—that is, conclude that the assumption of equal variances is justified.

NOTE

A similar test for equality of variances using samples of the same size from normal distributions is based on Cochran's statistic

$$R_{n,k} = \frac{s_{max}^2}{\sum\limits_{i=1}^{k} s_i^2} \qquad \textit{for k variances each having n} - 1 \textit{ degrees of freedom}$$

for which tables are available. In most cases the Hartley test and the Cochran test give the same result—however, the tests differ with respect to power against some alternatives. Clearly, the Hartley test is particularly useful for detecting lack of homogeneity due to one relatively large and one relatively small variance regardless of the other variances involved; the Cochran test is particularly useful for detecting lack of homogeneity due to one variance that is large compared to the other variances involved.

CASE 2

$H_0 : \sigma_1^2 = \sigma_2^2 = \sigma_k^2$

H_A: at least two variances are different

Assumptions:
    Independent random samples of sizes $n_1, n_2, \ldots, n_k$ from k normal populations having variances $\sigma_1^2, \sigma_2^2, \ldots, \sigma_k^2$.

Sample Statistics:

$$s_i^2 = \frac{\sum\limits_{j=1}^{n_i} (x_{ij} - \bar{x}_i)^2}{n_i - 1} \qquad \text{for } i = 1, 2, \ldots, k.$$

Sampling Distribution:

$$B = \frac{2.3026}{C} \left[(N - k) \log S^2 - \sum_{i=1}^{k} (n_i - 1) \log s_i^2 \right]$$

where

$$S^2 = \frac{\sum\limits_{i=1}^{k} (n_i - 1)s_i^2}{N - k} \qquad N = \sum_{i=1}^{k} n_i$$

and

$$C = 1 + \frac{1}{3(k - 1)} \left[\sum_{i=1}^{k} \frac{1}{n_i - 1} - \frac{1}{N - k} \right]$$

    B is Bartlett's statistic; for large samples this statistic has approximately the chi-square distribution with $k - 1$ degrees of freedom.

Critical Region:
    $B \geq \chi_{1-\alpha, n-1}^2$

NOTE:
In practice, it may not be necessary to compute C. Since $C > 1$, dividing $2.3026 \left[(N - k) \log S^2 - \sum\limits_{i=1}^{k} (n_i - 1) \log s_i^2 \right]$ by C will decrease its value and should be done only if the statistic falls in the critical region without such division.

EXAMPLE
An investigator has obtained 4 independent random samples from 4 presumably normal distributions and has computed the following sample variances:

$$s_1^2 = 103 \qquad n_1 = 10$$
$$s_2^2 = 86 \qquad n_2 = 12$$
$$s_3^2 = 416 \qquad n_3 = 8$$
$$s_4^2 = 212 \qquad n_4 = 9$$

Is the assumption of equal population variances justified at the 5 percent level of significance

$$H_0: \sigma_1^2 = \sigma_2^2 = \sigma_3^2 = \sigma_4^2$$

H_A: at least two variances are different

$$B = \frac{2.3026}{C} \left[(N - k) \log S^2 - \sum_{i=1}^{k} (n_i - 1) \log s_i^2 \right]$$

$$S^2 = \frac{(9)(103) + (11)(86) + 7(416) + 8(212)}{10 + 12 + 8 + 9 - 4} = \frac{6481}{35} = 185.17$$

$$(N - k) \log S^2 = 35(2.2676) = 79.3730$$

$$(n_1 - 1) \log s_1^2 = 9(2.0128) = 18.1152$$

$$(n_2 - 1) \log s_2^2 = 11(1.9345) = 21.2795$$

$$(n_3 - 1) \log s_3^2 = 7(2.6191) = 18.3337$$

$$(n_4 - 1) \log s_4^2 = 8(2.3263) = 18.6104$$

$$\sum_{i=1}^{4} (n_i - 1) \log s_i^2 = 76.3388$$

$$C = 1 + \frac{1}{3(3)} \left[\frac{1}{9} + \frac{1}{11} + \frac{1}{7} + \frac{1}{8} - \frac{1}{35} \right]$$

$$= 1 + \frac{.4412}{9} = 1.0490$$

$$B = \frac{2.3026}{1.0490} [79.3730 - 76.3388]$$

$$= 6.66$$

Critical region: $B \geq 7.815$

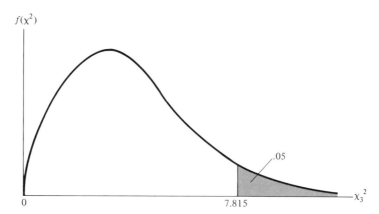

$f(\chi^2)$

.05

0   7.815 χ_3^2

So accept H_0—that is, the investigator is justified in assuming that the variances are equal.

NOTE:
Unfortunately, the usefulness of tests comparing variances is limited by the fact that these tests are extremely sensitive to nonnormality; since nonnormality of the parent population seriously affects the sampling distribution of the test statistic.

When a criterion for testing a hypothesis is derived, it is usually necessary for purposes of mathematical convenience to make simplifying assumptions, for example, normality of the population distribution. The resulting criterion should then be examined for "robustness" (i.e., for lack of sensitivity to these assumptions), in order to determine how literally the assumptions must be interpreted. This is usually done best empirically by taking random samples from distributions known not to be normal and by examining the corresponding sampling distribution of the statistic in question.

The usual tests for comparing means and variances are based on a number of assumptions; in particular, it is assumed that the parent populations are normal. However, in practice, usually little is known about the actual distributions of the populations from which samples are drawn and the tests are used as though the assumption of normality could be ignored. Fortunately, there is considerable empirical evidence that this practice is largely justified so far as tests comparing means are concerned, since these tests are remarkably insensitive ("robust") to general nonnormality of parent populations. Unfortunately, there is also considerable empirical evidence that this practice is not justified so far as tests comparing variances are concerned, since these tests do not share the property of robustness to nonnormality. Tests comparing variances are even more sensitive to nonnormality when more than two variances are involved; in fact, the Bartlett test is so sensitive to nonnormality that it is virtually a test for normality. Alternative tests for equality of variances, which are robust to nonnormality, depend on analysis of variance and are discussed in subsequent sections.

PROBLEMS

9.1 A random sample of boots worn by 60 combat soldiers in a desert region showed an average life of 1.16 yr with a standard deviation of 0.62 yr. Under standard conditions, these boots are known to have an average life of 1.33 yr. Is there reason to assert at a level of significance of 0.05 that use in the desert causes the mean life of such boots to decrease?

9.2 The IQs of 16 students from one area of a city showed a mean of 107 with a standard deviation of 10, while the IQs of 14 students from another area of the city showed a mean of 112 with a standard deviation of 8. Is there a significant difference between the IQs of the two groups at the .05 level?

9.3 Assume that the heights of 3000 male students at a university are normally distributed with a mean of 68.0 and a standard deviation of 3.0 in. If 80 samples of 25 students each are obtained, in how many samples would the mean be expected to lie between 66.8 and 68.3 in.?

9.4 The results of quality control tests on two manufacturing processes are:

> Process I: 1.5, 2.5, 3.5, 2.5
> Process II: 2.5, 3.0, 3.0, 4.0, 3.5, 2.0

Do these results allow the conclusion that the processes will yield different qualities? (Use 10 percent level of significance.)

9.5 From each of two normal and independent populations with identical means and with standard deviations of 6.40 and 7.20, respectively, a sample of 64 variates is drawn. Find the probability that the difference between the means of the samples exceeds .60 in absolute value.

9.6 The electric light bulbs of manufacturer A have a mean lifetime of 1400 hr with a standard deviation of 200 hr, while those of manufacturer B have a mean lifetime of

1200 hr with a standard deviation of 100 hr. If random samples of 125 bulbs of each brand are tested, what is the probability that brand A bulbs will have a mean lifetime that is at least 160 hr more than brand B?

9.7 Consumer Research Associates, Inc., is interested in comparing average yearly per family consumption of meat in rural and urban areas. In a random sample of 80 rural families average yearly consumption of meat is 170.6 lb with a standard deviation of 40; in a random sample of 125 urban families average yearly consumption of meat is 150.6 lb with a standard deviation of 25.

(a) Test the hypothesis that average yearly meat consumption per family is the same in urban and rural areas, 5 percent level of significance.

(b) Suppose that Consumer Research Associates, Inc., had been able to estimate the standard deviations 40 (for rural) and 25 (for urban) and had been interested in detecting a difference of 20 lb or more with probability 0.8 with a test at the 5 percent level of significance. How large should the sample be?

9.8 A random sample of 13 smokers of nonfilter cigarettes smoke an average of 150.5 cigarettes per week with a variance of $\frac{208}{9}$; a random sample of 14 smokers of filter cigarettes smoke an average of 141.5 cigarettes per week with a variance of $\frac{160}{3}$; the variance of the combined samples is $\frac{175}{3}$. The Cigarettes Manufacturers' Association suspects that heavier smokers favor nonfilter cigarettes. Assuming normality, do the data support that suspicion (5 percent level of significance)?

9.9 It is important in a certain industrial process that the average difference in pressure in the reactors of two successive stages does not exceed 5.5 lb/in^2. It is known from past records that the pressure in the first reactor has a variance of 6 lb/in^2. and the pressure in the second reactor has a variance of 9 lb/in^2. The quality control engineer has taken a random sample of 100 pressure readings from the first reactor and their average is 116.8 lb/in^2.; he has also taken a random sample of 144 pressure readings from the second reactor and their average is 109.9 lb/in^2. At the 5 percent level of significance, does the process appear to be in control in this respect?

9.10 It is suspected by a research technician that two populations of guinea pigs she intends to use for experimentation are not comparable with respect to average weight. She draws a random sample of 100 from one population and finds it to have a mean of 150 g and a variance of 1200 g; a random sample of 150 from the other population has a mean of 130 g and a variance of 600 g. At the 2 percent level of significance, what conclusion would she draw concerning the average weights of the two populations?

9.11 The Petro-Chem Company is considering two processes for refining petroleum. The engineering department claims that, from a standard input, process A produces an average of more than 25 gal more product than process B. The company statistician obtains data for a random sample of 25 batches from process B and a random sample of 10 batches from process A. The batches from process A have a mean of 512.2 gal and a standard deviation of 10 gal; the batches from process B have a mean of 480.8 gal and a standard deviation of 20 gal. At the 5 percent level of significance, is the engineering department's claim supported or refuted?

9.12 The following statistics were computed from the results of quality control tests on two manufacturing processes:

$$\bar{x}_1 = 116.7 \qquad s_1^2 = 140 \qquad n_1 = 10$$
$$\bar{x}_2 = 111.9 \qquad s_2^2 = 20 \qquad n_2 = 10$$

Do these results justify the conclusion (5 percent level of significance) that the processes will yield different qualities on the average?

9.13 The following statistics were computed from the results of blood counts on two groups of experimental animals:

$$\bar{x}_1 = 125.6 \qquad s_1^2 = 115 \qquad n_1 = 9$$
$$\bar{x}_2 = 113.1 \qquad s_2^2 = 92 \qquad n_2 = 16$$

Is there a significant difference in the blood counts (5 percent level)?

9.14 The Edison Thomas Company manufactures tungsten filaments for large industrial lamps. The quality control engineer is interested in the comparability of the average lengths of the filaments from two production lines. He obtains a random sample of filaments from production line 1, which have an average length of 25 cm and a variance of 14 cm. He obtains a random sample of filaments from production line 2, which have an average length of 23 cm and a variance of 10 cm. Do the data support or refute the quality control engineer's contention that the two production lines supply filaments having the same average length

(a) If he took a random sample of size 45 from production line 1 and a random sample of size 75 from production line 2

(b) If he took a random sample of size 10 from production line 1 and a random sample of size 8 from production line 2 (Use the 5 percent level of significance.)

9.15 The dressed weights of turkeys raised by grower A have a mean of 25 lb and a standard deviation of 1.6 lb. The dressed weights of turkeys raised by grower B have a mean of 23 lb and a standard deviation of 1.2 lb. If random samples of 100 dressed turkeys raised by each grower are weighed, what is the probability that the sample of turkeys from grower A will have an average weight at least 1.5 lb greater than the average weight of the sample of turkeys from grower B?

9.16 Manufacturer A claims that his snow blowers (which are cheaper) run at least as long on the average as those of manufacturer B before needing the first repair. A random sample of 5 blowers from manufacturer A ran an average of 125.4 hr with a standard deviation of 30 hr before needing repair; a random sample of 7 blowers from manufacturer B ran an average of 159.6 hr with a standard deviation of 10 hr before needing repair. On the basis of these data, is manufacturer A's claim justified at the 1 percent level of significance?

9.17 The president of Strongsteel Ltd. wants to purchase a new machine only if the tensile strength of the cable lengths it produces exceeds by at least 15 lb the tensile strength of the cable lengths produced by the machine now in use. While the new machine was in the factory for demonstration purposes he obtained a random sample of 9 cable lengths from the new machine and 16 cable lengths from the old machine. The cable lengths from the new machine had an average tensile strength of 135.2 lb with a variance of 21 lb; the cable lengths from the old machine had an average tensile strength of 116.7 lb with a variance of 44 lb. Should he buy the new machine (significance level .05)?

9.18 The Small Grocers Association claims that the average price of 5-lb canned hams is no more in neighborhood stores than in supermarkets. To check on this claim they have obtained a random sample of prices from 27 neighborhood stores and 24 supermarkets. The prices for the neighborhood stores have an average of $5.62 with a standard deviation of $.03; the prices for the supermarkets have an average of $5.59 with a standard deviation of $.08. Should the Small Grocers Association revise the claim (1 percent level)?

9.19 Galvanized Corporation is interested in comparing the average weight of cable wound on standard size spools by two different machines. A random sample of 21 spools wound on machine 1 has an average weight of 982.89 lb and a variance of 40 lb. A random sample of 6 spools wound on machine 2 has an average weight of 976.23 lb and a variance of 50 lb. On the basis of these data, do the two machines wind spools having the same average weight (5 percent level of significance)?

9.20 The quality control engineer for Super Manufacturing Company has specified that the standard deviation of the diameter of a particular type of washer should not exceed 3 mm. He obtains a random sample of 9 washers from a large batch delivered from the supplier and obtains the following measurements (in millimeters) for their diameters: 10, 12, 7, 6, 15, 9, 13, 10, 8. On the basis of these data does the batch of washers apparently meet specifications (5 percent level)?

9.21 A company wants to test whether the average tool life per sharpening of a cutting tool is at least 3000 pieces. If the tests on a random sample of 6 tools showed tool lives of 2970, 3020, 3005, 2900, 2940, and 2925 pieces, what conclusion will the company reach?

9.22 The production engineer of Plastics Incorporated is interested in comparing the average amount of sheet polythene produced per run by two reactors. He obtains random samples of 10 runs from the first machine and 12 runs from the second machine. The weights of the runs from the first machine had a mean of 126.8 lb and a variance of 36 lb. The weights of the runs from the second machine had a mean of 122.5 lb and a variance 24 lb. Assuming that the amounts are independently normally distributed, is this evidence (at the 5 percent level of significance) that the machines produce runs having different average weights?

9.23 The variance of independent random samples of sizes 30 and 15 from two normal populations are, respectively, 20 and 65. The investigator wished to decide mistakenly that the population variances are unequal with a probability not exceeding 0.02. What should she conclude about the homogeneity of variances of the two populations?

9.24 United Canners claims to buyers that the average weight of large (restaurant) size cans of peaches is at least 15 lb. A random sample of 5 cans of peaches are weighed and the following weights (in pounds) are obtained: 15.5, 14.9, 14.3, 15.4, 14.4. On the basis of this sample, does United Canners' claim appear to be justified?

9.25 The quality control engineer for Slick Oil, Inc. has agreed to accept oil drums either from manufacturer A or manufacturer B provided that the average diameters of the drums from the two manufacturers do not differ by more than 6 in. A random sample of 16 drums from manufacturer A shows an average diameter of 110.54 in. and a standard deviation of 3.6 in. A random sample of 36 from manufacturer B shows an average diameter of 103.17 in. and a standard deviation of 1.8 in. On the basis of these data, does the engineer's specifications seem to be met (5 percent level)?

9.26 The management of Electron-Tubes, Inc. wished to replace its equipment currently in use by new equipment only if the new equipment produces electronic tubes having an average length of life of at least 15 hr greater than that of the tubes produced by the current equipment. They obtain a random sample of 40 tubes produced by the new equipment (loaned to them for this purpose) and 150 tubes produced by the current equipment. The length of life of the tubes produced by the new equipment had an average of 123.5 hr and a standard deviation of 10 hr. The length of life of the tubes produced by the current equipment had an average of 103.2 hr and a standard deviation of 15 hr. On the basis of these data should the company replace their current equipment (1 percent level)?

9.27 The Tasty Fruit Growers Co. has developed a new apple that has a delicious flavor. However, since the apple was developed experimentally, its yielding capacities are unknown. They decided to test this important quality by planting the new apple adjacent to a standard apple in 8 orchards scattered about a region suitable for growing both varieties. When the trees are matured the yields in bushels are as follows:

ORCHARD NUMBER	NEW APPLE	STANDARD APPLE
1	13	12
2	14	16
3	19	17
4	10	9
5	15	16
6	14	12
7	12	10
8	11	8

Do these results indicate a different yield for the new apple as compared with the standard apple (5 percent level of significance)?

9.28 The Alure Perfume Co. is introducing two new scents ("Daytime Delight" and "Evening Madness") and has decided to investigate the effectiveness of two different methods of promotion for each—advertising posters throughout a store (method 1) and free samples given by pretty girls at the perfume counter of a store (method 2). The stores in which the company's perfume is sold are paired on the basis of average monthly perfume sales; a random sample of 8 pairs is chosen for "Daytime Delight" and a random sample of 12 stores is chosen for "Evening Madness," and weekly sales data below are obtained. One of the company's vice-presidents has a hunch that method 2 is more effective than method 1 and that the difference is greater for "Evening Madness" than for "Daytime Delight." Does she appear to be correct at the 5 percent level of significance?

Daytime Delight

STORE PAIR NUMBER	METHOD 2	METHOD 1
1	105	100
2	120	123
3	184	183
4	176	172
5	138	140
6	153	156
7	197	190
8	119	120

Evening Madness

STORE PAIR NUMBER	METHOD 2	METHOD 1
1	160	150
2	183	175
3	145	143
4	179	179
5	133	136
6	142	130
7	199	184
8	124	117
9	177	170
10	186	183
11	198	190
12	157	154

9.29 The Happy Super Market Co. is interested in comparing the costs of plane transportation and truck transportation of produce from their wareshouses to their stores. The company has two main warehouses located some distance apart, and the transportation department suspects that the more economical mode of transportation might differ for the two warehouses. For a random sample of 8 orders received at warehouse A and a random sample of 10 orders received at warehouse B, the costs of transportation by both plane and truck were determined, as given below. Do the warehouses appear to differ with regard to the most economical mode of transportation (10 percent level)?

WAREHOUSE A			WAREHOUSE B		
ORDER NUMBER	PLANE	TRUCK	ORDER NUMBER	PLANE	TRUCK
1	20	22	1	8	12
2	9	11	2	24	18
3	18	15	3	10	6
4	15	12	4	16	20
5	14	13	5	15	16
6	12	12	6	12	6
7	17	16	7	9	18
8	24	20	8	14	13
			9	20	23
			10	15	12

9.30 The Tweedy Sportswear Co. is considering computerizing the inventory control in all its many stores. The management has decided that this is justifiable financially if the average monthly reduction in sales lost as a result of poor inventory control is at least $1000 per store. In a random sample of 12 stores the clerks were asked to keep careful records for one month of all items (and their costs) that customers wished to see but which were unavailable as a result of having been sold out. Their inventory control was computerized on a trial basis in the 12 stores and clerks were again asked to keep records for one month of items unavailable as a result of having been sold out. The store totals (in thousands of dollars) of these "lost" sales are given below. Does computerizing the inventory control appear to be justifiable financially at the 1 percent level of significance? (Note: although not all items customers requested would actually have been sold if available, this was felt by management not to bias the experiment.)

STORE NUMBER	NONCOMPUTERIZED	COMPUTERIZED
1	100	96
2	182	182
3	169	167
4	174	171
5	126	127
6	118	119
7	135	134
8	193	193
9	147	140
10	181	180
11	155	151
12	107	103

9.31 Budget Counsellors, Inc. is interested in the discrepancy between husbands' and wives' view of the importance of clothes in the family budget and suspect that this discrepancy is greater for white collar families than for blue collar families. A random sample of 6 white collar families and a random sample of 9 blue collar families were chosen and the husbands and wives were asked (independently) what they thought their family budget should allot for clothes. On the basis of the data given below, are the suspicions comfirmed (1 percent level)?

FAMILY NUMBER	WHITE COLLAR		BLUE COLLAR	
	WIVES	HUSBANDS	WIVES	HUSBANDS
1	108	42	64	38
2	146	68	60	33
3	210	150	37	19
4	172	95	29	11
5	139	68	19	9
6	119	51	78	53
7			42	30
8			35	21
9			51	21

9.32 The Tree County School Board is interested in comparative incomes (after three years of work) of high school and college graduates in two communities in its jurisdiction—one community is essentially rural and the other is highly urban. Random samples of 8 high school and 8 college graduates were obtained from each community and the yearly salaries (in thousands of dollars) were as recorded below. Does the difference between average incomes of high school and college graduates appear to be greater in the urban community (5 percent level)?

RURAL		URBAN	
HIGH SCHOOL	COLLEGE	HIGH SCHOOL	COLLEGE
31	67	23	85
38	74	22	80
35	69	28	75
26	79	25	71
30	73	30	78
34	70	33	81
24	76	34	83
30	68	29	79

9.33 Dr. X-Ray is studying the effectiveness of two hormones (A and B) for producing tumors in mice. She has 9 pairs of mice for her experiment; each pair consists of litter-mates matched for weight. To one of each pair (at random) she administered a standard dose of drug A, to the other member of the pair she administered a standard dose of drug B. The weights (in milligrams) of the resulting tumors are given below. Using the average weights of the tumors as criteria, are the drugs equally effective in producing tumors (5 percent level of significance)?

PAIR	DRUG A	DRUG B
1	105	110
2	98	100
3	76	74
4	85	83
5	102	102
6	101	108
7	97	100
8	74	80
9	99	98

9.34 A company has put on an intensive (and expensive) advertising campaign and wants to evaluate the immediate results. The statistician associated with the project feels there was no major factor, other than the advertising campaign, which affected sales in an unusual manner during the period of interest. She obtains sales figures from a random sample of six of the company's distributors for the two weeks prior to the advertising campaign and for the two weeks following the advertising campaign. In order to be judged a success, the advertising campaign must have increased the average sales of the distributors by at least $32,500. On the basis of the data (given below in thousands of dollars), would the statistician conclude the campaign was or was not a success (5 percent level of significance)?

DISTRIBUTOR	BEFORE	AFTER
1	70	100
2	99	123
3	78	101
4	115	140
5	110	147
6	103	136

9.35 P. R., Inc. is interested in the effectiveness of two methods, A and B, of advertising coffee in urban and suburban supermarkets. The following data have been obtained (assume that the data represent sales appropriately adjusted for level prior to advertising):

URBAN		SUBURBAN	
METHOD A	METHOD B	METHOD A	METHOD B
20	18	19	19
22	20	19	18
19	16	16	15
17	12	20	18
23	21	22	21
19	15	18	17

The statistician at P. R., Inc. is told that method A is thought to be more effective than method B and that the average difference in effectiveness is believed to be the same for urban and suburban stores. On the basis of the data, what would the statistician conclude (5 percent level),

(a) If the data are obtained from four independent random samples of supermarkets
(b) If the data are obtained from two independent random samples of pairs of super-markets matched for size, type of customer, and so forth; one set of matched pairs of urban supermarkets and one set of matched pairs of suburban supermarkets

9.36 It has been observed that the first born of a pair of identical twins is heavier on the average at birth and through the school years. Sally Psychologist believes that the average weight of the first born of a pair of identical twins does not exceed the average weight of the second born by more than 5 lb when the twins are between the ages of of 45 and 60. The weights of a random sample of 10 pairs of identical twins between the ages of 45 and 60 are given below. Do these data tend to support or refute Sally's claim (5 percent level)?

FIRST BORN	SECOND BORN
152	143
189	170
172	176
145	146
159	150
195	186
125	130
168	160
179	163
182	172

CONTINGENCY TABLES AND GOODNESS OF FIT

10.1 INTRODUCTION

The hypotheses considered in the previous chapter concern parameters of continuous distributions. Tests of these hypotheses are based on sample statistics computed from data that are essentially continuous, although all measurements are discrete in practice. The hypotheses considered in the following sections concern parameters of discrete distributions; tests of these hypotheses are based on sample statistics whose values are computed from data that are essentially discrete. Note that, whereas data appropriate for testing hypotheses concerning parameters of continuous distributions consist of measurements of sample elements, data appropriate for testing hypotheses concerning parameters of discrete distributions consist of frequencies of sample elements falling into certain categories or classes. The former data are said to be measurable, the latter are said to be countable or categorical.

Frequently, in practice, the hypothesis of interest concerns the population frequencies or proportions of elements occurring in various classes or categories and the data obtained consist of the corresponding sample frequencies or proportions. For example, an investigator might be interested in the proportion of defectives resulting from a certain process or in comparing the proportions of defectives resulting from several different processes or in comparing the frequencies occurring in various categories with those predicted by independence or by some hypothesized distribution.

The following sections concern tests of hypotheses based on countable or categorical data. The population distributions most frequently assumed are the binomial distribution for two classes having population proportions π and $1 - \pi$, and the multinomial distribution for k classes having population proportions

$$\pi_1, \pi_2, \ldots, \pi_k; \qquad \sum_{i=1}^{k} \pi_i = 1.$$

10.2 HYPOTHESES CONCERNING PROPORTIONS

ONE SAMPLE

CASE

The hypotheses considered now concern the population proportion for one population:

$$H_0: \pi = \pi_0$$

$$H_A: \text{(a)} \ \pi > \pi_0$$
$$\text{(b)} \ \pi < \pi_0$$
$$\text{(c)} \ \pi \neq \pi_0$$

Assumptions:

Random sample of size n from a binomial population having proportion π_0 of the elements of interest.

Sample Statistic:

p, the sample proportion

Sampling Distribution:

Normal approximation if

$$n\pi_0 \geq 5, \qquad n(1 - \pi_0) \geq 5$$

$$Z = \frac{p - \pi_0}{\sqrt{\dfrac{\pi_0(1 - \pi_0)}{n}}} \quad \text{standard normal}$$

NOTE:

Equivalently, $Z = \dfrac{x - n\pi_0}{\sqrt{n\pi_0(1 - \pi_0)}}$

where $x = np$ is sample frequency

Critical Region:

(a) $Z \geq z_{1-\alpha}$
(b) $Z \leq z_{\alpha}$
(c) $Z \geq z_{1-\alpha/2}$ and $Z \leq z_{\alpha/2}$

EXAMPLE

From past experience we know that the proportion of cattle having a particular hereditary bone defect is .0005 if the prescribed breeding procedure is followed. This proportion increases if certain rules of the procedure are violated. In a random sample of 10,000 cattle there are 10 having the bone defect. Is this evidence (5 percent level of significance) that the prescribed procedure was violated?

$$H_0: \pi = .0005$$

$$H_A: \pi > .0005$$

$$z = \frac{p - \pi_0}{\sqrt{\dfrac{\pi_0(1 - \pi_0)}{n}}} = \frac{.0010 - .0005}{\sqrt{\dfrac{(.0005)(.9995)}{10,000}}} = \frac{.05}{.0224} = 2.23$$

Critical Region:
$z \geq 1.645$

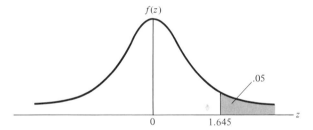

So reject H_0—that is, there is evidence the procedure was violated.

Whenever $n\pi_0 < 5$ or $n(1 - \pi_0) < 5$, the normal approximation for the sampling distribution of p is not appropriate and the binomial distribution should be used.

EXAMPLE
A motor company buys small carbon resistors in boxes of approximately 1000. (The resistors are not counted, but are sold by weight.) In the assembly line, all resistors are given a quick test for tolerance before being assembled as part of armatures. If the resistor does not meet the tolerance when tested, it is discarded. Because the discarded resistors may not be sent back to the manufacturer and because of the slowdown in the assembly line if too many resistors must be tested and replaced, the company will send back an entire box of resistors if they believe that 20 percent or more of the resistors are out of tolerance enough to be discarded. From each box, a random sample of 10 resistors is taken and tested. The company is anxious to avoid the error of sending back boxes that have less than 20 percent defective resistors. The α level is set at 5 percent. How many defective resistors out of the 10 will cause a conclusion of 20 percent or more defective and hence a return of the box?

$$H_0: \pi < 20 \text{ percent}$$

$$H_A: \pi \geq 20 \text{ percent}$$

$$n = 10; \quad \alpha = 5 \text{ percent}$$

Since $n\pi_0 = (10)(.20) = 2$, the normal approximation may not be used. Set up the following table to use the binomial distribution where x is the number of defectives found in a sample of 10 and $p = x/n$ is the proportion of defectives found.

x	$p = \dfrac{x}{n}$	$P(x) = P(p)$	$P(x \leq x_i) = P(p \leq p_i)$ CUMULATIVE	$P(x \geq x_i) = P(p \geq p_i)$ DECUMULATIVE
0	0	0.1074	0.1074	1.0000
1	0.1	0.2684	0.3758	0.8926
2	0.2	0.3020	0.6778	0.6242
3	0.3	0.2013	0.8791	0.3222
4	0.4	0.0881	0.9672	0.1209
5	0.5	0.0264	0.9936	0.0328
6	0.6	0.0055	0.9991	0.0064
7	0.7	0.0008	0.9999	0.0009
8	0.8	0.0001	1.0000	0.0001
9	0.9	0.0000	1.0000	0.0000
10	1.0	0.0000	1.0000	0.0000

The third column $P(x) = P(p)$ is obtained from the binomial table (Table B.1) for $n = 10$ and $\pi = .20$. From the last column in the table, the probability of getting 4 or more defectives is .1209 when $\pi = .20$, the probability of getting 5 or more defectives is .0328 when $\pi = .20$. Since the probability permitted by the α specified is .05 when $\pi = .20$ (actually .20 in the limit since $H_0: \pi < .20$), the box should be sent back if there are 5 or more defectives in the sample. From the decumulative column, we see that if boxes whose samples show 5 or more defectives are sent back, 3.28 percent of the time boxes whose proportion of defective is .20 will be sent back. Also, if boxes whose samples show 4 or more defectives are sent back 12.09 percent of the time, boxes whose proportion of defectives is .20 will be sent back. Since $\alpha = 5$ percent, the rule will be to send back boxes whose samples show 5 or more defectives.

MORE THAN ONE SAMPLE

CASE

The hypotheses in this section concern population proportions for more than one population.

$$H_0: \pi_1 = \pi_2 = \cdots = \pi_k = \pi_0$$

$$H_A: \pi_i\text{'s}, \qquad i = 1, 2, \ldots, k, \text{ not all equal}$$

Assumptions:

Independent random samples of sizes $n_1, n_2, \ldots, n_k$, respectively, from k binomial populations having proportions π_i of the elements of interest.

Sample Statistics:

$p_1, p_2, \ldots, p_k$, the sample proportions.

Sampling Distribution:

Chi-square approximation if

$$n_i\pi_0 \geq 5 \quad \text{and} \quad n_i(1 - \pi_0) \geq 5 \qquad \text{for } i = 1, 2, \ldots, k_{.2}$$

$$\chi^2 = \sum_{i=1}^{k} \frac{n_i(p_i - \hat{\pi}_0)^2}{\hat{\pi}_0(1 - \hat{\pi}_0)} = \sum_{i=1}^{k} \frac{(x_i - n_i\hat{\pi}_0)^2}{n_i\hat{\pi}_0(1 - \hat{\pi}_0)} \qquad \text{chi-square with d.f.} = k - 1$$

where

$$\hat{\pi}_0 = \frac{\displaystyle\sum_{i=1}^{k} n_i p_i}{\displaystyle\sum_{i=1}^{k} n_i}$$

NOTE:

$\hat{\pi}_0$ *is a sample estimate of the parameter* π_0. *If the value of* π_0 *is specified in the hypothesis, this value is used instead of* $\hat{\pi}_0$ *in the computation and d.f.* $= k$, *since no parameters are estimated.*

Critical Region:

$$\chi^2 \geq \chi^2_{1-\alpha,k-1}$$

An equivalent, generally more easily computed form of the chi-square statistic, is based on observed cell frequencies denoted by f_{ij} and expected cell frequencies denoted by e_{ij}. It is important to emphasize that the f_{ij} are the actual sample frequencies that occur and the e_{ij} are the expected values of the cell frequencies if the null hypothesis is true. Using this method of specifying the chi-square statistic, we have

$$\chi^2 = \sum_{i=1}^{2} \sum_{j=1}^{k} \frac{(f_{ij} - e_{ij})^2}{e_{ij}}$$

This statistic is computed from the sample data as defined in the following table:

Sample 1	Sample 2	$\cdots$	Sample k	
$f_{11} = x_1$ $e_{11} = n_1\hat{\pi}_0$	$f_{12} = x_2$ $e_{12} = n_2\hat{\pi}_0$	$\cdots$	$f_{1k} = x_k$ $e_{1k} = n_k\hat{\pi}_0$	$\sum\limits_{j=1}^{k} f_{ij} = \sum\limits_{k=1}^{k} x_j$
$f_{21} = n_1 - x_1$ $e_{21} = n_1(1 - \hat{\pi}_0)$	$f_{22} = n_2 - x_2$ $e_{22} = n_2(1 - \hat{\pi}_0)$	$\cdots$	$f_{2k} = n_k - x_k$ $e_{2k} = n_k(1 - \hat{\pi}_0)$	$\sum\limits_{j=1}^{k} f_{2j} = \sum\limits_{j=1}^{k} nj - \sum\limits_{j=1}^{k} x_j$
$f_{11} + f_{21} = n_1$	$f_{12} + f_{22} = n_2$	$\cdots$	$f_{1k} + f_{2k} = n_k$	$\sum\limits_{j=1}^{k} n_j$

In general,

$$f_{1j} = x_j \qquad f_{2j} = n_j - x_j$$
$$e_{1j} = n_j\hat{\pi}_0 \qquad e_{2j} = n_j(1 - \hat{\pi}_0)$$

The student who has difficulty in understanding the above table should review the discussion of summation in Chapter 4. Also, the following comments may help: When referring to the contents of a cell in a matrix or table with double subscripts, such as f_{ij}, the first subscript refers to the row and the second refers to the column. For example, f_{23} means the observed cell frequency in the cell found at the intersection of the second row and third column. In the above table there are only two rows—this results from the usual necessary binomial categorizations of observations regardless of the number of classes. The k columns result because there are now k samples with the multinomial instead of the one sample with the binomial. Each sample represents a class designated prior to sampling, and although there may be k classes, the outcome of an observation still falls into the binomial-type two categorizations. Note that there are n_1 observations in the first class, n_2 in the second, on up to n_k in the kth class. The letter n is used to indicate the total number of observations across all classes; it has been used throughout as the sample size.

$$\sum_{j=1}^{k} n_j = n_1 + n_2 + \ldots + n_k = n$$

EXAMPLE
A university counseling service is interested in whether the proportions of (eldest) sons taking up the occupations of their fathers are equal for doctors, lawyers, teachers, and ministers. The following data are obtained from independent random samples of doctors, lawyers, teachers, and ministers having sons. Test the hypothesis that the proportion of sons taking up the occupations of their fathers is the same for these four occupations (5 percent significance level).

	Doctors	Lawyers	Teachers	Ministers
Same Occupation	100	25	40	15
Different Occupation	100	75	110	35

$$H_0: \pi_1 = \pi_2 = \pi_3 = \pi_4 = \pi_0$$

$$H_A: \pi_i\text{'s, } i = 1, 2, 3, 4, \text{ not all equal}$$

100 $e_{11} = 72$	25 $e_{12} = 36$	40 $e_{13} = 54$	15 $e_{14} = 18$	$\sum\limits_{j=1}^{4} x_j = 180$
100 $e_{21} = 128$	25 $e_{22} = 64$	40 $e_{23} = 96$	15 $e_{24} = 32$	$\sum\limits_{j=1}^{4} (n_j - x_j) = 320$

$$n_1 = 200 \quad n_2 = 100 \quad n_3 = 150 \quad n_4 = 50 \quad \sum_{j=1}^{4} n_j = 500$$

$$\hat{\pi}_0 = \frac{\sum\limits_{j=1}^{4} x_j}{\sum\limits_{j=1}^{4} n_j} = \frac{180}{500} = \frac{9}{25}$$

$$1 - \hat{\pi}_0 = \frac{16}{25}$$

$$e_{1j} = n_j \hat{\pi}_0 = n_j \left(\frac{9}{25}\right)$$

for example,

$$e_{11} = 200 \left(\frac{9}{25}\right) = 72$$

$$e_{2j} = n_j (1 - \pi_0) = n_j \left(\frac{16}{25}\right)$$

for example,

$$e_{21} = 200 \left(\frac{16}{25}\right) = 128$$

$$\chi^2 = \sum_{i=1}^{2} \sum_{j=1}^{k} \frac{(f_{ij} - e_{ij})^2}{e_{ij}}$$

$$= \frac{(100 - 72)^2}{72} + \frac{(25 - 36)^2}{36} + \frac{(40 - 54)^2}{54} + \frac{(15 - 18)^2}{18} + \frac{(100 - 128)^2}{128}$$

$$+ \frac{(75 - 64)^2}{64} + \frac{(110 - 96)^2}{96} + \frac{(35 - 32)^2}{32}$$

$$= 10.89 + 3.33 + 3.63 + 0.50 + 6.13 + 1.89 + 2.04 + 0.28$$

$$= 28.69$$

Critical Region:

$\chi^2 \geq 7.8$,   *for $\alpha = 5$ percent and d.f. = 3*

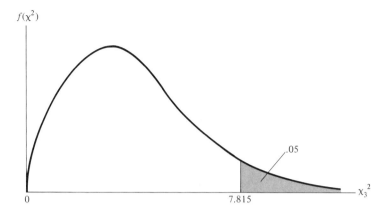

Therefore reject H_0 and conclude that the proportions of sons taking up the occupations of their fathers are not the same for all four occupations.

SPECIAL CASE: k = 2

The procedure discussed for the general case of testing the equality of k proportions can be used only to test against the alternative hypothesis that the proportions are not all equal. However, in the special case of two proportions, either of the one-sided alternative hypotheses $\pi_1 > \pi_2$ or $\pi_1 < \pi_2$ can be tested (rather than the two-sided alternative hypothesis $\pi_1 \neq \pi_2$). An equivalent procedure, based on the normal distribution rather than the chi-square distribution, provides a test against either of these two-sided alternatives.

$$H_0: \pi_1 = \pi_2$$

$$H_A: \text{(a)} \ \pi_1 > \pi_2$$
$$\text{(b)} \ \pi_1 < \pi_2 \left.\vphantom{\begin{matrix}1\\1\\1\end{matrix}}\right\} \quad \text{possible alternative hypotheses}$$
$$\text{(c)} \ \pi_1 \neq \pi_2$$

Assumptions:

Independent random samples of sizes n_1 and n_2 from two binomial populations having proportions π_1 and π_2 of the elements of interest.

Sample Statistics:

p_1 and p_2, the sample proportions

Sampling Distribution:

Approximately normal if

$$n_1 \pi_0 \geq 5 \qquad n_1(1 - \pi_0) \geq 5 \qquad n_2 \pi_0 \geq 5 \qquad n_2(1 - \pi_0) \geq 5$$

Z, the standard normal variate, is defined as follows:

$$Z = \frac{p_1 - p_2}{\sqrt{\hat{\pi}_0(1 - \hat{\pi}_0)\left(\dfrac{1}{n_1} + \dfrac{1}{n_2}\right)}} = \frac{n_2 x_1 - n_1 x_2}{\sqrt{\dfrac{n_1 n_2(x_1 + x_2)[(n_1 + n_2) - (x_1 + x_2)]}{n_1 + n_2}}} \sim N(0, 1)$$

where

$$\pi_0 = \frac{x_1 + x_2}{n_1 + n_2}$$

Critical Region:

(a) $Z \geq z_{1-\alpha}$

(b) $Z \leq z_\alpha$

(c) $Z \geq z_{1-\alpha/2}$ and $Z \leq z_{\alpha/2}$

NOTE:
It can be shown that the square of the Z statistic equals the χ^2 statistic (with 1 degree of freedom) and thus the tests are equivalent.

EXAMPLE
The registrar of a university is interested in whether the proportion of men and women who graduate without dropping out either permanently or temporarily is the same or whether it is (as alleged) lower for women. From independent random samples of the university's records, the registrar obtained the following data.

	Men	Women
Stayed	125	75
Dropped	25	25

What conclusion would be reached (1 percent level of significance)?

$$H_0: \pi_1 = \pi_2$$

$$H_A: \pi_1 < \pi_2 \text{ (1 denotes women, 2 denotes men)}$$

$$z = \frac{p_1 - p_2}{\sqrt{\hat{\pi}_0(1 - \hat{\pi}_0)\left(\frac{1}{n_1} + \frac{1}{n_2}\right)}} = \frac{\frac{3}{4} - \frac{5}{6}}{\sqrt{\left(\frac{4}{5}\right)\left(\frac{1}{5}\right)\left(\frac{1}{150} + \frac{1}{100}\right)}} = -1.614$$

Critical Region:
$z_{.01} \leq -2.326$

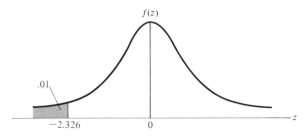

So accept H_0—that is, the proportions who drop out are the same. Alternatively,

125	75	200
$e_{11} = 120$	$e_{12} = 80$	
25	25	50
$e_{21} = 30$	$e_{22} = 20$	
150	100	250

$$\chi^2 = \sum_{i=1}^{2} \sum_{j=1}^{2} \frac{(f_{ij} - e_{ij})^2}{e_{ij}} = \frac{25}{120} + \frac{25}{80} + \frac{25}{30} + \frac{25}{20} = 2.604$$

In order to use the χ^2 statistic for a one-sided test, reject H_0 if the sign of $\pi_1 - \pi_2$ is as given by H_A and the computed $\chi^2 \geq \chi^2_{1-2\alpha}$. Note that an ordinary two-sided χ^2-test at level α is equivalent to a two-sided z-test at level α.

$$Z_\alpha^2 = Z_{1-\alpha}^2 = \chi^2_{1-2\alpha}$$

Critical Region:
   $\chi^2_{.98} \geq 5.4$

So accept H_0, as above.

   NOTE:

Computed Statistics:
   $z^2 = (-1.614)^2 \simeq 2.604 = \chi^2$

Critical Values:
   $z^2 = (-2.326)^2 \simeq 5.4 = \chi^2$

10.3 CONTINGENCY TABLES

A contingency table is a two-way table (having, in general, r rows and k columns) which summarizes the cross-classification of sample elements (persons, objects, and so forth) according to two characteristics. The investigator is interested in whether there is a relationship between the two characteristics, one represented by the rows and the other by the columns of the table. The null hypothesis is that the two characteristics are independent (i.e., there is no relationship between them). The argument is as follows: If the two characteristics are independent, then the expected frequency in any cell is the product of the total number of observations in the column containing the cell, times the total number of observations in the row containing the cell, divided by the total number of observations in the sample. That is, the expected frequency is the proportion of the total sample observations in the column times the proportion in the row divided by the total number of observations. If n is the total sample taken, and $\sum_{j=1}^{k} f_{1j}$ is the number of observations in the first row, and $\sum_{i=1}^{r} f_{i2}$ is the number of observations in the second column, then the expected frequency in cell (1, 2) is

$$\frac{\left(\sum_{j=1}^{k} f_{1j}\right)\left(\sum_{i=1}^{r} f_{i2}\right)}{\sum_{j=1}^{k} \sum_{i=1}^{r} f_{ij}} = \frac{\left(\sum_{j=1}^{k} f_{1j}\right)\left(\sum_{i=1}^{r} f_{i2}\right)}{n}$$

In probability terms—if the null hypothesis of independence is true, then the probability of a sample element being in a particular cell of the table is equal to the product of the probabilities of its being in the corresponding row and column of the table. The probability of being in a particular row, say the first row, is

$$\frac{\sum_{i=1}^{k} f_{1j}}{n} = p_{1.}$$

The probability of being in a particular column, say, the second column, is

$$\frac{\sum_{j=1}^{k} f_{i2}}{n} = p_{.2}$$

CASE

If the probability of being in any row is designated $p_{i.}$ and the probability of being in any column $p_{.j}$, then the expected frequency in cell (i, j) is $(p_{i.})(p_{.j})(n)$. The hypothesis of independence is tested on the basis of the chi-square statistic using these expected frequencies.

$$H_0: \pi_{ij} = \pi_{i.}\pi_{.j} \qquad \text{for } i = 1, 2, \ldots, r \text{ and } j = 1, 2, \ldots, k$$

$$H_A: \pi_{ij} \neq \pi_{i.}\pi_{.j} \qquad \text{for at least one pair of } i \text{ and } j$$

where π_{ij} is the probability that an element is in the ith row and jth column.

Assumptions:

Random sample of size n from a population having a multinomial distribution for each of the characteristics of interest.

Sample Statistics:

Observed frequencies in the rk cells of the contingency table

Sampling Distribution:

Approximately chi-square if $e_{ij} \geq 5$ for all i and j

$$\chi^2 = \sum_{i=1}^{r} \sum_{i=1}^{k} \frac{(f_{ij} - e_{ij})^2}{e_{ij}} \qquad \text{chi-square d.f.} = (r-1)(k-1)$$

Critical Region:

$$\chi^2 \geq \chi^2_{1-\alpha,(r-1)(k-1)}$$

f_{11}	f_{12}	$\cdots$	f_{1k}	$\sum_{j=1}^{k} f_{1j} = f_{1.}$
f_{21}	f_{22}	$\cdots$	f_{2k}	$\sum_{j=1}^{k} f_{2j} = f_{2.}$
$\vdots$		$\cdots$	$\vdots$	$\vdots$
f_{r1}	f_{r2}	$\cdots$	f_{rk}	$\sum_{j=1}^{k} f_{rj} = f_{r.}$
$\sum_{i=1}^{r} f_{i1} = f_{.1}$	$\sum_{i=1}^{r} f_{i2} = f_{.2}$	$\cdots$	$\sum_{i=1}^{r} f_{ik} = f_{.k}$	$\sum_{i=1}^{r} f_{i.} = \sum_{j=1}^{k} f_{.j} = \sum_{i=1}^{k} \sum_{j=1}^{k} f_{ij}$

Notice in the preceding matrix and in these definitions use of the notation

$$f_{i.} = \sum_{j=1}^{k} f_{ij} \qquad f_{.j} = \sum_{i=1}^{r} f_{ij}$$

In other words, $f_{1.}$ indicates the summing of observed frequencies across the first row; $f_{.1}$ indicates the summing of observed frequencies down the first column, and so forth. The dot (or dots) indicates the subscripts over which summation has been performed.

$$p_{i.} = \frac{f_{i.}}{n}$$

$$p_{.j} = \frac{f_{.j}}{n}$$

$$e_{ij} = p_{i.}p_{.j}n = \frac{(f_{i.})(f_{.j})}{n}$$

EXAMPLE
A production engineer is interested in whether there is a relationship between the method of manufacture of a certain type of polythene sheets (three methods are in use) and its transparency (graded as high, medium, or low after manufacture). From a random sample of 1000 sheets of polythene he obtained the following data:

	Method 1	Method 2	Method 3
High transparency	120	50	30
Medium transparency	200	70	80
Low transparency	180	80	190

Should the production engineer conclude that method of manufacture and transparency of polythene are or are not related on the basis of this sample (1 percent level of significance)?

$$H_0: \pi_{ij} = \pi_{i.}\pi_{.j} \qquad \text{for } i = 1, 2, 3 \text{ and } j = 1, 2, 3$$
$$H_A: \pi_{ij} \neq \pi_{i.}\pi_{.j} \qquad \text{for at least one pair of } i \text{ and } j$$

$$\chi^2 = \sum_{i=1}^{3} \sum_{j=1}^{3} \frac{(f_{ij} - e_{ij})^2}{e_{ij}} \qquad \text{d.f.} = 2 \times 2 = 4$$

$$= \frac{(120 - 100)^2}{100} + \frac{(50 - 40)^2}{40} + \frac{(30 - 60)^2}{60} + \frac{(200 - 175)^2}{175} + \frac{(70 - 70)^2}{70}$$

$$+ \frac{(80 - 105)^2}{105} + \frac{(180 - 225)^2}{225} + \frac{(80 - 90)^2}{90} + \frac{(190 - 135)^2}{135}$$

$$= 4 + 2.5 + 15 + 3.57 + 5.95 + 9 + 1.11 + 22.41$$

$$= 63.54$$

Critical Region:
$$\chi^2 \geq 13.28$$

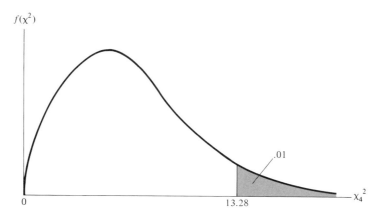

So reject H_0 and conclude that the method of manufacture and transparency are related.

	Method 1	Method 2	Method 3	
High	120 $e_{11} = 100$	50 $e_{12} = 40$	30 $e_{13} = 60$	200 $p_{1.} = 2$
Medium	200 $e_{21} = 175$	70 $e_{22} = 70$	80 $e_{32} = 105$	350 $p_{2.} = .35$
Low	180 $e_{31} = 225$	80 $e_{32} = 90$	190 $e_{33} = 135$	450 $p_{3.} = .45$
	500 $p_{.1} = .5$	200 $p_{.2} = .2$	300 $p_{.3} = .3$	1000

The χ^2 distribution is continuous while the sample data of interest are discrete. Nevertheless, it is appropriate to use the χ^2 distribution as an approximation provided the expected frequency in each cell is 5 or larger. When this is not the case, cells should be combined so that there is an expectation of 5 or more in each.

The problem of discontinuity is particularly troublesome for 2×2 contingency tables. Recall that an analogous situation was discussed in relation to the normal approximation for the binomial distribution: the more categories a discrete distribution has, the "less discrete" it is. The problem of discontinuity for 2×2 tables is partially solved by use of Yates correction, which essentially involves subtracting .5 from the absolute value of the difference between the observed and expected frequency for each cell. The formula for χ^2 including Yates correction is thus given by

$$\chi^2 = \sum_{j=1}^{k} \frac{(|f_j - e_j| - .5)^2}{e_j}$$

EXAMPLE

The American Medical Association claims that quite apart from flu, the proportion of people who have no colds during the winter months is higher for those who have flu shots than for those who do not have flu shots. On the basis of the following data, obtained from a random sample of 100 people who did and did not have flu shots, does the claim appear justified (5 percent level of significance)? The null hypothesis would be that of the people who have colds, there is no difference between the proportion who have had flu shots and those who have not. The hypothesis could also be stated in terms of independence of colds and flu shots—that is, having a cold is independent of having had a flu shot.

	No flu shot	Flu shot	
Colds	35 $e_{11} = 32$	5 $e_{12} = 8$	40
No colds	45 $e_{21} = 48$	15 $e_{22} = 12$	60
	80	20	100

Since this is a 2×2 table, using Yates correction, we have

$$\chi^2 = \sum_{j=1}^{k} \frac{(|f_j - e_j| - 0.5)^2}{e_j}$$

$$= \frac{(3 - 0.5)^2}{32} + \frac{(3 - 0.5)^2}{8} + \frac{(3 - 0.5)^2}{48} + \frac{(3 - 0.5)^2}{12} = 1.63$$

Critical Region:
$\chi^2 \geq 3.841$ for $\alpha = 5$ percent and d.f. $= 1$

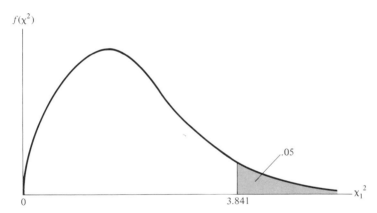

Based on this information, the claim of the AMA is not accepted.

10.4 GOODNESS OF FIT

An investigator is frequently interested in whether it is reasonable to assume that a set of obtained data represent a random sample from a particular theoretical distribution (binomial, normal, multinomial, etc.). He is interested in the goodness of fit of a sample frequency distribution to a theoretical frequency distribution. The expected frequencies assuming the null hypothesis are computed and the goodness of fit hypothesis is tested on the basis of the chi-square statistic, using these expected frequencies and the observed frequencies from the data. For the chi-square test to be appropriate, the expected frequency in each category must be at least 5; categories should be combined, if necessary, in order to satisfy this criterion.

CASE

H_0: sample data obtained from specified distribution

H_A: sample data not obtained from specified distribution

Assumptions:
Random sample from population of interest

Sample Statistics:
Observed frequencies in the k categories

Sampling Distribution:
Approximately chi-square if $e_j \geq 5$ for all $j = 1, 2, \ldots, k$

$$\chi^2 = \sum_{j=1}^{k} \frac{(f_j - e_j)^2}{e_n} \qquad \text{d.f.} = k - t - 1 \qquad \text{where } t \text{ is the number of independent parameters estimated from the data}$$

Critical Region:
$\chi^2 \geq \chi^2_{1-\alpha, k-t-1}$

EXAMPLE
According to a theory in physics, a particular reaction should provide particles of types α, β, γ, and δ in the ratio $9:10:6:5$. An experiment is conducted and the reaction produces 810 α-particles, 1040 β-particles, 660 γ-particles, and 490 δ-particles. Does the experiment tend to support or contradict the theory (1 percent level of significance)?

H_0: sample data from multinomial distribution with

$$\pi_1 = \frac{9}{30} = \frac{3}{10}$$

$$\pi_2 = \frac{10}{30} = \frac{1}{3}$$

$$\pi_3 = \frac{6}{30} = \frac{1}{5}$$

$$\pi_4 = \frac{5}{30} = \frac{1}{6}$$

H_A: sample data not from specified distribution

	OBSERVED	THEORETICAL
α-particles	810	$(9/30)(3000) = 900$
β-particles	1040	$(10/30)(3000) = 1000$
γ-particles	660	$(6/30)(3000) = 600$
δ-particles	490	$(5/30)(3000) = 500$
	3000	3000

$$\chi^2 = \frac{8100}{900} + \frac{1600}{1000} + \frac{3600}{600} + \frac{100}{500} \qquad \text{d.f.} = 3$$

$$= 9 + 1.6 + 6 + 0.2$$

$$= 16.8$$

Critical Region:
$\chi^2 \geq 11.345$

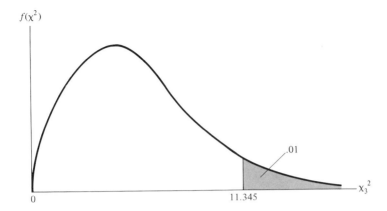

So reject H_0 and conclude that the data tend to refute the theory.

TESTING THE UNIFORM DISTRIBUTION

If an investigator observes different numbers in categories where he or she has the expectation that they are the same, what he or she wishes to determine is whether the differences are statistically significant. In this type of situation, where the theoretical ratio is $1:1:1 \ldots :1$, the hypothesis to test is that the observations are uniformly distributed across the categories. The expected number of observations in each category is equal to the total number of observations divided by the number of categories. Testing the goodness of fit of a uniform distribution (or of a distribution specified by a ratio as in the preceding example) is computationally simpler than testing the goodness of fit of a normal or Poisson distribution. In these cases it is more difficult to calculate the theoretical frequencies.

EXAMPLE

The sales report from a region in which Zuper Markets, Inc. operates eight stores indicates that 160 packages of Bim Bam Cereal were sold in one week. The regional sales manager speculates that the demand for Bim Bam is uniform (equal) for the eight stores. The expected frequency for each store would therefore be 20. The following table gives the actual and expected sales frequencies.

STORE	e_j EXPECTED FREQUENCY	f_j OBSERVED FREQUENCY
1	20	28
2	20	12
3	20	20
4	20	6
5	20	32
6	20	22
7	20	26
8	20	14

H_0: the demand is uniformly distributed

H_A: the demand is not uniformly distributed

$\alpha = 5$ percent   d.f. $= k - 1 = 7$

$$\chi^2 = \sum_{j=1}^{k} \frac{(f_i - e_j)^2}{e_j}$$

$$= \frac{64}{20} + \frac{64}{20} + 0 + \frac{196}{20} + \frac{144}{20} + \frac{4}{20} + \frac{36}{20} + \frac{36}{20}$$

$$= 27.2$$

Critical Region:
$\chi^2 \geq 14.067$

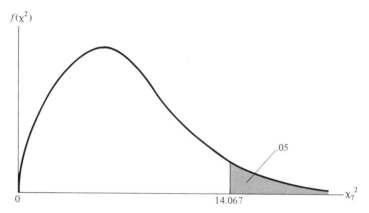

Therefore the hypothesis of uniformly distributed demand is rejected.

TESTING NORMALITY

Testing the goodness of fit for a continuous distribution, for example, the normal distribution, is computationally more involved than testing the goodness of fit for a discrete distribution. This additional computation is required to obtain the expected frequencies for the categorized distribution. One of the problems in testing goodness of fit for continuous distributions is in deciding what categories are to be used.

Either of two procedures may be used (under certain circumstances) for fitting a normal curve to observed data. In method A the data are categorized by percentiles; in method B the data are recorded in categories and are pooled if the number of expected observations in a category is too small. In both cases the sample mean and variance are computed as estimates of the corresponding population parameters, unless the population parameters are known.

METHOD A   FITTING BY PERCENTILES OF THE NORMAL DISTRIBUTION
This method is appropriate if the data are obtained without categorization.

Procedure:

1. Compute the sample mean, $\bar{x}$, and standard deviation, s.
2. Obtain values of $z = (x - \mu)/\sigma$ corresponding to the percentile points chosen; carry out the algebra to obtain x's (using $\mu = \bar{x}$ and $\sigma = s$).
3. Determine from the data the number of observations in each category, interpolating if necessary.
4. Compute χ^2 and test its significance; for k categories, degrees of freedom = $k - 3$, if $\bar{x}$ and s are estimated from the data.

NOTE:
Expected number for each category is n/k where n is the number of observations and k is the number of categories.

METHOD B   FITTING BY CATEGORIES ARISING FROM THE DATA
(GROUPING IF NECESSARY)
This method is appropriate if the data obtained are already categorized.

Procedure:
1. Compute the sample mean, $\bar{x}$, and standard deviation, s.
2. Obtain values of $z = (x - \mu)/\sigma$ corresponding to the end point of the categories (using $\mu = \bar{x}$ and $\sigma = s$); determine percentile values of z's.
3. Determine the proportion of the area under the normal curve corresponding to each category and multiply by the number of observations, n, to obtain the expected number for each category.
4. Compute χ^2 and test its significance; for k categories, degrees of freedom $= k - 3$ if $\bar{x}$ and s are estimated from the data.

Comments:
1. If k is the number of categories, degrees of freedom $= k - 1$ if μ and σ are known; degrees of freedom $= k - 3$ if μ and σ are estimated by $\bar{x}$ and s.
2. If the data are obtained without categorization, method A has certain advantages arising largely from the fact that the expected number of observations is the same for all categories—this can be shown to be a desirable state of affairs from the point of view of power of the test and also simplifies the computation of χ^2. However, if the obtained data are already categorized, method A may not be feasible due to the amount of interpolation required to determine the observed numbers for the categories—linear interpolation assumes a uniform distribution, which is obviously not compatible with the hypothesis of normality. In such cases, method B with appropriate grouping is more defensible.

EXAMPLE (METHOD A)
A random sample of 1000 members of a sports club open to members of the New York Stock Exchange has an average age of 45.55 years with a standard deviation of 10 years. The distribution of ages is given below, where age recorded is age at last birthday; for example, age 42 includes the interval 42 to 43. Using ten categories based on deciles, determine if this sample was obtained from a normal distribution (level of significance $= .01$)?

H_0: data are from a normal distribution

H_A: data are not from a normal distribution

AGES	FREQUENCY	AGES	FREQUENCY
< 30	10	45	64
30	11	46	58
31	13	47	51
32	14	48	47
33	15	49	44
34	17	50	38
35	23	51	35
36	31	52	30
37	34	53	25
38	43	54	21
39	43	55	16
40	40	56	15
41	46	57	13
42	51	58	12
43	50	59	12
44	57	60	11
		> 60	10

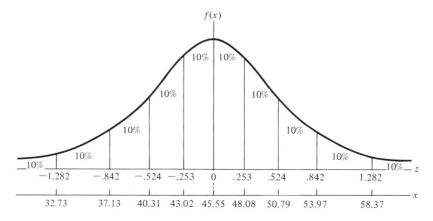

First, obtain from the normal table of areas the z values that divide the distribution into ten areas of 10 percent each. Starting from the mean, find in the table the z value giving the area of .10; interpolation is necessary. Then find the z value giving the area of .20, and so forth. Since the distribution is symmetric, once the z values for one side are found, the values for the other side are the same except for sign. The x values corresponding to the z values are calculated from the usual standard normal equation

$$z = \frac{x - \bar{x}}{s}$$

For example,

$$0.253 = \frac{x - 45.55}{10} \qquad x = 48.08$$

Both z and x values are given in the sketch above. As stated in the problem, the boundaries given are directed boundaries. For example, those ages given as 32 include all past 32 but not 33. Thus to calculate from the given data the number of individuals whose ages fall in the first class, take .73 of the frequency in the age class 32 and add to the frequencies in the age classes less than 32:

$$(.73)(14) + 13 + 11 + 10 = 44.22$$

The next class has boundaries 32.73 to 37.13, and frequencies:

$$(.27)(14) + 15 + 17 + 23 + 31 + (.13)(34) = 94.20$$

and so forth, resulting in the following table:

AGE CLASS	(f) OBSERVED FREQUENCY	(e) EXPECTED FREQUENCY
< 32.73	44.22	100
32.73–37.13	94.20	100
37.13–40.31	127.98	100
40.31–43.02	125.60	100
43.02–45.55	141.20	100
45.55–48.08	141.56	100
48.08–50.79	117.26	100
50.79–53.97	97.23	100
53.97–58.37	70.19	100
> 58.37	40.56	100
	1000	1000

$$\chi^2 = \sum_{j=1}^{k} \frac{(f_j - e_j)^2}{e_j} \qquad \text{d.f.} = 10 - 1 - 2 = 7$$

$$= \frac{(55.78)^2 + (5.80)^2 + (27.98)^2 + (25.60)^2 + (41.20)^2}{100}$$

$$\frac{+ (41.56)^2 + (17.26)^2 + (2.77)^2 + (29.81)^2 + (59.44)^2}{100}$$

$$= 127.35$$

Critical Region:
$$\chi^2 \geq 18.48 \quad \text{for } \alpha = 1 \text{ percent and d.f.} = 7$$

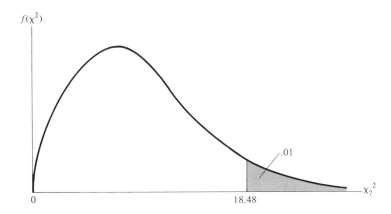

Therefore reject H_0; conclude the age information is not from a normal distribution.

EXAMPLE (METHOD B)
A market research firm is interested in whether the following random sample of sales figures came from a normal distribution.

SALES (000)	NUMBER OF STORES
less than $90	8
90–99	15
100–109	21
110–119	23
120–129	16
130–139	9
140 and above	8
	100

Before grouping, $\bar{x}$ and s were calculated as $\bar{x} = 113.5$ and $s = 16.65$. At the 5 percent level of significance, are the sample data apparently from a normal distribution?

$$H_0: \text{sample data are from a normal distribution}$$

$$H_A: \text{sample data are not from a normal distribution}$$

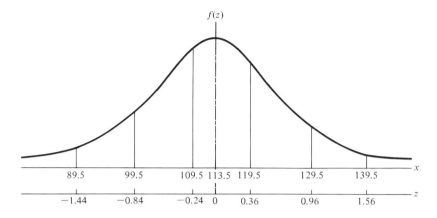

The z's are calculated in Table 10.1, p. 283.

$$\chi^2 = \sum_{j=1}^{k} \frac{(f_j - e_j)^2}{e_j} = \sum_{j=1}^{k} \frac{f_j^2}{e_j} - n \qquad \text{d.f.} = 7 - 2 - 1 = 4$$

$$= \frac{64}{7.49} + \frac{225}{12.56} + \frac{441}{20.47} + \frac{529}{23.54} + \frac{256}{19.09} + \frac{81}{10.91} + \frac{64}{5.94} - 100$$

$$= 8.54 + 17.91 + 21.54 + 22.47 + 13.41 + 7.42 + 10.77 - 100$$

$$= 2.06$$

Critical Region:
$$\chi^2 \geq 9.488$$

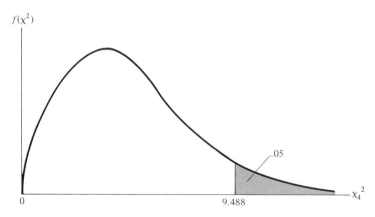

So accept H_0 and conclude that the sample data apparently are from a normal distribution.

TESTING THE POISSON DISTRIBUTION
EXAMPLE
One of the most famous examples of a Poisson Distribution fit is Bortkewitsch's study of deaths in the Prussian Cavalry from being kicked by a horse.* He collected data for 10 army corps for a period of 20 years in the nineteenth century

---

* Ladislaus von Bortkewitsch, *Das Gesetz der Kleinen Zahlen*, B. G. Teubner, Leipzig, Germany, 1898.

TABLE 10.1

(1) CLASS BOUNDARIES	(2) $\frac{x - \bar{x}}{}$ (UPPER CLASS BOUNDARY AS X)	(3) $z = \frac{x - \bar{x}}{s}$	(4) AREA TO LEFT OF z (UPPER CLASS BOUNDARY)	(5) AREA BETWEEN CLASS BOUNDARIES	(6) = (5)(n) EXPECTED FREQUENCY (e_j)	(7) OBSERVED FREQUENCY (f_j)
≤89.5	−24.0	−1.44	.0749	.0749	7.49	8
89.5–99.5	−14.0	−0.84	.2005	.1256	12.56	15
99.5–109.5	−4.0	−0.24	.4052	.2047	20.47	21
109.5–119.5*	+6.0	+0.36	.6406	.2354	23.54	23
119.5–129.5	+16.0	+0.96	.8315	.1909	19.09	16
129.5–139.5	+26.0	+1.56	.9406	.1091	10.91	9
≥139.5				.0594	5.94	8

* Note that this class falls on both sides of the sample mean; to get the area to the left of z, add .5 to the area corresponding to $z = +.36$. It is helpful when working this type of problem to draw the normal curve sketch and to label the class boundaries, the corresponding z values, and $\bar{x}$.

with these results:

DEATHS/CORPS YEAR	NUMBER OF CORPS YEARS	TOTAL DEATHS
0	109	0
1	65	65
2	22	44
3	3	9
4	1	4
5	0	0
	200	122

There are 200 observation intervals (10 corps for 20 years), containing 122 deaths. Hence the mean number of deaths per unit interval is $\lambda = 122/200 = .61$ deaths/corps year. Using 0.61 in the Poisson formula, we have

$$P(X) = e^{-.61} \frac{(.61)^x}{x!}$$

The probabilities of observing $0, 1, 2, 3, 4, 5$ deaths in an interval are as follows, with the corresponding expected frequencies:

X	$P(X)$	EXPECTED FREQUENCY $(200)\,P(X)$	OBSERVED FREQUENCY
0	0.55	110	109
1	0.32	64	65
2	0.10	20	22
3	0.02	4	3
4	0.003	0.6	1
5	0.0	0	0
		198.6	200

H_0: the sample data is from a Poisson distribution

H_A: the sample data is not from a Poisson distribution

$\alpha = 5$ percent

$$\chi^2 = \sum_{j=1}^{k} \frac{(f_j - e_j)^2}{e_j} \qquad \text{d.f.} = 6 - 1 - 1 = 4$$

$$= \frac{1}{110} + \frac{1}{64} + \frac{4}{20} + \frac{1}{4} + \frac{.16}{.6} + 0 = 1.04$$

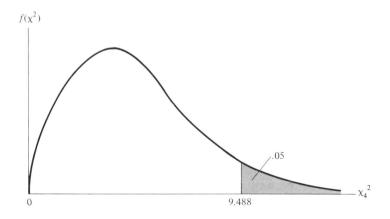

Critical Region:
$\chi^2 \geq 9.488$

Therefore accept the hypothesis that the sample information is from a Poisson distribution.

APPENDIX 10.1
SHORT-CUT FORMULAS FOR COMPUTING SPECIAL CASES OF CHI-SQUARE

The following short-cut formula for computing chi-square for a 2×2 table is sometimes convenient, particularly if the observed and/or expected frequencies involve decimals.

a	b	$a + b$
c	d	$c + d$

$a + c \quad b + d \qquad n$

$$\chi^2 = \frac{(ad - bc)^2 n}{(a + b)(c + d)(a + c)(b + d)} \qquad \text{d.f.} = 1$$

Yates correction for continuity: $\chi^2 = \dfrac{\left(|ad - bc| - \dfrac{n}{2}\right)^2 (n)}{(a + b)(c + d)(a + c)(b + d)}$

A short-cut computational method is also available for contingency tables that have only two rows (i.e., $2 \times k$) or only two columns (i.e., $k \times 2$) as follows:

a_1	a_2	$\cdots$	a_k	$\sum\limits_{i=1}^{k} a_i = A$
b_1	b_2	$\cdots$	b_k	$\sum\limits_{i=1}^{k} b_i = B$
n_1	n_2		n_k	$\sum\limits_{i=1}^{k} n_i = A + B = n$

or

a_1	b_1	n_1
a_2	b_2	n_2
$\vdots$	$\vdots$	
a_k	b_k	n_k
A	B	n

$$\chi^2 = \frac{\sum\limits_{i=1}^{k} \dfrac{a_i^2}{n_i} - \dfrac{A^2}{n}}{\dfrac{AB}{n^2}} - \frac{\sum\limits_{i=1}^{k} \dfrac{b_i^2}{n_i} - \dfrac{B^2}{n}}{\dfrac{AB}{n^2}} \qquad \text{d.f.} = k - 1$$

APPENDIX 10.2
DEGREES OF FREEDOM

The concept of degrees of freedom is very important in sampling distribution theory. Every sampling distribution has at least one parameter that depends at least partly on sample size and is referred to as *degrees of freedom*. That is, the exact form of a sampling distribution depends on its degrees of freedom.

Essentially, degrees of freedom always represent in some sense the number of observations that can be varied without violating some condition or constraint of the problem. In particular types of problems, degrees of freedom depend not only on sample size, but also on the number of ways the data are classified and the number of parameters estimated from the data.

The concept of degrees of freedom is frequently difficult to understand intuitively, although actually the term is quite descriptive. The following examples illustrate its meaning.

EXAMPLE
Consider the following set of 5 observations

$$
\begin{array}{r}
10 \\
15 \\
11 \\
7 \\
12 \\
\hline
\sum = 55 \\
\end{array}
$$
mean = 11

Suppose the mean and 4 observations are given, then the fifth observation is determined; that is, given the mean (or, equivalently, the sum) of n observations, it is possible to choose $n - 1$ of these in any of a large number of ways, but then the nth observation is determined. Thus the variance of n observations about their mean has $n - 1$ degrees of freedom, since only $n - 1$ observations are "free to vary" once the mean is specified. This corresponds to the degrees of freedom for the t distribution.

EXAMPLE
Consider the following contingency table,

4	3	10	17
10	6	5	21

14 9 15

and suppose the marginals are specified; then the values in two of the cells can be chosen in a number of ways, but the values in the other cells are then determined. Thus the table has two degrees of freedom. In general, a table with r rows and c columns has $(r - 1)(c - 1)$ degrees of freedom, corresponding to the rule for the degrees of freedom for a χ^2 test of contingency.

Similarly, since an F distribution is essentially the ratio of two chi-square distributions, it has the corresponding pair of degrees of freedom.

SUMMARY OF DEGREES OF FREEDOM FOR CHI-SQUARE

Testing equality of proportions: d.f. $= k - 1$ where k is the number of proportions tested for equality

Contingency tables: d.f. $- (r - 1)(k - 1)$ where r is the number of rows and k is the number of columns in the table

Goodness of fit: d.f. $= k - t - 1$ where k is the number of categories and t is the number of parameters estimated from the data

$$\text{General Rule: d.f.} = \begin{pmatrix} \text{number of terms} \\ \text{in chi-square} \\ \text{computation} \end{pmatrix} - \begin{pmatrix} \text{number of independent} \\ \text{parameters} \\ \text{estimated from data} \end{pmatrix} - 1$$

SUMMARY OF DEGREES OF FREEDOM FOR THE *F*

Testing equality of variances: d.f. $= n_1 - 1, n_2 - 1$; where n_1 is the sample size for the numerator and n_2 is the sample size for the denominator.

Analysis of variance: d.f. = degrees of freedom for the numerator sum of squares, degrees of freedom for the denominator sum of squares (see Chapter 12).

DEGREES OF FREEDOM FOR THE *t*

One sample: d.f. $= n - 1$, where n is the sample size.

Two sample: d.f. $= n_1 + n_2 - 2$, where n_1 and n_2 are the number of observations in the respective sample.

PROBLEMS

10.1 To test the hypothesis that a coin is fair (that is, heads and tails are equally likely), it is decided to accept the hypothesis if the number of heads in a single sample of 100 tosses is between 40 and 60 inclusive and to reject the hypothesis otherwise.
   (a) Find the probability of rejecting the hypothesis when it is correct.
   (b) Find the probability of accepting the hypothesis that the coin is fair if in fact the probability of heads is (i) .8, (ii) .6, (iii) .4.
   (c) Answer (a) and (b) if the test is such that the hypothesis is rejected only if the number of heads is greater than 60.

10.2 A manufacturer tells prospective buyers that the lock mechanisms he produces have a defective rate of only .002 (that is, .2 percent). The quality control engineer has been instructed to set up a test procedure which will reject lots meeting this criterion only 2.5 percent of the time; he has decided to test random samples of 2500 mechanisms (the lot size is 500,000).
   (a) For what sample proportion defective should lots be rejected?
   (b) Using this criterion, what is the probability of rejecting a lot having .001 defectives?
   (c) Using this criterion, what is the probability of accepting a lot having .004 defectives?

10.3 It is hypothesized that the proportion of people in a population who have brown eyes is .75 and it is decided to accept this hypothesis if in a random sample of 300 people between .70 and .80 do in fact have brown eyes, and reject the hypothesis otherwise.
   (a) Find the probability of rejecting the hypothesis when in fact it is true.
   (b) Find the probability of accepting the hypothesis when in fact the proportion of people having brown eyes is (i) .80, (ii) .60, (iii) .95, (iv) .72.

10.4 The Chatty Doll Company claims that only .5 percent of its talking dolls are defective. The company manufactures the dolls in lots of 100,000 and wants to set up a testing procedure which rejects a lot meeting this criterion only 2 percent of the time. If a random sample of 500 dolls is taken from each lot
(a) For what sample proportion defective should lots be rejected?
(b) Using this criterion, what is the probability of accepting a lot having .6 percent defective?
(c) Using this criterion, what is the probability of rejecting a lot having .4 percent defective? (assume the normal approximation is appropriate).

10.5 It is hypothesized that 80 percent or more of the rabbits in a large area of Australia have mixomitosis.
(a) At the 5 percent level of significance should the hypothesis be accepted or rejected (this particular area of Australia is literally overrun with rabbits, so you may consider their population infinite)?
(b) For the sample size of 196 state the rejection region in terms of the sample proportion having mixomitosis. (That is, what sample values of p lead to rejection of the hypothesis?)
(c) If, in fact, 64 percent of the rabbits have mixomitosis, what is the probability of (falsely) accepting the hypothesis?

10.6 (Use Normal approximation to solve this problem.) A manufacturer claims that at least 60 percent of the adults in a certain city have heard of his product. A random sample of 100 persons is taken and asked if they can identify the manufacturer's product. With alpha of 10 percent, how small would the sample percentage need to be to refute the manufacturer's claim?

10.7 (Do not use Normal approximation here.) A company will reject a shipment of gadgets if it is believed 10 percent or more are defective. The company is very anxious to avoid the error of rejecting a shipment that should not be rejected. Suppose $n = 10$, and the alpha level is $7\frac{1}{2}$ percent.
(a) Set up the null hypothesis.
(b) What is the minimum number of defectives in the sample that will cause rejection of a shipment?

10.8 A car salesman is interested in whether or not the proportions of Ford and Chevrolet owners who owned the same make car immediately before are the same. On the basis of the following data from the random samples of Ford and Chevrolet owners, what should he conclude (5 percent level of significance)?

	Ford	Chevrolet
Same	500	400
Different	500	800

10.9 The athletic director at Stanford claims that the proportion of Stanford students who regularly attend home football games is the same as the proportion of Notre Dame students who regularly attend home football games. The student paper claims that in fact this proportion is lower at Stanford and publishes the following data obtained from random samples of students at the two schools. At the 5 percent level of significance, does this evidence refute the athletic director's claim?

	Stanford	Notre Dame
Attend	350	400
Don't attend	250	100

10.10 From experience, it is known that 20 percent of a certain kind of seed germinate. If in an experiment, 60 out of 400 seeds germinate, is this evidence (1 percent level of significance) of poor germination?

10.11 Two groups, A and B, consist of 100 randomly chosen people each of whom have a certain disease. A serum is given to group A but not to group B (which is called the control group); otherwise the two groups are treated identically. In group A, 75 people recover and in group B, 65 people recover from the disease. At the 1 percent level of significance, is this evidence that the serum is effective?

10.12 The registrar of a university is interested in whether or not the proportions of juniors and seniors who have a 3.0 average or better are the same for engineering, commerce, and letters and science students. On the basis of the following data, from random samples of his files, what would the registrar conclude (5 percent level of significance)?

	Engineering	Commerce	Letters and Science
3.0 or above	70	60	70
below 3.0	280	190	330

10.13 A shopping center financial advisor is interested in whether or not the proportions of business done on credit are the same for the supermarket, department store, hardware store, and women's accessory shop in the center. From random samples of sales at the stores, he obtains the following data. At the 1 percent level of significance, what would he conclude?

	Super Market	Department Store	Hardware Store	Women's Accessories
Credit	50	100	15	75
Cash	50	50	35	25

10.14 A large commercial photography company purchased 100 flashbulbs of each of four brands A, B, C, and D. Of these sets of flashbulbs, 5, 10, 2, and 3, respectively, failed to light properly. Test the hypothesis that the true proportion of defective bulbs of each of the four brands is the same using the 5 percent level of significance.

10.15 Dr. X-Ray is concerned with the possible relationship between age of the patient and severity of German measles. He has taken a random sample of German measles sufferers and has classified each as old or young and as having a mild or severe case of the disease. His data are as follows:

	Mild	Severe
Young	240	360
Old	60	240

Dr. X-Ray's hypothesis is that the proportion of severe cases is the same for old as for young patients or, equivalently, that severity of the disease and age of the patient are independent.

(a) Give two formulas for testing this hypothesis using two *different* distributions and put the appropriate numbers into both formulas.

(b) Do the computations for *one* of the formulas and test the hypothesis (1 percent level of significance).

10.16 The Gigo Computing Center is making a study of the programs it runs in order to evaluate the advantages of time sharing. A random sample of 1000 programs has been classified as short (3 min or less) or long (more than 3 min) and as requiring or not requiring a tape to be called. The data are as follows:

	Tape	No Tape
Short	100	100
Long	300	500

(a) On the basis of these data, are length of program and use of tape independent (5 percent level)?

(b) One of the operators hypothesizes that the distribution of programs is as follows:

Short-tape	10%
Short-no tape	10%
Long-tape	33%
Long-no tape	47%

Do the data support or refute his hypothesis (5 percent level)?

10.17 A random sample of 1600 people living in a certain (well-defined) geographical region of Europe are observed with regard to hair color (light or dark) and eye color (blue or brown); the following data are obtained:

Light hair, blue eyes	210
Light hair, brown eyes	190
Dark hair, blue eyes	430
Dark hair, brown eyes	770
	1600

(a) Geneticist A has a theory that in this particular region $\frac{1}{8}$ of the people have light hair and blue eyes, $\frac{1}{8}$ have light hair and brown eyes, $\frac{1}{4}$ have dark hair and blue eyes, and $\frac{1}{2}$ have dark hair and brown eyes. On the basis of the obtained data, does he appear to be correct (5 percent level of significance)?

(b) Geneticist B has a theory that hair color and eye color are independent of each other for people in this region. On the basis of the obtained data, does he appear to be correct (5 percent level of significance)?

10.18 Amalgamated Auto Corporation is interested in the number of years its executives have been employed by the corporation and whether they completed the corporation's executive training program. The following data were obtained.

	Number of Years Employed		
	5	10	15
Training Program	30	15	5
No Training Program	10	25	15

(a) On the basis of these data, are number of years employed and completion of the training program independent (5 percent level of significance)

(b) The President of Amalgamated hypothesized the ratio 5:4:1:3:5:2 where the order is T-5, T-10, T-15, NT-5, NT-10, NT-15. (T and NT represent training program and no training program respectively, and 5, 10, and 15 represent years of employment). Do the data support his hypothesis (5 percent level of significance)?

10.19 In order to test whether or not it is balanced, a die is tossed 300 times. The tosses resulted in 52 ones, 54 twos, 45 threes, 60 fours, 46 fives and 43 sixes. At the 5 percent level of significance, is there evidence the die is not balanced?

10.20 A grower of Christmas trees is interested in whether or not their heights follow a normal distribution. From a random sample of 300 trees, he recorded the following data:

HEIGHT OF TREE (IN FEET)	NUMBER OF TREES
less than 1.5	11
1.5–2.4	26
2.5–3.4	46
3.5–4.4	60
4.5–5.4	62
5.5–6.4	50
6.5–7.4	30
7.5 or greater	15

The sample mean and variance (before grouping) are $\bar{x} = 4.5$ and $s = 1.75$. Based on this evidence, are the heights of the trees apparently normally distributed (5 percent level of significance)?

10.21 In three groups of people, randomly chosen from three different geographical regions, the distribution of hair color is as follows:

	RED HAIR	LIGHT HAIR	DARK HAIR
Group A	2	9	9
Group B	3	6	21
Group C	15	15	20

Do these data indicate (2 percent level of significance) that hair color is not dependent on geographical region?

10.22 Six coins are tossed together 512 times. The number of tosses on which 0, 1, 2, 3, 4, 5, and 6 heads occurred are, respectively, 10, 50, 115, 175, 120, 40, and 2. Is there evidence (5 percent level of significance) that not all six coins are fair?

10.23 On a particular proposal of national importance, a random sample of Democrats and Republicans cast votes as follows:

	FAVOR	OPPOSE	UNDECIDED
Democrats	85	78	37
Republicans	118	61	25

Are opinions on this issue and party affiliation apparently independent (5 percent level of significance)?

10.24 A manufacturer has contracted to sell several million light bulbs and has assured the buyer that $\frac{4}{5}$ of the light bulbs will have a life of at least 50 hours. In order to check on this, the quality control engineer obtains 25,000 random samples of 5 bulbs each over a period of several days and tests the bulbs. The numbers of samples in which 0, 1, 2, 3, 4, and 5 bulbs met the criterion of 50 hours were, respectively, 0, 150, 1300, 5150, 10,220, and 8180. Do these data support or refute the manudacturer's claim (5 percent level of significant)?

10.25 The evergreen trees in Sherwood Forest have been measured and the park commissioner wishes to know whether or not their heights follow a normal distribution. He has computed $\bar{x} = 4.45$ and $s = 1.78$ and has begun the job of fitting a normal distribution as follows:

HEIGHT OF TREE (IN FEET)	OBSERVED NO. OF TREES	EXPECTED NO. OF TREES
Less than 1.5	6	5
1.5–2.5	11	9
2.5–3.5	15	16
3.5–4.5	20	—
4.5–5.5	18	20
5.5–6.5	14	15
6.5–7.5	10	9
More than 7.5	6	—
	100	

(a) Find the two missing expected values.

(b) For testing goodness of fit, give the appropriate formula, substitute into it the appropriate numbers, and assume that the result is 2.26. State the conclusion, giving the critical value of the statistic on which it is based (use 5 percent level of significance).

10.26 Dr. I. Q. Ben has administered an intelligence test to 4,000 students and has had the results scored by his IBM machine. The predictive norms for this particular test were computed from a very large population of scores having a normal distribution with mean 100 and standard deviation 10. Dr. Ben wishes to determine whether the 4,000 test scores he obtained apparently also came from a normal distribution having mean 100 and standard deviation 10. He spilled eradicator on them and his work-sheet now looks as follows:

SCORE	NO. OBSERVED	Z	AREA LEFT OF Z	AREA	NO. EXPECTED
≤ 70.5	20	−2.95	.002		
70.5–80.5	90	−1.95	.026		
80.5–90.5	575		.853		
90.5–100.5	1282				
100.5–110.5	1450	1.05	.853		
110.5–120.5	499	2.05			
120.5–130.5	80	3.05	.999		
≥ 130.5	4				

(a) Fill in the missing numbers of Dr. Ben's worksheet.

(b) Test the hypotheses that the data came from a normal population with mean 100 and standard deviation 10. (Use the 5 percent level of significance and perform any pooling that is necessary.)

10.27 The statistician for Associated Airlines has been studying the data on airplane crashes over a period of nearly 35 years. He has complete data for exactly 400 months, which is tabulated as follows:

PLANE CRASHES	MONTHS
0	219
1	132
2	41
3	6
4	2
5	0
	400

As he looks at the data, the statistician feels that plane crashes might follow a Poisson distribution. Check the statistician's hunch by testing the goodness of fit of the above data to the Poisson.

SAMPLING DESIGNS

11.1 INTRODUCTION

Humankind's search for knowledge consists partly in attempting to generalize from specific instances in order to obtain general principles. A person observes individual occurrences and draws conclusions about the more general cases. A prominent contemporary philosopher, Karl Popper*, maintains that this is the essential process of man's acquisition of knowledge. Popper says that people observe, formulate a hypothesis, and then continually reformulate hypotheses based on further observations and that this process constitutes learning. This type of unplanned observation and of drawing premature conclusions is sometimes called "casual empiricism."

For example, a home owner might observe a neighbor having difficulty starting an XYZ power mower, and might say, "I'd never buy one of those XYZ mowers." Or a car owner might drive a set of tires for 30,000 miles and say, "Those Blat tires are really good." The main objection to such casual empiricism is that it is based on small samples, often on a sample of size one, and thus may be extremely misleading when used to generalize to the entire population.[†] In addition, there is the serious philosophical objection that what is observed may not be attributable to the presumed cause.

The purpose of *sampling design* is to obtain samples from which statistically valid generalizations may be made. Sampling design is often considered in the context of surveys or polls, but its application is much broader than the problems included in those areas. The analyses discussed in Chapters 7, 8, 9, and 10 assume properly obtained random samples, as do the analyses in subsequent chapters. The entire subject of statistical inference is concerned with making statements or drawing conclusions (with a probability attached) concerning a population on the basis of information obtained from a sample drawn from that population.

---

* K. R. Popper, *Conjectures and Refutations* (New York: Basic Books, 1965).
  † A noted scientist, Niels Abel, summed up this objection by stating, "Everywhere one observes the unfortunate habit of generalizing, without demonstration, from special cases."

This chapter discusses procedures for drawing samples so that inferences can be stated unambiguously in terms of probability, conveying the same meaning to all who consider them. These procedures involve *probability sampling*, that is, sampling in such a way that a theoretical sampling distribution is associated with the sample results and confidence limits can be placed on the estimates obtained.

Recall that a *population* is the set of all possible items, elements, or measurements about which one wishes to make an inference. These elements are referred to formally as *units* or *sampling units* and are specified in such a way that they are nonoverlapping and include the entire population. The *sampling frame* is an actual or hypothetical list of sampling units that permits the drawing of a probability sample from the population. A sampling frame need not always include the entire population, but it must permit access to that portion of the population under consideration. In this context, the portion of the population that the frame covers is referred to as the *sampled population*. The sampled population may be the total population of interest, called the target population, or it may be a subset of the target population.

Simple random sampling is defined as the selection of a sample from the population in such a way that every unit in the population has an equal chance of being included in the sample and every possible combination of units in the population has an equal chance of being included in the sample. This second condition should be noted carefully; it distinguishes simple random sampling from certain frequently used types of systematic sampling. For example, suppose a sample of names from a telephone book is obtained by taking, say, every twenty-fifth name starting at a randomly chosen place in the directory. This would not constitute a random sample; although the first condition is fulfilled, the second condition is not. For example, if the names of John A. Smith and John B. Smith were adjacent in the directory, both names could not appear in the same sample using the procedure of taking every twenty-fifth name.

A sample consisting of every twenty-fifth unit, or obtained using some similar rule, is called a *systematic sample*. Two other types of samples frequently used are: (1) *judgment sample*, in which the investigator uses his or her knowledge of the population, rather than a random process, as the basis for selecting the sample and (2) *convenience sample*, in which the investigator includes in the sample those elements of the population that are convenient or readily accessible.

It should be noted that probability statements concerning the population can be made only when sampling is based on a random process. By definition, probability sampling is based on a random process, and all probability sampling designs are based on some variant of simple random sampling.

Frequently, there is confusion concerning the meaning of the terms "random sampling" and "probability sampling." Technically these terms are synonomous and refer to any method of sampling in which the selection of the sample depends on a random (probabilistic or stochastic) process. The confusion arises because random sampling is often used in practice to refer to simple random sampling. Simple random sampling requires that every unit in the population has an equal probability of selection, independent of the other units included in the sample. Random or probability sampling requires only that selection of the sample is based on a probabilistic or random process rather than on judgment or convenience.

There are many situations in which simple random sampling is inefficient, inconvenient or impossible, and some other method of random or probability sampling should be used. Two sampling methods that are frequently useful are *stratified sampling* and *cluster sampling*. In both of these methods the population to be sampled is partitioned. In stratified sampling, partitioning is on the basis of some characteristic thought to be relevant, for example, age in a survey of buying patterns. In cluster sampling, partitioning is on the basis of some characteristic related to location or accessibility, for example, geographical area in a survey of household characteristics. These two types of sampling are discussed in some detail later in the chapter. The basic difference between them is that in stratified sampling the population is divided on the basis of some known characteristics, while in cluster sampling the population is divided on the basis of convenience or the necessity of covering the entire population. In many situations, even if a list of the population is not available, it may still be possible to divide the relevant area into sections, for example, to obtain a (cluster) sampling frame.

In stratified sampling, units are selected randomly from each stratum; an equal number may be selected from each stratum or the sample sizes may be proportional to the numbers of units in the respective strata. In any case, stratified sampling is appropriate only if the variable of interest is relatively homogeneous within strata and heterogeneous among strata. In cluster sampling, clusters are randomly selected and then all or a random sample of the units in each of these clusters is included in the final sample. Cluster sampling may be used almost as a matter of necessity in order to construct a sampling frame; it is appropriate from a sampling design point of view when the variable of interest is relatively heterogeneous within clusters and homogeneous among clusters.

Two sampling procedures that are frequently used are *double sampling* (or *multiphase sampling*) and *sequential sampling*. In double sampling, a large preliminary sample is selected to obtain information that will aid in the selection of a smaller subsample from which the estimate of primary interest is obtained. If such a procedure involves more than two steps, it is called multiphase sampling. Sequential sampling is often used in acceptance sampling for quality control. Essentially, the decision to accept, reject, or continue sampling is made after every observation. Discussions of multiphase and sequential sampling can be found in several of the references in Appendix A.

11.2 SAMPLING DESIGNS OF SURVEYS

A statistical survey attempts to estimate population parameters whose values are required for some particular purpose. The sampling design for conducting the survey is an important part of the design of the survey. The general questions involved in survey design are: (1) what information is the investigator attempting to obtain and (2) for what purpose does he or she intend to use the information when it is obtained. The survey design specifies the sampling units, the target and sampled populations, the sampling frame, the methods by which the data are to be collected and processed, the error to be tolerated, and the disposition of the results.

Essentially, a sampling design consists of three parts or procedures: (1) the procedure to be used for selecting the units in the sample; (2) the procedure to be used in calculating the sample statistics, their standard error and their bias; and (3) the procedure to be used in calculating the estimates that would be obtained for the sample statistics if complete coverage of the frame were obtained in the same way that the sample was obtained—this provides a basis for separating sampling and nonsampling errors and their effects.

There are two general considerations in sample design—the accuracy and the precision of the measurements involved. The accuracy of a measurement is the closeness with which the measurement approximates the true value of the parameter being measured; that is, accuracy is concerned with bias of measurement. The precision of a measurement concerns its repeatability or reliability; precision is concerned with the variance of measurement. The smaller the standard error of a statistic, the more precise it is as an estimate of the corresponding parameter. When reporting estimates based on sampling, standard errors should be included. One of the main purposes of sample design is to obtain precision at minimum cost. One sample design is said to be more *efficient* than another if it provides the same precision at less cost or provides greater precision at a given cost.

From a statistical point of view, bias is a property of an estimator; however, it is quite common practice, especially in nontechnical writing, to refer to a sample as being biased. In this context, biased generally means that the sample is not a random sample from the population of interest and presumably is not representative of the population and that estimates obtained from the sample are thus not appropriate for the population. As noted, statistical (probability) statements can be made concerning sample estimates only when such estimates are based on data obtained from a random sample of the population of interest. For example, suppose a polling organization is interested in determining voter preference for candidates for national office. If the sample were obtained from those attending the national convention of one of the political parties, any information obtained obviously would be of no value with respect to determining general voter preferences. In this case the sample might be said to be a biased sample of the population, meaning that it is not a random sample and furthermore is manifestly not representative of the population.

This meaning of bias must be distinguished carefully from its meaning when applied to estimators. A biased estimator is an estimator based on a random sample but having an expected value (over its sampling distribution) that is not equal to the true value of the parameter of which it is an estimator. As already stated, biasedness or unbiasedness in this sense is a statistical property of estimators based on random sampling and is not attributable to incorrect sampling procedures. There is no *a priori* intuitive basis for determining whether an estimator is biased or unbiased; this property depends entirely on the estimator's mathematical expectation. For example, the "intuitively unbiased" estimator of variance is, in fact, biased. A general discussion of bias and other properties of estimators, which is applicable to estimators used for sample designs, is given in Chapter 7.

Accuracy and precision are a reflection or an indication of the two types of measurement error—sampling error and nonsampling error. Sampling error arises from the random variation that is inherent in the nature of things. To eliminate sampling error it would be necessary to examine all the units in the population. Nonsampling error, on the other hand, occurs as a result of mistakes in design or procedure and the biases of measurement that result. Errors due to bias are sometimes referred to as systematic or persistent errors and are not attributable to random sampling. Although sampling error decreases as sample size increases, bias generally is independent of sample size.

11.3 SIMPLE RANDOM SAMPLING

In the discussion of conditional probability in Chapter 4, we mentioned that a random sample from 66 companies could be selected using a table of random numbers rather than using the procedure of placing 66 numbered balls in an urn and drawing out a sample. In practice, it is preferable to use random numbers rather than balls in an urn, both for convenience and because it is impossible actually to determine when balls in an urn are "thoroughly mixed" so that each ball has an "equally likely" chance of being drawn.

In 1969 the administration of the draft in the United States was changed to a lottery system and on December 1, 1969, a drawing was made of the 366 days in a year (one for February 29 in leap year) corresponding to birthdays of potential draftees. The order in which the days were drawn indicated the order in which draft calls were to be made. This drawing was performed on television from a large urnlike apparatus containing capsules (rather than balls), one for each day of the year. The ordering of the 366 days drawn is given in Table 11.1.

After the lottery-drawn numbers were published, the question was raised concerning whether or not the drawing was random. Some people claimed that the procedure used would not generate randomness. The procedure was approximately as follows: the capsules for January were placed in a large square box and pushed to one side by a cardboard divider. The February capsules were then placed in the box and scraped into the January capsules. The same method was used for subsequent months. Thus the January capsules were mixed eleven times, the February capsules were mixed ten times, and so on until the December capsules were mixed only once. After all capsules were placed in the box, it was turned over a few times. Before the drawing was made, the capsules were poured from the box into a bowl. The capsules were not stirred after they were poured into the bowl.

The specific challenge concerning the randomness of the numbers arises from a calculation of the average number for each month. The average for a given month is computed by adding the numbers for each day of a year and by dividing by the number of days in a year of the given months. Therefore each month could be expected to have an average of 181 (31-day month), 186 (30-day month), or 193 (29-day month). These averages are calculated as follows:

$$1 + 2 + 3 + \cdots + 366 = 67161$$

$$\frac{67161}{(12)(29)} = 193 \qquad \frac{67161}{(12)(30)} = 186 \qquad \frac{67161}{(12)(31)} = 181$$

TABLE 11.1   ORDERING OF DAYS OF THE
YEAR BY DRAFT LOTTERY, DECEMBER 1, 1969

1. Sept. 14	47. Nov. 27	93. July 1	139. March 6
2. April 24	48. Aug. 8	94. Oct. 28	140. Jan. 18
3. Dec. 30	49. Sept. 3	95. Dec. 24	141. Aug. 18
4. Feb. 14	50. July 7	96. Dec. 16	142. Aug. 12
5. Oct. 18	51. Nov. 7	97. Nov. 8	143. Nov. 17
6. Sept. 6	52. Jan. 25	98. July 17	144. Feb. 2
7. Oct. 26	53. Dec. 22	99. Nov. 29	145. Aug. 4
8. Sept. 7	54. Aug. 5	100. Dec. 31	146. Nov. 18
9. Nov. 22	55. May 16	101. Jan. 5	147. April 7
10. Dec. 6	56. Dec. 5	102. Aug. 15	148. April 16
11. Aug. 31	57. Feb. 23	103. May 30	149. Sept. 25
12. Dec. 7	58. Jan. 19	104. June 19	150. Feb. 11
13. July 8	59. Jan. 24	105. Dec. 8	151. Sept. 29
14. April 11	60. June 21	106. Aug. 9	152. Feb. 13
15. July 12	61. Aug. 29	107. Nov. 16	153. July 22
16. Dec. 29	62. April 21	108. March 1	154. Aug. 17
17. Jan. 15	63. Sept. 20	109. June 23	155. May 6
18. Sept. 26	64. June 27	110. June 6	156. Nov. 21
19. Nov. 1	65. May 10	111. Aug. 1	157. Dec. 3
20. June 4	66. Nov. 12	112. May 17	158. Sept. 11
21. Aug. 10	67. July 25	113. Sept. 15	159. Jan. 2
22. June 26	68. Feb. 12	114. Aug. 6	160. Sept. 22
23. July 24	69. June 13	115. July 3	161. Sept. 2
24. Oct. 5	70. Dec. 21	116. Aug. 23	162. Dec. 23
25. Feb. 19	71. Sept. 10	117. Oct. 22	163. Dec. 13
26. Dec. 14	72. Oct. 12	118. Jan. 23	164. Jan. 30
27. July 21	73. June 17	119. Sept. 23	165. Dec. 4
28. June 5	74. April 27	120. July 16	166. March 16
29. March 2	75. May 19	121. Jan. 16	167. Aug. 28
30. March 31	76. Nov. 6	122. March 7	168. Aug. 7
31. May 24	77. Jan. 28	123. Dec. 28	169. March 15
32. April 1	78. Dec. 27	124. April 13	170. March 26
33. March 17	79. Oct. 31	125. Oct. 2	171. Oct. 15
34. Nov. 2	80. Nov. 9	126. Nov. 13	172. July 23
35. May 7	81. April 4	127. Nov. 14	173. Dec. 26
36. Aug. 24	82. Sept. 5	128. Dec. 18	174. Nov. 30
37. May 11	83. April 3	129. Dec. 1	175. Sept. 13
38. Oct. 30	84. Dec. 25	130. May 15	176. Oct. 25
39. Dec. 11	85. June 7	131. Nov. 15	177. Sept. 19
40. May 3	86. Feb. 1	132. Nov. 25	178. May 14
41. Dec. 10	87. Oct. 6	133. May 12	179. Feb. 25
42. July 13	88. July 28	134. June 11	180. June 15
43. Dec. 9	89. Feb. 15	135. Dec. 20	181. Feb. 8
44. Aug. 16	90. April 18	136. March 11	182. Nov. 23
45. Aug. 2	91. Feb. 7	137. June 25	183. May 20
46. Nov. 11	92. Jan. 26	138. Oct. 13	184. Sept. 8

The actual averages for the months, calculated from Table 11.1, are as follows:

January	201
February	203
March	226
April	204
May	208

TABLE 11.1 ORDERING OF DAYS OF THE
YEAR BY DRAFT LOTTERY, DECEMBER 1, 1969 (continued)

185. Nov. 20	231. April 14	277. July 9	323. March 10
186. Jan. 21	232. Sept. 4	278. May 18	324. Aug. 11
187. July 20	233. Sept. 27	279. July 4	325. Jan. 10
188. July 5	234. Oct. 7	280. Jan. 20	326. May 22
189. Feb. 17	235. Jan. 17	281. Nov. 28	327. July 6
190. July 18	236. Feb. 24	282. Nov. 10	328. Dec. 2
191. April 29	237. Oct. 11	283. Oct. 8	329. Jan. 11
192. Oct. 20	238. Jan. 14	284. July 10	330. May 1
193. July 31	239. March 20	285. Feb. 29	331. July 14
194. Jan. 9	240. Dec. 19	286. Aug. 25	332. March 18
195. Sept. 24	241. Oct. 19	287. July 30	333. Aug. 30
196. Oct. 24	242. Sept. 12	288. Oct. 17	334. March 21
197. May 9	243. Oct. 21	289. July 27	335. June 9
198. Aug. 14	244. Oct. 3	290. Feb. 22	336. April 19
199. Jan. 8	245. Aug. 26	291. Aug. 21	337. Jan. 22
200. March 19	246. Sept. 18	292. Feb. 18	338. Feb. 9
201. Oct. 23	247. June 22	293. March 5	339. Aug. 22
202. Oct. 4	248. July 11	294. Oct. 14	340. April 26
203. Nov. 19	249. June 1	295. May 13	341. June 18
204. Sept. 21	250. May 21	296. May 27	342. Oct. 9
205. Feb. 27	251. Jan. 3	297. Feb. 3	343. March 25
206. June 10	252. April 23	298. May 2	344. Aug. 20
207. Sept. 16	253. April 6	299. Feb. 28	345. April 20
208. April 30	254. Oct. 16	300. March 12	346. April 12
209. June 30	255. Sept. 17	301. June 3	347. Feb. 6
210. Feb. 4	256. March 23	302. Feb. 20	348. Nov. 3
211. Jan. 31	257. Sept. 28	303. July 26	349. Jan. 29
212. Feb. 16	258. March 24	304. Dec. 17	350. July 2
213. March 8	259. March 13	305. Jan. 1	351. April 25
214. Feb. 5	260. April 17	306. Jan. 7	352. Aug. 27
215. Jan. 4	261. Aug. 3	307. Aug. 13	353. June 29
216. Feb. 10	262. April 28	308. May 28	354. March 14
217. March 30	263. Sept. 9	309. Nov. 26	355. Jan. 27
218. April 10	264. Oct. 27	310. Nov. 5	356. June 14
219. April 9	265. March 22	311. Aug. 19	357. May 26
220. Oct. 10	266. Nov. 4	312. April 8	358. June 24
221. Jan. 12	267. March 3	313. May 31	359. Oct. 1
222. June 28	268. March 27	314. Dec. 12	360. June 20
223. March 28	269. April 5	315. Sept. 30	361. May 25
224. Jan. 6	270. July 29	316. April 22	362. March 29
225. Sept. 1	271. April 2	317. March 9	363. Feb. 21
226. May 29	272. June 12	318. Jan. 13	364. May 5
227. July 19	273. April 15	319. May 23	365. Feb. 26
228. June 2	274. June 16	320. Dec. 15	366. June 8
229. Oct. 29	275. March 4	321. May 8	
230. Nov. 24	276. May 4	322. July 5	

June	196
July	180
August	173
September	157
October	182
November	149
December	122

TABLE 11.2 ORDERING OF DAYS OF THE YEAR BY RANDOM NUMBERS

1. April 9	47. May 18	93. Feb. 24	139. April 8
2. March 24	48. March 26	94. July 28	140. April 14
3. May 7	49. Dec. 31	95. June 30	141. May 12
4. Nov. 5	50. July 29	96. Nov. 18	142. Feb. 21
5. April 27	51. May 24	97. Jan. 29	143. Aug. 3
6. April 6	52. Jan. 1	98. Oct. 23	144. Jan. 4
7. May 4	53. Aug. 28	99. Oct. 12	145. July 30
8. June 2	54. March 11	100. Jan. 21	146. Sept. 27
9. Aug. 22	55. Oct. 3	101. July 10	147. March 15
10. Feb. 13	56. June 28	102. Nov. 30	148. Feb. 3
11. Jan. 5	57. Dec. 17	103. Dec. 25	149. Dec. 8
12. Dec. 24	58. June 5	104. Jan. 15	150. March 1
13. Nov. 16	59. March 12	105. April 10	151. May 15
14. July 13	60. Feb. 20	106. Aug. 16	152. Nov. 29
15. Nov. 26	61. Nov. 20	107. Dec. 19	153. March 28
16. July 4	62. Oct. 14	108. May 10	154. Jan. 11
17. April 25	63. Aug. 12	109. Aug. 23	155. Sept. 15
18. Aug. 13	64. Sept. 26	110. Dec. 7	156. July 23
19. July 8	65. Dec. 9	111. Dec. 22	157. Aug. 26
20. March 19	66. July 26	112. Oct. 9	158. Jan. 2
21. July 19	67. Sept. 2	113. March 14	159. Dec. 29
22. Feb. 19	68. April 26	114. Sept. 7	160. Aug. 29
23. Aug. 30	69. June 16	115. May 2	161. Jan. 3
24. June 10	70. Aug. 27	116. April 11	162. Dec. 5
25. Feb. 17	71. May 14	117. Jan. 24	163. Feb. 22
26. March 4	72. April 2	118. Jan. 17	164. May 21
27. Aug. 24	73. Dec. 26	119. June 19	165. June 9
28. March 20	74. Oct. 17	120. Feb. 6	166. July 18
29. Dec. 11	75. Oct. 2	121. June 13	167. Dec. 15
30. Nov. 28	76. March 27	122. June 18	168. May 31
31. Oct. 20	77. Dec. 3	123. Sept. 4	169. March 8
32. Sept. 3	78. Sept. 24	124. March 18	170. Nov. 24
33. Sept. 9	79. April 17	125. July 21	171. Nov. 9
34. Sept. 8	80. Oct. 10	126. Jan. 25	172. April 3
35. July 31	81. Oct. 8	127. March 17	173. Sept. 1
36. Jan. 20	82. March 5	128. Jan. 7	174. Dec. 2
37. July 12	83. Nov. 17	129. April 28	175. Feb. 11
38. Sept. 5	84. Feb. 1	130. July 24	176. May 1
39. Feb. 28	85. June 1	131. June 7	177. July 25
40. May 19	86. March 3	132. July 2	178. Oct. 18
41. May 22	87. Jan. 13	133. Jan. 30	179. Oct. 31
42. July 17	88. May 23	134. Sept. 18	180. Oct. 4
43. Feb. 9	89. Aug. 31	135. May 9	181. Sept. 29
44. Nov. 14	90. Feb. 10	136. July 5	182. Aug. 5
45. July 20	91. Oct. 1	137. Sept. 6	183. March 29
46. Oct. 29	92. Nov. 15	138. Feb. 23	184. Aug. 9

The results of one attempt at using a table of random numbers instead of a mixing urn approach to order the 366 days are given in Table 11.2. The method used was the following: every day was assigned a number, starting with January 1 as 001 and ending with December 31 as 366. Then the numbers in a table of random numbers (Table B.14) were divided into columns of three digits each, and the columns were read from top to bottom beginning at the left and proceed-

TABLE 11.2 ORDERING OF DAYS OF THE YEAR BY RANDOM NUMBERS (*con't.*)

185. Oct. 13	231. Nov. 23	277. Nov. 7	323. Jan. 22
186. April 12	232. July 27	278. June 3	324. July 22
187. Jan. 14	233. April 22	279. Feb. 7	325. June 26
188. Dec. 1	234. June 21	280. Oct. 15	326. May 8
189. April 7	235. Dec. 20	281. May 28	327. Dec. 21
190. Sept. 12	236. July 9	282. June 12	328. Aug. 6
191. April 19	237. Feb. 16	283. Sept. 22	329. Feb. 2
192. Dec. 6	238. March 13	284. Aug. 17	330. June 6
193. Jan. 26	239. Dec. 27	285. April 29	331. Nov. 6
194. Sept. 14	240. Jan. 28	286. Nov. 11	332. Aug. 18
195. Nov. 12	241. May 20	287. Feb. 4	333. April 30
196. May 3	242. Oct. 24	288. Aug. 19	334. Feb. 29
197. Oct. 21	243. Oct. 7	289. Sept. 13	335. June 27
198. Nov. 13	244. Oct. 26	290. Dec. 30	336. Nov. 4
199. Aug. 20	245. April 21	291. Dec. 28	337. April 23
200. March 7	246. Feb. 5	292. March 31	338. Oct. 25
201. Dec. 23	247. April 20	293. June 11	339. Oct. 6
202. June 20	248. Sept. 1	294. July 6	340. April 16
203. July 3	249. Nov. 22	295. April 13	341. April 15
204. Oct. 19	250. Aug. 7	296. Oct. 28	342. Jan. 10
205. March 6	251. Oct. 16	297. Aug. 4	343. Feb. 12
206. June 23	252. Aug. 2	298. June 17	344. Aug. 11
207. June 29	253. Sept. 19	299. Nov. 25	345. March 22
208. May 26	254. Feb. 8	300. Nov. 19	346. Aug. 25
209. Nov. 8	255. July 14	301. March 30	347. Aug. 1
210. March 25	256. May 5	302. April 24	348. April 18
211. March 9	257. Oct. 11	303. Jan 27	349. July 11
212. Feb. 14	258. April 4	304. Jan. 8	350. Feb. 26
213. June 25	259. Sept. 21	305. Dec. 4	351. May 6
214. Feb. 27	260. Nov. 2	306. Sept. 17	352. Aug. 10
215. Nov. 3	261. Aug. 15	307. Dec. 12	353. May 30
216. Dec. 13	262. Aug. 8	308. July 1	354. March 10
217. Oct. 22	263. March 21	309. Nov. 21	355. Sept. 25
218. March 2	264. Sept. 16	310. Nov. 10	356. Feb. 8
219. Jan. 6	265. Dec. 16	311. Oct. 5	357. Sept. 11
220. Dec. 10	266. Oct. 27	312. Jan. 31	358. Aug. 14
221. Jan. 12	267. March 16	313. May 27	359. May 25
222. Jan. 18	268. Dec. 18	314. Dec. 14	360. June 15
223. Feb. 25	269. Jan. 9	315. April 1	361. June 14
224. Sept. 20	270. July 16	316. Oct. 30	362. Jan. 16
225. June 22	271. March 23	317. Nov. 27	363. Sept. 28
226. Aug. 21	272. May 29	318. May 16	364. April 5
227. June 24	273. Nov. 1	319. June 8	365. July 7
228. May 13	274. Feb. 15	320. May 11	366. Sept. 30
229. July 15	275. June 4	321. Sept. 23	
230. Jan. 19	276. Jan. 23	322. May 17	

ing to the right. Only numbers between 001 and 366 were recorded; any number appearing more than once was ignored after the first occurrence. For example, the first five numbers in the first column of digits are: 100, 375, 084, 990, 128. The second and fourth numbers are greater than 366 and therefore are ignored. The first number, 100, corresponds to April 9; the third, 084, corresponds to March 24; the fifth, 128, corresponds to May 7. Hence these are the first three dates in the ordering of the 366 days. This method was continued until all the numbers were obtained; the sequence in which they occurred provides a random ordering. This method is somewhat tedious, because considerable repetition of numbers occurs before all 366 numbers are recorded. As a comparison with the averages of the December 1, 1969 lottery drawing, the averages of the random numbers for each month from Table 11.2 are:

January	185
February	182
March	162
April	195
May	189
June	201
July	156
August	198
September	193
October	182
November	188
December	176

The chi-square test can be used to investigate for the two tables the extent to which the observed averages deviate from the expected averages. (See Chapter 10.)

From Table 11.1:

Month	Observed (f_i)	Expected (e_i)	$(f_i - e_j)^2$
January	201	181	400
February	203	193	100
March	226	181	2025
April	204	186	324
May	208	181	729
June	196	186	100
July	180	181	1
August	173	181	64
September	157	186	841
October	182	181	1
November	149	186	1369
December	122	181	3481

$$\chi^2 = \sum_{j=1}^{k} \frac{(f_j - e_j)^2}{e_j} \qquad \text{d.f.} = k - 1 = 11$$

$$= \frac{400 + 2025 + 729 + 1 + 64 + 1 + 3481}{181} + \frac{324 + 100 + 841 + 1369}{186} + \frac{100}{193} = 51.69$$

The value of $\chi^2 = 51.69$ for 11 degrees of freedom is significant beyond any tabulated level. This indicates rejection of the hypothesis that the observed values are a random sample from the expected (nearly uniform) distribution.

From Table 11.2:

Month	Observed (f_j)	Expected (e_j)	$(f_j - e_j)^2$
January	185	181	16
February	182	193	121
March	162	181	361
April	195	186	81
May	189	181	64
June	201	186	225
July	156	181	625
August	198	181	289
September	193	186	49
October	182	181	1
November	188	186	4
December	176	181	25

$$\chi^2 = \sum_{j=1}^{k} \frac{(f_i - e_j)^2}{e_j} \qquad \text{d.f.} = k - 1 = 11$$

$$= \frac{16 + 361 + 64 + 625 + 289 + 1 + 25}{181} + \frac{81 + 225 + 49 + 4}{186} + \frac{121}{193} = 10.10$$

The tabled value of χ^2 for 11 degrees of freedom and $\alpha = .05$ is 19.675. Therefore the hypothesis that the observed averages are a random sample from the distribution of expected averages is accepted at the 5 percent level of significance.

On the basis of the two chi-square tests just discussed, there is some reason to believe that the draft lottery drawing on December 1, 1969 was not random.

Although the procedure for obtaining Table 11.2 involved ordering rather than sampling, it illustrates the procedures for sampling using a table of random numbers. If it is possible to assign a number to each unit in the population, the type of procedure outlined ensures that a random order and thus a random sample is obtained from the population. Suppose that the population contains 1000 units and a sample of 40 is to be drawn. Each unit is given a number from 0001 to 1000. The random number table is divided into columns of fours and read as in the above example or, alternatively, each group of four is read across in a row, as in reading the lines in a book; that is, groups of four digits are read from left to right across the rows proceeding to the bottom. Numbers greater than 1000 are ignored, as are repetitions. The units corresponding to the first 40 numbers between 0001 and 1000 constitute the random sample of 40 units.

Sometimes there is a question concerning where to start in a random number table. The starting place is unimportant, except that if an investigator always starts in the same place, the same sequence of numbers will occur. When the same random number table is used frequently, this problem is easily solved by having a large table* and selecting the starting place randomly. For example, a dollar bill may be examined to determine the first three digits of its serial number. The first digit can be taken as the page, the second as the column, the

---

* Such as The Rand Corporation, *A Million Random Digits with 100,000 Normal Deviates* (Glencoe, Illinois: Free Press, 1955). The source of the pages in Table B.14 in this book. Table B.14 does not contain all of the pages necessary to order the days of the year as described.

third as the row, at which to start in the table. The starting place will be somewhere between pages 1 and 9, columns 1 and 9, and rows 1 and 9, if this method is used.

The procedure discussed in this section is referred to as simple or unrestricted random sampling. As mentioned in the introduction, various types of restricted random sampling are also used. Most of these are based on methods of stratified sampling or cluster sampling, which are now discussed.

11.4 STRATIFIED SAMPLING

The efficiency of a sample design frequently can be improved by using stratified sampling; this is the case when the population can be categorized on the basis of some known characteristic and the estimator of interest has different standard errors for the different categories or strata. In stratified sampling, sampling takes place randomly from each stratum and the results are combined into a weighted estimate for the population. The purpose of stratification is to reduce the standard error of an estimator and thus obtain a more precise estimate than would be obtained by simple random sampling. Suppose an investigator wishes to estimate the average consumption of beer by persons 18 and over in a college town. Before obtaining a sample, it would be wise to stratify, for example, by separating the townspeople from the college students and then stratifying both of these groups again by sex and age.

The division of a population into strata should be on the basis that the elements within each stratum are more similar than the elements of the population as a whole. That is, the population should be somewhat heterogeneous while each stratum is relatively homogeneous. If the entire population is fairly homogeneous, stratification in unwarranted. Note that stratification also is unwarranted unless the variable used for stratification is related to the parameter being estimated.

Suppose a population is divided into p strata, with N_i elements in the ith stratum and a total of N elements in the population, so that $\sum_{i=1}^{p} N_i = N$. Let n_i be the size of the sample from the ith stratum, so that the total sample size is $n = \sum n_i$. Let $\bar{x}_i$ be the sample mean of the ith stratum and s_i be the sample standard deviation of the ith stratum. The weight of the ith stratum is given by

$$w_i = \frac{N_i}{N}$$

The sample mean of the ith stratum is

$$\bar{x}_i = \frac{\sum_j x_{ij}}{n_i} \qquad \text{where } j = 1, 2, \ldots, n_i$$

and the standard error of the mean of the ith stratum is

$$s_{\bar{x}} = \frac{s_i}{\sqrt{n_i}}$$

The estimates of the population mean and its standard error are obtained from the strata means and strata standard errors as follows:

$$\bar{x} = \sum_i \bar{w}_i \bar{x}_i$$

$$s_{\bar{x}} = \sqrt{\sum_i w_i^2 s^2 \bar{x}_i}$$

The finite population correction should be used if the sample includes 10 percent or more of the population.

EXAMPLE
Suppose a financial analyst wishes to estimate the mean capital budgeting expenditure made during the past year by a large conglomerate firm. There is reason to believe that expenditures made in the category of "expansion" are considerably higher than those made in the category of "replacement and modernization." Separate files are maintained for the two expenditure categories and the analyst estimates that 800 capital expenditures were made during the year in the "expansion" category and 2400 in the "replacement and modernization" category. The analyst obtains a random sample of 36 reports from the first category and a random sample of 100 reports from the second category. The results are as follows:

STRATUM NUMBER (i)	NUMBER IN STRATUM (N_i)	SAMPLE SIZE (n_i)	SAMPLE MEAN ($\bar{x}_i$)	SAMPLE STANDARD DEVIATION (s_i)
1	800	36	$200,000	$24,000
2	2400	100	50,000	5,000

The weights for the strata are:

$$w_1 = \frac{N_1}{N} = \frac{800}{3200} = \frac{1}{4}$$

$$w_2 = \frac{N_2}{N} = \frac{2400}{3200} = \frac{3}{4}$$

The estimate of the population mean is

$$\bar{x} = \sum_i w_i \bar{x}_i = \left(\frac{1}{4}\right)(200,000) + \left(\frac{3}{4}\right)(50,000) = \$87,500$$

The sample standard errors of the means are:

$$s_{\bar{x}_1} = \frac{s_1}{\sqrt{n_1}} = \frac{24,000}{\sqrt{36}} = \$4000$$

$$s_{\bar{x}_2} = \frac{s_2}{\sqrt{n_2}} = \frac{5000}{\sqrt{100}} = \$500$$

The estimate of the standard error of the mean for the population is

$$s_{\bar{x}} = \sqrt{\sum_i w_i^2 s_{\bar{x}_i}^2} = \sqrt{\left(\frac{1}{4}\right)^2 (4000)^2 + \left(\frac{3}{4}\right)^2 (500)^2} = \sqrt{1,140,800} = \$1070$$

When nothing is known about the variability within strata, the size of the sample from a stratum relative to the total sample size should be in the same ratio as the number of elements in the stratum to the total number of elements in the population. This is known as *proportionate stratification*. The sample size for each stratum in the previous example is approximately proportional to the number of elements in that stratum.

Disproportionate sampling, or sampling in which the sample sizes are not proportional to the stratum sizes, may be done on the basis of judgment or on the basis of known or expected variability within strata. Disproportionate sampling is appropriate when a stratum (or strata) influences the estimate more than its size, relative to the total, would indicate. In particular, it is preferable to have sample sizes not proportional to strata sizes when variability is known to differ within the strata. The most efficient sample sizes in this case are determined by what is referred to as optimum allocation. In general, the most efficient sample sizes for the strata are those that result in the smallest standard error of the mean for a given total sample size. To achieve optimum allocation, we make the sample size for each stratum proportional to the product of the number of elements in the stratum and the standard deviation of the relevant variable in the stratum:

$$n_i = n \left(\frac{N_i s_i}{\sum_i N_i s_i} \right)$$

The difficulty in optimum allocation, and in applying this formula, is determination of the s_i. How can these values be known in advance of sampling from the various strata? The answer is that they must be known from sampling in these or similar strata at a previous time, or the investigator must be able to make reasonable estimates based on his or her experience. If nothing is known in advance about the s_i, an investigator can do no better than proportionate sampling.

EXAMPLE
In Ajax County there are three types of farms: squire farms that are big, poor farms that are small, and intermediate farms that are intermediate in size. There are $N_i = 120$, $N_2 = 480$, and $N_3 = 240$ farms of each type, respectively. An investigator wishes to estimate the total hog population of Ajax County. Some of the squire farms have very large hog populations, and some raise no hogs at all. Nearly all the poor farms are thought to raise some hogs, but none of them have enough acres to raise very many hogs. Nearly all of the intermediate farms raise some hogs and some of them raise quite a large number of hogs. Hence the variation in the number of hogs is highest on the squire farms, lowest on the poor farms, and presumably in between on the intermediate farms. Based on previous research in similar counties, the investigator estimates the standard deviation of the hog population in each category to be, respectively, $s_1 = 150$, $s_2 = 20$, $s_3 = 60$. Using the optimum allocation formula, we find the sample size

from each stratum to be as follows, assuming a total sample of 100 farms:

$$n_i = n \left(\frac{N_i s_i}{\sum_i N_i s_i} \right)$$

$$n_1 = (100) \frac{120(150)}{120(150) + 480(20) + 240(60)} = (100) \frac{18,000}{42,000} = 43$$

$$n_2 = (100) \frac{480(30)}{120(150) + 480(20) + 240(60)} = (100) \frac{9,600}{42,000} = 23$$

$$n_3 = (100) \frac{240(60)}{120(150) + 480(20) + 240(60)} = (100) \frac{14,400}{42,000} = 34$$

The weights for the strata are:

$$w_1 = \frac{N_1}{N} = \frac{120}{840} = \frac{1}{7}$$

$$w_2 = \frac{N_2}{N} = \frac{480}{840} = \frac{4}{7}$$

$$w_3 = \frac{N_3}{N} = \frac{240}{840} = \frac{2}{7}$$

The sample means for the strata are:

$$\bar{x}_1 = 500 \qquad \bar{x}_2 = 60 \qquad \bar{x}_3 = 140$$

The estimate of the total hog population for the county therefore is:

$$\sum x = \bar{x}_1 n_1 + \bar{x}_2 n_2 + \bar{x}_3 n_3 = 120(500) + 480(60) + 240(140) = 122,400$$

The strata standard errors are as follows, assuming that the s_i calculated are the same as those estimated: (Note that the finite population correction $(N_i - n_i)/(N_i - 1)$ is appropriate.)

$$s_{\bar{x}_1} = \frac{s_1}{\sqrt{n_1}} \sqrt{\frac{N_1 - n_1}{N_1 - 1}} = \frac{150}{\sqrt{43}} \sqrt{\frac{120 - 43}{120 - 1}} = 11$$

$$s_{\bar{x}_2} = \frac{s_2}{\sqrt{n_2}} \sqrt{\frac{N_2 - n_2}{N_2 - 1}} = \frac{20}{\sqrt{480}} \sqrt{\frac{480 - 23}{479}} = .9$$

$$s_{\bar{x}_3} = \frac{s_3}{\sqrt{n_3}} \sqrt{\frac{N_3 - n_3}{N_3 - 1}} = \frac{60}{240} \sqrt{\frac{240 - 34}{239}} = 3.8$$

The mean hog population of a farm in Ajax County is

$$\bar{x} = \sum w_i \bar{x}_i = \left(\frac{1}{7} \right)(500) + \left(\frac{4}{7} \right)(60) + \left(\frac{2}{7} \right)(140) = 145.7 \simeq 146.$$

The estimate of the standard error of the population mean is

$$s_{\bar{x}} = \sqrt{\sum w_i^2 s_{\bar{x}_i}^2} = \sqrt{\left(\frac{1}{7} \right)^2 (11)^2 + \left(\frac{4}{7} \right)^2 (.9)^2 + \left(\frac{2}{7} \right)^2 (3.8)^2} = 2.3$$

11.5 CLUSTER SAMPLING

In *cluster sampling* the population is divided into clusters on the basis of groups or areas and a sample of clusters is obtained. This procedure frequently is used because of the inability of the investigator to identify each of the units in the population. In many situations cluster sampling provides a method of gaining access to all the units in the population to be sampled. In some cases, formation of clusters provides the only sampling frame. Once the clusters have been specified, a random sample of clusters is drawn, and from that sample a sample of the elements is obtained. Cluster sampling is sometimes referred to as area sampling if the clusters are geographical areas.

One of the most frequently cited examples of the use of cluster sampling involves estimating some characteristic of a city's population, such as average household expenditure on consumer durables. A complete list of households in a large city is not available, except in some unusual circumstance, and is very difficult and costly to construct accurately. One alternative to such a list is to number all the blocks in the city and obtain a random sample of the blocks. Each household in the selected blocks could then be questioned, or a random sample of households in each block could be questioned. In this case the blocks are referred to as *primary sampling units* and the households are referred to as *sampling elements* or *secondary units*. If a random sample of households is taken from the sample of blocks, the procedure is referred to as *two-stage sampling*.

Let C be the number of clusters in the population from which a random sample of c clusters is drawn. Each cluster contains N_i units; $\sum N_i = N$, where N is the total number of units in the population. Typically, the values of N_i and N are not known. Let x_{ij} denote the jth observation in the ith cluster and assume that every sampling unit in each of the c sampled clusters is observed. Then an unbiased estimate of the population total is given by

$$\hat{T} = \frac{C}{c} \sum_{i=1}^{c} \sum_{j=1}^{N_i} x_{ij} = \frac{C}{c} \sum_{i=1}^{c} T_i$$

where $T_i = \sum_{j=1}^{N_i} x_{ij}$ is the observed total for the ith cluster and C/c is the inverse of the sampling proportion. The estimated variance of $\hat{T}$ is

$$s_{\hat{T}}^2 = \frac{C(C - c)}{c(c - 1)} \sum_{i=1}^{c} (T_i - \bar{T})^2$$

where $\bar{T} = \sum_{i=1}^{c} T_i/c$ is the average of the observed cluster totals. The estimated number of sampling units in the population is given by

$$\hat{N} = \frac{C}{c} \sum_{i=1}^{c} N_i$$

and the estimated variance of $\hat{N}$ is

$$s_{\hat{N}}^2 = \frac{C(C - c)}{c(c - 1)} \sum_{i=1}^{c} (N_i - \bar{N})^2$$

where $\bar{N} = \sum_{i=1}^{c} N_i/c$ is the average number of sampling units in the observed clusters.

The population average value of x per sample unit is estimated by

$$\hat{\bar{x}} = \frac{\hat{T}}{\hat{N}}$$

which is a ratio of two other estimators and has estimated variance*

$$s_{\hat{\bar{x}}}^2 = \frac{C(C-c)}{c(c-1)\bar{N}^2}\left(\sum_{i=1}^{c} T_i^2 + \hat{\bar{x}}^2 \sum_{i=1}^{c} N_i^2 - 2\hat{\bar{x}} \sum_{i=1}^{c} T_i N_i\right)$$

EXAMPLE
Outstanding invoices in a number of file drawers were divided by a ruler into 60 groups, each having approximately the same number of invoices. Six groups were selected at random and examined completely. The firm's accountant is interested in estimating the total amount of outstanding invoices, the total number of outstanding invoices, and the average amount per outstanding invoice. The following sample data were obtained:

GROUP NUMBER (i)	NUMBER IN THE GROUP (N_i)	TOTAL AMOUNT OF INVOICES IN THE GROUP (T_i)
19	106	$6,450
26	98	7,225
04	99	7,025
51	105	6,850
17	103	6,650
42	107	6,600
	$N = 618$	$\sum T_i = 40,800$

$$\hat{T} = \frac{C}{c}\sum_{i=1}^{c} x_{ij} = \frac{C}{c}\sum_{i=1}^{c} T_i = \frac{60}{6}(40,800) = \$408,000$$

$$s_{\hat{T}}^2 = \frac{C(C-c)}{c(c-1)}\sum_{i=1}^{c}(T_i - \bar{T})^2$$

$$= \frac{(60)(54)}{6(5)}[(-350)^2 + (425)^2 + (225)^2 + (50)^2 + (-150)^2 + (-200)^2]$$

$$= (108)(418,750)$$

$$= 45,225,000$$

$$s_{\hat{T}} = 6,724.95$$

$$\hat{N} = \frac{C}{c}\sum_{i=1}^{c} N_i = \frac{60}{6}(618) = 6,180$$

* Derivations of formulas for variances in this section are given in Appendix 11.1.

$$s_{\hat{N}}^2 = \frac{C(C-c)}{c(c-1)} \sum_{i=1}^{c} (N_i - \bar{N})^2$$

$$= \frac{60(54)}{6(5)} [(3)^2 + (-5)^2 + (-4)^2 + (2)^2 + (0)^2 + (4)^2]$$

$$= (108)(70)$$

$$= 7,560$$

$$s_{\hat{N}} = 67,00$$

$$\hat{\bar{x}} = \frac{\hat{T}}{\hat{N}} = \frac{408,000}{6,180} = 66.02$$

$$s_{\hat{\bar{x}}}^2 = \frac{C(C-c)}{c(c-1)\bar{N}^2} \left(\sum_{i=1}^{c} T_i^2 + \hat{\bar{x}}^2 \sum_{i=1}^{c} N_i^2 - 2\hat{\bar{x}} \sum_{i=1}^{c} T_i N_i \right)$$

$$= \frac{(60)(54)}{(6)(5)(103)^2} [277,858,750 + (66.02)^2(63,724) - 2(66.02)(4,197,625)]$$

$$= \frac{(108)(1,522,225)}{(103)^2}$$

$$= 15,222$$

$$s_{\hat{\bar{x}}} = 123.4$$

Thus, summarizing the results, we see that the estimate of the total amount of invoices outstanding is $408,000 with a standard error of estimate of $6,724.95; the estimated number of invoices outstanding is 6,180 with a standard error of estimate of 87.00; the estimate of the average amount per invoice outstanding is $66.02 with a standard error of estimate of $123.40.

Note that $s_{\hat{T}}^2$, the variance of the estimator $\hat{T}$ of the population total, is based on variation among cluster totals. This variance of the estimated population total is large if the clusters vary considerably in size, even if all the observations are identical. When cluster sampling is used, the structure of the sampling frame should be such that the clusters are as small as possible and consist of approximately the same number of sampling units.

In many cases it is impossible or impractical to make the clusters equal in size; this is particularly likely to occur for area sampling. When clusters cannot be made equal in size, it is frequently possible to sample clusters with probabilities proportional to size; the probability that the ith cluster will be drawn into the frame is then

$$P_i = \frac{N_i}{N} \quad \text{and} \quad \sum_{i=1}^{c} p_i = \sum_{i=1}^{c} \frac{N_i}{N} = 1$$

Even if the values of the N_i are not known, there may be data available that are at least approximately proportional to the cluster sizes; for example, census data for households or areas on maps.

For sampling with replacement the population total is estimated by

$$\hat{T}_p = \frac{1}{c} \sum_{i=1}^{c} \frac{T_i}{P_i}$$

with a variance of

$$\sigma_{\hat{T}_p}^{\ 2} = \sum_{i=1}^{c} \frac{\left[\dfrac{T_i}{P_i} - \hat{T}_p\right]^2}{c(c-1)}$$

Note that the summations over the c sampled clusters may include duplicated sample clusters due to sampling with replacement. Such clusters are treated computationally in the same way as the other clusters.

In many cases it is more efficient to subsample from the clusters using a two-stage design. Formulas for estimators and their standard errors for these designs can be found in the references in Appendix A.

11.6 SUMMARY

The basic concepts of sampling are discussed in this chapter. Most of the examples are in the area of survey design, although the concepts of population, sampling units, sampling frame, and probability sampling also apply to experimental situations. The differences among random samples, systematic samples, judgment samples, and convenience samples are considered with respect to the inferences that can be made on the basis of each type of sample. Sampling designs based on simple random sampling, stratified sampling, and cluster sampling are discussed and illustrated. Multiphase sampling and sequential sampling are defined and the reader is referred to more advanced texts for details and examples of these methods.

APPENDIX 11.1
DERIVATION OF CLUSTER SAMPLING VARIANCE

For sampling without replacement it can be shown that if the proportion n/N is sampled,

$$\text{var}(y) = \left(1 - \frac{n}{N}\right) n s_y^{\ 2}$$

Thus if

$$\hat{T} = \frac{C}{c} \sum_{i=1}^{c} T_i \quad \text{where } T_i = \sum_{j=1}^{N_i} x_{ij} \quad \text{and } \bar{T} = \frac{\displaystyle\sum_{i=1}^{c} T_i}{c}$$

then

$$s_{\hat{T}}^{\ 2} = \frac{C^2}{c^2}\left(1 - \frac{c}{C}\right) c \sum_{i=1}^{c} \frac{(T_i - \bar{T})^2}{c-1}$$

$$= \frac{C^2}{c^2}\left(\frac{C-c}{C}\right) \frac{c}{c-1} \sum_{i=1}^{c} (T_i - \bar{T})^2$$

$$= \frac{C(C-c)}{c(c-1)} \sum_{i=1}^{c} (T_i - \bar{T})^2$$

Similarly, if

$$\hat{N} = \frac{C}{c} \sum_{i=1}^{c} N_i \quad \text{and} \quad \bar{N} = \frac{\sum_{i=1}^{c} N_i}{c} -$$

then

$$s_{\hat{N}}^2 = \frac{C(C-c)}{c(c-1)} \sum_{i=1}^{c} (N_i - \bar{N})^2$$

It can be shown that the variance of the ratio of two random variables $r = y/x$ is given approximately by

$$\text{var}(r) = \frac{1}{x^2} \left[\text{var}(y) + r^2 \, \text{var}(x) - 2r \, \text{cov}(y, x) \right]$$

As a rule of thumb, this approximation is satisfactory if the coefficient of variation of x is less than 0.2. (See Kish [Appendix A], p. 207 for derivation of the formula and justification of the rule of thumb.) Thus the variance of $\bar{x} = \hat{T}/\hat{N}$ (where $\hat{T} = C\bar{T}$ and $\hat{N} = C\bar{N}$) is given by

$$
\begin{aligned}
\text{var}\left(\hat{\bar{x}} = \frac{1}{\bar{N}^2} \left[\text{var}(\hat{T}) + \frac{\hat{T}^2}{\hat{N}^2} \text{var}(\hat{N}) = 2\frac{\hat{T}}{\hat{N}} \, \text{cov}(\hat{T}, \hat{N}) \right] \right. \\
= \frac{1}{\bar{N}^2} \left[\frac{C(C-c)}{c(c-1)} \right] \left[\sum_{i=1}^{c} (T_i - \bar{T})^2 + \frac{\hat{T}^2}{\hat{N}^2} \sum_{i=1}^{c} (N_i - \bar{N})^2 \right. \\
\left. - 2\frac{\hat{T}}{\hat{N}} \sum_{i=1}^{c} (T_i - \bar{T})(N_i - \bar{N}) \right] \\
= \frac{1}{\bar{N}^2} \left[\frac{C(C-c)}{c(c-1)} \right] \left[\sum_{i=1}^{c} T_i^2 + c\bar{T}^2 + \frac{\hat{T}^2}{\hat{N}^2} \sum N_i^2 + \frac{\hat{T}^2}{\hat{N}^2}(c\bar{N}^2) \right. \\
\left. - 2\frac{\hat{T}}{\hat{N}} \sum T_i N_i - 2\frac{\hat{T}}{\hat{N}} \left(-\sum_{i=1}^{c} T_i \bar{N} - \sum_{i=1}^{c} N_i \bar{T} + c\bar{N}\bar{T} \right) \right] \\
= \frac{C(C-c)}{c(c-1)\bar{N}^2} \left[\sum_{i=1}^{c} T_i^2 + \frac{\hat{T}^2}{\hat{N}^2} \sum N_i^2 - 2\frac{\hat{T}}{\hat{N}} \sum T_i N_i \right. \\
\left. + c\bar{T}^2 + c\bar{T}^2 - 2c\bar{T}^2 \right] \\
= \frac{C(C-c)}{c(c-1)\bar{N}^2} \left[\sum_{i=1}^{c} T_i^2 + \hat{\bar{x}}^2 \sum N_i^2 - 2\hat{\bar{x}} \sum T_i N_i \right]
\end{aligned}
$$

PROBLEMS

11.1 There are 25 students in Professor Huckster's advertising class. Each student has prepared a paper on the advertising program of a retailer in the city. The professor has allocated only 5 class periods for presentation of the papers, and each paper will require an entire period. The professor decides to choose those who will be asked to present their papers in class by a random process. Also, the date on which each student will give his or her paper will be randomly determined. The dates for the papers are as follows:

May 3, 10, 17, 24, and 31.

Using random numbers in Table 13.14, select a sample of 5 students from 25 and determine when each will deliver his or her paper.

11.2 It is desired to estimate the number of persons 65 years of age and older who have moved from other states to Tucson, Arizona, during the past year. The city has a total of 36 precincts, of which 6 have been surveyed. The precinct numbers and the number of persons in this category are as follows:

PRECINCT NUMBER	NUMBER OF PERSONS
4	22
7	13
18	25
24	12
29	30
35	8

Estimate the total number of such persons and obtain the standard error of this estimate.

11.3 Because of complaints from wholesalers, a sample inventory audit of Gigantic Hardware Manufacturing is to be made. GHM is supposed to be able to supply 40,000 different items. The purpose of the audit is to estimate the percent of GHM inventory items that are out of stock. Management wishes the percentage estimate to be accurate within ± 5 percent, with 95 percent confidence. Based on previous total inventory counts, the percentage of out-of-stock items is believed to lie between a minimum of 10 percent and a maximum of 20 percent. Given this information and the assumption that p (percentage of out-of-stock items) is normally distributed, what sample size should be chosen to be on the conservative side? Assuming a sample size of 256, calculate the confidence interval for p.

11.4 Newly admitted students of Cornstalk College have been stratified by the admissions officer into 5 economic classifications based on the occupations of their parents. She wishes to make some estimates concerning the amount of money these students are likely to spend during the next academic year in the city in which C.C. is located. The first column below gives the stratum number and the second gives the number of students in the stratum. Since the admissions officer knows nothing about the variability of expenditures within strata, she uses proportional allocation. After the samples have been taken, the admissions officer estimates the expenditures that each student will make in the city by means of an interview. The sample sizes, sample means, and sample standard deviations are as follows:

STRATUM	NUMBER OF STRATUM	SAMPLE SIZE	SAMPLE MEANS	SAMPLE STANDARD DERIVATION
1	50	10	$4000	$2000
2	250	50	3000	1000
3	300	60	2000	800
4	150	30	1500	600
5	50	10	1000	500

From this information estimate
(a) The average expenditures per student
(b) The standard error of the expenditures per student
(c) The total expenditures of the incoming students

11.5 An investigator wishes to estimate the total portfolio value for persons doing business with the SPLAM Bokerage Company. He separates the customers into two categories: active accounts and inactive accounts. There are 640 acounts in the first group and 260 in the second group. Based on some previous work, the investigator knows that the variation in portfolio value in the first group of accounts is higher than in the second group. In fact, he believes the standard deviation in the first will be about $10,000 and in the second about $4000. Assuming a total sample of 60 accounts, how should the sample be allocated between the two groups of accounts? If the sample mean portfolio value is $40,000 for the active accounts and $80,000 for the inactive accounts, estimate the total portfolio value of all accounts.

11.6 Criticize the following sampling procedures:

A. To estimate the number of pre school children in Sawbuck City, an investigator obtained a random sample of children in the city's only school and asked each child how many brothers and sisters he or she had who were too young to be in school.

B. To estimate the number of persons from eastern states entering southern Arizona, every tenth car on the Interstate Highway leading into that part of the state was stopped. The license plate on the vehicle was noted and the occupants counted. This was done every Saturday for a month.

C. To estimate the deer population in a well-defined area of Elk County, Pennsylvania, a large group of foresters, game wardens, and local sportsmen walked through the area abreast, about 20 yards apart, and counted the deer sighted.

ANALYSIS OF VARIANCE

12.1 INTRODUCTION

Analysis of variance is a statistical technique for analyzing observations or measurements that depend, in general, on the effects of several variables operating simultaneously. The purpose of analysis of variance is to determine which of the variables have significant effects and to estimate those effects. This is accomplished by partitioning the total variance in the observations into variances attributable to the different experimental variables and to random variation or error. The measurements or observations are usually obtained in an experimental context, for example, marketing research, industrial production, or agricultural crop yield studies. In some cases, observations are obtained in a nonexperimental context, for example, many types of economic and financial data can be analyzed using analysis of variance.

Analysis of variance is based on assumptions about both the model representing the observations and the variables affecting them. Analysis of variance also requires assumptions concerning the procedures used in obtaining the observations and thus involves problems of experimental design.

Experimental design is concerned with specifying the type and number of observations to be obtained and the experimental variables to be considered, as well as with the structure of an experiment and the randomization procedure used to prevent bias in the results. The purpose of experimental design is to obtain maximum information for minimum cost. Almost always, more precise conclusions can be drawn if careful thought is given to designing an experiment before data are obtained. When the design of an experiment has been considered and specified, a mathematical model that describes the experiment can be stated. From the mathematical model, the appropriate analysis, referred to as analysis of variance, follows. It is essentially a partitioning of the total variance of the observations into components representing variability due to different sources. Under appropriate assumptions, these components can be tested for statistical significance.

12.2 MATHEMATICAL MODEL FOR ANALYSIS OF VARIANCE

The analysis of variance model assumes that the relationship between the observed and experimental variables is linear. Suppose that there are n observations or measurements of the random variable Y denoted by $Y_1, Y_2, \ldots, Y_n$. Furthermore, assume that these random observations can be expressed as linear combinations of p unknown quantities $\beta_1, \beta_2, \ldots, \beta_p$ plus errors ε_1, $\varepsilon_2, \ldots, \varepsilon_n$, as follows:

$$Y_i = x_{1i}\beta_1 + x_{2i}\beta_2 + \cdots + x_{pi}\beta_p + \varepsilon_i, \qquad i = 1, 2, \ldots, n$$

The x_{ji}'s are known constant coefficients and the β_j's represent sources of variance that are of interest to the investigator in studying the phenomena underlying the observations. The purpose of analysis of variance is to make inferences about the ε_i's and some of the β_j's.

It is always assumed that the random variables ε_i have mean zero and it is usually assumed, in addition, that the ε_i are independent of each other and have the same variance σ^2:

$$E[\varepsilon_{ij}] = 0$$

$$E[\varepsilon_i\varepsilon_j] = \begin{cases} 0 & \text{if } i \neq j \\ \sigma^2 & \text{if } i = j \end{cases}$$

For purposes of making inferences, the ε_i usually are also assumed to be normally distributed.

Analysis of variance can be defined as a set of statistical methods for analyzing observations assumed to be of this structure. The coefficients x_{ji} are values of indicator variables and indicate the presence or absence of the effects β_j in the experimental conditions under which the observations are taken. If the x_{ji} are values assumed by continuous variables rather than by indicator variables, the model is appropriate for regression analysis (see Chapters 13 and 14).*

In general, the purpose of an experiment is to investigate the effects of independent or experimental variables on an observed or response variable. Any induced or selected variation in experimental procedures or conditions whose effect is to be observed is referred to as a *factor* or *treatment*. Each factor or treatment occurs at two or more different *levels*. Level is a general term referring to the characteristic or amount that defines or designates a particular category or classification of a factor.

The treatments may represent different (quantitative) amounts or degrees of a single variable, for example, tensile strength of cables, square feet of display area, debt ratio of a bond, and so forth; or they may represent different complex (qualitative) combinations of several variables that may or may not be speci-

---

* If some of the x_{ji} are values of indicator variables and some are values of continuous variables, referred to in this case as concomitant variables, the model is appropriate for covariance analysis. This model is discussed in more advanced texts. (See Appendix A.)

fically identified, for example, type of advertising or promotion, industry classi-fication, packaging design, and so forth. Thus the levels of a treatment or experimental variable may be quantitative or qualitative. The basic model for analysis of variance makes no distinction between quantitative and qualitative treatment variables, although some additional analyses are available for quantitative, but not for qualitative, treatment variables.

FIXED, RANDOM, AND MIXED MODELS

The effects β_j may be unknown constants, referred to as parameters, or they may be random variables. This corresponds to the fact that in any experiment the levels of a treatment variable may be fixed or random, depending on whether the investigator is interested in the particular levels used or in generalizing to a population of which those levels are a random sample.

In practice, quantitative variables are usually set at fixed levels in order to be sure that the range of interest is covered adequately—to avoid the possi-bility that random selection might result in a set of levels not representative of the relevant range of the variable. It is particularly important for the levels used to be representative of the relevant range of a variable when its effects may be nonlinear, since in this case the results for various parts of the range may be quite different.

Some sources of variance are clearly not of interest at fixed levels. For example, in most experiments particular operators, factories, vats, test tubes, experimental animals, and apparatus of various types are of interest only as an indication of the variability due to such factors. Thus they are appropriately selected at random.

Whether the levels of a treatment variable are to be considered fixed or random should be decided before data are collected. If levels are to be considered random, they should be chosen at random from all possible levels concerning which the investigator wishes to generalize.

A model in which all the β_j (treatment levels) are known constants, that is, all treatment levels are fixed, is referred to as a *fixed-effects model, parametric model*, or *model I*. A model in which all the β_j are random variables is referred to as a *random-effects model, components of variance model*, or *model II*. A model in which at least one β_j is an unknown constant and at least one β_j is a random variable is referred to as a *mixed model*.

One of the most important aspects of analysis of variance is the deter-mination of appropriate tests of significance. These tests of significance are not always intuitively appealing at first. This can be attributed, at least partly, to the fact that hypotheses are stated in terms of means, and the corresponding tests of significance are in terms of estimated variances. Keep in mind that the hypothesis that the variance among parameters is zero is equivalent to the hypothesis that the parameters are equal. This is the basic logic of tests of signi-ficance in analysis of variance, although the situation is complicated somewhat since variability assumed to be present even in the absence of experimental treatments must be taken into account.

Tests of significance in analysis of variance are based on ratios of mean squares. These ratios have the F distribution under the assumptions of the analysis of variance model. The mean square for a treatment or factor is an estimate of the variance in the response variable that is attributable to that treatment of factor. The significance of a factor is tested by comparing its mean square with the mean square estimated on the assumption that only random variation is present.

The appropriate tests of significance (ratios of mean squares) for various sources of variance are determined on the basis of the expected mean squares. Recall that a parameter, say μ, is estimated by a statistic, say $\bar{X}$, which has a known distribution under the null hypothesis; this distribution is the basis of tests of significance concerning μ. Similarly, the mean squares computed in analysis of variance are estimates of expected mean squares; these estimates have sampling distributions that form the basis of tests of significance concerning the corresponding sources of variance. The expected mean squares for analysis of variance are obtained by writing the model in terms of variances and taking expectations. For more complex models, this procedure is rather tedious algebraically; its details can be found in any analysis of variance text. (See Appendix A.)

In many cases the expected mean squares, and thus the appropriate tests of significance, differ for fixed, random, and mixed models. For the simple models discussed in this chapter, the appropriate tests of significance are generally the same for fixed, random, and mixed models; exceptions are discussed whenever they occur.

12.3 RESIDUAL VARIATION

The results of experiments are influenced not only by the effects of factors of interest to the investigator, but also by extraneous variation which tends to obscure or distort these effects. This extraneous variation is referred to as *residual variation* or *experimental error*. The mathematical model for analysis of variance includes an error term and assumptions concerning its properties.

Several sources of residual variation in an experiment include inherent variability in the experimental material, variation due to factors not included in the experimental design, and failure to standardize the experimental procedures. In practice, even in a very carefully designed and executed experiment, some residual variation is present. Whether it is possible in principle to eliminate residual variation completely or whether an inherent irreducible randomness is present in nature is essentially a philosophical question. In any case, inferences can be made in the presence of residual variation, provided that the assumptions of the model are satisfied. The effects on statistical inference when the assumptions of the model are violated are discussed below.

The amount of material to which a treatment is applied to obtain a single measurement of the response variable is an *experimental unit*. An experimental unit might be, for example, a grocery store, a person shopping for shoes, a TV tube, a production run of light bulbs, a group of consumers, and so forth. It is characteristic of such units that they respond differently even to apparently identical treatments; these differences in response contribute to the experimental error.

Randomly assigning experimental units to treatments or procedures is referred to as *randomization*. The purpose of randomization is to avoid bias, due to variability in the experimental units, that might otherwise be associated with the presence of experimental error. Randomization ensures that a treatment is not systematically either favored or handicapped by some extraneous source of variation, known or unknown. Although the result of any specific randomization may favor a particular treatment or treatments, this occurs only to an extent that is accounted for in the corresponding probability statements for significance tests and confidence limits.

Randomization is advisable, even when there is no reason to think that serious bias will result from not doing so. In more complicated experiments, randomization may be appropriate at several stages of the experimental procedure. It should be noted that failure to randomize at any stage of an experiment may introduce bias, unless either the variation introduced at that stage is negligible or the experiment effectively randomizes itself.

The method of randomization appropriate for a particular experiment is determined by the experimental design. Suitable restrictions on the randomization and an appropriate experimental design permit the investigator to use any available knowledge that will increase the precision of the experiment.

ACCURACY AND PRECISION

In designing and interpreting experiments the difference between accuracy and precision of measurements should be considered. The *accuracy* of a measurement is the closeness with which the measurement approaches the true value of whatever is being measured; accuracy is concerned with the bias of a measurement, that is, the difference between its observed and expected value. The *precision* of a measurement is its repeatability or reliability; precision is concerned with the variance of a measurement.

Although the proper use of randomization effectively eliminates bias associated with the assignment of experimental units to treatments, it does not affect bias resulting from the method of measurement itself. Physical measurements may be biased as a result of badly designed or maintained equipment. Measurements obtained from questionnaires or other types of reporting forms are frequently biased by the statements of the questions or by the types of responses required; this type of bias may be more difficult to detect and correct. With respect to measurement bias, the accuracy of an experiment can usually be improved only by a refinement or change in the experimental technique. Although such considerations are not properly the concern of the statistical design of experiments, their importance should be emphasized when a statistician discusses experimental design with an investigator.

Refining the experimental technique may improve the precision, as well as the accuracy, of an experiment. The precision of an experiment can also be improved by increasing the size of the experiment, either by providing more replications (repetitions) or by including additional treatments. In some cases the experimental material can be used in such a way that the effects of its variability are reduced and the precision of the experiment is improved. This can be accomplished, for example, by carefully selecting the material, by obtaining additional measurements that provide information about the material, or by grouping experimental units so that the units to which the different experimental treatments are applied are closely comparable.

The considerations involved in increasing the accuracy of an experiment differ in each case; the investigator must attempt to increase accuracy using his knowledge of the experimental treatments and the difficulties that might arise in measuring their effects.

The precision of an experiment is a more general and abstract problem with which the statistical design of experiments is concerned. The effects of replication, the use of additional measurements, and the grouping of experimental units can be estimated by statistical procedures. Analysis of variance provides an estimate of the experimental error and thus an estimate of the precision with which the treatment effects are estimated. Note, however, that analysis of variance estimates the treatment effects as they are measured in the experiment—if the measurements of the treatment effects are biased, analysis of variance can neither estimate nor correct for this bias.

12.4 ASSUMPTIONS OF ANALYSIS OF VARIANCE

Analysis of variance is based on the following assumptions: (1) statistical independence of the errors, (2) equality of variance (homoskedasticity) of the errors, and (3) normality of the errors and normality of the random effects in the models in which they appear. These assumptions are made for mathematical convenience in deriving results; unfortunately, they may be violated in practical applications of analysis of variance.

The assumption of statistical independence of the errors is often reasonable. If randomization is properly performed and care is taken so that measurements or observations are made separately and if one measurement or observation does not affect a later measurement or observation, then independence of the errors can reasonably be assumed. However, in practice there are situations in which measurements or observations, particularly those involving considerable judgment, may not be independent; for example, there may be a tendency on the part of the person recording observations to balance "high" observations by "low" observations.

The homoskedasticity assumption is particularly important since estimation of this (presumed common) variance is an essential feature of analysis of variance. The assumption of homoskedasticity is probably much less likely to be valid than the assumption of statistical independence of errors. For example, it is quite likely that different methods of measurement or observation will produce variations of different magnitudes, and higher expected values are often associated with higher variance.

Normality of errors and of random effects is probably the assumption that is least likely to be valid in practice. If measurements or observations are restricted in their range, for example, to positive values or to integer values, then clearly the assumption of normality is not justified. At best, the normality assumption is met only more or less approximately. As discussed below, the effects of violating the assumption of normality are not usually too serious, unless the deviation from normality is extreme.

A statistical method is said to be *robust* with respect to an assumption used in its derivation if the corresponding inferences are not seriously invalidated by violation of the assumption. The robustness of statistical methods is of great practical importance, since it is never possible to be certain that assumptions are satisfied and often there is good reason to think that they are not satisfied, even approximately.

The effects of violating the assumptions of analysis of variance can be summarized as follows:

1. Nonnormality has little effect on inferences concerning means, but in most cases has serious effects on inferences concerning variances of random effects.
2. Inequality of variances has little effect on inferences concerning means based on the same number of observations, but serious effects on inferences concerning means based on unequal numbers of observations.
3. Nonindependence of the observations can have serious effects on inferences about means.

The effects of violating the assumptions of analysis of variance are discussed in detail later.

In the following sections a number of basic experimental designs are detailed and brief descriptions of additional designs, with references to appropriate sources for further material, are presented.

The experimental design or structure of an experiment is described by the factors included and by the way in which the levels of the various factors are combined. A research study should be planned from the point of view of translating the questions or problems involved in the study into an appropriate experimental design. This requires careful consideration of the factors to be included in the investigation and the levels at which they are to be measured. The investigator should not underestimate the importance of these decisions and should also carefully consider whether the levels of each factor are to be fixed or random, since this affects the conclusions drawn from the results. Unfortunately, an extremely complicated and costly experiment may give very precise answers to questions in which the investigator has no interest. In order to avoid such results and the erroneous conclusions likely to be drawn from them, we must understand the types of questions and problems for which particular experimental designs are appropriate.

12.5 COMPLETELY RANDOMIZED DESIGNS

The simplest type of experimental design corresponds to the problem of comparing several levels of a single factor. The levels of this factor may be quantitative or qualitative, fixed or random. An experiment or investigation that concerns only one factor is referred to as a *single-factor experiment*, or a *one-way classification*. If the experimental units are assigned to factor levels completely at random, the design is referred to as a *completely randomized design*.

This simplest type of experimental design can be regarded as a straightforward generalization of the problem of comparing two means. Tests of the hypothesis $H_0: \mu_1 = \mu_2$, where μ_1 and μ_2 are the means of two independent normal populations, are considered in Chapter 9. When an investigator is interested in comparing the means of k populations, where $k > 2$, the hypothesis can be stated as

$$H_0: \mu_1 = \mu_2 = \cdots = \mu_k$$

A completely randomized design is appropriate for obtaining data to test this hypothesis; if the assumptions of normality and homogeneity of variance are satisfied, the hypothesis can then be tested on the basis of the analysis of variance F statistic.

In the case of two populations the hypothesis $H_0: \mu_1 = \mu_2$ can be tested against the two-sided alternative hypothesis $H_A: \mu_1 \neq \mu_2$ or against either of the one-sided alternative hypotheses $H_A: \mu_1 > \mu_2$ or $H_A: \mu_1 < \mu_2$. Thus if the null hypothesis is rejected, something is inferred concerning the relationship between the two population means.

If the null hypothesis $H_0: \mu_1 = \mu_2 = \cdots = \mu_k$ is rejected, then the only inference possible is that the population means are not all equal. The investigator is almost always interested in analyzing the data further in order to determine in more detail the nature of the differences among the population means. Several of these analyses which involve linear contrasts among the means and the circumstances in which they are valid are discussed subsequently.

There are many problems for which a completely randomized design is appropriate—for example, the comparison of the effectiveness of different types of advertising, the wearability of different types of tire tread, the output of different production lines, the effect of different incentives on the morale of factory workers, and so forth.

In the first example, advertising is the factor and the different types of advertising are the levels, which would probably be qualitative and considered to be fixed. Some measure of effectiveness, presumably in terms of sales, is the response or observed variable. In the second example, tire tread is the factor and different types of tread are the levels, which might be either qualitative or quantitative and would probably be considered fixed. Some measure of wearability is the response or observed variable. In the third example, production lines is the factor and the various production lines are the levels, which would probably be quantitative and considered to be random. Some measure of output is the response or observed variable. In the fourth example, incentives is the factor and the various incentives are the levels, which might be either qualitative or quantitative, fixed, or random. Some measure of morale is the response or observed variable.

In a completely randomized design the experimental units are assigned at random to the factor or treatment levels. Clearly, if the experimental units vary with respect to some characteristic relevant to the experimental observations, the results of the experiment will not be precise, although random assignment assures that they are unbiased. For example, if the stores available for investigating the effects of various advertising campaigns differ considerably with regard to number or type of customers, the results obtained using a completely randomized design might involve such a large proportion of extraneous variation that they would be useless. On the other hand, the raw material available for different production lines might be sufficiently homogeneous so that this source of extraneous variation could safely be ignored.

The mathematical model for a completely randomized design is

$$Y_{ij} = \mu + \beta_j + \varepsilon_{ij}$$

where Y_{ij} represents the ith observation ($i = 1, 2, \ldots, n_j$) on the jth level ($j = 1, 2, \ldots, k$); μ represents a common effect or overall mean for the whole experiment, β_j represents the effect of the jth level, and ε_{ij} represents the random error for the ith observation on the jth level.*

The errors are assumed to be normally and independently distributed with mean zero and homogeneous variance. That is,

$$E[\varepsilon_{ij}] = 0 \qquad \text{for all } i \text{ and } j$$

$$\text{var}(\varepsilon_{ij}) = \sigma^2 \qquad \text{for all } i \text{ and } j$$

ε_{ij}'s mutually independent and normally distributed

If the levels are fixed, $\beta_1, \beta_2, \ldots, \beta_k$ are fixed parameters and it is assumed that their expectation is zero:

$$\sum_{j=1}^{k} n_j \beta_j = 0$$

If the k levels are chosen at random, the β_j's are assumed to be normally and independently distributed with mean zero and variance σ^2. The β_j's are also assumed to be independent of the ε_{ij}'s.

The basic analysis of variance for a single-factor completely randomized experiment consists primarily of a test of the hypothesis

$$H_0: \beta_j = 0 \qquad \text{for all } j = 1, \ldots, k$$

or, equivalently,

$$H_0: \mu_1 = \mu_2 = \cdots = \mu_k$$

where the μ_j are the means for the k levels. If this hypothesis is true, there is no factor effect and each observation, Y_{ij}, consists of the common experimental effect, μ, and a random error, ε_{ij}.

In a one-way (single-factor) analysis of variance, there are assumed to be k populations, each representing one factor level. Independent random samples of n_j observations, $j = 1, 2, \ldots, k$ are obtained from the k populations. These observations can be represented by the format in Table 12.1.

TABLE 12.1 FORMAT FOR
ONE-WAY ANALYSIS OF VARIANCE

Factor Level: $j =$	1	2	$\cdots$	k
	Y_{11}	Y_{12}	$\cdots$	Y_{1k}
	Y_{21}	Y_{22}	$\cdots$	Y_{2k}
	$\vdots$	$\vdots$		$\vdots$
	$Y_{n_1 1}$	$Y_{n_2 2}$	$\cdots$	$Y_{n_k k}$
	$\mu_{.1}$	$\mu_{.2}$	$\cdots$	$\mu_{.k}$

---

* The observations, denoted in the general discussion by $Y_1, Y_2, \ldots, Y_n$, and the errors, denoted in the general discussion by $\varepsilon_1, \varepsilon_2, \ldots, \varepsilon_n$, are written more precisely here with double subscripts to indicate both the observation and the treatment level for which they were obtained.

where $E(Y_{i1}) = \mu_{.1}, \ldots, E(Y_{ik}) = \mu_{.k}$ represent the means for the k populations and $E(Y_{ij}) = \mu$ is the mean over all k populations. It is assumed that each of the k populations is infinite or sampling is with replacement, effectively making the populations infinite.

In the model $Y_{ij} = \mu + \beta_j + \varepsilon_{ij}$ the effect β_j can be indicated by $\mu_{.j} - \mu$ and the error ε_{ij} can be indicated by $Y_{ij} - \mu_{.j}$; the model can then be written as the identity

$$Y_{ij} - \mu = (\mu_{.j} - \mu) + (Y_{ij} - \mu_{.j})$$

The parameters μ and $\mu_{.j}$ are estimated, respectively, by $\bar{y}_{..}$ and $\bar{y}_{.j}$ which are obtained from the data as shown in Table 12.2. Note that $\bar{y}_{..}$ is the overall mean of the observations; $\bar{y}_{.j}$ is the mean of the observations on the jth factor level and is referred to as the jth group mean.

TABLE 12.2 FORMAT FOR
MEANS AND TOTALS (ONE-WAY ANOVA)

Factor Level: $j =$	1	2	$\cdots$	k	
	y_{11}	y_{12}	$\cdots$	y_{1k}	
	y_{21}	y_{22}	$\cdots$	y_{2k}	
	$\vdots$	$\vdots$		$\vdots$	
	$y_{n_1 1}$	$y_{n_2 2}$		$y_{n_k k}$	
Totals:	$T_{.1}$	$T_{.2}$	$\cdots$	$T_{.k}$	$T_{..}$
Number:	n_1	n_2	$\cdots$	n_k	N
Means:	$\bar{y}_{.1}$	$\bar{y}_{.2}$	$\cdots$	$\bar{y}_{.k}$	$\bar{y}_{..}$

---

where

$$T_{.j} = \sum_{i=1}^{n_j} y_{ij} \qquad j = 1, 2, \ldots, k$$

$$\bar{y}_{.j} = \frac{T_{.j}}{n_j} \qquad j = 1, 2, \ldots, k$$

$$T_{..} = \sum_{j=1}^{k} \sum_{i=1}^{n_j} y_{ij} = \sum_{j=1}^{k} T_{.j}$$

$$N = \sum_{j=1}^{k} n_j$$

$$\bar{y}_{..} = \frac{\sum_{j=1}^{k} \sum_{i=1}^{n_j} y_{ij}}{N} = \frac{\sum_{j=1}^{k} n_j \bar{y}_{.j}}{N}$$

The sample identity corresponding to the model (population) identity given above is

$$y_{ij} - \bar{y}_{..} = (\bar{y}_{.j} - \bar{y}_{..}) + (y_{ij} - \bar{y}_{.j})$$

If both sides of the equation are squared and summed over i and j, then

$$\sum_{j=1}^{k} \sum_{i=1}^{n_j} [(\bar{y}_{.j} - \bar{y}_{..}) + (y_{ij} - \bar{y}_{.j})]^2$$

$$= \sum_{j=1}^{k} \sum_{i=1}^{n_j} (\bar{y}_{.j} - \bar{y}_{..})^2 + \sum_{j=1}^{k} \sum_{i=1}^{n_j} (y_{ij} - \bar{y}_{.j})^2$$

(see Appendix 12.1 for the proof), and thus

$$\sum_{j=1}^{k} \sum_{i=1}^{n_j} (y_{ij} - \bar{y}_{..})^2 = \sum_{j=1}^{k} \sum_{i=1}^{n_j} (\bar{y}_{.j} - \bar{y}_{..})^2 + \sum_{j=1}^{k} \sum_{i=1}^{n_j} (y_{ij} - \bar{y}_{.j})^2$$

These sums of squares can be identified as follows:

$$\begin{pmatrix} \text{sum of squared} \\ \text{deviations about} \\ \text{overall mean} \end{pmatrix} = \begin{pmatrix} \text{sum of squared} \\ \text{deviations of} \\ \text{group means} \\ \text{about overall mean} \end{pmatrix} + \begin{pmatrix} \text{sum of squared} \\ \text{deviations of} \\ \text{observations about} \\ \text{group means} \end{pmatrix}$$

or, as is frequently written,

$$\text{SS}_{\text{Total}} = \text{SS}_{\text{Between groups}} + \text{SS}_{\text{Within groups}}$$

The corresponding degrees of freedom for these sums of squares are:

$$(N - 1) = (k - 1) + (N - k)$$

The mean square for a source of variance is equal to the corresponding sum of squares divided by its degrees of freedom. Thus

$$\text{MS}_{\text{Total}} = \text{SS}_{\text{Total}}/(N - 1)$$

$$\text{MS}_{\text{Between groups}} = \text{SS}_{\text{Between groups}}/(k - 1)$$

$$\text{MS}_{\text{Within groups}} = \text{SS}_{\text{Within groups}}/(N - k)$$

MS_{Total} estimates the variance attributable to differences among all the elements of the population; $\text{MS}_{\text{Between groups}}$ estimates the variance attributable to differences between groups; and $\text{MS}_{\text{Within groups}}$ estimates the variance attributable to differences among the elements within groups.

Note that although the sums of squares and the degrees of freedom are additive, the mean squares are not.

$$\text{MS}_{\text{Total}} \neq \text{MS}_{\text{Between groups}} + \text{MS}_{\text{Within groups}}$$

The within groups sum of squares and mean square are sometimes referred to as the residual or error sum of squares and mean square, respectively. Analysis of variance partitions the total sum of squares into parts attributable to different sources of variance. In a one-way classification the only source of variance introduced or manipulated by the investigator is that due to differences among groups or treatments. The variance within groups is thus essentially residual or error variance which is not attributable to differences among groups.

In a completely randomized design, the factor levels are sometimes referred to as treatment groups, treatments or groups. This terminology reflects the fact that a completely randomized design is frequently used for investigating effects not generally thought of as representing different levels of a factor; it is also suggestive of the appropriate comparison between mean squares. Variability between the groups (levels) that is large relative to variability within the groups (levels) indicates differences among the groups.

More precisely, under the assumptions that the ε_{ij} are independently normally distributed with zero mean and common variance, it can be shown that the ratio

$$F = \frac{\text{between groups mean square}}{\text{within groups mean square}}$$

has the F distribution with degrees of freedom $k - 1$ and $N - k$, when the null hypothesis is true.

A between groups mean square that is large relative to the within groups mean square indicates the existence of differences among groups or factor levels; the hypothesis

$$H_0: \mu_1 = \mu_2 = \cdots = \mu_k$$

is rejected when $F > F_{1-\alpha, k-1, N-k}$ for a test whose significance level is α. This test is appropriate whether the model is fixed or random.

The analysis of variance for a completely randomized design is summarized in Table 12.3. The computation of sums of squares is usually simplified if they are expanded and then written in terms of totals. These computational formulas are included in Table 12.3.

The computational steps for analyzing data from a completely randomized design are as follows:

1. List the observations by columns in the treatment groups to which they belong.
2. Obtain the sum $T_{.j} = \sum_{i=1}^{n_j} y_{ij}$ for each column and the sum $T_{..} = \sum_{i=1}^{n_j} \sum_{j=1}^{k} y_{ij}$ for the whole experiment.
3. Square each of the observations and obtain the sum of their squares $\sum_{i=1}^{n_j} \sum_{j=1}^{k} y_{ij}^2$.
4. Square each of the column totals obtained in (2), divide by the number of observations in the total, and obtain the sum $\sum_{j=1}^{k} T_{.j}^2/n_j$ for all the columns.
5. Square the total of the observations for the experiment and divide by the total number of observations $T_{..}^2/N$.

TABLE 12.3   ANALYSIS OF VARIANCE: COMPLETELY RANDOMIZED DESIGN

Source of Variation	d.f.	Sum of Squares (SS)	Mean Squares (MS)
Between groups (treatments)	$k - 1$	$\sum\limits_{j=1}^{k} n_j(\bar{y}_{.j} - \bar{y}_{..})^2$ $= \sum\limits_{j=1}^{k} \dfrac{T_{.j}^2}{n_j} - \dfrac{T_{..}^2}{N}$	$SS_{\text{Between}}/(k - 1)$
Within groups (residual or error)	$N - k$	$\sum\limits_{j=1}^{k} \sum\limits_{i=1}^{n_j} (y_{ij} - \bar{y}_{.j})^2$ $= \sum\limits_{j=1}^{k} \sum\limits_{i=1}^{n_j} y_{ij} - \sum\limits_{j=1}^{k} \dfrac{T_{.j}^2}{n_j}$	$SS_{\text{Within}}/(N - k)$
Total	$N - 1$	$\sum\limits_{j=1}^{k} \sum\limits_{i-1}^{n_j} (y_{ij} - \bar{y}_{..})^2$ $= \sum\limits_{j=1}^{k} \sum\limits_{i=1}^{n_j} y_{ij} - \dfrac{T_{..}^2}{N}$	

6. Obtain the sum of squares total from the computations in (3) and (5)

$$\sum_{j=1}^{k} \sum_{i=1}^{n_j} y_{ij} - \frac{T_{..}^2}{N}$$

7. Obtain the sum of squares between groups from the computations in (4) and (5)

$$\sum_{j=1}^{k} \frac{T_{.j}^2}{n_j} - \frac{T_{..}^2}{N}$$

8. Obtain the sum of squares within groups from the computations in (3) and (4)

$$\sum_{j=1}^{k} \sum_{i=1}^{n_j} y_{ij} - \sum_{j=1}^{k} \frac{T_{.j}^2}{n_j}$$

and divide by $N - k$ to obtain the mean square within groups.
9. Compute F by dividing the mean square between groups by the mean square within groups.
10. Compare the F computed in (9) with the F having degrees of freedom $k - 1$ and $N - k$ for the desired significance level.

EXAMPLE
A market analyst is interested in whether average rates of return differ for bonds having different ratings (AA, A, and BBB). Rates of return for the preceding year are recorded for random samples of 10 bonds from each rating class, as given below. Is there evidence that bonds having different ratings have different average rates of return?

	BOND RATING	
AA	A	BBB
6.2	5.4	5.2
7.8	5.8	6.3
9.5	6.9	5.9
8.9	7.5	7.5
7.9	8.4	6.4
8.7	9.1	5.8
7.4	7.8	7.2
6.9	5.6	6.5
9.1	6.5	5.6
9.4	6.7	5.5

(Step 2) $T_{.1} = 81.8$ $T_{.2} = 69.7$ $T_{.3} = 61.9$ $T_{..} = 213.4$

(Step 3) $\sum\limits_{i=1}^{n_j} \sum\limits_{j=1}^{k} y_{ij}^2 = (6.2)^2 + (7.8)^2 + (9.5)^2 + \cdots + (5.5)^2$

$$= 1597.44$$

(Step 4) $\sum\limits_{j=1}^{k} \dfrac{T_{.j}^2}{n_j} = \dfrac{(81.8)^2 + (69.7)^2 + (61.9)^2}{10} = \dfrac{15380.94}{10}$

$$= 1538.094$$

(Step 5) $\dfrac{T_{..}^2}{N} = \dfrac{(213.4)^2}{30} = \dfrac{45539.56}{30}$

$$= 1517.985$$

(Step 6) $SS_{Total} = \sum\limits_{i=1}^{n_j} \sum\limits_{j=1}^{k} y_{ij}^2 - \dfrac{T_{..}^2}{n}$

$$= 1597.44 - 1517.985 = 79.455$$

(Step 7) $SS_{Between} = \sum\limits_{j=1}^{k} \dfrac{T_{.j}^2}{n_j} - \dfrac{T_{..}^2}{N}$

$$= 1538.094 - 1517.985 = 20.109$$

$MS_{Within} = \dfrac{20.109}{2} = 10.055$

(Step 8) $SS_{Within} = \sum\limits_{j=1}^{k} \sum\limits_{i=1}^{n_j} y_{ij}^2 - \sum\limits_{j=1}^{k} \dfrac{T_{.j}^2}{n_j}$

$$= 1597.44 - 1538.094 = 59.346$$

$MS_{Within} = \dfrac{59.346}{27} = 2.198$

(Step 9) $F = \dfrac{MS_{Between}}{MS_{Within}} = \dfrac{10.055}{2.198} = 4.57$

(Step 10) $F_{2,27,.05} = 3.35$

The computed $F = 4.57$ is significant at the .05 level.

SOURCE OF VARIATION	d.f.	SUM OF SQUARES	MEAN SQUARE
Between bonds	2	20.109	10.055
Within bonds	27	59.346	2.198
Total	29	79.455	

The hypothesis tested is that the average rates of return are equal for bonds having different ratings.

$$H_0: \mu_{AA} = \mu_A = \mu_{BBB}$$

The critical value of F at the 5 percent level is 3.35 and the computed value of F is 4.57. Therefore reject the hypothesis and conclude that average rates of return differ for bonds having different ratings.

HYPOTHESES CONCERNING LINEAR CONTRASTS

As previously mentioned, an investigator is usually interested in testing hypotheses other than the hypothesis of equal means. Thus if the hypothesis of equal means is rejected, then the investigator usually wishes to determine the nature of the differences among groups or treatments. If the hypothesis of equality of two means is tested, then there are only three possible alternative hypotheses: $\mu_1 \neq \mu_2, \mu_1 > \mu_2$ and $\mu_1 < \mu_2$. The alternative of particular interest to the investigator is specified as part of the test of equality of means. Thus, in the case of two means, if the hypothesis of equality is rejected, then the general nature of the difference is indicated by the alternative hypothesis.

When more than two means are involved, there are a number of possible alternative hypotheses. For example, if the hypothesis $H_0: \mu_1 = \mu_2 = \mu_3 = \mu_4 = \mu_5$ is rejected, then the inequality among means may be attributable, for instance, to the fact that $\mu_1 = \mu_2 = \mu_3 = \mu_4$ but μ_5 is much larger (or smaller) than μ_1, μ_2, μ_3 and μ_4; or to the fact that $\mu_1 = \mu_2 = \mu_3$ and $\mu_4 = \mu_5$, but μ_1, μ_2 and μ_3 are larger (or smaller) than μ_4 and μ_5, or to any of a number of other alternatives. The analysis of variance F-test for the equality of means cannot be used as a basis for accepting any particular alternative; thus additional analyses are required in order to explore the nature of the differences among more than two means. In many cases investigators may have a particular alternative or several alternatives in which they are interested. In other cases they may not be able to specify alternatives before conducting the experiment and may wish instead to test alternatives suggested by the data.

Thus investigators may formulate hypotheses concerning differences among means before obtaining the data, particularly if they are reasonably certain that all the means are not equal; or they may formulate hypotheses after obtaining and examining the data. Hypotheses concerning differences among treatment means are usually stated in the form of linear contrasts among the means. Three methods for testing hypotheses concerning linear contrasts of treatment means are described in this section.

The method of orthogonal contrasts is used to test hypotheses concerning a set of orthogonal contrasts, which must be formulated prior to obtaining the data. The *Tukey method* (T-method) is used primarily to test hypotheses concerning contrasts involving two population means; it requires equal sample sizes, but permits the contrasts to be formulated after the data are obtained. The

Tukey method provides a confidence interval which holds simultaneously for every possible contrast between or among population means. Note that although the Tukey method actually can be used for contrasts involving any number of population means, in practice it is seldom used except for the special case of two means. The Scheffé method (S-method) is used to test hypotheses concerning all possible contrasts involving any number of population means; the Scheffé method permits the contrasts to be formulated after the data are obtained and does not require equal sample sizes.

ORTHOGONAL LINEAR CONTRASTS

A *linear combination* of the population means $\mu_1, \mu_2, \ldots, \mu_k$ is given by $L = c_1\mu_1 + \cdots + c_k\mu_k = \sum_{j=1}^{k} c_j\mu_j$. If $\sum_{j=1}^{k} c_j = 0$, then the linear combination is a linear *contrast*. If the contrasts

$$L_1 = c_{11}\mu_1 + c_{21}\mu_1 + \cdots + c_{k1}\mu_k$$

and

$$L_2 = c_{12}\mu_1 + c_{22}\mu_2 + \cdots + c_{k2}\mu_k$$

are such that

$$c_{11}c_{12} + c_{21}c_{22} + \cdots + c_{k1}c_{k2} = \sum_{j=1}^{k} c_{j1}c_{j2} = 0$$

then L_1 and L_2 are *orthogonal contrasts*.

In the preceding example investigators might be interested in the following contrasts:

$$L_1 = \tfrac{1}{2}(\mu_{AA} + \mu_A) - \mu_{BBB}$$

and

$$L_2 = \mu_{AA} - \mu_A$$

which correspond to the hypotheses

$$H_1: \frac{\mu_{AA} - \mu_A}{2} = \mu_{BBB}$$

and

$$H_2: \mu_{AA} = \mu_A$$

These contrasts are orthgonal, since

$$\sum_{j=1}^{r} c_{j1}c_{j2} = \left(\frac{1}{2}\right)(1) - \left(\frac{1}{2}\right)(1) - (1)(0) = 0$$

If $L = c_1\mu_1 + \cdots + c_k\mu_k$, then the hypothesis $L = L_0$ can be tested using

$$\hat{L} = c_1\bar{x}_{.1} + c_2\bar{x}_{.2} + \cdots + c_k\bar{x}_{.k}$$

and either the t statistic

$$t_{n-k} = \frac{\hat{L} - L_0}{\sqrt{MS_W \sum_{j=1}^{k} \frac{c_j^2}{n_j}}}$$

or the F statistic

$$F_{1,n-k} = \frac{(\hat{L} - L_0)^2}{MS_W \left(\sum_{j=1}^{k} \frac{c_j^2}{n_j} \right)}$$

The most frequent situation occurs when $n_j = n$ for all j and the hypothesis is $H_0: L = 0$. For this case, multiplying numerator and denominator by n^2, we obtain

$$F = \frac{\left[\sum_{j=1}^{k} c_j T_{.j} \right]^2}{MS_W \left[n \sum_{j=1}^{k} c_j^2 \right]}$$

where $T_{.j} = n_j \bar{x}_{.j}$
    The quantity

$$D^2 = \frac{\left[\sum_{j=1}^{k} c_j T_{.j} \right]^2}{n \sum_{j=1}^{k} c_j^2}$$

is a sum of squares with which 1 degree of freedom is associated. The degrees of freedom associated with the sum of squares for treatments is $k - 1$. If $L_1, L_2, \ldots, L_{k-1}$ are mutually orthogonal contrasts, each of which is equal to zero by hypothesis and if $D_1, D_2, \ldots, D_{k-1}$ are the sums of squares associated, respectively, with these contrasts, then

$$D_1 + D_2 + \cdots + D_{k-1} = SS_{Treatments}$$

Note that a set of orthogonal contrasts is not unique; note also that if there are k treatments, sets of at most $k - 1$ mutually orthogonal contrasts can be specified.

EXAMPLE
Consider the orthogonal contrasts given above

$$L_1 = \frac{1}{2}(\mu_{AA} + \mu_A) - \mu_{BBB}$$

and

$$L_2 = \mu_{AA} - \mu_A$$

and the hypotheses $L_1 = 0$ and $L_2 = 0$. The sums of squares corresponding to these contrasts are:

$$D_1{}^2 = \frac{\left[\sum\limits_{j=1}^{k} c_j T_{.j}\right]^2}{n \sum\limits_{j=1}^{k} c_j{}^2}$$

$$= \frac{[(\tfrac{1}{2})(81.8) + (\tfrac{1}{2})(69.7) - (1)(61.9)]^2}{10(\tfrac{1}{4} + \tfrac{1}{4} + 1)}$$

$$= \frac{(13.85)^2}{15}$$

$$= 12.7882$$

and

$$D_2{}^2 = \frac{\left[\sum\limits_{j=1}^{k} c_j T_{.j}\right]^2}{n \sum\limits_{j=1}^{k} c_j{}^2}$$

$$= \frac{[(1)(81.8) - (1)(69.7)]^2}{10(1 + 1)}$$

$$= \frac{(146.41)^2}{20}$$

$$= 7.3205$$

Note that

$$D_1{}^2 + D_2{}^2 = 20.109 = SS_{\text{Treatments}}$$

The analysis of variance table can be written with the treatment sum of squares partitioned into the sums of squares corresponding to a set of orthogonal contrasts among the treatments. Thus if $D_1, D_2, \ldots, D_{k-1}$ are the sums of squares associated, respectively, with mutually orthogonal linear contrasts among k treatment means, then the treatment sum of squares can be partitioned into $D_1, D_2, \ldots, D_{k-1}$, each having one degree of freedom. Thus the $k - 1$ degrees of freedom for treatments is partitioned into 1 degree of freedom for each of $k - 1$ sums of squares. This partitioning is shown for $k = 3$ in the following table.

SOURCE OF VARIATION	d.f.	SUM OF SQUARES	MEAN SQUARE
Between bonds (treatments)	2	20.109	10.055
L_1	1	12.7882	12.7882
L_2	1	7.3205	7.3205
Within bonds (residual)	27	59.346	2.198
Total	29	79.455	

The hypothesis

$$H_0: L_1 = 0$$

can be tested against the alternative hypothesis

$$H_A: L_1 \neq 0$$

using the F statistic

$$F_{1,27} = \frac{D_1{}^2}{MS_W} = \frac{12.7882}{2.198} = 5.82$$

Since $F_{1,27,.05} = 4.21$, $H_0 : L_1 = 0$ is rejected and the alternative hypothesis $H_A : L_1 \neq 0$ is accepted. The hypothesis

$$H_0 : L_2 = 0$$

can be tested against the alternative hypothesis

$$H_A : L_2 \neq 0$$

using the F statistic

$$F_{1,27} = \frac{D_2{}^2}{MS_W} = \frac{7.3025}{2.198} = 3.32$$

Since $F_{1,27,.95} = 4.21$, $H_0 : L_1 = 0$ is accepted.

If each of the hypotheses $L_1 = 0$ and $L_2 = 0$ is tested at the 5 percent level, the significance level for the experiment is not equal to .05 but is larger than .05. The sum of the two significance levels, .1, is an upper bound of this significance level; its exact value is not easy to determine.

Several comments should be made concerning the method of orthogonal contrasts. As noted, hypotheses concerning contrasts must be formulated before conducting an experiment. They can concern any contrasts of interest and need not concern only orthogonal contrasts. However, sums of squares associated with linear contrasts add to the sum of squares for treatments only if the contrasts are orthogonal.

Whenever several hypotheses are tested, the problem of determining the significance level for the experiment as a whole is very complicated. Clearly, if the probability of rejecting a true hypothesis is α for each of several tests, then the probability of rejecting at least one true hypothesis when several tests are conducted is considerably greater than α. In fact, if enough tests are conducted, then it is likely that one or more hypotheses will be rejected, even if all the hypotheses are true. The sum of the significance levels of all the tests is an upper bound to the probability of rejecting one or more hypotheses when all of them are true; the exact probability, that is, the significance level of the overall experiment, is difficult to determine.

TUKEY'S METHOD FOR MULTIPLE COMPARISONS

The usual procedure in analysis of variance is to test the hypothesis that all treatment means are equal; usually no particular concern is given to other hypotheses, unless the hypothesis of equal means is rejected. Then the investigator generally looks for contrasts that are responsible—that is, he attempts to determine the nature of the differences among the means. For this purpose, it is useful to have a procedure that permits selection of contrasts after the data have been obtained and the hypothesis of equal means has been tested and with which a known level of significance is associated. One such procedure, referred to as the T-method, has been developed by Tukey.

Let L be a contrast that is estimated by $\hat{L}$ and let all k samples be of size n. Tukey has shown that, with probability $1 - \alpha$,

$$\hat{L} - T\sqrt{MS_W}\left(\frac{1}{2}\sum_{j=1}^{k}|c_j|\right) \leq L \leq \hat{L} + T\sqrt{MS_W}\left(\frac{1}{2}\sum_{j=1}^{k}|c_j|\right)$$

holds simultaneously for *every* possible contrast that can be constructed. The statistic T is defined as follows:

$$T = \frac{1}{\sqrt{n}}\, q_{1-\alpha,k,N-k}$$

where $q_{1-\alpha,k,N-k}$ is the point exceeded with probability $1 - \alpha$ in the distribution of the studentized range (Table B.15).*

The T-method was originally designed for contrasts involving two means and is seldom used in practice except for this case. For contrasts involving two means

$$\frac{1}{2}\sum_{j=1}^{k}|c_j| = \frac{1}{2}(|1| + |-1|) = 1$$

and, with probability $1 - \alpha$, the intervals

$$(\bar{x}_{.j} - \bar{x}_{.j'}) - T\sqrt{MS_W} \leq \mu_j - \mu_{j'} \leq (\bar{x}_{.j} - \bar{x}_{.j'})T\sqrt{MS_W}$$

include all $k(k-1)/2$ differences of pairs of means $\mu_j - \mu_{j'}$, where j and j' refer to any two treatment means. When the interval computed for a contrast L does not include zero, L is significant and the hypothesis $\mu_j = \mu_{j'}$ is rejected. Thus, in order to determine which means apparently differ from each other, the corresponding differences between the sample means are compared with

$$T\sqrt{MS_W} = q_{1-\alpha,k,n-k}\sqrt{\frac{MS_W}{n}}$$

EXAMPLE
For the data given above,

$$\bar{x}_{.1} = 8.18 \qquad \bar{x}_{.2} = 6.97 \qquad \bar{x}_{.3} = 21.34$$

$$q_{.95,3,27}\sqrt{\frac{MS_W}{n}} = 3.51\sqrt{\frac{2.198}{10}} = 1.646$$

There are $(3)(2)/2 = 3$ differences between pairs of means. It is convenient to arrange these differences in a table, with the means ordered from largest to smallest and the largest differences listed first.

---

* If $Y_1, \ldots, Y_n$ are independently $N(\mu, \sigma^2)$ and s^2 is an unbiased estimate of σ^2 based on v degrees of freedom, then

$$q_{n,v} = \frac{\max Y - \min Y}{s}$$

is the studentized range.

	$\bar{x}_{.j}$	$\bar{x}_{.j} - \bar{x}_{.3}$	$\bar{x}_{.j} - \bar{x}_{.2}$
$\bar{x}_{.1}$	8.18	1.99	1.21
$\bar{x}_{.2}$	6.97	0.78	
$\bar{x}_{.3}$	6.19		

Thus, comparing the differences in this table with the critical value 1.646, the hypothesis $H_0 \colon \mu_1 = \mu_3$ is rejected and the hypotheses $H_0 \colon \mu_1 = \mu_2$ and $H_0 \colon \mu_2 = \mu_3$ are accepted. Thus there seems to be a difference between the average rates of return of AA and BBB bonds; apparently there is no difference between the average rates of return of AA and A bonds nor the average rates of return of A and BBB bonds. The probability is .05 that one or more of these conclusions is incorrect.

More complicated contrasts can also be tested using the T-method. For example, the contrast

$$L = \frac{1}{2}(\mu_1 + \mu_2) - \mu_3$$

can be tested as follows:

$$\frac{1}{2} \sum_{j=1}^{3} |c_j| = \frac{1}{2}\left(\frac{1}{2} + \frac{1}{2} + 1\right) = 1$$

and thus

$$T\sqrt{\mathrm{MS_W}} \frac{1}{2} \sum_{j=1}^{3} |c_j| = 1.646$$

and the 95 percent confidence interval for $\frac{1}{2}(\mu_1 + \mu_2) - \mu_3$ is given by

$$\hat{L} - T\sqrt{\mathrm{MS_W}} \frac{1}{2} \sum_{j=1}^{k} |c_j| \le L \le \hat{L} + T\sqrt{\mathrm{MS_W}} \frac{1}{2} \sum_{j=1}^{k} |c_j|$$

$$1.385 - 1.646 \le L \le 1.385 + 1.646$$

$$-0.261 \le L \le 3.031$$

Since this interval includes zero, the contrast $L = \frac{1}{2}(\mu_1 + \mu_2) - \mu_3$ is not significantly different from zero.

Other contrasts involving three means can be tested simultaneously with probability $\alpha = .05$ that one or more of the conclusions is incorrect. The other two contrasts between three means are:

$$L' = \frac{1}{2}(\mu_1 + \mu_3) - \mu_2$$

and

$$L'' = \frac{1}{2}(\mu_2 + \mu_3) - \mu_1$$

Since $L' = 0.215$ and $L'' = -1.6$,

$$-1.431 \le L' \le 1.861$$

and

$$-3.246 \le L'' \le 0.046$$

Neither of these contrasts is significantly different from zero.

SCHEFFÉ'S METHOD FOR MULTIPLE COMPARISONS

Another procedure for testing multiple comparisons, the S-method, has been proposed by Scheffé. Unlike the T-method, the S-method can be used when the sample sizes are not equal; in addition, more is known about the statistical properties of the S-method.

Scheffé has shown that, with probability $1 - \alpha$, all possible contrasts are included in the set of intervals

$$\hat{L} - S\hat{\sigma}_{\hat{L}} \leq L \leq \hat{L} + S\hat{\sigma}_{\hat{L}}$$

where

$$S^2 = (k - 1)F_{1-\alpha, k-1, N-k}$$

and

$$\hat{\sigma}^2_{\hat{L}} = \mathrm{MS_W}\left(\sum_{j=1}^{k} \frac{c_j}{n_j}\right)$$

Again, the probability is α that one or more incorrect inferences will be made.

For the above example,

$$S^2 = (2)F_{.95, 2, 27}$$
$$= (2)(3.35)$$
$$= 6.70$$
$$S = 2.588$$

For contrasts involving two means

$$\hat{\sigma}^2_{\hat{L}} = (2.198)\left(\frac{1 + 1}{10}\right)$$
$$= .440$$
$$\hat{\sigma}_{\hat{L}} = .663$$

and

$$S\hat{\sigma}_{\hat{L}} = 1.716$$

Thus, for the S-method, differences between two sample means are compared with 1.716 rather than with 1.646 as in the T-method. In this example the contrast $\bar{x}_{.1} - \bar{x}_{.3} = 1.99$ is significant using either the T-method or the S-method. In general, for contrasts involving two means and equal sample sizes the T-method gives shorter confidence intervals and thus finds more differences significant than the S-method.

For contrasts of the form $L = \frac{1}{2}(\mu_1 + \mu_2) - \mu_3$

$$\hat{\sigma}^2_{\hat{L}} = (2.198)\left(\frac{\frac{1}{4} + \frac{1}{4} + 1}{10}\right)$$
$$= .330$$
$$\hat{\sigma}_{\hat{L}} = .574$$

and $S\hat{\sigma}_{\hat{L}} = (2.588)(.574) = 1.486$. The S-method thus provides confidence intervals

$$\hat{L} - 1.486 \leq L \leq \hat{L} + 1.486$$

for contrasts involving three means, while the T-method provides confidence intervals

$$\hat{L} - 1.646 \leq L \leq \hat{L} + 1.646$$

for these contrasts. For more complex contrasts the S-method generally provides shorter intervals than the T-method.

An interesting property of the S-method is that whenever the hypothesis of equal means is rejected using the F-test, one or more of the intervals obtained using the S-method will not cover zero. That is, whenever $H_0 : \mu_1 = \cdots = \mu_k$ is rejected, the S-method will identify at least one significant contrast. The S-method has an additional advantage in that it is known to be affected very little if the assumptions of normality and equal variances are not satisfied.

Note that a confidence interval for any specified contrast L can be obtained on the basis of the t-distribution, just as a sum of squares and F statistic can be associated with any contrast. With probability $1 - \alpha$, the contrast L is included in the interval

$$\hat{L} - t_{1-\alpha/2, N-r}\hat{\sigma}_{\hat{L}} \leq L \leq \hat{L} + t_{1-\alpha/2, N-r}\hat{\sigma}_{\hat{L}}$$

For example, for the contrast $L = \frac{1}{2}(\mu_1 + \mu_2) - \mu_3$

$$\hat{L} - (2.052)(.574) \leq L \leq \hat{L} + (2.052)(.574)$$

$$\hat{L} - 1.178 \leq L \leq \hat{L} + 1.178$$

which is a much shorter interval than that provided by either the T-method or the S-method. This is reasonable, since the interval is constructed for one particular contrast and the intervals constructed in the T-method or the S-method are for all possible contrasts. Note that the interval based on the t-statistic is appropriate only if one particular contrast is specified in advance as the only contrast to be tested.

This is unlikely to be the case in practice; as discussed, the usual procedure is to test the hypothesis of equal means and if this hypothesis is rejected, then to look for a contrast or contrasts to which the inequality is attributable. It is very difficult to determine the significance levels associated with several intervals based on the t-statistic. Thus, in most situations, intervals given by the T-method or the S-method should be used, since (1) these intervals can be computed for contrasts selected after the data are obtained and the hypothesis of equal means is rejected and (2) an exact probability can be associated with these intervals.

UNEQUAL SAMPLE SIZES

One of the advantages of a one-way classification or completely randomized design is the relative ease with which unequal numbers of observations for the groups can be handled in the analysis; there is no theoretical difficulty and the computations are only slightly more difficult than for equal numbers of observations. For some more complicated designs, valid analysis for unequal numbers of observations (or for "missing observations") is computationally difficult, or even impossible in theory.

The following example illustrates the analysis of variance for a completely randomized design with unequal numbers of observations in the groups.

EXAMPLE

The manager of Amphibious, Inc., assigned 14 sales people at random to three districts; 3 to District A, 6 to District B, and 5 to District C. (The number of sales people for each district was determined on the basis of population). Sales for each of the sales people for the first quarter were recorded in tens of thousands of dollars, as given below. Does there seem to be a difference among districts with respect to sales potential?

DISTRICT A	DISTRICT B	DISTRICT C	
10	5	4	
13	4	3	
9	6	5	
	3	2	
	2	3	
	5		
$T_{.j}$ 32	25	17	$T_{..} = 74$
n_j 3	6	5	$N = 14$
$\bar{x}_{.j}$ 10.67	4.17	3.40	

$$SS_{Total} = \sum_{j=1}^{k} \sum_{i=1}^{n_j} y_{ij}^2 - \frac{T_{..}^2}{N} = 528 - \frac{(74)^2}{14} = 136.86$$

$$SS_{Between} = \sum_{j=1}^{k} \frac{T_{.j}^2}{n_j} - \frac{T_{..}^2}{N} = \frac{(32)^2}{3} + \frac{(25)^2}{6} + \frac{(17)^2}{5} - \frac{(74)^2}{14}$$

$$= 112.16$$

$$SS_{Within} = SS_{Total} - SS_{Between} = 136.86 - 112.16 = 24.70$$

SOURCE OF VARIATION	d.f.	SUM OF SQUARES	MEAN SQUARE
Between districts	2	112.16	56.08
Within districts	11	24.70	2.25
Total	13	136.86	

$$F_{2,11} = \frac{56.08}{2.25} = 24.92$$

Since $F_{2,11,.99} = 7.20$, the hypothesis is rejected and it is concluded that apparently the three districts differ in sales potential. Based on an examination of the sample means, the manager might decide to compare district A with districts B and C. Since the sample sizes are unequal, the T-method cannot be used. For contrasts involving two means, the 95 percent Scheffé intervals are given by

$$\hat{L} - S\hat{\sigma}_L \leq L \leq \hat{L} + S\hat{\sigma}_L$$

where

$$S^2 = (2)F_{.95,2,12}$$
$$= (2)(3.89) = 7.78$$
$$S = 2.789$$

and

$$\hat{\sigma}_{\hat{L}} = MS_W \sum_{j=1}^{3} \frac{c_j^2}{n_j}$$

$$= 2.25 \sum_{j=1}^{3} \frac{c_j^2}{n_j}$$

$$\text{For } A \text{ and } B \quad \sum_{j=1}^{3} \frac{c_j^2}{n_j} = \frac{1}{8} + \frac{1}{6} = .5$$

$$\text{For } A \text{ and } C \quad \sum_{j=1}^{3} \frac{c_j^2}{n_j} = \frac{1}{3} + \frac{1}{5} = .533$$

$$\text{For } B \text{ and } C \quad \sum_{j=1}^{3} \frac{c_j^2}{n_j} = \frac{1}{6} + \frac{1}{5} = .367$$

Thus the intervals for contrasts of two means are as follows:

$$(\bar{x}_A - \bar{x}_B) - S\hat{\sigma}_{\hat{L}} \leq \mu_A - \mu_B \leq (\bar{x}_A - \bar{x}_B) + S\hat{\sigma}_{\hat{L}}$$

$$6.5 - (2.789)(1.061) \leq \mu_A - \mu_B \leq 6.5 + (2.789)(1.061)$$

$$6.5 - 2.959 \leq \mu_A - \mu_B \leq 6.5 + 2.959$$

$$3.541 \leq \mu_A - \mu_B \leq 9.459$$

This interval does not include zero, so the contrast $\mu_A - \mu_B$ is significant.

$$(\bar{x}_A - \bar{x}_C) - S\hat{\sigma}_{\hat{L}} \leq \mu_A - \mu_C \leq (\bar{x}_A - \bar{x}_C) + S\hat{\sigma}_{\hat{L}}$$

$$7.27 - (2.789)(1.095) \leq \mu_A - \mu_C \leq 7.27 + (2.2789)(1.095)$$

$$7.27 - 3.054 \leq \mu_A - \mu_C \leq 7.27 + 3.054$$

$$4.276 \leq \mu_A - \mu_C \leq 10.324$$

This interval does not include zero, so the contrast $\mu_A - \mu_C$ is significant.

$$(\bar{x}_B - \bar{x}_C) - S\hat{\sigma}_{\hat{L}} \leq \mu_B - \mu_C \leq (\bar{x}_B - \bar{x}_C) + S\hat{\sigma}_{\hat{L}}$$

$$0.77 - (2.789)(.909) \leq \mu_B - \mu_C \leq 0.77 + (2.789)(.909)$$

$$0.77 - 2.535 \leq \mu_B - \mu_C \leq 0.77 + 2.535$$

$$-1.765 \leq \mu_B - \mu_C \leq 3.305$$

This interval includes zero, so the contrast $\mu_B - \mu_C$ is not significant.
The contrast $L = \mu_A - \frac{1}{2}(\mu_B + \mu_C)$ can be tested as follows:

$$\hat{L} - S\hat{\sigma}_{\hat{L}} \leq L \leq \hat{L} + S\hat{\sigma}_{\hat{L}}$$

where

$$S = 2.789$$

$$\hat{\sigma}_{\hat{L}}^2 = MS_W \sum_{j=1}^{3} \frac{c_j}{n_j} = (2.25)\left(\frac{1}{3} + \frac{1}{24} + \frac{1}{20}\right)$$

$$= (2.25)(.425) = .956$$

$$\hat{\sigma}_{\hat{L}} = .978$$

$$6.885 - (2.789)(.978) \leq L \leq 6.885 + (2.789)(.978)$$

$$6.885 - 2.728 \leq L \leq 6.885 + 2.728$$

$$4.157 \leq L \leq 9.613$$

This interval does not include zero, so the contrast $L = \mu_A - \frac{1}{2}(\mu_B + \mu_C)$ is not significant.

Thus the differences among districts with respect to sales potential apparently are attributable to the differences between district A and districts B and C; districts B and C apparently do not differ significantly.

12.6 RANDOMIZED BLOCK DESIGNS

In a completely randomized design, experimental units are assigned at random to treatments; any relevant lack of homogeneity among experimental units increases the error mean square and decreases the precision of the experiment in detecting differences among treatments. If the experimental units are heterogeneous but can be divided into fairly homogeneous groups (referred to as blocks), then the precision of the experiment can be improved considerably. If the number of experimental units in each block is equal to the number of treatments or factor levels and if the experimental units within each block are assigned at random, one to each treatment, then the experimental design is referred to as a *randomized complete block design* or a *randomized block design*.

The purpose of blocking is to reduce a source of variation that is not of interest in order to increase the precision of measurement of treatment effects that are of interest. Blocking is essentially an extension of the method of pairing or matching to more than two treatments. When a source of extraneous variation is removed from an experiment, it is more likely that (true) differences among treatments will be detected; that is, effective blocking reduces the probability of type II error. Effective blocking frequently results in a dramatic increase in precision. However, if the experimental units are not sufficiently heterogeneous or if the blocking variable is not relevant to the treatments being investigated, then a randomized block design decreases the degrees of freedom for error with no compensating increase in precision. Since each block contains one observation on each treatment, a block can be considered to be a complete experiment; thus a block is sometimes referred to as a *replicate* or *replication* of an experiment.

Suppose, for example, that a manufacturer of electric golf carts is interested in the resistance to wear of four brands of ball bearings. He decides to test the bearings under actual conditions of use, rather than in a laboratory simulation. Four golf carts are used in the experiment. Wear is measured by decrease in weight of the ball bearing after 50 hours of use. One ball bearing is required for each wheel of a golf cart; 16 ball bearings (4 of each brand) are used in the experiment.

One possibility is to assign brands at random to golf carts, using 4 ball bearings of a given brand on each golf cart. An assignment of this type is shown below. The 4 golf carts are denoted by I, II, III, and IV and the 4 brands of ball bearings are denoted by A, B, C, and D.

	CARTS			
	I	II	III	IV
Brand of ball bearing	B	D	A	C
	B	D	A	C
	B	D	A	C
	B	D	A	C

The disadvantages of this design are apparent, since differences in brands are indistinguishable from differences in golf carts or their drivers or the routes over which they travel. Such a design is said to be *completely confounded*, since the factor of interest (brands of ball bearings) is indistinguishable from another source of variance (e.g., differences in golf carts) in the analysis.

In order to randomize differences in golf carts over the brands of ball bearings, we can use a completely randomized design. In this case the 16 ball bearings are assigned at random to golf carts; for example, the assignment might be as shown below:

	CARTS			
	I	II	III	IV
Brand of ball bearing	D	C	C	A
	B	A	C	D
	A	D	C	A
	B	B	B	D

The appropriate model is

$$y_{ij} = \mu + \tau_j + \varepsilon_{ij} \qquad i, j = 1, 2, 3, 4$$

where y_{ij} represents the ith observation on the jth treatment and τ_j is the jth treatment effect.

A completely randomized design is clearly preferable to a completely confounded design. However, a completely randomized design has some obvious disadvantages in this example; the completely randomized design randomizes or averages out variation among carts, but it does not remove this source of variance from the error mean square and thus does not increase the precision of the experiment. Since the main purpose of experimental design is to decrease experimental error, a design that removes variation due to golf carts from variation due to error is clearly preferable to a completely randomized design; a randomized block design accomplishes this.

For a randomized block design, each golf cart is considered to be a block of 4 experimental units (wheels); one ball bearing of each brand is used on each golf cart, assigned at random to a particular wheel. Randomization is thus said to be restricted within blocks. The model for a randomized complete block design is

$$y_{ij} = \mu + \beta_i + \tau_j + \varepsilon_{ij} \qquad i, j = 1, 2, 3, 4$$

where β_i represents the effect of blocking, τ_j represents the jth treatment effect, and y_{ij} represents the observation on the jth treatment in the ith block.

The analysis of variance for a randomized complete block design is a two-way analysis of variance, since both treatment and block effects can be estimated and removed from the error mean square. The observations can be represented by the format in Table 12.4.

TABLE 12.4 FORMAT FOR
TWO-WAY ANALYSIS OF VARIANCE

Treatment: $j=$	1	2	$\cdots$	k	
Block: $i=1$	y_{11}	y_{12}	$\cdots$	y_{1k}	$T_{1.}$
2	y_{21}	y_{22}	$\cdots$	y_{2k}	$T_{2.}$
$\vdots$	$\vdots$	$\vdots$		$\vdots$	$\vdots$
n	y_{n1}	y_{n2}	$\cdots$	y_{nk}	$T_{n.}$
	$T_{.1}$	$T_{.2}$	$\cdots$	$T_{.k}$	$T_{..}$

As noted, the model for this randomized complete block design is

$$y_{ij} = \mu + \beta_i + \tau_j + \varepsilon_{ij}$$

or, as an identity for all y_{ij},

$$y_{ij} = \mu + (\mu_{i.} - \mu) + (\mu_{.j} - \mu) + (y_{ij} - \mu_{i.} - \mu_{.j} + \mu)$$

where $\mu_{i.}$ is the mean of block i and $\mu_{.j}$ is the mean of treatment j. The corresponding sample model is

$$y_{ij} - \bar{y}_{..} = (\bar{y}_{i.} - \bar{y}_{..}) + (\bar{y}_{.j} - \bar{y}_{..}) + (y_{ij} - \bar{y}_{i.} - \bar{y}_{.j} + \bar{y}_{..})$$

If both sides of the equation are squared and summed over i and j,

$$\sum_{i=1}^{n} \sum_{j=1}^{k} (y_{ij} - \bar{y}_{..})^2 = \sum_{i=1}^{n} \sum_{j=1}^{k} (\bar{y}_{i.} - \bar{y}_{..})^2 + \sum_{i=1}^{n} \sum_{j=1}^{k} (\bar{y}_{.j} - \bar{y}_{..})^2$$
$$+ \sum_{i=1}^{n} \sum_{j=1}^{k} (y_{ij} - \bar{y}_{i.} - \bar{y}_{.j} + \bar{y}_{..})^2$$

The cross-product terms sum to zero and

$$\text{SS}_{\text{Total}} = \text{SS}_{\text{Blocks}} + \text{SS}_{\text{Treatments}} + \text{SS}_{\text{Error}}$$

The corresponding degrees of freedom for these sums of squares are

$$(nk - 1) = (n - 1) + (k - 1) + (n - 1)(k - 1)$$

Two sources of variance introduced or manipulated by the investigator in a randomized complete block design are blocks and treatments. Under the assumption that the ε_{ij} are independently normally distributed with zero mean and common variance, both of these sources of variance can be tested for significance. Note, however, that variance due to blocks and variance due to treatments differ with respect to interpretation. The purpose of blocking is to increase the precision of estimating the treatment differences; a test of the significance of differences among treatments is the main purpose of the design. In this sense, an experiment for which a randomized block design is used is a single-factor experiment. A test of significance for blocks is essentially a test of the appropriateness of the randomized block design. If differences due to

TABLE 12.5   ANALYSIS OF VARIANCE: RANDOMIZED BLOCK DESIGN

Source of Variation	d.f.	SS	MS
Between blocks	$n - 1$	$\sum_{i=1}^{n} \dfrac{T_{i.}^{2}}{k} - \dfrac{T_{..}^{2}}{nk}$	$SS_{Blocks}/(n - 1)$
Between treatments	$k - 1$	$\sum_{j=1}^{k} \dfrac{T_{.j}^{2}}{n} - \dfrac{T_{..}^{2}}{nk}$	$SS_{Treatments}/(k - 1)$
Error	$(n - 1)(k - 1)$	$\sum_{i=1}^{n} \sum_{j=1}^{k} y_{ij}^{2} - \sum_{i=1}^{n} \dfrac{T_{i.}^{2}}{k}$ $- \sum_{j=1}^{n} \dfrac{T_{.j}^{2}}{n} + \dfrac{T_{..}^{2}}{nk}$	$SS_{Error}/(n - 1)(k - 1)$
Total	$nk - 1$	$\sum_{i=1}^{n} \sum_{j=1}^{k} y_{ij}^{2} - \dfrac{T_{..}^{2}}{nk}$	

blocks are not significant, blocking is not very effective in reducing error variance, either because the blocking variable is not appropriate or because the experimental material is not divided effectively into blocks. When the variance due to blocks is not significant, an investigator frequently is tempted to ignore this source of variance, in order to increase the degrees of freedom for estimating error; that is, he may want to "pool" or add together the sum of squares for blocks and the sum of squares for error and also add their respective degrees of freedom. This procedure is not justifiable theoretically; when randomization is restricted to take account of blocks, there is no way of estimating the error involved in ignoring this restriction for purposes of analysis.

The hypothesis of equal treatment means

$$H_0 : \mu_{.1} = \mu_{.2} = \cdots = \mu_{.k}$$

is tested using the F ratio

$$F = \frac{MS_{Treatments}}{MS_{Error}} \qquad \text{d.f. } (k - 1), (n - 1)(k - 1)$$

and H_0 is rejected if $F \geq F_{1-\alpha}$.

The hypothesis of equal block means

$$H_B : \mu_{1.} = \mu_{2.} = \cdots = \mu_{n.}$$

is tested using the F ratio

$$F = \frac{MS_{Blocks}}{MS_{Error}} \qquad \text{d.f. } (n - 1), (n - 1)(k - 1)$$

and H_B is rejected if $F \geq F_{1-\alpha}$.

The analysis of variance for a randomized complete block design is summarized in Table 12.5; computational formulas are given for obtaining sums of squares.

HYPOTHESES CONCERNING LINEAR CONTRASTS

The methods of testing linear contrasts which are discussed for completely randomized designs in Section 12.5 are applicable to randomized block designs, with a few modifications. In randomized block designs, contrasts involve the treatment means, $\mu_{.j}$, $j = 1, 2, \ldots, k$ or, equivalently, the treatment effects $T_{.j}, j = 1, 2, \ldots, k$.

Contrasts are of the form

$$L = c_1 \mu_{.1} + c_2 \mu_{.2} + \cdots + c_k \mu_{.k} = \sum_{j=1}^{k} c_j \mu_{.j}$$

$$= c_1 \beta_1 + c_2 \beta_2 + \cdots + c_k \beta_k = \sum_{j=1}^{k} c_j \beta_j$$

where $\sum_{j=1}^{k} c_j = 0$ and the estimate of L is given by

$$\hat{L} = c_1 \bar{x}_{.1} + c_2 \bar{x}_{.2} + \cdots + c_k \bar{x}_{.k} = \sum_{j=1}^{k} c_j \bar{x}_{.j}$$

For a randomized blocks design, $\mathrm{MS_E}$, the mean square error, replaces $\mathrm{MS_W}$ and $(n-1)(k-1)$ replaces $N-k$ for degrees of freedom. Thus, for the T-method,

$$T = \frac{1}{\sqrt{n}} q_{1-\alpha,k,(n-1)(k-1)}$$

and for the S-method,

$$S^2 = (k-1)F_{1-\alpha,k-1,(n-1)(k-1)}$$

The appropriate use and advantages and disadvantages of these procedures for randomized blocks designs are the same as those discussed for completely randomized designs.

EFFICIENCY OF A RANDOMIZED BLOCK DESIGN

A randomized block design is used for the purpose of reducing extraneous variation and thus increasing the precision with which the treatment effects are estimated. In particular, blocking reduces the error mean square without excessively reducing its degrees of freedom. For purposes of planning future experiments, it is useful to evaluate the effectiveness or efficiency of blocking after an experiment has been analyzed.

Complete randomization and randomized blocks can be compared by estimating the sum of squares error associated with each design and then comparing them. In general, efficiency is defined as a ratio of variances. The relative efficiency of randomized blocks with respect to complete randomization is estimated by

$$\hat{E}(\mathrm{RB}|\mathrm{CR}) = \frac{(n-1)\mathrm{MS_B} + n(k-1)\mathrm{MS_E}}{(nk-1)\mathrm{MS_E}}$$

where all quantities in the ratio can be obtained from the ANOVA table for the randomized block design. A relative efficiency greater than 1, say 1.5, may be interpreted to mean that the randomized block design is 150 percent as efficient as complete randomization. A relative efficiency less than 1 indicates that the randomized block design is less efficient than complete randomization, probably as a result of loss of degrees of freedom.

EXAMPLE
Suppose that for the example of the golf carts and ball bearings, the following data are obtained (observations are recorded as milligrams decrease in weight and then coded* by subtracting 30 mg.) Is there evidence that the brands of ball bearings differ in resistance to wear? Does the evidence indicate that blocking was effective. (1 percent level)?

| | | \multicolumn{4}{c|}{BRAND} | | | | |
		A	B	C	D	$T_{i.}$
	I	2	3	1	0	6
Golf Cart	II	4	-1	1	-1	3
	III	1	-2	-3	-2	-6
	IV	1	-5	-4	-3	-11
	$T_{.j}$	8	-5	-5	-6	$T_{..} = -8$
$\sum_{i=1}^{4} y_{ij}^2$		22	39	27	14	$\sum_{j=1}^{4} \sum_{i=1}^{4} y_{ij}^2 = 102$

$$SS_{Total} = \sum_{j=1}^{4} \sum_{i=1}^{4} y_{ij}^2 - \frac{T_{..}^2}{16} = 102 - \frac{(-8)^2}{16} = 98$$

$$SS_{Treatments} = \sum_{j=1}^{4} \frac{T_{.j}^2}{4} - \frac{T_{..}^2}{16} = \frac{(8)^2 + (-5)^2 + (-5)^2 + (-6)^2}{4} - \frac{(-8)^2}{16} = 33.50$$

$$SS_{Blocks} = \sum_{i=1}^{4} \frac{T_{i.}^2}{4} - \frac{T_{..}^2}{16} = \frac{(6)^2 + (3)^2 + (-6)^2 + (-11)^2}{4} - \frac{(-8)^2}{16} = 46.50$$

$$SS_{Error} = SS_{Total} - SS_{Treatments} - SS \qquad = 98 - 3.50 - 46.50 = 18$$

SOURCE OF VARIATION	d.f.	SS	MS	F
Brands (treatments)	3	33.50	11.17	5.58
Carts (blocks)	3	46.50	15.50	7.75
Error	9	18.00	2.00	
Total	15	98.00		

$$H_0: \mu_{.1} = \mu_{.2} = \mu_{.3} = \mu_{.4}$$

$$F_{3,9} = \frac{11.17}{2.00} = 5.58$$

$$F_{.99,3,9} = 6.99 \quad (\text{Table B.11})$$

* Coding, by subtracting a constant from each observation, does not affect the analysis of variance F-tests and frequently makes the computations less cumbersome.

So accept H_0 and conclude that the four brands of ball bearings apparently do not differ with respect to resistance to wear.

$$H_B: \mu_{1.} = \mu_{2.} = \mu_{3.} = \mu_{4.}$$

$$F_{3,9} = \frac{15.50}{2.00} = 7.75$$

$$F_{.99,3,9} = 6.99 \quad \text{(Table B.11)}$$

So reject H_B and conclude that apparently there are differences among the golf carts, and thus blocking seems appropriate and effective.

Since the hypothesis of equal treatment means was accepted, linear contrasts among treatment means are not appropriate.

For these data the relative efficiency of randomized blocks with respect to complete randomization can be estimated as follows:

$$\hat{E}(RB|CR) = \frac{(n-1)MS_B + n(k-1)MS_E}{(nk-1)MS_E}$$

$$= \frac{(3)(15.50) + (4)(3)(2.00)}{(15)(2.00)}$$

$$= \frac{70.50}{30}$$

$$= 2.35$$

Thus a randomized blocks design is approximately 235 percent as efficient as complete randomization for this example, and the investigator would probably continue to use this type of design rather than complete randomization in future similar experiments.

EXAMPLE

The Hidden Valley Construction Company is interested in whether landscaping increases the sales price of a property by an amount that exceeds the cost of the landscaping. The company is developing five different subdivisions in Megalopolis. In each subdivision the sales manager chose three lots that were similar in size, location, surrounding property, and so forth and that had identical houses built on them. In each subdivision he chose, at random, one property that was left without any landscaping, one for which some (but not much) landscaping was done, and one for which extensive landscaping was done. The sales prices, minus the costs of landscaping, follows. What conclusions can be drawn on the basis of these data?

		LANDSCAPING			
		NONE	SOME	EXTENSIVE	$T_{i.}$
	1	43.1	42.3	45.6	131.0
	2	27.6	29.3	35.1	92.0
Subdivision	3	27.0	26.5	29.3	82.8
	4	31.0	32.3	34.5	97.8
	5	21.3	20.2	25.6	67.1
	$T_{.j}$	150.0	150.6	170.1	$T_{..} = 470.7$

$$SS_{Total} = \sum_{j=1}^{3} \sum_{i=1}^{5} y_{ij}^2 - \frac{T_{..}^2}{(5)(3)}$$

$$= (45.6)^2 + (35.1)^2 + \cdots + (21.3)^2 - \frac{(470.7)^2}{15}$$

$$= 15579.89 - 14770.57 = 809.32$$

$$SS_{Treatments} = \sum_{j=1}^{3} \frac{T_{.j}^2}{5} - \frac{T_{..}^2}{(5)(3)}$$

$$= \frac{(150.0)^2 + (150.6)^2 + (170.1)^2}{5} - \frac{(470.7)^2}{15}$$

$$= 14822.87 - 14770.57 = 52.30$$

$$SS_{Blocks} = \sum_{i=1}^{5} \frac{T_{i.}^2}{3} - \frac{T_{..}^2}{(5)(3)}$$

$$= \frac{(131.0)^2 + (92.0)^2 + (82.8)^2 + (97.8)^2 + (67.1)^2}{3} - \frac{(470.7)^2}{15}$$

$$= 15516.03 - 14770.57 = 745.46$$

$$SS_{Error} = SS_{Total} - SS_{Treatments} - SS_{Blocks}$$
$$= 809.32 - 52.30 - 745.46 = 11.56$$

SOURCE OF VARIATION	d.f.	SS	MS	F
Landscaping (treatments)	2	52.30	26.15	17.55
Subdivision (blocks)	4	745.46	186.36	125.07
Error	8	11.56	1.49	
Total	14			

$$H_0: \mu_{.1} = \mu_{.2} = \mu_{.3}$$

$$F_{2,8} = \frac{26.15}{1.49} = 17.55$$

$$F_{.99,2,8} = 8.65 \text{ (Table B.11)}$$

So reject H_0 and conclude that sales price varies with landscaping more than accounted for by the cost of the landscaping.

$$H_B: \mu_{1.} = \mu_{2.} = \mu_{3.} = \mu_{4.} = \mu_{5.}$$

$$F_{4,8} = \frac{186.36}{1.49} = 125.07$$

$$F_{.99,4,8} = 3.84 \text{ (Table B.11)}$$

So reject H_B and conclude that sales price apparently varies among subdivisions, and thus this is an appropriate blocking variable.

After examining the sample means, the investigator might decide to test the contrast between extensive landscaping and none or some landscaping. Since the sample sizes are equal, both the T-method and the S-method are appropriate.

For the T-method, contrasts involving two means are compared with

$$T\sqrt{\mathrm{MS_E}}\left(\frac{1}{2}\sum_{j=1}^{3}|c_j|\right) = \left(\frac{1}{\sqrt{5}}\right)(q_{.95,3,8})(\sqrt{1.49})(1)$$

$$= \left(\frac{1}{\sqrt{5}}\right)(4.0)(\sqrt{1.49})$$

$$= 2.205$$

Thus the contrasts $\mu_3 - \mu_1$ and $\mu_3 - \mu_2$ are significant; the contrast $\mu_2 - \mu_1$ is not significant.

For the S-method, contrasts involving two means are compared with

$$S\sqrt{\mathrm{MS_E}}\left(\sqrt{\sum_{j=1}^{3}\frac{c_j^2}{n_j}}\right) = \sqrt{(2)(F_{.95,2,8})(1.49)\left(\frac{2}{5}\right)}$$

$$= \sqrt{(2)(4.46)(1.49)(.4)}$$

$$= 2.306$$

Again, the contrasts $\mu_3 - \mu_1$ and $\mu_3 - \mu_2$ are significant; the contrasts $\mu_2 - \mu_1$ is not significant. Thus apparently "extensive" landscaping is worthwhile, but "some" landscaping is not.

The efficiency of randomized blocks with respect to complete randomization can be estimated as follows:

$$\hat{\mathrm{E}}(\mathrm{RB}|\mathrm{CR}) = \frac{(n-1)\mathrm{MS_B} + n(k-1)\mathrm{MS_E}}{(nk-1)\mathrm{MS_E}}$$

$$= \frac{(4)(186.36) + (5)(2)(1.49)}{(14)(1.49)}$$

$$= \frac{760.34}{20.86}$$

$$= 36.45$$

Therefore the efficiency of a randomized block design relative to complete randomization is approximately 3645 percent for this example, and the investigator would probably continue to use this type of design rather than complete randomization in future similar experiments.

12.7 LATIN SQUARE DESIGNS

In many situations, an investigator wishes to remove two or more sources of extraneous variation from the experimental error of an experiment. In the golf cart example the investigator might suspect, for example, that wheel position (right front, left front, right rear, or left rear) affects the wear on ball bearings. If feasible, he could rotate the ball bearings during the experiment so that this effect would be balanced over the brands for each golf cart. Alternatively, the experiment can be designed so that each brand is used not only once on each golf cart, but also once in each position.

A design in which each treatment occurs once and only once in each row and once and only once in each column is a *Latin square design*. The rows and columns, respectively, represent levels of the two sources of extraneous variation. This design, as the name implies, must be a square; that is, the number of levels for each extraneous variable and the number of levels for treatments must be the same. A particular square of the required size can be constructed using a table of random numbers or chosen at random from tables of Latin Squares.*

The following is an example of a 4 × 4 Latin Square design. The Roman and Arabic numerals represent, respectively, levels of two sources of extraneous variation. The letters A, B, C, and D represent the treatments.

	I	II	III	IV
1	B	C	D	A
2	A	B	C	D
3	D	A	B	C
4	C	D	A	B

A Latin Square design is appropriate whenever the investigator is interested in one factor (or one set of treatments) and wishes to take account of two sources of extraneous variation, provided that all three variables have the same number of levels. For example, (extraneous) variations in temperature and pressure may affect the clarity of polythene produced by several different processes, (extraneous) variations in operators and material may affect the speed attained with several types of calculating machines, (extraneous) variations in store type and day of the week may affect the amount of sales corresponding to various types of promotion, and so forth.

The model for a Latin Square design is

$$y_{ijk} = \mu + R_i + C_k + \tau_j + \varepsilon_{ijk}$$

where μ is the overall mean, R_i represents the ith row effect, C_k represents the kth column effect, τ_j represents the jth treatment effect, and ε_{ijk} represents the (normally and independently distributed) random error; i, j, $k = 1, 2, \ldots, t$. The total number of observations is $N = t^2$.

The basic partitioning of the sums of squares for a Latin Square design is

$$SS_{Total} = SS_{Rows} + SS_{Columns} + SS_{Treatments} + SS_{Error}$$

with corresponding degrees of freedom

$$(t^2 - 1) = (t - 1) + (t - 1) + (t - 1) + (t - 1)(t - 2)$$

Assuming that the ε_{ij} are independently normally distributed with zero mean and common variance, we can test for significance the variances due to rows, columns and treatments. The interpretation of the tests of significance for rows and columns in a Latin Square design is similar to that of the test of significance for blocks in a randomized block design. As is the case for a randomized block design, a test for the significance of treatment differences is

* R. A. Fisher, and Frank Yates, *Statistical Tables.* (New York: Hafner, 1957).

the main purpose of a Latin Square design. Note that in practice a Latin Square design is sometimes used when the variance represented by rows and/or columns is of interest as a treatment effect. This is a legitimate and, in many cases, useful application; however, since Latin Square designs were devised originally for estimating (and eliminating from the experimental error) two sources of extraneous variation, they are usually discussed in this context.

Under the usual assumptions concerning the error distribution, the hypothesis of equal treatment means

$$H_0: \mu_{.1.} = \mu_{.2.} = \cdots = \mu_{.t.}$$

is tested using the F ratio

$$F = \frac{MS_{\text{Treatments}}}{MS_{\text{Error}}} \qquad \text{d.f. } (t-1), (t-1)(t-2)$$

The hypothesis of equal row means

$$H_R: \mu_{1..} = \mu_{2..} = \cdots = \mu_{t..}$$

is tested using the F ratio

$$F = \frac{MS_{\text{Rows}}}{MS_{\text{Error}}} \qquad \text{d.f. } (t-1), (t-1)(t-2)$$

The hypothesis of equal column means

$$H_C: H_{..1} = \mu_{..2} = \cdots = \mu_{..T}$$

is tested using the F ratio

$$F = \frac{MS_{\text{Columns}}}{MS_{\text{Error}}} \qquad \text{d.f. } (t-1), (t-1)(t-2)$$

The analysis of variance for a Latin Square design is summarized in Table 12.6; computational formulas are given for computing sums of squares. The notation used is an extension of that for randomized block designs, with a dot subscript indicating that the corresponding variable has been summed. For example, $T_{i..}$ indicates the ith total obtained by summing over j and k.

$$T_{i..} = \sum_j \sum_k x_{ijk}$$

$$T_{.j.} = \sum_i \sum_k x_{ijk}$$

$$T_{..k} = \sum_i \sum_j x_{ijk}$$

$$T_{...} = \sum_i \sum_j \sum_k x_{ijk}$$

TABLE 12.6 ANALYSIS OF VARIANCE: LATIN SQUARE DESIGN

Source of Variation	d.f.	SS	MS
Rows	$t - 1$	$\sum_{i=1}^{t} \dfrac{T_{i..}^2}{t} - \dfrac{T_{...}^2}{N}$	$SS_{Rows}/(t-1)$
Columns	$t - 1$	$\sum_{k=1}^{t} \dfrac{T_{..k}^2}{t} - \dfrac{T_{...}^2}{N}$	$SS_{Columns}/(t-1)$
Treatments	$t - 1$	$\sum_{j=1}^{t} \dfrac{T_{.j.}^2}{t} - \dfrac{T_{...}^2}{N}$	$SS_{Treatments}/(t-1)$
Error	$(t-1)(t-2)$	$\sum_{i=1}^{t}\sum_{j=1}^{t}\sum_{k=1}^{t} y_{ijk}^2 - \sum_{i=1}^{t} \dfrac{T_{i..}^2}{t}$ $- \sum_{j=1}^{t} \dfrac{T_{.j.}^2}{t} - \sum_{k=1}^{t} \dfrac{T_{..k}^2}{t} + \dfrac{2T_{...}^2}{N}$	$SS_{Error}/(t-1)(t-2)$
Total	$t - 1$	$\sum_{i=1}^{t}\sum_{j=1}^{t}\sum_{k=1}^{t} y_{ijk}^2 - \dfrac{T_{...}^2}{N}$	

HYPOTHESES CONCERNING LINEAR CONTRASTS

Again, the method of testing linear contrasts which is discussed in Section 12.5 is applicable with a few modifications. In a Latin Square design, contrasts are of the form

$$L = c_1\mu_{..1} + c_2\mu_{..2} + \cdots + c_t\mu_{..t} = \sum_{j=1}^{t} c_j\mu_{..j}$$

$$= c_1\tau_1 + c_2\tau_2 + \cdots + c_t\tau_t = \sum_{j=1}^{t} c_j\tau_j$$

where

$$\sum_{j=1}^{t} c_j = 0 \quad \text{and} \quad \hat{L} = c_1\bar{x}_{..1} + c_2\bar{x}_{..2} + \cdots + c_t\bar{x}_{..t} = \sum_{j=1}^{t} c_j\bar{x}_{..j}$$

For a Latin Square design, MS_E, the mean square error, replaces MS_W and $(t-1)(t-2)$ replaces $N - k$ for degrees of freedom. In addition, n is replaced by t. Thus for the T-method

$$T = \frac{1}{\sqrt{t}} q_{1-\alpha,t,(t-1)(t-2)}$$

and for the S-method

$$S^2 = (t-1)F_{1-\alpha;t-1,(t-1)(t-2)}$$

EFFICIENCY OF A LATIN SQUARE DESIGN

Estimates of efficiency, similar to those for a randomized block design, can be obtained for a Latin Square design.

The relative efficiency of a Latin Square design with respect to complete randomization is estimated by

$$\hat{E}(\text{LS}|\text{CR}) = \frac{\text{MS}_R + \text{MS}_C + (t-1)\text{MS}_E}{(t+1)\text{MS}_E}$$

An estimate of the relative efficiency of a Latin Square design with respect to a randomized block design is given by

$$\hat{E}(\text{LS}|\text{RB}) = \frac{\text{MS}_C + (t-1)\text{MS}_E}{t\text{MS}_E}$$

when the row variable of the square is the block variable and by

$$\hat{E}(\text{LS}|\text{RB}) = \frac{\text{MS}_R + (t-1)\text{MS}_E}{t\text{MS}_E}$$

when the column variable of the square is the block variable.

EXAMPLE
Suppose that the data in the randomized complete block example concerning golf carts and ball bearings were obtained using the Latin Square design, with rows representing wheel positions (in the order right front, left front, right rear, left rear), columns representing golf carts, and treatments representing brands of ball bearings. Is there evidence that the brands of ball bearing differ in resistance to wear? Does the evidence indicate that accounting for variance due to wheel position and/or golf cart was effective?

		I	II	III	IV	$T_{i..}$
	1	$B = 3$	$C = 1$	$D = -2$	$A = 1$	3
Wheel	2	$A = 2$	$B = -1$	$C = -3$	$D = -3$	-5
position	3	$D = 0$	$A = 4$	$B = -2$	$C = -4$	-2
	4	$C = 1$	$D = -1$	$A = 1$	$B = -5$	-4
	$T_{.j.}$	6	3	-6	-11	$T_{...} = -8$
	$\sum\limits_{i=1}^{t}\sum\limits_{j=1}^{t} y_{ijk}^2$	14	19	18	51	$\sum\limits_{i=1}^{t}\sum\limits_{j=1}^{t}\sum\limits_{k=1}^{t} y_{ijk}^2 = 102$

$$\text{SS}_{\text{Total}} = \sum_{i=1}^{4}\sum_{j=1}^{4}\sum_{k=1}^{4} y_{ijk}^2 - \frac{T_{...}^2}{16} = 102 - \frac{(-8)^2}{16} = 98$$

(as for randomized block design)

$$\text{SS}_{\text{Rows}} = \sum_{i=1}^{4} \frac{T_{i..}^2}{4} - \frac{T_{...}^2}{16} =: \frac{(3)^2 + (-5)^2 + (-2)^2 + (-4)^2}{4} - \frac{(-8)^2}{16} = 9.50$$

$$\text{SS}_{\text{Columns}} = \sum_{k=1}^{4} \frac{T_{..k}^2}{4} - \frac{T_{...}^2}{16} = \frac{(6)^2 + (3)^2 + (-6)^2 + (-11)^2}{4} - \frac{(-8)^2}{16} = 46.50$$

(same as $\text{SS}_{\text{Blocks}}$ in randomized block design)

$$\text{SS}_{\text{Treatments}} = \sum_{j=1}^{4} \frac{T_{.j.}^2}{4} - \frac{T_{...}^2}{16} = \frac{(8)^2 + (-5)^2 + (-5)^2 + (-6)^2}{4} - \frac{(-8)^2}{16} = 33.50$$

(as for randomized block design)

$$SS_{Error} = SS_{Total} - SS_{Rows} - SS_{Columns} - SS_{Treatments}$$
$$= 98 - 9.50 - 46.50 - 33.50 = 8.50$$

SOURCE OF VARIATION	d.f.	SS	MS
Brands (treatments)	3	33.50	11.17
Carts (columns)	3	46.50	15.50
Positions (rows)	3	9.50	3.17
Error	6	8.50	1.42
Total	15	98.00	

$$H_0 : \mu_{.1.} = \mu_{.2.} = \mu_{.3.} = \mu_{.4.}$$

$$F_{3,6} = \frac{11.17}{1.42} = 7.87$$

$$F_{.95,3,6} = 4.76 \quad \text{(Table B.11)}$$

so reject H_0 and conclude (as also concluded above) that brands of ball bearings apparently differ with respect to resistance to wear.

$$H_R : \mu_{1..} = \mu_{2..} = \mu_{3..} = \mu_{4..}$$

$$F_{3,6} = \frac{3.17}{1.42} = 2.23$$

$$F_{.95,3,6} = 4.76 \quad \text{(Table B.11)}$$

So accept H_R and conclude that wheel position apparently does not affect resistance to wear.

$$H_C : \mu_{..1} = \mu_{..2} = \mu_{..3} = \mu_{..4}$$

$$F_{3,6} = \frac{15.50}{1.42} = 10.92$$

$$F_{.99,3,6} = 9.78 \quad \text{(Table B.11)}$$

So reject H_C and conclude (as already concluded) that resistance to wear apparently is affected by the golf cart used. The apparent differences among brands can be examined using the T-method or the S-method.

$$\bar{x}_A = 2$$

$$\bar{x}_B = -1.25$$

$$\bar{x}_C = -1.25$$

$$\bar{x}_D = -1.5$$

Differences between pairs of means are compared, for the T-method, with

$$\frac{1}{\sqrt{k}} q_{1-\alpha,t,(t-1)(t-2)} \sqrt{MS_E} \left(\frac{1}{2} \sum_{j=1}^{k} |c_j| \right) = \left(\frac{1}{\sqrt{4}} \right) (4.90)(\sqrt{1.42})(1)$$

$$= 2.920$$

and, for the S-method, with

$$\sqrt{(t-1)(F_{1-\alpha,t-1,(t-1)(t-2)})(\text{MS}_E)\left(\sum_{j=1}^{k}\frac{c_j^2}{n_j}\right)}$$

$$= \sqrt{(3)(4.76)(1.42)(.5)}$$

$$= 3.184$$

Thus, on the basis of either the T-method or the S-method, brand A ball bearings have significantly greater resistance to wear than brands B, C, and D. There are no significant differences in resistance to wear among brands B, C, and D.

For this example the relative efficiency of a Latin Square design with respect to complete randomization can be estimated as follows:

$$\hat{E}(\text{LS}|\text{CR}) = \frac{\text{MS}_R + \text{MS}_C + (t-1)\text{MS}_E}{(t-1)\text{MS}_E}$$

$$= \frac{3.17 + 15.50 + (3)(1.42)}{(5)(1.42)}$$

$$= \frac{22.93}{7.10}$$

$$= 3.23$$

Thus the relative efficiency of a Latin Square design with respect to complete randomization is approximately 323 percent for this example.

The estimated relative efficiency of a Latin Square design with respect to a randomized block design with the row variable used for blocking is

$$\hat{E}(\text{LS}|\text{RB}) = \frac{\text{MS}_C + (t-1)\text{MS}_E}{t\text{MS}_E}$$

$$= \frac{15.50 + (3)(1.42)}{(4)(1.42)}$$

$$= \frac{19.76}{5.68}$$

$$= 3.48$$

When the column variable is used for blocking, the estimated relative efficiency is

$$\hat{E}(\text{LS}|\text{RB}) = \frac{\text{MS}_R + (t-1)\text{MS}_E}{t\text{MS}_E}$$

$$= \frac{3.17 + (3)(1.42)}{(4)(1.42)}$$

$$= \frac{7.43}{5.68}$$

$$= 1.31$$

Thus for this example a Latin Square design is approximately 348 percent as efficient as a randomized block design if the row variable is used for blocking and approximately 131 percent as efficient as a randomized block design if the column variable is used for blocking.

For future similar experiments, therefore, the investigator would probably continue to use a Latin Square design.

Since the effect of wheel position on resistance to wear is not significant, the investigator might be tempted to ignore this source of variation in order to increase the degrees of freedom for estimating error. That is, he might "pool" or add together the sum of squares for positions and the sum of squares for error and also add their respective degrees of freedom. As noted, this procedure is not justifiable theoretically – randomization is restricted to take into account the effect of position and there is no way of estimating the possible error involved in accepting the hypothesis of no position effect and in ignoring the corresponding restriction on randomization.

EXAMPLE

The Efficient-Girl Secretarial Agency is developing new materials for the typing test given to all applicants for secretarial jobs. Three different letters (identified as A, B, and C) are being considered for part of the test, and the agency wishes to determine whether they are of equal difficulty. Three students type each of the letters using three different typewriters. The times required (in minutes) are given in the following table. Are the letters apparently of equal difficulty?

		STUDENTS			
		1	2	3	$T_{i..}$
	I	$A = 3.8$	$B = 4.2$	$C = 5.6$	13.6
TYPEWRITERS	II	$C = 3.6$	$A = 3.9$	$B = 5.4$	12.9
	III	$B = 4.0$	$C = 3.5$	$A = 5.7$	13.2
	$T_{.j.}$	11.4	11.6	16.7	$T_{...} = 39.7$

Total for $A = 13.4$
Total for $B = 13.6$
Total for $C = 12.7$

$$\text{SS}_{\text{Total}} = \sum_{i=1}^{3} \sum_{j=1}^{3} \sum_{k=1}^{3} y_{ijk}^2 - \frac{T_{...}^2}{9}$$

$$= (3.8)^2 + (3.6)^2 + \cdots + (5.7)^2 - \frac{(39.7)^2}{9}$$

$$= 181.51 - 175.12 = 6.39$$

$$\text{SS}_{\text{Rows}} = \sum_{i=1}^{3} \frac{T_{i..}^2}{3} - \frac{T_{...}^2}{9} = \frac{(13.6)^2 + (12.9)^2 + (13.2)^2}{3} - \frac{(39.7)^2}{9}$$

$$= 175.20 - 175.12 = 0.08$$

$$\text{SS}_{\text{Columns}} = \sum_{k=1}^{3} \frac{T_{..k}^2}{3} - \frac{T_{...}^2}{9} = \frac{(11.4)^2 + (11.6)^2 + (16.7)^2}{3} - \frac{(39.7)^2}{9}$$

$$= 181.14 - 175.12 = 6.02$$

$$\text{SS}_{\text{Treatments}} = \sum_{j=1}^{3} \frac{T_{.j.}^2}{3} - \frac{T_{...}^2}{9} = \frac{(13.4)^2 + (13.6)^2 + (12.7)^2}{3} - \frac{(39.7)^2}{9}$$

$$= 175.27 - 175.12 = 0.15$$

$$\text{SS}_{\text{Error}} = \text{SS}_{\text{Total}} - \text{SS}_{\text{Rows}} - \text{SS}_{\text{Columns}} - \text{SS}_{\text{Treatments}}$$

$$= 6.39 - 0.08 - 6.02 - 0.15 = 0.14$$

SOURCE OF VARIATION	d.f.	SS	MS	F
Letters (treatments)	2	0.15	0.075	1.07
Students (columns)	2	6.02	3.01	43.00
Typewriter (rows)	2	0.08	0.04	<1
Error	2	0.14	0.07	
Total	8			

$$H_0: \mu_{.1.} = \mu_{.2.} = \mu_{.3.}$$

$$F_{2,2} = \frac{0.075}{0.07} = 1.07$$

$$F_{.95,2,2} = 19.0 \quad \text{(Table B.11)}$$

So accept H_0 and conclude that the letters do not vary in difficulty.

$$H_B: \mu_{..1} = \mu_{..2} = \mu_{..3}$$

$$F_{2,2} = \frac{0.04}{0.07} < 1$$

So accept H_B and conclude that the typewriters do not vary in speed.

$$H_C: \mu_{1..} = \mu_{2..} = \mu_{3..}$$

$$F_{2,2} = \frac{3.01}{0.07} = 43.00$$

$$F_{.95,2,2} = 19.0 \quad \text{(Table B.11)}$$

So reject H_C and conclude that the students vary in speed of typing.

Since the hypothesis of equal treatment means was accepted, linear contrasts among treatment means are not appropriate.

For this example the estimated relative efficiency of a Latin Square design with respect to complete randomization is

$$\hat{E}(LS \mid CR) = \frac{MS_R + MS_C + (t-1)MS_E}{(t+1)MS_E}$$

$$= \frac{.04 + 3.01 + (2)(.07)}{4(.07)}$$

$$= \frac{3.19}{.28}$$

$$= 11.39$$

Thus the relative efficiency of a Latin Square design with respect to complete randomization is approximately 1139 percent for this example.

The estimated relative efficiency of a Latin Square design with respect to a randomized block design with the row variable used for blocking is

$$\hat{E}(LS \mid CR) = \frac{MS_C + (t-1)MS_E}{tMS_E}$$

$$= \frac{3.01 + (2)(.07)}{(3)(.07)}$$

$$= \frac{3.15}{.21}$$

$$= 15.00$$

When the column variable is used for blocking, the estimated relative efficiency is

$$\hat{E}(LS|RB) = \frac{MS_R + (t - 1)MS_E}{tMS_E}$$

$$= \frac{.04 + (2)(.07)}{(3)(.07)}$$

$$= \frac{.18}{.21}$$

$$= .86$$

Thus for this example a Latin Square design is approximately 1500 percent as efficient as a randomized block design if the row variable is used for blocking and approximately 86 percent as efficient as a randomized block design if the column variable is used for blocking. This result is consistent with the significant difference among column means and the nonsignificant difference among row means.

For future similar experiments, therefore, the investigator might choose to use a randomized block design with students as the blocking variable.

12.8 FACTORIAL EXPERIMENTS

In a factorial experiment the effects of several factors are investigated simultaneously and the treatments consist of all possible combinations of levels of the factors. For example, if an investigator is interested in four methods of advertising three different products, a factorial experiment would involve $4 \times 3 = 12$ treatments; such an experiment is referred to as 4×3 factorial experiment. Similarly, if four methods of advertising three different products are investigated in two types of stores, a $4 \times 3 \times 2$ factorial experiment is appropriate.

A *factorial experiment* is defined as an experiment in which the treatments consist of all combinations of factor levels. Factorial experiments can be performed using various designs—complete randomization, randomized blocks, Latin Squares, and so forth—in the same way as other experiments.

Note that the interactions and other treatment effects in a factorial experiment are obtained by partitioning the sum of squares for treatments and are not removed from the error term, as in the case of blocking variables. Thus the analysis of factorial experiments, using any design, proceeds in the usual way, with the exception that the treatments sum of squares is partitioned into main effects and interactions.

The advantages of factorial experiments depend on the purpose of the experiment. Factorial experiments are particularly appropriate in exploratory research when the investigator is interested in determining the effects of several factors over a specified range. In such research, factorial experiments require less time and experimental material than would be needed to conduct separate experiments for investigating each factor; in addition, factorial experiments provide information concerning interactions among factors.

On the other hand, if considerable information has already been accumulated or if the investigator is concerned with very specific objectives, it might be more advantageous to concentrate efforts on a single factor or on a few combinations of factors. In particular, if the investigator's purpose is to find the combination of factor levels that will produce a maximum response, a well-planned series of single-factor experiments may be more efficient than a large factorial experiment.

MATHEMATICAL MODEL FOR A
COMPLETELY RANDOMIZED DESIGN

The mathematical model for a factorial experiment depends on the type of design used. For a two-factor factorial experiment with n observations per cell (i.e., n observations per treatment combination), run as a completely randomized design, the mathematical model is

$$y_{ijk} = \mu + A_i + B_j + AB_{ij} + \varepsilon_{k(ij)}$$

where A and B represent the two factors and AB represents their interaction; there are $i = 1, 2, \ldots, a$ levels of factor A, $j = 1, 2, \ldots, b$ levels of factor B and $k = 1, 2, \ldots, n$ observations per cell. Note that $\varepsilon_{k(ij)}$ represents the random error within the cell (i, j). In terms of population means

$$y_{ijk} - \mu_{...} = (\mu_{i..} - \mu_{...}) + (\mu_{.j.} - \mu_{...}) + (\mu_{ij.} - \mu_{i..} - \mu_{.j.} + \mu_{...}) + (y_{ijk} - \mu_{ij.})$$

where μ_{ij} represents the population mean of the cell or treatment combination (i, j). As in previous discussions, a dot subscript indicates a variable over which summation has been performed.

NOTE
The interaction term can be justified heuristically by subtracting the A and B effects from the cell effect as follows:

$$(\mu_{ij.} - \mu_{...}) - (\mu_{i..} - \mu_{...}) - (\mu_{.j.} - \mu_{...}) = \mu_{ij.} - \mu_{i..} - \mu_{.j.} + \mu_{...}$$

The corresponding sample model is

$$(y_{ijk} - \bar{y}_{...}) = (\bar{y}_{i..} - \bar{y}_{...}) + (\bar{y}_{.j.} - \bar{y}_{...}) + (\bar{y}_{ij.} - \bar{y}_{i..} - \bar{y}_{.j.} + \bar{y}_{...}) + (y_{ijk} - \bar{y}_{ij.})$$

If both sides of the equation are squared and summed over i, j, and k, the cross-product sums are zero and

$$\sum_{i=1}^{a} \sum_{j=1}^{b} \sum_{k=1}^{n} (y_{ijk} - \bar{y}_{...})^2 = \sum_{i=1}^{a} \sum_{j=1}^{b} \sum_{k=1}^{n} (\bar{y}_{i..} - \bar{y}_{...})^2 + \sum_{i=1}^{a} \sum_{j=1}^{b} \sum_{k=1}^{n} (\bar{y}_{.j.} - \bar{y}_{...})^2$$

$$+ \sum_{i=1}^{a} \sum_{j=1}^{b} \sum_{k=1}^{n} (\bar{y}_{ij.} - \bar{y}_{i..} - \bar{y}_{.j.} + \bar{y}_{...})^2$$

$$+ \sum_{i=1}^{a} \sum_{j=1}^{b} \sum_{k=1}^{n} (y_{ijk} - \bar{y}_{ij.})^2$$

or

$$SS_{Total} = SS_A + SS_B + SS_{AB} + SS_{Error}$$

with corresponding degrees of freedom

$$(abn - 1) = (a - 1) + (b - 1) + (a - 1)(b - 1) + ab(n - 1)$$

The interaction between A and B can be estimated only if there is more than one observation for each cell, that is, for each treatment combination. If $n = 1$, there are no degrees of freedom for estimating interaction; in this case the partitioning of the sums of squares is

$$SS_{Total} = SS_A + SS_B + SS_{Error}$$

where the error term actually includes any interaction between A and B. This model then corresponds formally to the model for a randomized block design, although different randomization is required for the corresponding experiments.

Under the usual assumptions concerning the error distribution, hypotheses concerning the treatments and their interaction can be tested as follows: The hypothesis

$$H_1: \mu_{1..} = \mu_{2..} = \cdots = \mu_{a..} \qquad \text{(Factor A has no effect)}$$

is tested using the F ratio

$$F = \frac{MS_A}{MS_{Error}} \qquad \text{d.f. } (a - 1), ab(n - 1)$$

The hypothesis

$$H_2: \mu_{.1.} = \mu_{.2.} = \cdots = \mu_{.b.} \qquad \text{(Factor B has no effect)}$$

is tested using the F ratio

$$F = \frac{MS_B}{MS_{Error}} \qquad \text{d.f. } (b - 1), ab(n - 1)$$

and the hypothesis

$$H_3: \mu_{ij.} - \mu_{i..} - \mu_{.j.} + \mu_{...} = 0 \qquad \text{for all } i, j \qquad \text{(no interaction)}$$

is tested using the F ratio

$$F = \frac{MS_{AB}}{MS_{Error}} \qquad \text{d.f. } (a - 1)(b - 1), ab(n - 1)$$

The preceding discussion of two-factor factorial experiments assumes that both A and B are fixed effects. If this is not the case; that is, if A or B or both A and B are random effects, then the appropriate F-tests are not as given above. Note that for the completely randomized, randomized blocks, and Latin Square designs discussed previously, the expected mean squares indicate that the F-tests do not change when treatment effects are random rather than fixed, although interpretation of the results does change.

The appropriate F-tests for a two-factor, completely randomized experiment when A or B or both are random are summarized in Table 12.7.

TABLE 12.7 F-TESTS FOR TWO-FACTOR
COMPLETELY RANDOMIZED EXPERIMENTS

	A and B Fixed	A Fixed B Random	A Random B Fixed	A and B Random
A	MS_A/MS_{Error}	MS_A/MS_{AB}	MS_A/MS_{Error}	MS_A/MS_{AB}
B	MS_B/MS_{Error}	MS_B/MS_{Error}	MS_A/MS_{AB}	MS_B/MS_{AB}
AB	MS_{AB}/MS_{Error}	MS_{AB}/MS_{Error}	MS_{AB}/MS_{Error}	MS_{AB}/MS_{Error}

TABLE 12.8 ANALYSIS OF VARIANCE:
TWO-FACTOR FACTORIAL EXPERIMENT, COMPLETELY RANDOMIZED DESIGN

Source of Variation	d.f.	SS	MS
Factor A	$a-1$	$\sum_{i=1}^{a} \dfrac{T_{i..}^2}{bn} - \dfrac{T_{...}^2}{abn}$	$SS_A/(a-1)$
Factor B	$b-1$	$\sum_{j=1}^{b} \dfrac{T_{.j.}^2}{an} - \dfrac{T_{...}^2}{abn}$	$SS_B/(b-1)$
A × B interaction	$(a-1)(b-1)$	$\sum_{i=1}^{a}\sum_{j=1}^{b} \dfrac{T_{ij.}^2}{n} - \sum_{i=1}^{a} \dfrac{T_{i..}^2}{bn}$ $-\sum_{j=1}^{b} \dfrac{T_{.j.}^2}{an} + \dfrac{T_{...}^2}{abn}$	$SS_{AB}/(a-1)(b-1)$
Error	$ab(n-1)$	$\sum_{i=1}^{a}\sum_{j=1}^{b}\sum_{k=1}^{n} y_{ijk}^2$ $-\sum_{i=1}^{a}\sum_{j=1}^{b} \dfrac{T_{ij.}^2}{n}$	$SS_{Error}/ab(n-1)$

The analysis of variance for a two-factor factorial experiment (fixed effects) using a completely randomized design is summarized in Table 12.8; computational formulas are given for obtaining sums of squares.

NOTE
The sum of squares for error can be written

$$SS_{Error} = \sum_{i=1}^{a}\sum_{j=1}^{b}\left[\sum_{k=1}^{n} y_{ijk}^2 - \frac{T_{ij.}^2}{n}\right]$$

which emphasizes the fact that the sum of squares for error is the pooled sums of squares within cells. The interaction sums of squares can be obtained as

$$SS_{AB} = SS_{Cells} - SS_A - SS_B$$

$$SS_{AB} = \left(\sum_{i=1}^{a}\sum_{j=1}^{b}\frac{T_{ij.}^2}{n} - \frac{T_{...}^2}{abn}\right) - \left(\sum_{i=1}^{a}\frac{T_{i..}^2}{bn} - \frac{T_{...}^2}{abn}\right) - \left(\sum_{j=1}^{b}\frac{T_{.j.}^2}{an} - \frac{T_{...}^2}{abn}\right)$$

EXAMPLE
An investigator is interested in the effectiveness of four different types of advertising or promotion on the sales of three different products (floor wax,

toothpaste, and detergent) and has chosen a random sample of 48 supermarkets for the experiment. The four types of advertising or promotion are: coupons mailed to the supermarket's list of customers cleared to cash checks; premium (large comb) attached to the product; samples sent to homes of the supermarket's list of customers cleared to cash checks; special display of product at sale price. The 12 combinations of the 4 methods of advertising or promotion and the 3 products are assigned at random to the supermarkets, each combination being used in 4 supermarkets. Thus the experiment is run using a completely randomized design. The data obtained, recorded as percent change in sales volume and then coded, follow. Compute and interpret the appropriate analysis of variance.

	FLOOR WAX	*TOOTHPASTE*	*DETERGENT*	$T_{i..}$
COUPON	-1 -1 -1 $\sum = -3$ 0 $\sum^2 = 3$	2 1 1 $\sum = 5$ 1 $\sum^2 = 7$	1 1 3 $\sum = 7$ 2 $\sum^2 = 15$	9
PREMIUM	0 0 -1 $\sum = -2$ -1 $\sum^2 = 2$	3 5 4 $\sum = 16$ 4 $\sum^2 = 66$	5 5 6 $\sum = 2$ 4 $\sum^2 = 102$	34
SAMPLE	-1 -2 -1 $\sum = -4$ 0 $\sum^2 = 6$	-1 1 -1 $\sum = -2$ -1 $\sum^2 = 4$	0 0 -1 $\sum = -1$ 0 $\sum^2 = 1$	-7
SALE	-1 0 0 $\sum = 0$ 1 $\sum^2 = 2$	2 2 2 $\sum = 7$ 1 $\sum^2 = 13$	3 2 2 $\sum = 8$ 1 $\sum^2 = 18$	15
$T_{.j.}$	-9	26	34	$T_{...} = 51$

$$SS_{Total} = \sum_{i=1}^{a} \sum_{j=1}^{b} \sum_{k=1}^{n} y_{ijk}^2 - \frac{T_{...}^2}{abn}$$

$$= 239 - 54.19 = 184.81$$

$$SS_A = \sum_{i=1}^{a} \frac{T_{i..}^2}{bn} - \frac{T_{...}^2}{abn} = \frac{(9)^2 + (34)^2 + (-7)^2 + (15)^2}{12} - \frac{(51)^2}{48}$$

$$= 125.92 - 54.19 = 71.73$$

$$SS_B = \sum_{j=1}^{b} \frac{T_{.j.}^2}{an} - \frac{T_{...}^2}{abn} = \frac{(-9)^2 + (26)^2 + (34)^2}{16} - \frac{(51)^2}{48}$$

$$= 119.56 - 54.19 = 65.37$$

$$SS_{AB} = \sum_{i=1}^{a} \sum_{j=1}^{b} \frac{T_{ij.}^2}{n} - \sum_{i=1}^{a} \frac{T_{i..}^2}{bn} - \sum_{j=1}^{b} \frac{T_{.j.}^2}{an} + \frac{T_{...}^2}{abn}$$

$$= \frac{(-3)^2 + (5)^2 + \cdots + (7)^2 + (8)^2}{4} - 125.92 - 119.56 + 54.19$$

$$= 219.25 - 125.92 - 119.56 + 54.19 = 27.96$$

$$SS_{Error} = \sum_{i=1}^{a} \sum_{j=1}^{b} \sum_{k=1}^{n} y_{ijk}^2 - \sum_{i=1}^{a} \sum_{j=1}^{b} \frac{T_{ij.}^2}{n}$$

$$= 239 - 219.25 - 19.75$$

SOURCE OF VARIATION	d.f.	SS	MS	F
Advertising (A)	3	71.73	23.91	43.47
Products (B)	2	65.37	32.69	59.44
Advertising × products (AB)	6	27.96	4.66	8.47
Error	36	19.75	0.55	
Total	47	184.81		

Methods of advertising or promotion and products and their interaction are
highly significant. Linear contrasts among types of advertising or linear contrasts
among products can be tested using the preceding methods for a completely
randomized design. However, the interaction of types of advertising and products
probably would be of more interest. This interaction indicates that the most
effective type of advertising might depend on the product involved. The interaction
of types of advertising and products can be analyzed graphically, as discussed
in the following section.

ANALYSIS OF INTERACTIONS

If an interaction is significant, representing the interaction graphically
will usually help the investigator to interpret it. One of the factors is represented
on the x-axis and the data (responses) are represented in the y-axis, with one
set of plotted points for each level of the other factor. Using the coded values
of the data above, for example, we can represent the interaction of method of
advertising or promotion and products graphically as in Figure 12.1. If there
were no interaction, then the lines on the graph would be parallel; clearly, the
effectiveness of the premium method for toothpaste and detergent, but not for
floor wax, is largely responsible for the significant interaction.

Similarly, the interaction could be represented with products on the x-
axis and sets of points for each method of advertising or promotion, as in

Figure 12.1 Graphical representation of interaction

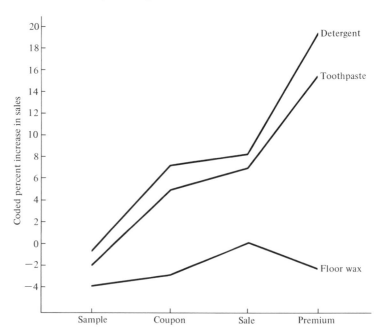

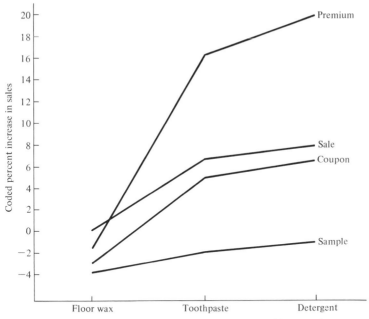

Figure 12.2   Alternative graphical representation of interaction

Figure 12.2. This representation again indicates the effectiveness of premiums for advertising toothpaste and detergent.

Factorial experiments with three (or more) factors are analyzed by an extension of the methods for analyzing two-factor experiments. For example, the mathematical model for a three-factor factorial experiment using a completely randomized design is

$$Y_{ijk\ell} = \mu + A_i + B_j + C_k + AB_{ij} + AC_{ik} + BC_{jk} + ABC_{ijk} + \varepsilon_{\ell(ijk)}$$

12.9 SUMMARY

Three of the simplest types of experimental designs are discussed in this chapter: completely randomized designs, randomized block designs and Latin Square designs. There are numerous other experimental designs—Graeco–Latin Squares, Youden squares, randomized incomplete blocks, split-plot designs, partially confounded designs, fractionally replicated designs, and so forth. All experimental designs have the same purpose—increasing the precision of estimating the effects of interest by decreasing extraneous variation—but differ with regard to the situations in which they are appropriate. Before choosing an experimental design, an investigator must thoroughly understand the nature of the experimental variables and the likely sources of extraneous variation. Several books that are virtually encyclopedias of experimental designs (see Appendix A) can be used to help determine the appropriate experimental design.

PROBLEMS

12.1 A doctor is interested in the relative effectiveness of 3 proposed treatments for leukemia. He administers each treatment to 10 randomly selected leukemia patients and records the change in a particular measure of red blood cells. The data, after coding and initial computations, are given below. Is there evidence that the treatments differ in effectiveness (5 percent level)?

Computational information:

$$\sum Y_1 = 50 \qquad \sum Y_1^2 = 300$$
$$\sum Y_2 = 30 \qquad \sum Y_2^2 = 200$$
$$\sum Y_3 = 70 \qquad \sum Y_3^2 = 600$$

(where the subscripts 1, 2, and 3 represent treatment groups)
Use both the T-method and the S-method to analyze differences in treatments in more detail.

12.2 From a random work sample for each of three operators, 5 observations of the time required to do a standard bookkeeping entry were made. Using the following data, perform an analysis of variance and, at the 5 percent level of significance, test the hypothesis that the average times required are equal.

TIME REQUIRED (IN MINUTES)		
OPERATOR A	OPERATOR B	OPERATOR C
9	4	8
10	9	6
7	8	7
6	9	6
8	5	8

12.3 A market analyst is interested in the relationship between length of time in business and annual profits of real estate firms. Blocks of firms of 4 different sizes were determined; from each block a firm was randomly selected for each of 5 lengths of time in business. In recording the data, size (block) was represented by rows and length of time in business (treatment) was represented by columns. Thus x_{ij} represents the profits of the company in the ith size classification ($i = 1, \ldots, 4$) and the jth time classification ($j = 1, \ldots, 5$). Using the following statistics, perform an analysis of variance, stating and testing the appropriate hypotheses.

Computational information:

$$\sum_{i=1}^{4} T_{i.}^2 = 1450 \qquad \sum_{j=1}^{5} T_{.j}^2 = 2740$$

$$\sum_{i=1}^{4} \sum_{j=1}^{5} y_{ij}^2 = 1010 \qquad \frac{T_{..}^2}{20} = 200$$

State and test hypotheses concerning contrasts among means, if appropriate.

12.4 A doctor is interested in the relative effectiveness of four drugs proposed for raising blood counts in patients suffering from pernicious anemia. He administered each drug to 10 randomly selected anemia patients and obtained data concerning their blood counts. The data, after coding and some calculations, are summarized below:

$$\sum Y_1 = 10 \qquad (\sum Y_1)^2 = 100 \qquad \sum Y_1^2 = 20$$
$$\sum Y_2 = 40 \qquad (\sum Y_2)^2 = 1{,}600 \qquad \sum Y_2^2 = 200$$
$$\sum Y_3 = 80 \qquad (\sum Y_3)^2 = 6{,}400 \qquad \sum Y_3^2 = 800$$
$$\sum Y_4 = 50 \qquad (\sum Y_4)^2 = 2{,}500 \qquad \sum Y_4^2 = 400$$
$$\sum = 180 \qquad \sum = 10{,}600 \qquad \sum = 1{,}420$$

(where the subscripts 1, 2, 3, 4 represent the 4 drugs)

(a) Is there evidence that the drugs differ in effectiveness (1 percent level)?

(b) Test the contrast T_3 versus T_1, T_2, and T_4 (assumed to have been specified in advance).

12.5 A 4×4 Latin Square experiment was designed to investigate the relationship between the number of statistics courses completed (treatments) and computing speed. Brand of calculator was the row variable and time of day was the column variable. The (coded) data obtained are as follows:

A	B	C	D		2	3	5	2
B	C	D	A		3	1	1	3
C	D	A	B		1	1	2	2
D	A	B	C		4	1	2	3

(a) Perform an analysis of variance (5 percent level) testing the hypothesis of equal treatment means.

(b) Determine the relative efficiency of the Latin Square design used with respect to a completely randomized design and with respect to each of the possible randomized block designs.

12.6 A 3×3 Latin Square experiment was designed to investigate the relation of types of advertising (treatment) to sales of cigarettes. The row variable is day of the week; the column variable is season of the year. The following (coded) data were obtained:

		SEASON					
	B	A	C		2	2	2
DAY	A	C	B		4	4	7
	C	B	A		1	10	1

(a) Perform an analysis of variance (5 percent level).

(b) Determine the relative efficiency of the Latin Square design with respect to complete randomization.

12.7 The Plyboard Polythene Company is interested in the possible effects of the pressure under which polythene sheeting is produced on its strength. Random samples from batches run under low, medium, and high pressure have been measured for strength with the results given below. State the appropriate hypotehsis and the conclusion from testing it (1 percent level).

LOW	MEDIUM	HIGH
4	15	5
7	13	6
6	12	7
5	12	6
8		6

12.8 Three types of TV advertising appeals, price (A), quality (B), and convenience (C) were investigated for a gourmet frozen food using a Latin Square design with time of day as the row variable and day of the week as the column variable. The coded sales data are:

$A = 0$	$C = 2$	$B = 5$
$B = -2$	$A = 0$	$C = 0$
$C = -5$	$B = -4$	$A = -6$

(a) Compute the appropriate analysis of variance.
(b) Determine the efficiency of the Latin Square design with respect to complete randomization and with respect to a randomized block design using the row variable for blocking.

12.9 Data are obtained using a 3×3 Latin Square design, as follows:

$$
\begin{array}{ccc}
A = 4 & B = 0 & C = -4 \\
B = 2 & C = -2 & A = 6 \\
C = -6 & A = 8 & B = 1
\end{array}
$$

(a) Complete the following ANOVA table.

SOURCE OF VARIANCE	d.f.	SS	MS	F
Rows	☐		☐	☐
Columns	☐	☐	☐	☐
Treatments	☐	☐	☐	☐
Error	☐	☐	☐	
Total	☐			

(b) State and test the appropriate hypotheses.
(c) Show that the following contrasts are orthogonal and test their significance (assuming they were specified *a priori*):

$$
T_A \text{ and } T_B \text{ vs. } T_C
$$

$$
T_A \text{ vs. } T_B
$$

(d) Determine the estimated efficiency of the Latin Square design relative to a completely randomized design for this experiment.

12.10 An experiment has been performed using the following 4×4 Latin Square and obtaining the following data:

D A C B	10	3	2	7	$T_{1.} = 22$
A C B D	5	4	6	8	$T_{2.} = 23$
B D A C	9	6	2	3	$T_{3.} = 20$
C B D A	3	8	6	4	$T_{4.} = 21$
	$T_{.1} = 27$	$T_{.2} = 21$	$T_{.3} = 16$	$T_{.4} = 22$	$T_{..} = 86$

Computational information:

$$
\sum_{i=1}^{4} T_{i.}^{2} = 1854 \qquad \sum_{i=1}^{4} \sum_{j=1}^{4} y_{ij}^{2} = 558
$$

$$
\sum_{j=1}^{4} T_{.j}^{2} = 1910 \qquad T_{..}^{2} = 7396
$$

(a) Give the appropriate analysis of variance, filling in all entries.
(b) Obtain the expected error mean square using a randomized block design eliminating the row variable; which design is more efficient on this basis?

12.11 A 2×3 two-factor factorial design (A at 3 levels, B at 2 levels) was used for an experiment and the data below were obtained.

	b_1	b_2	
a_1	10 8 10 10 $T_{11.} = 38$	5 4 6 5 $T_{12.} = 20$	$T_{1..} = 58$
a_2	6 8 7 5 $T_{21.} = 26$	8 6 6 6 $T_{22.} = 26$	$T_{2..} = 52$
a_3	4 5 5 4 $T_{31.} = 18$	8 7 7 6 $T_{32.} = 28$	$T_{3..} = 46$
	$T_{.1.} = 82$	$T_{.2.} = 74$	$T_{...} = 156$

Computational information:

$$\sum_{j=1}^{2} T_{.j.}^2 = 12200 \qquad \sum_{i=1}^{3} T_{i..}^2 = 8184 \qquad \sum_{i=1}^{3} \sum_{j=1}^{2} \sum_{k=1}^{4} y_{ijk}^2 = 1092$$

$$\sum_{i=1}^{3} \sum_{j=1}^{2} T_{ij.}^2 = 4304 \qquad T_{...}^2 = 24336$$

(a) Assuming a fixed-effects model, fill in the missing entries in the following analysis of variance table and state and test the appropriate hypotheses.

SOURCE OF VARIATION	d.f.	SS	MS	F
B	☐	2.667	☐	☐
A	☐	☐	☐	☐
AB	☐	☐	☐	☐
Error	☐	☐	☐	☐
Total	☐	78	☐	☐

(b) The experimenter wishes to test the following contrasts (specified in advance)

$$2\mu_{A_1} - \mu_{A_2} - \mu_{A_3}$$

$$\mu_{A_2} - \mu_{A_3}$$

Show that these contrasts are orthogonal, compute the appropriate sums of squares, and test the significance of each contrast.

12.12 An experimenter is interested in the relation of temperature (5 different degrees, represented as rows) and pressure (6 different degrees, represented as columns) to the quality of polythene produced in a certain type of reactor. The experiment involves a measurement at each of 30 combinations of temperature and pressure for each of 4 randomly selected reactors. Perform an analysis of variance (5% level).

Computational information:

$$\frac{1}{24} \sum_i T_{i..}^2 = 1960 \qquad \frac{1}{4} \sum_i \sum_j T_{ij.}^2 = 4160$$

$$\frac{1}{20} \sum_j T_{.j.}^2 = 1600$$

$$\sum_i \sum_j \sum_k y_{ijk}^2 = 12305 \qquad \frac{T_{...}^2}{120} = 1000$$

12.13 An experiment is performed to investigate the relation of type of training (5 types represented as the row variable) and amount of previous experience (4 amounts represented as the column variable) to dollar sales volume. Six salespeople, randomly selected, are used in each of the 20 groups.
(a) Perform an analysis of variance assuming that only the amounts of previous experience used in the experiment are of interest (1% level).
(b) Perform an analysis of variance assuming that the amounts of previous experience used in the experiment are a random sample of those of interest (1% level).

Computational information:

$$\frac{1}{24} \sum_{i=1}^{5} T_{i..}^2 = 980 \qquad \frac{1}{30} \sum_{j=1}^{4} T_{.j.}^2 = 1550$$

$$\frac{1}{6} \sum_{i=1}^{5} \sum_{j=1}^{4} T_{ij.}^2 = 3230 \qquad \sum\sum\sum y_{ijk}^2 = 4230$$

$$\frac{T_{...}^2}{120} = 500$$

12.14 Suppose that an analysis of variance (5 × 4 factorial, completely randomized design) involves 5 rows, 4 columns, and 3 observations per cell and that the following totals have been obtained.

ROW TOTALS   COLUMN TOTALS

$T_{1..} = 18$ $T_{.1.} = 16$
$T_{2..} = 20$ $T_{.2.} = 18$
$T_{3..} = 10$ $T_{.3.} = 20$
$T_{4..} = 14$ $T_{.4.} = 14$
$T_{5..} = 6$

The appropriate error mean square is .9 based on 40 degrees of freedom.
(a) Using the appropriate F statistic, test the significance of the row total contrast $T_{1..}$ and $T_{2..}$ vs. $T_{3..}$, $T_{4..}$ and $T_{5..}$ (specified *a priori*).
(b) Test the significance of the column total contrast $T_{.1.}$ and $T_{.4.}$ vs. $T_{.2.}$ (specified *a priori*).
(c) Perform the same tests using the S-method.

12.15 An experiment was designed to investigate the effects of type of store (drug, grocery, or variety, represented by rows) and location (5 large cities, represented by columns) on dollar sales volume of Gooey-Crunch candy. Four randomly selected stores were used for each of the 15 store–location combinations. An analysis of variance was computed as follows:

SOURCE OF VARIANCE	d.f.	SS	MS
Rows (store type)	2	40	20
Column (cities)	4	160	40
Interaction R × C	8	160	20
Error (observations/cells)	45	180	4
Total	59	540	

Perform the appropriate tests of significance (5 percent level) assuming in both cases that the three store types were specifically of interest

(a) When the five cities were specifically of interest

(b) When the five cities were a random sample from a much larger population of cities of interest.

12.16 Suppose that an analysis of variance involves 6 rows, 5 columns and 2 observations per cell and that the following totals have been obtained.

ROW TOTALS   COLUMN TOTALS

$T_{1.} = 10$ $T_{.1} = 20$
$T_{2.} = 15$ $T_{.2} = 25$
$T_{3.} = 17$ $T_{.3} = 30$
$T_{4.} = 19$ $T_{.4} = 15$
$T_{5.} = 14$ $T_{.5} = 25$
$T_{6.} = 11$

The appropriate error mean square is 0.2 based on 30 degrees of freedom. Using the appropriate F statistic,

(a) Test the significance of the row total contrast $T_{1.}$, $T_{3.}$, and $T_{4.}$ versus $T_{2.}$, $T_{5.}$, and $T_{6.}$ (1 percent level).

(b) Test the significance of the column total contrast $T_{.1}$ versus $T_{.2}$, $T_{.3}$, and $T_{.5}$ (1 percent level).

12.17 For the data in problem 17,

(a) Obtain confidence intervals for contrasts involving two means and for contrasts involving three means using the T-method and the S-method

(b) Obtain confidence intervals for contrasts involving four means, five means and six means using the S-method

(c) State conclusions drawn from (a) and (b)

12.18 Data are obtained for a two-factor factorial experiment with A at levels α_1, α_2, and α_3 and B at levels β_1, β_2, β_3, β_4, β_5, and β_6. For each treatment combination, 4 observations are obtained using a completely randomized design. The results are summarized as follows:

$$\frac{1}{24} \sum_{i=1}^{3} T_{i..}^2 = 100$$

$$\frac{1}{12} \sum_{j=1}^{6} T_{.j.}^2 = 140$$

$$\frac{1}{4} \sum_{i=1}^{3} \sum_{j=1}^{6} T_{ij.}^2 = 240$$

$$\sum_{i=1}^{3} \sum_{j=1}^{6} \sum_{k=1}^{4} y_{ijk}^2 = 510$$

$$\frac{T^2}{72} = 60$$

(a) Obtain the ANOVA table.

(b) State and test the appropriate hypotheses assuming a fixed effects model.

(c) State and test the appropriate hypotheses assuming A is random and B is fixed.

BIVARIATE REGRESSION AND CORRELATION

13.1 INTRODUCTION

This chapter concerns the analysis of relationships between two variables. Analyses of relationships among more than two variables are discussed in the following chapters.

The data analyzed in business and economics are of two general types—cross-section and time series. *Cross-section data* consist of observations for a sample from a population defined at a specified time or for a specified (short) time period. For example, cross-section data might be obtained for a random sample of firms, individuals, regions, machines, and so forth. *Time series data* consists of observations for a sample obtained over a period of time. For example, unemployment rate, GNP, Standard & Poors average, business failures, and so forth are usually observed at various points over time. Some descriptive analyses of this type of data are discussed in Chapter 3.

Any variable that is studied as a function of time or is observed over time is said to be a time series variable. A *time series variables* assumes values corresponding to the points in time when the variable is observed. A *cross-section variable* assumes values corresponding to the different sample elements for which it is observed.

The analyses discussed in this chapter and in Chapter 14 are appropriate for either cross-section or time series data, provided the assumptions of the analyses are satisfied. However, there are special problems involved both in meeting the assumptions and in interpreting the results of analyses of time series data. The simplest type of time series analysis consists of the study of the values of a variable over time; that is, as a function of time. There is no analogous analysis of cross-section data. More complicated economic analyses involve the study of relationships between or among variables measured at different points in time; virtually all econometric studies are of this type. These time series analyses are analogous to cross-section analyses with observations obtained at different points in time rather than for different individuals, firms, or other sample elements.

Analysis of time series data was one of the first, and still is one of the most frequent, types of economic analysis. This emphasis is largely the result of economists' interest in forecasting future values of economic variables and indices. In this chapter, regression analysis is discussed in general terms applicable to both cross-section and time series data; some of the special problems involved in the analysis of time series data are discussed in a section on time series regression.

13.2 REGRESSION AND CORRELATION

In some problems involving the relationship between two variables, one variable is considered to predict or account for the other variable; these problems are analyzed using regression. In other problems the relationship between two variables is considered simultaneously and symmetrically. For these problems correlation analysis is appropriate.

The purpose of *bivariate regression analysis* is to establish a relationship between a dependent variable and an independent variable so that the former can be predicted from or accounted for by the latter. The purpose of *bivariate correlation analysis* is to measure the degree of association or mutual variation between two variables; correlation does not provide a functional relationship for predicting one variable from the other. In particular, a given correlation coefficient is consistent with an infinite number of regression curves–it measures the strength of the relationship between two variables but does indicate its functional form. Neither regression nor correlation implies causation.

Regression analysis, and the corresponding correlation analysis, consists of two logically and computationally distinct aspects: (1) determination of the appropriate mathematical form of the relationship – for example, linear, quadratic, exponential, or logarithmic and (2) estimation of the parameters of the relationship.

In the following sections the computational and inferential procedures for bivariate linear regression and correlation analysis are discussed first. The problem of determining the appropriate form of the relationship is then considered; a variety of mathematical forms and the estimation of their parameters are mentioned. This order clarifies the discussion, since some types of nonlinear regression analysis are based on the methods of linear regression analysis. However, in practice, the first consideration is always the determination of the appropriate mathematical form of the regression. Linear regression is important, not only because many relationships are approximately of this form, but also because linear equations frequently provide good approximations to more complex relationships that are difficult to handle mathematically. However, linearity should not be assumed automatically to be the appropriate mathematical form for a regression equation.

The general approriateness of a linear equation for expressing the relationship between two variables can be determined from a *scatter plot* or *scatter diagram* of the points corresponding to the pairs of values of two variables X and Y. For example, of the scatter plots in Figure 13.1, the data corresponding to (b) and (d) can be described quite accurately by appropriate linear equations; the data corresponding to (a) and (e) cannot be described adequately by linear

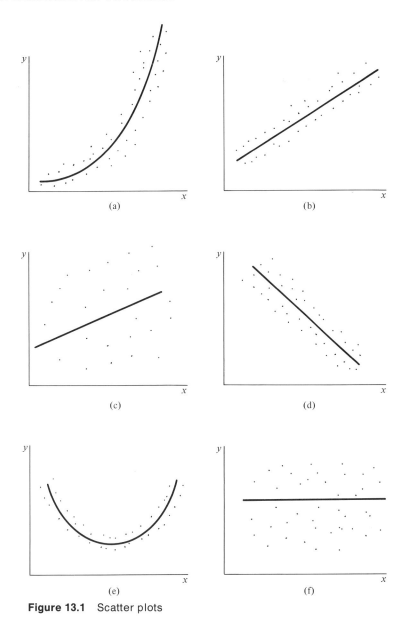

Figure 13.1 Scatter plots

equations ((a) requires an exponential equation, (e) requires a parabolic equation); and the data corresponding to (c) and (f) cannot be described very well by any type of equation, but a linear equation appears to be at least as appropriate as any other form.

Note that a scatter plot is primarily used to determine the appropriateness of a particular type of equation for describing the data. The approximate "goodness of fit" of the equation is also apparent from a scatter plot, for example, the fit in (b) and (d) is quite good compared with the fit in (c) and (f) in Figure 13.1. However, "goodness of fit" can and should be defined and computed precisely.

13.3 ESTIMATING A LINEAR REGRESSION EQUATION

In many cases the purpose of an investigation is to develop a relationship for predicting one variable from another. For example, an investigator may wish to predict quantity demanded from price, profits from overall economic conditions, brand share from attitude survey results, tensile strength from purity of steel, and so forth. Note again that causality is not implied by any of these predictive relationships.

A number of mathematical relationships between variables have been discovered in the physical and biological sciences. Some of these relationships, for example, the laws of motion and the laws relating pressure and temperature of gases, have known for many years; others, for example, the relationships basic to the theory of relativity, are of more recent origin. Although there are many difficulties in the social, economic, and behavioral sciences that are not present in the physical and biological sciences, relationships for predicting one variable from another have been and are being discovered in these areas also. For example, under certain conditions, demand for a commodity can be predicted from its price.

Of course, it is desirable to be able to predict one variable exactly from knowledge of another variable; this is frequently possible in the physical and biological sciences, at least within errors of measurement, but is only rarely realistic in the economic and behavioral sciences. In these latter areas, predictions are therefore in terms of average or expected values. This difference is important in interpreting functional relationships and will be discussed in some detail in later sections.

The problem of bivariate regression is that of estimating or predicting the average or expected value of one variable from the observed value of another variable. In this section it is assumed that the relationship between the two variables is linear; that is, the relationship between X and Y is of the form

$$Y_i = \beta_0 + \beta_1 X_i + \varepsilon_i \qquad i = 1, 2, \ldots, n$$

where ε_i represents a random error term and β_0 and β_1 are the regression parameters or coefficients that are to be estimated. The Y_i are observable random variables; the X_i are observable mathematical (nonrandom) variables; the errors, ε_i, are independent, unobservable, normally distributed, random variables with mean zero and variance σ^2 and are independent of X_i. The Y_i are referred to as *dependent* or *response variables;* the X_i are referred to as *independent* or *regressor variables.* The values of X_i are specified and the corresponding values of the random variable Y_i are observed. These specifications define a *simple linear model.* (See Figure 13.2.)

There are three generally recognized sources of error in most regression problems: (1) specification error, arising from the omission of one or more relevant independent variables; (2) sampling error, arising from random variation of the observations around their expectation; and (3) measurement error, arising from lack of precision in measuring the variables. All of these sources contribute to the error term represented by ε. Every effort should be made to specify the model correctly and to control measurement error. Sampling error is inherent and, ideally, is the only type of error that should occur. Statistically,

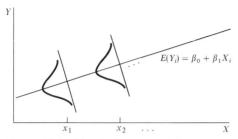

Figure 13.2 Simple linear model

specification error and measurement error result in estimation problems that are not easily handled and thus are not discussed here. (See references in Appendix A for further consideration of these problems.) Subsequent discussions assume that the ε_i consist only of sampling error.

In order to facilitate confidence interval estimation and testing of hypotheses using regression, we assume the errors are normally distributed and have the same variance for each value of X. The values of Y are then normally distributed with the same variance for each value of X. The Y values are observable and the assumption that they are normally distributed for each value of X can sometimes be justified on the basis of either prior knowledge or the sample data. Since the errors cannot be observed directly, knowledge of their distribution is less easily obtained. If it can be assumed that the errors are additive and that they can be partitioned into a large number of independent terms, then the central limit theorem applies and the normality assumption is justified. However, existence of the conditions for the central limit theorem to apply is difficult to verify in practice.

Thus although normality of the error distribution is assumed at various points in subsequent sections, this assumption is not always justifiable and its appropriateness should be tested; methods for testing normality are discussed later.

METHOD OF LEAST SQUARES

If a scatter plot or other analysis indicates that a set of data can be described adequately by a linear equation, the next step is to determine the particular linear relationship that provides the best possible fit in some precisely specified sense. An infinite number of straight lines can be drawn on any scatter plot and a number of these usually provide a reasonably good fit to the data. If any one line passes through *all* the data points, there is no problem in deciding that it provides the best fit. Unfortunately, this is rarely the case in practice and a criterion for choosing the best of less than perfect fits is required. This is done by estimating β_0 (the intercept) and β_1 (the slope) of the regression line that best fits the data, since these two parameters uniquely determine a straight line. The method of least squares is usually used for fitting straight lines (and other types of curves) to data. This method is based on the criterion of minimizing the sum of the squared errors of prediction, that is, the sum of the squares of the observed values minus the corresponding predicted values. Least squares is an intuitively appealing criterion and has various desirable statistical properties.

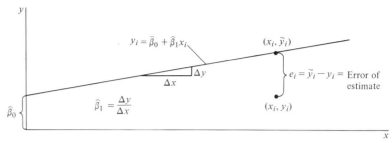

Figure 13.3 Graphical representation of the method of least squares

Graphically, the method of least squares minimizes the sum of squares of the vertical distances from the data points to the regression line. (See Figure 13.3). Since the fitted regression line is used to predict Y from X, the predicted value of Y is the point on the regression line corresponding to the given value of X. Thus the vertical distance (or deviation) from a data point to the regression line represents the error of prediction (positive or negative) for this observation, and minimizing the sum of squares of these distances minimizes the sum of the squared errors of prediction. (Recall from Chapter 2 that the sum of the deviations about their mean is zero. Squared deviations are used for the same reasons as in computing variances.)

By convention, Y denotes the predicted (dependent) variable and X denotes the regressor (independent) variable on which predictions are based. If X were predicted by Y, the sum of squares of the horizontal deviations from the regression line would be minimized.

The method of least squares, when used for fitting a linear regression equation to observed data, is essentially a procedure for estimating the two parameters, the slope and the intercept, of the line that best fits the observed data. This method guarantees that $\sum e_i^2$ is a minimum and $\sum e_i = 0$, where $e_i = \hat{y}_i - y_i$ is the error of estimate in predicting y_i.

If the observed pairs of sample data are denoted by

$$(x_1, y_1), (x_2, y_2), \ldots, (x_n, y_n)$$

then the least squares regression line is given by

$$\tilde{y}_i = \hat{\beta}_0 + \hat{\beta}_1 x_i$$

or, equivalently,*

$$\tilde{y}_i = \bar{y} + \hat{\beta}_1 (x_i - \bar{x})$$

---

* This can be shown as follows:

$$\tilde{y}_i = \hat{\beta}_0 + \hat{\beta}_1 x_i$$

$$\sum \tilde{y}_i = \sum \hat{\beta}_0 + \hat{\beta}_1 \sum x_i$$

$$\frac{\sum \tilde{y}_i}{n} = \frac{n\hat{\beta}_0}{n} + \frac{\hat{\beta}_1 \sum x_i}{n}$$

$$\bar{y} = \hat{\beta}_0 + \hat{\beta}_1 \bar{x}$$

$$\tilde{y}_i - \bar{y} = \hat{\beta}_1 (x_i - \bar{x}) \qquad \text{(subtracting the fourth equation from the first equation)}$$

$$\tilde{y}_i = \bar{y} + \hat{\beta}_1 (x_i - \bar{x})$$

where $\tilde{y}$ denotes the predicted value of y and the least squares estimates $\hat{\beta}_0$ and $\hat{\beta}_1$ are such that

$$\sum_{i=1}^{n} (y_i - \tilde{y}_i)^2 = \sum_{i=1}^{n} [y_i - (\hat{\beta}_0 + \hat{\beta}_1 x_i)]^2 = \sum e_i^2$$

is minimized. The most straightforward derivation of the estimation equations for $\hat{\beta}_0$ and $\hat{\beta}_1$ depends on calculus and is given in Appendix 13.1. The resulting estimates are

$$\hat{\beta}_1 = \frac{n \sum_{i=1}^{n} x_i y_i - \left(\sum_{i=1}^{n} x_i\right)\left(\sum_{i=1}^{n} y_i\right)}{n \sum_{i=1}^{n} x_i^2 - \left(\sum_{i=1}^{n} x_i\right)^2}$$

$$\hat{\beta}_0 = \frac{\sum_{i=1}^{n} y_i - \hat{\beta}_1 \sum_{i=1}^{n} x_i}{n}$$

If β_0 and β_1 are estimated by $\hat{\beta}_0$ and $\hat{\beta}_1$ using these equations, then the corresponding line is the best linear fit to the points $(x_1, y_1), (x_2, y_2), \ldots, (x_n, y_n)$, in the sense of minimizing the sum of squared distances from the points to the line.

COMPUTATIONAL EXAMPLE

The following data are used for illustrating the equations given in the next several sections. The independent variable, X, is the number of years since an account executive in a large brokerage firm completed the firm's training program. The dependent variable, Y, is the account executive's current annual income from commissions in thousands of dollars. It is assumed in subsequent analyses of these data that the independence and normality assumptions of the simple linear model are satisfied.

X (No. of Years)	Y (Current Annual Commissions)
3	2.8
3	2.3
3	2.0
3	1.5
6	5.0
6	4.2
6	3.8
9	6.6
9	6.0
9	5.8
9	5.5
12	8.9
12	8.1
12	7.4
15	10.5
15	9.7

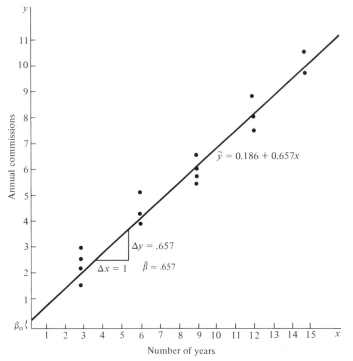

Figure 13.4 Scatter plot for example data

These data are plotted in a scatter diagram in Figure 13.4. A linear relationship appears to be appropriate.

The following summary computations can be verified for these data.

$$\sum_{i=1}^{4} y_i \big| (x = 3) = 8.6 \qquad \sum_{i=1}^{3} y_i \big| (x = 6) = 13.0 \qquad \sum_{i=1}^{4} y_i \big| (x = 9) = 23.9$$

$$\sum_{i=1}^{3} y_i \big| (x = 12) = 24.4 \qquad \sum_{i=1}^{2} y_i \big| (x = 15) = 20.2$$

$$\sum_{i=1}^{16} y_i = 90.1; \bar{y} = 5.63 \qquad \sum_{i=1}^{16} x_i = 132; \bar{x} = 8.25$$

$$\sum_{i=1}^{16} x_i^2 = 4(3)^2 + 3(6)^2 + \cdots + 2(15)^2 = 1350$$

$$\sum_{i=1}^{16} y_i^2 = (2.8)^2 + (2.3)^2 + \cdots + (9.7)^2 = 623.83$$

$$\sum_{i=1}^{16} x_i y_i = 3(8.6) + 6(13.0) + \cdots + 15(20.2) = 914.7$$

The regression parameters can be estimated as follows:

$$\hat{\beta}_1 = \frac{n \sum\limits_{i=1}^{n} x_i y_i - \left(\sum\limits_{i=1}^{n} x_i\right)\left(\sum\limits_{i=1}^{n} y_i\right)}{n \sum\limits_{i=1}^{n} x_i^2 - \left(\sum\limits_{i=1}^{n} x_i\right)^2}$$

$$= \frac{(16)(914.7) - (132)(90.1)}{(16)(1350) - (132)^2} = .657$$

$$\hat{\beta}_0 = \frac{\sum\limits_{i=1}^{n} y_i - \hat{\beta}_1 \sum\limits_{i=1}^{n} x_i}{n}$$

$$= \frac{(90.1) - (.66)(132)}{16} = .186$$

and the least squares estimated regression equation is given by

$$\tilde{y} = .186 + .657x$$

This line is drawn on the scatter plot in Figure 13.4.

Note that the predicted values of y_i, the $\tilde{y}_i$'s, are predicted average or expected values of Y. Even if β_0 and β_1 are known exactly, the prediction of Y from X is only a prediction of the average or expected value of Y, since the Y's are assumed to have a (normal) distribution for each value of X. This is in constrast to the usual assumption in the physical and biological sciences that a relationship is exact and that variations are caused by error of measurement or, perhaps, by failure to meet some condition(s) necessary for the relationship to hold. This difference is primarily due to the nature of the subject matter; light rays, atoms, cells, and some of the other objects of physical and biological research can be depended on to behave in a predictable way under specified conditions. Human beings, who are usually either directly or indirectly the objects of social, economic, and behavioral research, are much less dependable— they are inherently variable and prediction of averages is usually the best the investigator can hope to do. For example, if the quality of a product is known, the average price a group of consumers would pay for it could be predicted, but predicting the price an individual consumer would pay would be very difficult.

13.4 EVALUATING A REGRESSION EQUATION

The method of least squares can be applied to any set of paired observations to obtain a linear regression equation. However, tests of hypotheses concerning the regression coefficients and confidence intervals for regression coefficients and predicted values require assumptions concerning the variables and the error term. As mentioned, in the standard regression model the values of X are fixed and thus are the same for all possible samples; the population mean μ_x is equal to the sample mean $\bar{x}$. The Y's, however, are random variables; they are normally distributed for each X with means $E(Y|X)$, denoted by $\mu_{y \cdot x}$, which

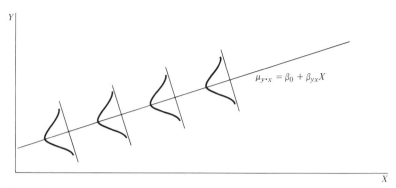

Figure 13.5 Population model for linear regression

depend on X; the y's have the same variance σ^2 for each value of X. (See Figure 13.5.) The population regression equation is thus

$$\mu_{y \cdot x} = \beta_0 + \beta_1 x$$

and the corresponding least squares sample regression equation is

$$\tilde{y}_i = \hat{\beta}_0 + \hat{\beta}_1 x_i$$

where $\hat{\beta}_0$ and $\hat{\beta}_1$ are the least squares estimates of β_0 and β_1 and are computed as shown on p. 377.*

A regression equation is of little use without some measure of its appropriateness or goodness of fit for the data, since estimates of β_0 and β_1 can be obtained even when the relationship $Y_i = \beta_0 + \beta_1 X_i + \varepsilon_i$ is grossly inappropriate. The goodness of fit of a regression line can be evaluated using the following analyses: (1) estimation of confidence intervals for β_0, β_1, $\mu_{y \cdot x}$, and σ^2; (2) tests of significance and (3) test for linearity of regression. Note that these analyses are based on the assumption of normality and independence of the error distributions, resulting in normality of the conditional distributions of Y.

CONFIDENCE INTERVAL ESTIMATION

Under the assumption that the variance of the errors, σ^2, is the same for all values of X and that the regression is linear, an unbiased estimate of σ^2 is given by

$$s_{y \cdot x}^2 = \frac{1}{n-2} \sum_{i=1}^{n} (y_i - \tilde{y}_i)^2$$

$$= \frac{1}{n-2} \sum_{i=1}^{n} [y_i - (\hat{\beta}_0 + \hat{\beta}_1 x_i)]^2$$

---

* The population regression equation can also be written in the form $\mu_{y \cdot x} = \mu_y + \beta_1(X - \mu_x)$; the corresponding sample regression equation is $\tilde{y}_i = \bar{y} + \hat{\beta}_1(x - \bar{x})$. These equations express the regression in terms of deviations of x and y from their means.

or, equivalently,

$$s_{y \cdot x}^2 = \frac{n-1}{n-2}(s_y^2 - \hat{\beta}_1^2 s_x^2)$$

$$= \frac{\sum\limits_{i=1}^{n} y_i - \hat{\beta}_0 \sum\limits_{i=1}^{n} y_i - \hat{\beta}_1 \sum\limits_{i=1}^{n} x_i y_i}{n-2}$$

Where s_x^2 and s_y^2 are the variances of the observed X and Y values, respectively. Note that $s_{y \cdot x}^2$ is a function of the sum of squared deviations of sample points from the estimated regression line. Thus $s_{y \cdot x}^2$ is essentially the variance of the sample errors of estimation; $s_{y \cdot x}$ is referred to as the *standard error of estimate*. The divisor is $n-2$ since one degree of freedom is lost for each of the two parameters, β_0 and β_1, estimated from the data. Thus, given $n-2$ values of $y_i - \tilde{y}_i$ and the estimating equations for β_0 and β_1, the remaining two values of $e_i = y_i - \tilde{y}_i$ can be obtained.

Under the additional assumption of normality, the following sampling distributions can be derived.

$$\frac{\beta_0 - \hat{\beta}_0}{s_{\hat{\beta}_0}} \sim t \text{ with } n - 2 \text{ degrees of freedom}$$

where

$$s_{\hat{\beta}_0} = \frac{s\sqrt{\sum x^2}}{s_x \sqrt{n(n-1)}}$$

$$\frac{\beta_1 - \hat{\beta}_1}{s_{\hat{\beta}_0}} \sim t \text{ with } n - 2 \text{ degrees of freedom}$$

where

$$s_{\hat{\beta}_1} = \frac{s}{s_x \sqrt{n-1}}$$

$$\frac{(n-2)s^2}{\sigma^2} \sim \chi^2 \text{ with } n - 2 \text{ degrees of freedom}$$

These sampling distributions can be used as the basis for testing hypotheses or obtaining confidence interval estimates for β_0, β_1, and σ^2. $100(1 - \alpha)$ percent confidence intervals for β_0, β_1, and σ^2 are given by

$$\hat{\beta}_0 + t_{\alpha/2} s_{\hat{\beta}_0} < \beta_0 < \hat{\beta}_0 + t_{1-\alpha/2} s_{\hat{\beta}_0}$$

$$\hat{\beta}_1 + t_{\alpha/2} s_{\hat{\beta}_1} < \beta_1 < \hat{\beta}_1 + t_{1-\alpha/2} s_{\hat{\beta}_1}$$

$$\frac{(n-2)s_{y \cdot x}^2}{\chi_{1-\alpha/2}^2} < \sigma^2 < \frac{(n-2)s_{y \cdot x}^2}{\chi_{\alpha/2}^2}$$

where $t_{\alpha/2}$ and $t_{1-\alpha/2}$ denote the $\alpha/2$ and $1 - \alpha/2$ percentage points, respectively, of the t distribution with $n - 2$ degrees of freedom and $\chi_{\alpha/2}^2$ and $\chi_{1-\alpha/2}^2$ denote the $\alpha/2$ and $(1 - \alpha/2)$ percentage points, respectively, of the χ^2 distribution with $n - 2$ degrees of freedom.

For the example data given

$$s_x{}^2 = \frac{\sum\limits_{i=1}^{n} x_i{}^2 - \dfrac{\left(\sum\limits_{i=1}^{n} x_i\right)^2}{n}}{n-1} = \frac{1350 - \dfrac{(132)^2}{16}}{15} = \frac{261}{15} = 17.40$$

$$s_x = 4.17$$

$$s_y{}^2 = \frac{\sum\limits_{i=1}^{n} y_i{}^2 - \dfrac{\left(\sum\limits_{i=1}^{n} y_i\right)^2}{n}}{n-1} = \frac{623.83 - \dfrac{(90.1)^2}{16}}{15} = \frac{116.45}{15} = 7.76$$

$$s_y = 2.79$$

$$s_{y \cdot x}^2 = \frac{n-1}{n-2}(s_y{}^2 - \hat{\beta}_1 s_x{}^2) = \frac{15}{14}\left[(7.76) - (.657)^2(17.40)\right] = \frac{15}{14}(.26) = .28$$

$$s_{y \cdot x} = .53$$

$$s_{\hat{\beta}_0} = \frac{s\sqrt{\sum x^2}}{s_x\sqrt{n(n-1)}} = \frac{.53\sqrt{1350}}{4.17\sqrt{(16)(15)}} = .31$$

$$s_{\hat{\beta}_1} = \frac{s_{y \cdot x}}{s_x\sqrt{n-1}} = \frac{.53}{(4.17)(3.87)} = .033$$

A 95 percent confidence interval for β_0 is given by

$$\hat{\beta}_0 + t_{\alpha/2}s_{\hat{\beta}_0} < \beta_0 < \hat{\beta}_0 + t_{1-\alpha/2}s_{\hat{\beta}_0}$$
$$.186 - (2.145)(.31) < \beta_0 < .186 + (2.145)(.31)$$
$$-.479 < \beta_0 < .851$$

A 95 percent confidence interval for β_1 is given by

$$\hat{\beta}_1 + t_{\alpha/2}s_{\hat{\beta}_1} < \beta_1 < \hat{\beta}_1 + t_{1-\alpha/2}s_{\hat{\beta}_1}$$
$$.657 - (2.145)(.033) < \beta_1 < .657 + (2.145)(.033)$$
$$.586 < \beta_1 < .728$$

A 95 percent confidence interval for σ^2 is given by

$$\frac{(n-2)s_{y \cdot x}^2}{\chi_{1-\alpha/2}^2} < \sigma^2 < \frac{(n-2)s_{y \cdot x}^2}{\chi_{\alpha/2}^2}$$
$$\frac{(.14)(.28)}{26.12} < \sigma^2 < \frac{(.14)(.28)}{5.62}$$
$$.15 < \sigma^2 < .70$$

Under the same assumptions, the following $100(1 - \alpha)$ percent confidence interval can be obtained for the expected value or mean of the Y values for a given X value:

$$\bar{y}_x + t_{\alpha/2} s_{y \cdot x} \sqrt{\frac{1}{n} + \frac{(x - \bar{x})^2}{(n - 1)s_x^2}} < \mu_{y \cdot x} < \bar{y}_x + t_{1 - \alpha/2} s_{y \cdot x} \sqrt{\frac{1}{n} + \frac{(x - \bar{x})^2}{(n - 1)s_x^2}}$$

where $t_{\alpha/2}$ and $t_{1-\alpha/2}$ represent the $\alpha/2$ and $(1 - \alpha/2)$ percentage points, respectively, of the t distribution with $n - 2$ degrees of freedom and $\bar{y}_x$ is the average of the sample values of Y for a given value of X.

Note that this confidence interval is for the mean or expected value of Y for a given X and is not appropriate for individual observations; the confidence interval below is appropriate for single observations.

As shown in Figure 13.6, the confidence interval is narrowest at $(\bar{x}, \bar{y})$ and fans out in both directions. The point $(\bar{x}, \bar{y})$ may be thought of as the pivot or fulcrum; the farther away an observation is from this point, the wider is the associated confidence interval:

$$\bar{y}_x + t_{\alpha/2} s_{y \cdot x} \sqrt{1 + \frac{1}{n} + \frac{(x - \bar{x})^2}{(n - 1)s_x^2}} < Y < \bar{y}_x + t_{1 - \alpha/2} s_{y \cdot x} \sqrt{1 + \frac{1}{n} + \frac{(x - \bar{x})^2}{(n - 1)s_x^2}}$$

where $t_{\alpha/2}$, $t_{1-\alpha/2}$, and $\bar{y}_x$ are defined as already stated. Note that this is the appropriate interval for an individual observation and that, as should be expected, it is wider than the corresponding interval for the mean. For the example data given (p. 377) if $x = 6$,

$$\tilde{y} = \hat{\beta}_0 + \hat{\beta}_1 x = .186 + (.657)(6) = 4.13$$

A 95 percent confidence interval for $\mu_{y \cdot x}$, the expected value of y when $x = 6$ is given by

$$\bar{y}_x + t_{\alpha/2} s_{y \cdot x} \sqrt{\frac{1}{n} + \frac{(x - \bar{x})^2}{(n - 1)s_x^2}} < \mu_{y \cdot x} < \bar{y}_x + t_{1 - \alpha/2} \sqrt{\frac{1}{n} + \frac{(x - \bar{x})^2}{(n - 1)s_x^2}}$$

$$4.33 - (2.145)(.53) \sqrt{\frac{1}{16} + \frac{(6 - 8.25)^2}{(15)(17.40)}} < \mu_{y \cdot x}$$

$$< 4.33 + (2.145)(.53) \sqrt{\frac{1}{16} + \frac{(6 - 8.25)^2}{(15)(17.40)}}$$

$$4.33 - .29 < \mu_{y \cdot x} < 4.33 + .29$$

$$4.04 < \mu_{y \cdot x} < 4.62$$

Thus for $x = 6$ the expected value of y is between 4.04 and 4.62 with confidence .95.

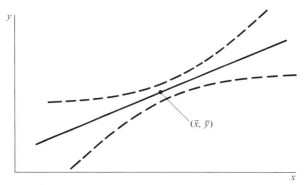

Figure 13.6   Confidence interval for $\mu_{y \cdot x}$

A 95 percent confidence interval for an individual value of y when $x = 6$ is given by

$$\bar{y}_x + t_{\alpha/2} s_{y \cdot x} \sqrt{1 + \frac{1}{n} + \frac{(x - \bar{x})^2}{(n - 1){s_x}^2}}$$

$$< Y < \bar{y}_x + t_{1 - \alpha/2} s_{y \cdot x} \sqrt{1 + \frac{1}{n} + \frac{(x - \bar{x})^2}{(n - 1){s_x}^2}}$$

$$4.33 - (2.145)(.53) \sqrt{1 + \frac{1}{16} + \frac{(6 - 8.25)^2}{(15)(17.40)}}$$

$$< Y < 4.33 + (2.145)(.53) \sqrt{1 + \frac{1}{16} + \frac{(6 - 8.25)^2}{(15)(17.40)}}$$

$$4.33 - 1.04 < Y < 4.33 + 1.04$$

$$3.29 < Y < 5.37$$

Thus for $x = 6$ the value of an individual Y observation is between 3.29 and 5.37 with confidence .95.

TEST FOR SIGNIFICANCE OF LINEAR REGRESSION

The purpose of linear regression analysis is to estimate the parameters of a linear relationship between two variables and to use this relationship as a basis for predictions. If two variables were in fact independent (not related), estimating the parameters of a linear relationship between them would be inappropriate. Thus the hypothesis that the value of Y is independent of the value of X is frequently tested as an indication of the appropriateness of regression analysis. In linear regression, if the expectation of Y is the same for each value of X—that is, if X and Y are independent—then $\beta_1 = 0$. If the hypothesis $\beta_1 = 0$ is rejected, then this is evidence that Y is in fact linearly dependent on X. Under the appropriate normality assumptions,

$$\frac{(\beta_1 - \hat{\beta}_1) s_x \sqrt{n - 1}}{s_{y \cdot x}} \sim t \text{ with } n - 2 \text{ degrees of freedom.}$$

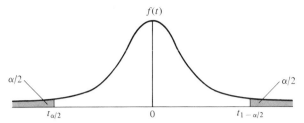

Figure 13.7 *t* distribution

Thus for the hypothesis

$$H_0 : \beta_1 = 0$$

$$H_A : \beta_1 \neq 0$$

the test statistic is

$$t_{n-2} = \frac{\hat{\beta}_1 s_x \sqrt{n-1}}{s_{y \cdot x}}$$

and the critical regions are determined in the usual way, as shown in Figure 13.7. Note again that the hypothesis $H_0 : \beta_1 = 0$, if true, implies that Y is linearly independent of X and thus that linear regression analysis is inappropriate. Rejection of this hypothesis leads to the conclusion that linear regression analysis is appropriate.

For the example data

$$t = \frac{\hat{\beta}_1 s_x \sqrt{n-1}}{s_{y \cdot x}}$$

$$= \frac{(1657)(4.17)\sqrt{15}}{.53} = 20.01$$

which is significant at the .01 Level.

Essentially, this same test can be obtained from other considerations, which are perhaps more appealing intuitively. It is possible to partition the total sum of squared deviations of the Y values from their sample mean into a part explained by regression on X and a part not explained by regression on X. This partitioning or breakdown of the sum of squares can be used as the basis for a test of the hypothesis $\beta_1 = 0$. It can be shown (see Appendix 13.2) that

$$\sum_{i=1}^{n} (y_i - \bar{y})^2 = \sum_{i=1}^{n} (y_i - \tilde{y}_i)^2 + \sum_{i=1}^{n} (\tilde{y}_i - \bar{y})^2$$

$$\begin{pmatrix} \text{sum of squares} \\ \text{total variance} \end{pmatrix} = \begin{pmatrix} \text{sum of squares} \\ \text{variance not} \\ \text{due to regression} \\ \text{on } X \end{pmatrix} + \begin{pmatrix} \text{sum of squares} \\ \text{variance due to} \\ \text{regression on } X \end{pmatrix}$$

Equivalently, this equation can be written:

$$(n-1)s_y^2 = (n-2)s_{y \cdot x}^2 + (n-1)\hat{\beta}_1^2 s_x^2$$

TABLE 13.1   ANALYSIS OF VARIANCE FOR REGRESSION

Source of Variance	Degrees of Freedom (d.f.)	Sum of Squares (SS)	Mean Square (MS)
Regression	1	$\hat{\beta}_1^2(n-1)s_x^2$	$\hat{\beta}_1^2(n-1)s_x^2$
Deviation about regression	$n-2$	$(n-2)s_{y\cdot x}^2$	$s_{y\cdot x}^2$
Total	$n-1$	$(n-1)s_y^2$	

This partitioned sum of squares is summarized in the analysis of variance in Table 13.1 and is shown graphically in Figure 13.8.

The hypothesis $\beta_1 = 0$ can be tested on the basis of the F distribution, since

$$\frac{\hat{\beta}_1^2(n-1)s_x^2}{s_{y\cdot x}^2} \sim F \text{ with 1 and } n-2 \text{ degrees of freedom.}$$

Note that this F statistic is the square of the t statistic already; given as discussed in Chapter 5, if a statistic is distributed as t with v degrees of freedom, its square is distributed as F with 1 and v degrees of freedom.

For the example data, the analysis of variance is given in Table 13.2. Note that $401.57 \approx (20.01)^2$, the difference being attributable to rounding.

Figure 13.8   Graphical representation of partitioning of sums of squares

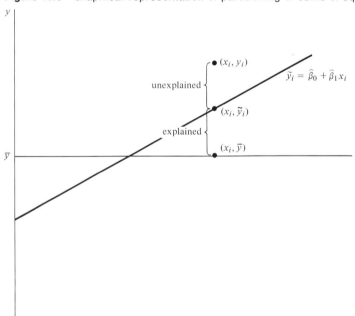

TABLE 13.2 ANALYSIS OF VARIANCE FOR REGRESSION (EXAMPLE DATA)			
Source of Variance	d.f.	SS	MS
Regression	1	112.44	112.44
Deviation about regression	15	3.96	0.28
Total	16	116.40	

$$F = \frac{\text{Regression MS}}{\text{Deviation MS}} = \frac{112.44}{0.28} = 401.57$$

The proportion of the total variance of Y which is accounted for by regression on X is given by

$$\frac{\sum\limits_{i=1}^{n} (\tilde{y}_i - \bar{y})^2}{\sum\limits_{i=1}^{n} (y_i - \bar{y})^2} = \frac{\text{SS}_{\text{Regression}}}{\text{SS}_{\text{Total}}}$$

This ratio is denoted by r^2, the square of the correlation coefficient, r, which is a measure of the degree or strength of the linear relationship between two variables, discussed in the next section. It can be shown that

$$\frac{r\sqrt{n-2}}{\sqrt{1-r^2}} \sim t \text{ with } n-2 \text{ degrees of freedom}$$

or, equivalently,

$$\frac{(n-2)r^2}{1-r^2} \sim F \text{ with 1 and } n-2 \text{ degrees of freedom.}$$

These sampling distributions are algebraic equivalents of those given in terms of β_1^2, s_x^2, and $s_{y \cdot x}^2$.

For the example data

$$r^2 = \frac{\sum\limits_{i=1}^{n} (\tilde{y}_i - \bar{y})^2}{\sum\limits_{i=1}^{n} (y_i - \bar{y})^2} = \frac{\text{SS}_{\text{Regression}}}{\text{SS}_{\text{Total}}}$$

$$= \frac{112.44}{116.40} = .966$$

Thus 96.6 percent of the variation in y is accounted for by regression on x.

TABLE 13.3 ANALYSIS OF VARIANCE
FOR TEST OF LINEARITY OF REGRESSION

Source of Variance	d.f.	SS
Regression	1	$\hat{\beta}_1{}^2(n-1)s_x{}^2$
About regression	$k-2$	$\sum\limits_{j=1}^{k} n_j(n_j-1)s_{\bar{y}_j}{}^2 - \hat{\beta}_1{}^2(n-1)s_x{}^2$
Within groups	$n-k$	$(n-1)s_y{}^2 - \sum\limits_{j=1}^{k} n_j(n_j-1)s_{\bar{y}_j}{}^2$
Total	$n-1$	$(n-1)s_y{}^2$

TEST FOR LINEARITY OF REGRESSION

The preceding sections discuss estimation and evaluation procedures for linear regression analysis. As noted previously, the relationship between two variables may be nonlinear. A scatter plot of the data may indicate the possibility of a nonlinear relationship; the significance of the nonlinearity can be determined using an analysis of variance test. This test is based on a comparison of the variance of the sample means of Y (for each value of X) about the regression line with the variance of the Y values within groups (for each value of X). The test indicates the possible presence of nonlinearity, but not the functional form of the nonlinear relationship. The determination of the functional form and the estimation of its parameters are discussed in a subsequent section. Table 13.3 is a summary of the computations required for the test of linearity.

Note that k is the number of groups, that is, the number of values of X. The following computing formulas, based on the same algebraic manipulations as the computing formulas for variances, are usually more convenient than those in Table 13.3.

$$\sum_{j=1}^{k} n_j(n_j-1)s_{\bar{y}_j}{}^2 = \sum_{j=1}^{k} n_j(\bar{y}_j - \bar{y})^2 = \sum_{j=1}^{k} \frac{T_j{}^2}{n_j} - \frac{T^2}{n}$$

Where T_j is $\sum y_i$ for specified x_j; n_j is the number of observations for which x is x_j; $T = \sum_{j=1}^{k} T_j$ and $n = \sum_{j=1}^{k} n_j$. The sum of squares about regression is obtained by subtraction.

Under the appropriate assumptions, if the regression is linear,

$$\frac{\text{MS About regression}}{\text{MS Within groups}} \sim F \text{ with } k-2 \text{ and } n-k \text{ degrees of freedom;}$$

that is,

$$\frac{\sum\limits_{j=1}^{k} n_j(n_j-1)s_{\bar{y}_j}{}^2 - \hat{\beta}_1{}^2(n-1)s_x{}^2}{(n-1)s_{\bar{y}}{}^2 - \sum\limits_{j=1}^{k} n_j(n_j-1)s_{\bar{y}_j}{}^2} \cdot \frac{n-k}{k-2} \sim F_{k-2,n-k}$$

Note that each of these terms corresponds to a particular source of variance in the values of Y.

$$\sum_{j=1}^{k} n_j(n_j - 1)s_{\bar{y}_j}^2$$
  Represents the variance or deviations of the sample means of Y (for each value of X) from the overall mean of Y

$$\hat{\beta}_1^2(n - 1)s_x^2$$
  Represents the variance in Y attributable to regression on X

$$(n - 1)s_y^2$$
  Represents the variance of the values of Y about their overall mean

Denote the ratio

$$\frac{\sum_{j=1}^{k} n_j(n_j - 1)s_{\bar{y}_j}^2}{(n - 1)s_y^2}$$

by E^2; this is referred to as the *correlation ratio* and is the proportion of the total variance in Y that is attributable to differences in the means of Y for different values of X. It is an appropriate measure of association (not necessarily linear) between Y and X only if the numbers of observations for each value of X are reasonably large. Note, in particular, that if there is only one observation of Y for each value of X, the correlation ratio is 1, regardless of the relationship (or lack of it) between X and Y. Interpretation of regression analysis in the case of only one observation for each value of X is discussed in the next section.

The F ratio for the test of linearity can be stated in terms of r^2 and E^2 as follows:

$$F = \frac{E^2 - r^2}{1 - E^2} \cdot \frac{n - k}{k - 2}$$

This is the ratio of the variance in Y that is attributable to nonlinear association with X divided by the variance in Y that is not attributable to association (linear or nonlinear) with X. The hypothesis of linearity is rejected for sufficiently large values of F.

A test for association (not necessarily linear) between X and Y is based on the following ratio which has the F distribution with $k - 1$ and $n - k$ degrees of freedom under the hypothesis of independence:

$$F = \frac{E_{yx}^2}{1 - E_{yx}^2} \cdot \frac{n - k}{k - 1}$$

The hypothesis of independence between X and Y is rejected for sufficiently large values of F.

TABLE 13.4 ANALYSIS OF VARIANCE
FOR TEST OF LINEARITY OF REGRESSION
(EXAMPLE DATA)

Source of Variance	d.f.	SS
Regression	1	112.44
About regression	3	0.27
Within groups	11	3.69
Total	15	116.40

For the example data the analysis of variance table for the test of linearity is given in Table 13.4. Note that

$$\sum_{j=1}^{k} n_j(n_j - 1)s_{\bar{y}_j}^2 = \sum_{j=1}^{k} \frac{T_j^2}{n_j} - \frac{T^2}{n}$$

$$= \frac{(8.6)^2}{4} + \frac{(13.0)^2}{3} + \frac{(23.9)^2}{4} + \frac{(24.4)^2}{3} + \frac{(20.2)^2}{3} - \frac{(90.1)^2}{16}$$

$$= 112.71$$

The other terms in the sums of squares can be obtained from the analysis of variance table for regression, Table 13.2. For the test of linearity

$$F = \frac{\text{MS About regression}}{\text{MS Within groups}}$$

$$= \frac{0.27}{3.69} \cdot \frac{11}{3} = .27$$

and the hypothesis of linearity is not rejected. For the test of independence

$$E_{yx}^2 = \frac{\sum_{j=1}^{k} n_j(n_j - 1)s_{\bar{y}_j}^2}{(n-1)s_y^2} = \frac{112.71}{116.40} = .968$$

$$F = \frac{E_{yx}^2}{1 - E_{yx}^2} \cdot \frac{n-k}{k-1} = \frac{.968}{.032} \cdot \frac{11}{4} = 83.19$$

and the hypothesis of independence is rejected for $\alpha = .01$. Thus the data indicate that y is linearly related to x.

13.5 LINEAR CORRELATION

In linear regression analysis, Y is assumed to be normally distributed for each value of X and the values of X are fixed. The data are assumed to be obtained by taking a random sample of Y values for each fixed value of X. In linear correlation analysis, X and Y are assumed to be jointly normally distributed random variables. The data are assumed to be obtained by taking a random sample of pairs of values of X and Y.

If X and Y have a bivariate normal distribution, the distribution of Y for each value of X is normal with constant variance and the marginal distribution of X for each value of Y is normal with constant variance. In practice, regression analysis is used in such situations, although, strictly speaking, even in this case inferences should be made in terms of fixed values of X. That is, independent random samples of Y should be obtained for each fixed value of X; data should not be obtained as random samples of pairs of values of X and Y.

On the other hand, correlation analysis is used in practice when the regression model is strictly appropriate, that is, when the values of X are fixed. In this context, correlation analysis frequently is used as a basis for selecting variables to be used in regression analysis. As noted in Section 13.3, the square of the correlation coefficient (the coefficient of determination) is a measure of the goodness of fit of the corresponding regression line.

Thus although regression and correlation are based on somewhat different assumptions and were developed for different purposes, there are many situations in which both are used. In particular, if random sampling from a bivariate normal population can be assumed, both analyses are appropriate if regression analysis is interpreted in terms of fixed values of X. The example data in this section are assumed to have been obtained by random sampling from a bivariate normal population.

Correlation analysis has two distinct interpretations, corresponding to the two purposes for which it is used: (1) correlation analysis provides an estimate of the degree of (linear) association between two variables considered symmetrically and (2) correlation analysis provides an estimate of the goodness of fit of a linear regression equation.

Most of the following discussion of correlation analysis is in terms of the goodness of fit of a linear regression equation. However, it should be noted that, even in that case, correlation analysis is symmetric. That is, in correlation analysis the two variables are treated identically rather than considering one the independent variable and one the dependent variable, as in regression.

GOODNESS OF FIT OF A
LINEAR REGRESSION EQUATION

The method of least squares provides the best linear regression equation for predicting one variable from another, based on the criterion of minimizing the sum of the squared errors of prediction. Correlation analysis provides a measure of the strength of the linear association or relationship between two variables; thus the correlation coefficient is a measure of the goodness of fit of the linear regression equation. The population correlation coefficient (sometimes referred to as the Pearson product-moment correlation coefficient) is denoted by ρ, where $\rho = \sigma_{xy}/\sigma_x\sigma_y$. The sample correlation coefficient, denoted by r, is $r = \text{cov}(x, y)/s_x s_y$ where $\text{cov}(x, y)$, s_x and s_y are the sample values corresponding to σ_{xy}, σ_x and σ_y. The sample correlation coefficient can be obtained from an analysis of variance, as follows:

$$\sum_{i=1}^{n} (y_i - \bar{y})^2 = \sum_{i=1}^{n} (\tilde{y}_i - \bar{y})^2 + \sum_{i=1}^{n} (y_i - \tilde{y}_i)^2$$

$$SS_{Total} = SS_{Regression} + SS_{Error}$$

$$(n-1)s_y^2 = (n-1)\hat{\beta}_1^2 s_x^2 + (n-2)s_{y \cdot x}^2$$

$$(n-1)s_y^2 = (n-1)r^2 s_y^2 + (n-1)(1-r^2)s_y^2$$

And thus

$$r = \pm \sqrt{1 - \frac{\sum_{i=1}^{n} (y_i - \tilde{y}_i)^2}{\sum_{i=1}^{n} (y_i - \bar{y})^2}} = \pm \sqrt{\frac{\sum_{i=1}^{n} (\tilde{y}_i - \bar{y})^2}{\sum_{i=1}^{n} (y_i - \bar{y})^2}}$$

Note that $\sum_{i=1}^{n} (y_i - \bar{y})^2$ is the sum of the squares of the deviations of the sample y_i's from their mean $\bar{y}$ (total sum of squares, denoted by SST); $\sum_{i=1}^{n} (\tilde{y}_i - \bar{y})^2$ is the sum of the squares of the deviations of the predicted y_i's from the sample mean $\bar{y}$ (regression sum of squares, denoted by SSR); $\sum_{i=1}^{n} (y_i - \tilde{y}_i)^2$ is the sum of the squares of the errors of prediction (error sum of squares, denoted SSE), the quantity minimized by the method of least squares. If the linear association between X and Y is weak and thus the fit of the regression line is poor, r is close to zero; if the linear association between X and Y is strong and thus the fit of the regression line is good, r is close to 1 in absolute value. (Note that as SSR → SST, $r \rightarrow \pm 1$).

The most convenient formula for computing r from sample data is

$$r = \frac{n \sum_{i=1}^{n} x_i y_i - \left(\sum_{i=1}^{n} x_i \right)\left(\sum_{i=1}^{n} y_i \right)}{\sqrt{\left[n \sum_{i=1}^{n} x_i^2 - \left(\sum_{i=1}^{n} x_i \right)^2 \right]\left[n \sum_{i=1}^{n} y_i^2 - \left(\sum_{i=1}^{n} y_i \right)^2 \right]}}$$

Note that this formula is symmetric in x and y; the correlation coefficient r is a measure of the (symmetric) linear relationship between X and Y, without regard to which variable is used to predict the other.

Consider a sample from a bivariate normal distribution of X and Y for which the regression of y on x

$$\tilde{y} = \hat{\beta}_0 + \hat{\beta}_1 x$$

is computed and the regression of x on y

$$\tilde{x} = \hat{\beta}_0' + \hat{\beta}_1' y$$

is also computed. For purposes of the following discussion, denote $\hat{\beta}_1$ by $\hat{\beta}_{yx}$ and $\hat{\beta}_1'$ by $\hat{\beta}_{xy}$.

By definition of least squares estimation,

$$\hat{\beta}_{yx} = \frac{n \sum_{i=1}^{n} x_i y_i - \left(\sum_{i=1}^{n} x_i\right)\left(\sum_{i=1}^{n} y_i\right)}{n \sum_{i=1}^{n} x_i^{2} - \left(\sum_{i=1}^{n} x_i\right)^{2}}$$

Similarly,

$$\hat{\beta}_{xy} = \frac{n \sum_{i=1}^{n} x_i y_i - \left(\sum_{i=1}^{n} x_i\right)\left(\sum_{i=1}^{n} y_i\right)}{n \sum_{i=1}^{n} y_i^{2} - \left(\sum_{i=1}^{n} y_i\right)^{2}}$$

Thus the regression coefficients $\hat{\beta}_{yx}$ and $\hat{\beta}_{xy}$ are related to the correlation coefficient r as follows:

$$r_{xy} = r_{yx} = \hat{\beta}_{yx}\hat{\beta}_{xy}$$

$$\hat{\beta}_{yx} = r_{xy} \frac{s_y}{s_x}$$

$$\hat{\beta}_{xy} = r_{xy} \frac{s_x}{s_y}$$

Note that $\hat{\beta}_{yx}$, $\hat{\beta}_{xy}$ and r_{xy} all have the same sign (see Figure 13.8). A positive sign for $\hat{\beta}_{yx}$, $\hat{\beta}_{xy}$ and r_{xy} indicates that X and Y are directly related; that is, as one variable increases, the other variable also increases. Geometrically, a direct relationship between X and Y is associated with an upward sloping regression line. A negative sign for $\hat{\beta}_{yx}$, $\hat{\beta}_{xy}$ and r_{xy} indicates that X and Y are inversely related. An inverse relationship between X and Y is associated with a downward sloping regression line; that is, as one variable increases, the other variable decreases. The regression line of y on x and the regression line of x on y intersect at the point $(\bar{x}, \bar{y})$; if $s_y > s_x$, $|\hat{\beta}_{yx}| > |\hat{\beta}_{xy}|$ and if $s_x > s_y$, $|\hat{\beta}_{xy}| > |\hat{\beta}_{yx}|$. (See Figure 13.9.)

Figure 13.9 Regression lines

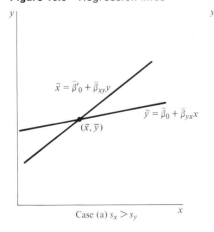

Case (a) $s_x > s_y$

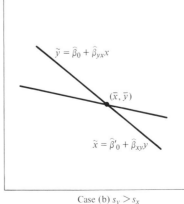

Case (b) $s_y > s_x$

The relationship between regression and correlation is also apparent in the formula for the standard error of estimate of y on x. In the population, the standard error can be written

$$\sigma_{y \cdot x} = \sigma_y \sqrt{1 - \rho^2}$$

and its sample estimate is given by

$$s_{y \cdot x} = s_y \sqrt{\frac{n - 1}{n - 2} (1 - r^2)}$$

Recall that in Section 13.3, $s_{y \cdot x}$ was given by

$$s_{y \cdot x} = \sqrt{\frac{n - 1}{n - 2} (s_y - \hat{\beta}_1^2 s_x^2)}$$

which is equivalent to the equation given above, since $\hat{\beta}_1^2 s_x^2 = r^2 s_y^2$.

EXAMPLE
The following data have been obtained for a random sample of ten electronics firms, where X represents age of the company in years and Y represents annual sales in millions of dollars. Determine the correlation between age of the firm and sales.

X	Y
3	2.5
10	6
5	2.5
6	3.5
12	6
15	6.5
9	6
2	1.5
9	5.5
7	5

$$\sum_{i=1}^{10} x_i = 78 \qquad \sum_{i=1}^{10} y_i = 45 \qquad \sum_{i=1}^{10} x_i^2 = 754 \qquad \sum_{i=1}^{10} y_i^2 = 232.5 \qquad \sum_{i=1}^{10} x_i y_i = 412$$

$$r = \frac{n \sum_{i=1}^{n} x_i y_i - \left(\sum_{i=1}^{n} x_i \right) \left(\sum_{i=1}^{n} y_i \right)}{\sqrt{\left[n \sum_{i=1}^{n} x_i^2 - \left(\sum_{i=1}^{n} x_i \right)^2 \right] \left[n \sum_{i=1}^{n} y_i^2 - \left(\sum_{i=1}^{n} y_i \right)^2 \right]}}$$

$$= \frac{(10)(412) - (78)(45)}{[(10)(754) - (78)^2][(10)(232.5) - (45)^2]}$$

$$= \frac{610}{\sqrt{(1456)(300)}} = .923$$

and

$$r^2 = .852$$

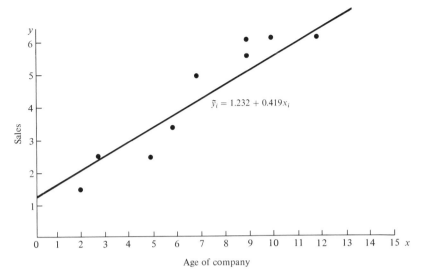

Note that for these data

$$\hat{\beta}_1 = \frac{610}{1456} = .419$$

$$\hat{\beta}_0 = 4.5 - (.419)(7.8) = 1.232$$

Thus the least squares regression line is given by $\tilde{y} = 1.232 + .419x$, as shown.

COEFFICIENT OF DETERMINATION

Correlation analysis provides a measure of the strength or degree of the linear relationship between two variables; the correlation coefficient is a measure of the goodness of fit of the regression line for predicting one variable from the other. Clearly, if $r = 0$, then knowledge of X is of no use in predicting Y. However, if r is not very close to 0 or 1, then its interpretation is not so clear. The *coefficient of determination*, r^2, is useful in interpreting an intermediate value of r, since it is the proportion of the total variance in y that is attributable to variance in x or, equivalently, the proportion of the total variance in X that is attributable to variance in Y. As noted in Section 13.3,

$$\sum_{i=1}^{n} (y_i - \bar{y})^2 = \sum_{i=1}^{n} (y_i - \tilde{y})^2 + \sum_{i=1}^{n} (\tilde{y}_i - \bar{y}_i)^2$$

$$\begin{pmatrix} \text{sum of squares} \\ \text{total variance} \end{pmatrix} = \begin{pmatrix} \text{sum of squares} \\ \text{variance not} \\ \text{attributable} \\ \text{to } X \end{pmatrix} + \begin{pmatrix} \text{sum of squares} \\ \text{variance} \\ \text{attributable} \\ \text{to } X \end{pmatrix}$$

$$(n-1)s_y^2 = (n-1)r^2 s_y^2 + (n-1)(1-r^2)s_y^2$$

and thus the proportion of the total variance of Y due to variance in X is

$$\frac{\sum_{i=1}^{n} (\tilde{y}_i - \bar{y})^2}{\sum_{i=1}^{n} (y_i - \bar{y})^2}$$

which is equal to r^2.

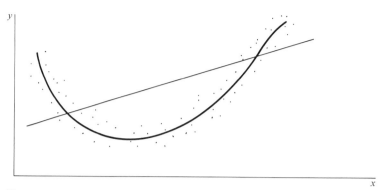

Figure 13.10 Nonlinear scatter plot

Unfortunately, it is tempting to associate with correlation analysis various interpretations and implications that cannot be justified and may be misleading. The following cautions concern two quite common sources of misinterpretation.

CAUTIONS
1. *The correlation coefficient r measures the strenth only of linear relationships. A low value of r thus does not necessarily indicate a weak relationship between X and Y, but only that there is not a strong linear relationship between X and Y. (See Figure 13.10.)*
2. *Correlation between X and Y (a relatively high value of r) may be attributable to any of the following:*
   (a) X causes Y
   (b) Y causes X
   (c) Both X and Y are related to Z (spurious correlation)
   (d) Random sampling error
   The probability that the correlation between X and Y is attributable to sampling error can be assessed on the basis of statistical tests of significance. The question concerning causation can be answered only on the basis of knowledge of the nature of the relationship between X and Y. Even a perfect correlation, r = ±1, does not prove (nor disprove) causation.

13.6 INFERENCES CONCERNING CORRELATION COEFFICIENTS

If the assumption of normality of the errors is satisfied, the sample correlation coefficient can be used as a test of the significance of linear regression (see Section 13.3). If $\rho = 0$, which implies that $\beta = 0$, then the sampling distribution of r is symmetric and

$$\frac{r\sqrt{n-2}}{\sqrt{1-r^2}} \sim t \text{ with } n-2 \text{ degrees of freedom}$$

Unfortunately, if $\rho \neq 0$, the distribution of r is complicated and depends on the value of ρ, being quite skewed if ρ is close to $+1$ and degenerate if $\rho = \pm1$. Although the exact distribution of r is unmanageable if $\rho \neq 0$, a function of r, denoted by z_r and defined by

$$z_r = \frac{1}{2} \ln\left(\frac{1 + r}{1 - r}\right)$$

is approximately normally distributed; its standard deviation depends explicitly on the unknown value of ρ, but is given approximately by

$$\sigma_{z_r} = \frac{1}{\sqrt{n - 3}}$$

For $n \geq 50$ this approximation is very accurate.

The function z_r is referred to as *Fisher's z transformation*; some of its values are given in Table B.16. Tests of hypotheses and confidence interval estimation for correlation coefficients are obtained by transforming r and ρ to z_r and ζ, respectively, proceeding as usual with the computations, and then transforming back to r and ρ.

TESTS OF SIGNIFICANCE

For testing the hypothesis

$$H_0: \rho = \rho_0$$

where ρ_0 is any specified value, the appropriate sampling distribution is given by

$$(z_r - \zeta)\sqrt{n - 3} \sim N(0, 1)$$

where

$$z_r = \frac{1}{2} \ln\left(\frac{1 + r}{1 - r}\right)$$

$$\zeta = \frac{1}{2} \ln\left(\frac{1 + \rho_0}{1 - \rho_0}\right)$$

and z_r and ζ can be obtained from Table B.10.

EXAMPLE

For the example data, test the hypothesis

$$H_0: \rho = .50$$

$$H_A: \rho > .50$$

at the 5 percent level of significance.

The hypothesized value of ρ is not zero and the Fisher z transformation (Table B.16) must be used.

$$z = (z_r - \zeta)\sqrt{n - 3}$$

$$z_r = \frac{1}{2} \ln\left(\frac{1 + .923}{1 - .923}\right) = 1.60983 \qquad \text{(from Table B.16)}$$

$$= \frac{1}{2} \ln\left(\frac{1 + .50}{1 - .50}\right) = 0.54931 \qquad \text{(from Table B.16)}$$

and thus

$$z = (1.60983 - .54931)\sqrt{7}$$

$$= (1.06052)(2.6458)$$

$$= 2.805$$

$$z_{.95} = 1.645$$

So reject H_0 and conclude that the alternative hypothesis $\rho > .50$ is supported.

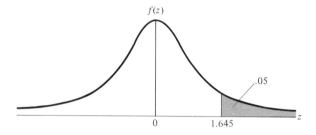

CONFIDENCE INTERVAL ESTIMATION

Using the approximate sampling distribution $(z_r - \zeta)\sqrt{n-3} \sim N(0,1)$, we get the $(1 - \alpha)$ percent confidence interval for ζ

$$z_r + \frac{z_1}{\sqrt{n-3}} < \zeta < z_r + \frac{z_2}{\sqrt{n-3}}$$

where z_1 is the $(\alpha/2)$th percentile of the standard normal distribution and z_2 is the $(1 - \alpha/2)$th percentile of the standard normal distribution. When this confidence interval for ζ has been obtained, the corresponding interval for ρ is computed by performing an inverse Fisher z transformation on the confidence limits for ζ.

EXAMPLE

Obtain a 95 percent confidence interval estimate for ρ based on the example data.

$$z_r + \frac{z_1}{\sqrt{n-3}} < \zeta < z_r + \frac{z_2}{\sqrt{n-3}}$$

$$1.60983 - \frac{1.96}{\sqrt{7}} < \zeta < 1.60983 + \frac{1.96}{\sqrt{7}}$$

$$1.60983 - .74074 < \zeta < 1.60983 + .74074$$

$$.86909 < \zeta < 2.35057$$

and, using Table B.16, we have

$$.70 < \rho < .98$$

Thus $.70 < \rho < .98$ with confidence .95.

TEST OF SIGNIFICANCE OF THE DIFFERENCE BETWEEN TWO CORRELATION COEFFICIENTS

In some problems an investigator is interested in determining whether two correlation coefficients, obtained for independent random samples, differ by a specified amount. For testing the hypothesis

$$H_0 : \rho_1 - \rho_2 = \delta$$

where ρ_1 is the correlation of X and Y for one bivariate normal population and ρ_2 is the correlation of X and Y for another (independent) bivariate normal population, the appropriate sampling distribution is given by

$$\frac{(z_{r_1} - z_{r_2}) - (\zeta_1 - \zeta_2)}{\sqrt{\dfrac{1}{n_1 - 3} + \dfrac{1}{n_2 - 3}}} \sim N(0, 1)$$

where z_{r_1}, z_{r_2}, ζ_1, and ζ_2 are the transformed values of r_1, r_2, ρ_1, and ρ_2, respectively. For the special case $H_0 : \rho_1 - \rho_2 = 0$ or, equivalently, $\rho_1 = \rho_2$, this sampling distribution becomes

$$\frac{z_{r_1} - z_{r_2}}{\sqrt{\dfrac{1}{n_1 - 3} + \dfrac{1}{n_2 - 3}}} \sim N(0, 1)$$

EXAMPLE
The correlation between price and sales volume of a random sample of 15 brands of cigarettes is $-.32$ and the correlation between price and sales volume of a random sample of 10 brands of cigars is $-.76$. Test the hypothesis that the two corresponding population correlation coefficients are equal (.10 level of significance).

$$z = \frac{z_{r_1} - z_{r_2}}{\sqrt{\dfrac{1}{n_1 - 3} + \dfrac{1}{n_2 - 3}}}$$

$$z = \frac{(-.33165) - (-.99621)}{\sqrt{\frac{1}{12} + \frac{1}{7}}}$$

$$z = \frac{.66456}{\sqrt{.22619}} = \frac{.66456}{.4756} = 1.40$$

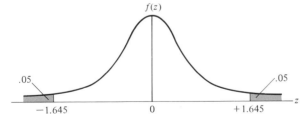

Since $z_{\alpha/2} = -1.645$ and $z_{1-\alpha/2} = +1.645$, the null hypothesis $\rho_1 = \rho_2$ is accepted. Note that it is very difficult to reject the hypothesis $\rho_1 = \rho_2$ when the sample sizes are so small.

TEST OF SIGNIFICANCE INVOLVING
THREE CORRELATION COEFFICIENTS
If X, Y, and Z are three variables that are jointly distributed so that Z is normal with the same variance for each set of values of X and Y, then the

hypothesis

$$H_0: \rho_{xz} = \rho_{yz}$$

can be tested using the following sampling distribution.

$$(r_{xz} - r_{yz}) \sqrt{\frac{(n-3)(1 + r_{xy})}{2(1 - r_{xy}^2 - r_{xz}^2 - r_{yz}^2 + 2r_{xy}r_{xz}r_{yz})}} \sim t \text{ with } n - 3 \text{ degrees of freedom}$$

EXAMPLE

For a random sample of 60 observations (quarterly data) the following correlations have been obtained where X represents long-term interest rate (average of high and low for the quarter), Y represents rate of increase of money supply, and Z represents rate of inflation:

$$r_{xy} = .70 \qquad r_{xz} = .75 \qquad r_{yz} = .60$$

Test the hypothesis that $\rho_{xz} = \rho_{yz}$ at the 5 percent level of significance.

$$t = (r_{xz} - r_{yz}) \sqrt{\frac{(n-3)(1 + r_{xy})}{2(1 - r_{xy}^2 - r_{xz}^2 - r_{yz}^2 + 2r_{xy}r_{xz}r_{yz})}}$$

$$= (.75 - .60) \sqrt{\frac{(57)(1 + .70)}{2[1 - .49 - .5626 - .36 + 2(.70)(.75)(.60)]}}$$

$$= (.15) \sqrt{\frac{96.9}{.435}}$$

$$= (.15) \sqrt{222.7586}$$

$$= (.15)(14.92)$$

$$= 2.24$$

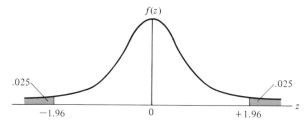

Therefore reject the hypothesis $\rho_{xz} = \rho_{yz}$ and conclude that $\rho_{xz} \neq \rho_{yz}$, that is, the correlation between long term interest rate and rate of inflation is not equal to the correlation between rate of increase of the money supply and rate of inflation.

13.7 VIOLATIONS OF MODEL ASSUMPTIONS

The regression model and the analyses previously discussed are based on the following assumptions: (1) the regressor variable is linearly related to the dependent or response variable and is independent of the errors; (2) the errors are normally distributed for each value of X, with expected value zero and the same variance σ^2 (homogeneity of variance); (3) the errors are pairwise un-correlated (serially independent).

When the assumptions of the regression model are violated, the results of the analysis should be interpreted with caution. When the errors have nonhomogeneous variances or are not serially independent, the estimated regression coefficients are not precise; that is, they have large variances. Thus predictions based on the estimated model are unreliable and can be grossly in error. If the assumptions of linearity of the model or normality of the errors are not satisfied, then the coefficients may be biased and totally inappropriate as a basis for predictions.

A test for linearity is given in Section 13.4 and the problem of detecting nonlinearity is discussed in the context of nonlinear regression in Section 13.8. The model assumptions concerning the error distributions are tested on the basis of the observed sample errors or residuals from the least squares regression equation. If the least squares equation is $\tilde{y}_i = \hat{\beta}_0 + \hat{\beta}_1 x_i$ and the observations are (x_i, y_i), $i = 1, 2, \ldots, n$, then the residuals or estimated errors are $e_i = y_i - \tilde{y}_i$, $i = 1, 2, \ldots, n$. (Note that $\sum_{i=1}^{n} e_i = 0$.) The standard tests for normality (Chapter 10) and equality of variances (Chapter 9) can be applied to the residuals. However, these tests require several observations of Y for each value of X, and in many situations only one observation of X is available for each value of Y. In practice, the validity of the assumptions of normality and homogeneity of variance are frequently evaluated somewhat subjectively on the basis of plots of the residuals. The assumption of pairwise independence of the errors can also be evaluated on the basis of plots of the residuals. A statistical test for this assumption, which is based on the Durbin–Watson statistic, is discussed below.

PLOTS OF RESIDUALS

A scatter plot of the observations (x_i, y_i) with a plot of the least squares regression equation frequently indicates possible violations of the assumptions concerning the errors. However, it is usually more convenient to examine the residuals by plotting them for the different values of X. Checking for violations of assumptions by examining residuals is not exact and depends considerably on judgment. Homogeneity of variance is suspected if the variance of the residuals seems to depend on the value of X, nonindependence is suspected if the residuals seem to have a pattern with respect to the value of X, and non-normality is suspected if the distribution of the residuals seems not to be unimodal or seems to be asymmetric or skewed.

Consider, for example, the plots in Figure 13.11. In these examples there are six values of X with multiple observations of Y for each value of X. The residuals in plot (a) seem to satisfy the assumptions of normality, homogeneity of variance, and independence. In plot (b) the residuals seem to satisfy the assumptions of normality and independence, but not the assumption of homogeneity of variance; the variance of the residuals appears to increase as X increases. The residuals in plot (c) seem to satisfy the assumptions of normality and homogeneity of variance, but not the assumption of independence; the residuals associated with small and large values of X are negative, those associated with intermediate values of X are positive. In plot (d) the residuals seem to satisfy the assumption of normality, but not the assumptions of homogeneity of variance or independence; both the residuals and their variances appear to increase as X increases.

The residuals shown in Figure 13.11 could be used as a basis for statistical tests of the regression model assumptions, since there are multiple observations of Y for each value of X. One of the advantages of plotting residuals is that this

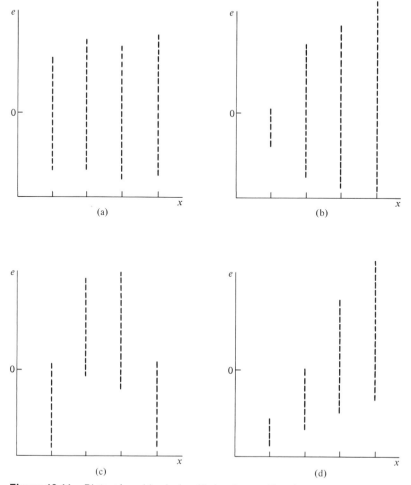

Figure 13.11 Plots of residuals (multiple observations)

method of evaluating the validity of model assumptions can be used even when multiple observations of Y are not available for each value of X. For example, consider the plots in Figure 13.12, which indicate the same types of violations of assumptions as those in Figure 13.11. In plot (a) the assumptions seem to be satisfied, in plot (b) the assumption of homogeneity of variance seems to be violated, in plot (c) the assumption of independence seems to be violated, and in plot (d) the assumptions of homogeneity of variance and independence seem to be violated.

EFFECTS OF VIOLATING ASSUMPTIONS

When the assumptions of the regression model are violated, the correction procedure usually involves transformation of the variables. Various types of linearizing transformations are discussed in Section 13.8. If the variances of the errors increase as the value of X increases, homogeneity of the error variances can sometimes be achieved by a square root transformation. Similarly, a first-difference transformation sometimes corrects pairwise correlation of the

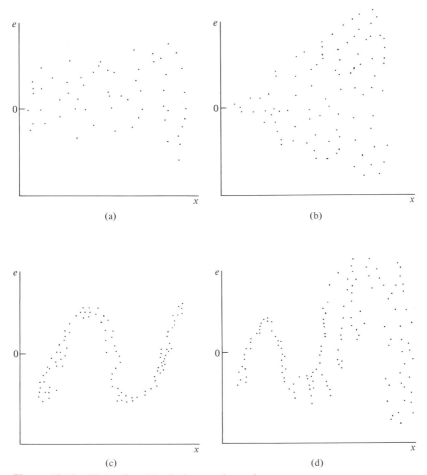

Figure 13.12 Plots of residuals (general case)

errors, as discussed below. Various transformations are also available for correcting nonnormality of the errors. Detailed consideration of the choice of the appropriate transformation to correct violations of model assumptions is beyond the scope of the present discussion.

Pairwise correlation of the errors is discussed in the following section in the context of time series regression. The validity of the assumption of pairwise independence of the errors can be tested statistically. The most common test of this assumption, which is based on the Durbin–Watson statistic, is discussed.

13.8 TIME SERIES REGRESSION

As mentioned earlier, time series regression involves the analysis of observations of one variable at various points in time or the analysis of observations of several variables at various points in time. In the former case, time is the regressor variable; an example of this type of time series regression is given later. An example of the second type, involving the observation of several variables over time, is given in Chapter 14.

There are several characteristics of time series regression that distinguish it from regression analysis in general. The sample size for time series regression is frequently relatively small and increasing it is usually beyond the control of the investigator. For example, economic data are usually recorded at fixed intervals (monthly, quarterly, annually, and so forth) and records are available only for a given length of time. In addition, some of the available observations may be inappropriate for certain analyses; for example, many economic analyses omit data for war years or other atypical periods.

In time series regression there is almost always only one observation of the dependent variable for each observation of the independent variable. The sense in which these observations represent a random sample from a defined population usually is not very clear.

The question of seasonal adjustment frequently arises for time series regression, and some of the procedures for seasonal adjustment are discussed in Chapter 3. There is a difference of opinion regarding the appropriateness of adjusting time series data; details of this discussion can be found in econometrics texts.

There are also statistical problems that occur frequently and almost exclusively when time series data are used for regression analysis. These problems are associated with pairwise correlation of the errors. The regression model assumes that the errors, ε_i and ε_j, associated with the ith and jth observations, are uncorrelated. When the observations have a natural sequential order, as in time series, correlation between the errors is referred to as *autocorrelation* or *serial correlation*.

When data are obtained using standard random sampling techniques, the assumption of pairwise independent errors is reasonable; the data have no natural sequential order and there is usually no reason to expect correlated errors. However, when the data are obtained in a natural temporal or spatial order, for example, when time is the regressor variable, autocorrelation of the errors frequently occurs. Adjacent residuals tend to be similar for either temporally or spatially ordered observations. In particular, successive residuals in economic time series tend to be positively correlated—that is, large positive errors are followed by other positive errors and large negative errors are followed by other negative errors. This occurs because observations sampled from adjacent or nearby areas or locations tend to have residuals that are correlated, since they are affected by similar external conditions. Other patterns of autocorrelation may also occur. For example, for quarterly data the errors for any given quarter may be positively correlated, not the errors for adjacent observations. In practice, autocorrelation for most time series is positive; it is difficult to describe a realistic example involving negative autocorrelation. Autocorrelation may also occur because a relevant regressor variable is omitted from the equation, discussed further in Chapter 14.

There are several statistical tests for the presence of autocorrelation. The most frequently used of these, based on the Durbin–Watson statistic, is discussed next.

THE DURBIN–WATSON TEST

The Durbin–Watson test is based on the assumption that if the errors are not independent, then their dependence is of the form of a first-order autoregressive series. Thus the errors are assumed to be of the form

$$\varepsilon_i = \rho \varepsilon_{i-1} + u_i \quad |\rho| < 1, \qquad i = 1, 2, \ldots, n$$

where μ_i is normally distributed with zero mean and constant variance; that is, μ_i, $i = 1, 2, \ldots, n$, satisfies the assumptions of the regression model. In many cases in practice, the correlation structure of ε_i is much more complex than first-order autocorrelation. First-order autocorrelation may sometimes be taken as an approximation to the actual, more complex error structure. However, it should be noted that the Durbin–Watson test is designed to test only first-order autocorrelation, not any higher order dependence among the errors.

The Dubin–Watson statistic is defined as follows:

$$d = \frac{\sum_{i=2}^{n} (e_i - e_{i-1})^2}{\sum_{i=2}^{n} e_i^2}$$

where $e_i = y_i - \tilde{y}_i$. This statistic is used for testing the hypothesis

$$H_0: \rho = 0$$

against the appropriate alternative hypothesis, which usually is $H_A: \rho > 0$, as stated. Note that if $\rho = 0$, then $e_i = u_i$, and thus the errors are uncorrelated.

Since

$$d = \frac{\sum_{i=2}^{n} (e_i - e_{i-1})^2}{\sum_{i=2}^{n} e_i^2} = \frac{\sum_{i=2}^{n} e_i^2 - 2 \sum_{i=2}^{n} e_i e_{i-1} - \sum_{i=2}^{n} e_{i-1}^2}{\sum_{i=1}^{n} e_i^2}$$

and the parameter ρ is estimated by $\hat{\rho}$ where

$$\hat{\rho} = \frac{\sum_{i=2}^{n} e_i e_{i-1}}{\sum_{i=2}^{n} e_i^2}$$

there is the following approximate relationship between d and $\hat{\rho}$

$$d \approx 2(1 - \hat{\rho})$$

Note that, approximately,

$$\text{if } \hat{\rho} = -1, d = 4$$

$$\text{if } \hat{\rho} = 0, d = 2$$

$$\text{if } \hat{\rho} = 1, d = 0$$

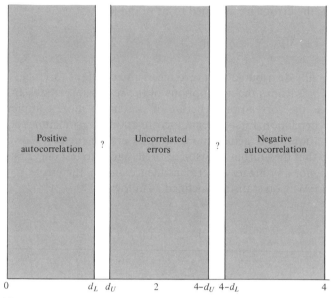

Figure 13.13 Durbin–Watson test

Thus d has the range 0 to 4 and d is close to 2 when $\rho = 0$; relatively small values of d indicate positive autocorrelation and relatively large values of d indicate negative autocorrelation.

    Unfortunately, the distribution of d is not independent of the data matrix. However, bounds that hold for all data matrices have been obtained and tabled. These bounds, denoted by d_L and d_U, are used to test the hypothesis $H_0: \rho = 0$ against the appropriate alternative hypothesis. The interpretation of the values of the Durbin–Watson statistic is summarized in Figure 13.13. For example, if

$$H_0: \rho = 0 \quad \text{(uncorrelated errors)}$$

$$H_A: \rho > 0 \quad \text{(positive autocorrelation)}$$

then the Durbin–Watson test is as follows:

$$\text{if } d < d_L, \text{ reject } H_0$$

$$\text{if } d > d_U, \text{ do not reject } H_0$$

$$\text{if } d_U < d < d_L, \text{ test is inconclusive}$$

If a test for negative autocorrelation is required, the values $4 - d_U$ and $4 - d_L$ are used rather than d_L and d_U. The question marks in Figure 13.13 indicate the ranges of values of d for which the Durbin–Watson test is inconclusive. These ranges decrease as the sample size increases and are small for reasonably large values of n. Values of d_L and d_U are given for $\alpha = .05$ and $\alpha = .01$ in Table B.17.

    As noted, correlated errors result in estimation problems. When the errors are correlated, the regression coefficient estimates are not minimum variance and their variances may be underestimated. Thus statistical tests and confidence intervals may be distorted. In some cases a transformation will remove or decrease autocorrelation of the errors. If the Durbin–Watson statistic is significant, the regression equation is sometimes reestimated using the transformed variables $y_i - \hat{\rho}y_{i-1}$ and $x_i - \hat{\rho}x_{i-1}$, where $\hat{\rho}$ is the estimate of ρ obtained from the residuals

$$\hat{\rho} = \frac{\sum\limits_{i=2}^{n} e_i e_{i-1}}{\sum\limits_{i=2}^{n} e_i^2}$$

If the residuals from this transformed regression are autocorrelated, the procedure can be iterated, although this is usually not necessary.[†] The difficulties in interpreting regression analyses of tranformed data are discussed in the next section.

The following example illustrates time series regression analysis with time as the regressor variable. There is one observation of the dependent variable for each observation of the regressor variable. Computation and interpretation of the Durbin–Watson statistic are illustrated.

EXAMPLE
An economist is interested in the variation in business gross retained earnings from 1950 to 1969. He obtained the following time series (Economic Report of the President, 1970).

BUSINESS GROSS RETAINED
EARNINGS (BILLIONS OF DOLLARS)

YEAR	TIME PERIOD x_i	y_i
1950	1	29.4
1951	2	33.1
1952	3	35.1
1953	4	36.1
1954	5	39.2
1955	6	46.3
1956	7	47.3
1957	8	49.8
1958	9	49.4
1959	10	56.8
1960	11	56.8
1961	12	58.7
1962	13	66.3
1963	14	68.8
1964	15	76.2
1965	16	84.7
1966	17	91.3
1967	18	93.3
1968	19	96.7
1969	20	98.6

[†] It can be shown that if $y_i = \beta_0 + \beta_1 x_i + \varepsilon_i$ where $\varepsilon_i = \rho\varepsilon_{i-1} + u_t$, then $y_i - \rho y_{i-1} = \beta_0^* + \beta_1^*(x_i - x_{i-1}) + \varepsilon_i^*$ where ε_i^* are uncorrelated.

Determine the linear relationship between business gross retained earnings and time.

To simplify computations, we take the time variable as having integer values from 1 to 20. Note that the correlation coefficient is not changed if either or both variables are transformed linearly. This linear transformation of the time variable consists of subtracting 1949 from each of the original values. The least squares estimates are:

$$\hat{\beta}_1 = \frac{n \sum_{i=1}^{n} x_i y_i - \left(\sum_{i=1}^{n} x_i \right) \left(\sum_{i=1}^{n} y_i \right)}{n \sum_{i=1}^{n} x_i^2 - \left(\sum_{i=1}^{n} x_i \right)^2}$$

$$= \frac{(20)(15254.5) - (210)(1213.9)}{(20)(2870) - (210)^2} = \frac{50171}{13300} = 3.77$$

$$\hat{\beta}_0 = \frac{\sum_{i=1}^{n} y_i - \hat{\beta}_1 \sum_{i=1}^{n} x_i}{n} = \frac{1213.9 - (3.77)(210)}{20} = \frac{422.2}{20} = 21.11$$

Thus the estimated regression equation is $\hat{y}_i = 21.11 + 3.77 x_i$. The associated correlation coefficient is

$$r = \frac{n \sum_{i=1}^{n} x_i y_i - \left(\sum_{i=1}^{n} x_i \right) \left(\sum_{i=1}^{n} y_i \right)}{\sqrt{\left[n \sum_{i=1}^{n} x_i^2 - \left(\sum_{i=1}^{n} x_i \right)^2 \right] \left[n \sum_{i=1}^{n} y_i^2 - \left(\sum_{i=1}^{n} y_i \right)^2 \right]}}$$

$$= \frac{50171}{\sqrt{(13300)(195536.2)}} = 0.98$$

Since

$$s_x^2 = \frac{\sum_{i=1}^{n} x_i^2 - \dfrac{\left(\sum_{i=1}^{n} x_i \right)^2}{n}}{n - 1} = \frac{2870 - \dfrac{(210)^2}{20}}{19} = 35$$

and

$$s_y^2 = \frac{\sum_{i=1}^{n} y_i^2 - \dfrac{\sum_{i=1}^{n} y_i^2}{n}}{n - 1} = \frac{83454.47 - \dfrac{(1213.9)^2}{20}}{19} = 514.57$$

the standard error of estimate is

$$s_{y \cdot x}^2 = \frac{n - 1}{n - 2} s_y^2 - \hat{\beta}_1^2 s_x^2 = \frac{19}{18} [514.57 - (14.2129)(35)] = 18.07$$

$$s_{y \cdot x} = 4.25$$

In this problem there is no reason for testing the hypothesis $H_0 : \beta_0 = 0$, since the intercept clearly is not zero and its value is of no particular interest

Significance of the regression can be tested as follows:

$$H_0: \beta_1 = 0 \Leftrightarrow \rho = 0$$

$$H_A: \beta_1 \neq 0 \Leftrightarrow \rho \neq 0$$

$$t = \frac{\hat{\beta}_1}{s_{\hat{\beta}_1}}$$

where

$$s_{\beta_1} = \frac{s_{y \cdot x}}{s_x \sqrt{n-1}}$$

thus

$$t = \frac{\hat{\beta}_1 s_x \sqrt{n-1}}{s_{y \cdot x}} = \frac{(3.77)(5.92)\sqrt{19}}{4.25} = 22.89$$

$t_{18,.99} = 2.552$, so there is a significant linear relation between business gross retained earnings and time; more specifically, business gross retained earnings increase linearly with time.

Alternatively, the hypothesis of linear relationship can be tested using

$$t_{n-2} = \frac{r\sqrt{n-2}}{\sqrt{1-r^2}}$$

or since

$$t_{n-2}^2 = F_{1,n-2}$$

$$F_{1,n-2} = \frac{r^2(n-2)}{1-r^2}$$

The analysis of variance for regression, which also provides a test for linear relationship, is given as

SOURCE OF VARIANCE	d.f.	SS	MS	F
Regression	1	9451.57	9451.57	523.05
Deviation about regression	18	325.26	18.07	
Total	19	9776.83		

Note that, except for rounding differences, $F = 523.05$ is equal to $t^2 = (22.89)^2$

In the above analysis of variance table,

$$SS_{Regression} = \beta_1^2(n-1)s_x^2 = (3.77)^2(19)(35) = 9451.57$$

$$SS_{Deviations} = (n-2)s_{y \cdot x}^2 = (18)(18.07) = 325.26$$

$$SS_{Total} = (n-1)s_y^2 = (19)(514.47) = 9776.83$$

The observations (x_i, y_i) and the least squares equation $\tilde{y}_i = 21.11 + 3.77x_j$ are shown in Figure 13.14. There is clearly a linear relationship between X and Y. The residuals from regression appear to be autocorrelated, since there are runs of positive and then negative residuals. The values x_i. the predicted values

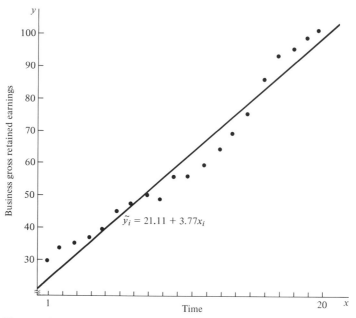

Figure 13.14   Scatter plot and least squares regression line (example data)

values $\tilde{y}_i$ and the residuals e_i are:

x_i	$\tilde{y}_i$	e_i
1	24.88	4.52
2	28.65	4.45
3	32.42	2.68
4	36.19	−0.09
5	39.96	−0.76
6	43.73	2.57
7	47.50	−0.20
8	51.27	−1.47
9	55.04	−5.64
10	58.81	−2.01
11	62.58	−5.78
12	66.35	−7.65
13	70.12	−3.82
14	73.89	−5.09
15	77.66	−1.46
16	81.43	3.27
17	85.20	6.10
18	88.97	4.33
19	92.74	3.96
20	96.51	2.09

A plot of x_i and e_i (see Figure 13.15) also indicates autocorrelation of the residuals. Note that there is a pattern, looking approximately U-shaped, rather than a random scatter of the residuals.

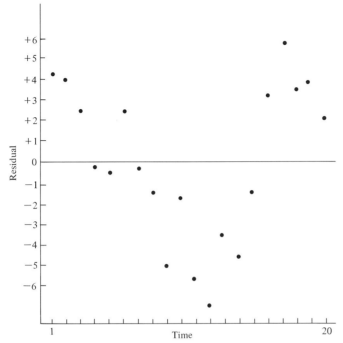

Figure 13.15 Plot of residuals (example data)

The Durbin–Watson statistic is

$$d = \frac{\sum\limits_{i=2}^{20} (e_i - e_{i-1})^2}{\sum\limits_{i=2}^{20} e_i^2} = \frac{146.5171}{293.4906} = .499$$

For one regressor variable ($k = 1$), 20 observations ($n = 20$) and $\alpha = .01$, $d_L = .95$ and $d_U = 1.15$ (Table B.17). Since $d < d_L$, the hypothesis of uncorrelated errors is rejected and the alternative hypothesis of positive autocorrelation is accepted. In practice, the data probably would be transformed; however, that analysis is beyond the scope of the present discussion.

13.9 TRANSFORMATIONS TO LINEARIZE DATA

The classical linear regression model assumes that the dependent or response variable Y is a linear function of the independent or regressor variable X. In order to satisfy this assumption, we sometimes perform the analysis on transformed variables. A regression model is linear when the parameters of the model occur linearly. For example, because the β's occur linearly, each of the following models is linear:

$$y_i = \beta_0 + \beta_1 x_i + \varepsilon_i$$

$$y_i = \beta_0 + \beta_1 x_i + \beta_2 x_i^2 + \varepsilon_i$$

$$y_i = \beta_0 + \beta_1 \sqrt{x_i} + \varepsilon_i$$

$$y_i = \beta_0 + \beta_1 \log x_i + \varepsilon_i$$

However,

$$y_i = \beta_0 + e^{\beta_1 x_i} + \varepsilon_i$$

and

$$y_i = \beta_0 + \log \beta_1 x_i + \varepsilon_i$$

are not linear models because β_1 does not occur linearly.

If a model is linear, that is, linear in the parameters, the standard least squares procedures can be applied using transformed variables. Transformations of the variables may be indicated either by theoretical considerations or by examination of a scatter plot of the observations. In many cases a scatter plot suggests several possible transformations. Regression analyses can be performed using each of these transformations and the results can then be compared as a basis for choosing the most appropriate transformation. Because the results of regression analyses are attributable partly to sampling variability, comparisons of the results of analyses using different transformations of the data are somewhat subjective, as illustrated in the examples later in the section.

As previously discussed, transformations of the data may be used because the model in terms of the original variables violates one or more of the standard assumptions concerning the errors as well as because the model is not linear in the original variables. There are several difficulties inherent in analyses of transformed data, whatever the reason for the transformation. These difficulties include:

1. It is not possible to list all the possible transformations of the variables. There are several logarithmic, exponential, and reciprocal transformations that provide reasonable flexibility for most data; however, there is no assurance that the "best" transformation has been chosen for a given set of data.
2. A transformation used for the purpose of satisfying one assumption of the model may result in violation of another assumption. The most frequently violated model assumptions probably are homogeneity of the error variances and linearity of the model. In practice, the transformations chosen to provide homogeneity of the error variances usually improve normality of the error distributions as well. The situation is not always so convenient with respect to transformations to achieve linearity of the data. Any transformation used to linearize the data also transforms the distribution of the errors and may result in violation of one or more of the assumptions of the model. Residuals from a regression using transformed data should be examined to check on the validity of the model assumptions.

3. Regression coefficients obtained using transformed variables are least squares estimates (and thus minimum variance unbiased) only for the transformed variables, not for the original variables. Inferences based on regression models for transformed variables apply to the transformed variables, not to the original variables.

Several of the transformations most commonly used to linearize data are given in Table 13.5; corresponding graphs are given in Figure 13.16. When nonlinearity is observed in the scatter plot of (x_i, y_i), one of these transformations may be chosen to respresent the data. Note that there are several apparently simple nonlinear models that cannot be linearized by transformation of the variables. For example, in addition to the two examples previously given, the models

$$y_i = \beta_0 + \beta_1^{x_i} + \varepsilon_i$$

and

$$y_i = e^{\beta_0 x_i} + e^{\beta_1 x_i} + \varepsilon_i$$

cannot be linearized. These strictly nonlinear models—models that cannot be linearized by transformation of the variables—require different, usually iterative, methods for estimating the parameters. Such methods are not discussed here.

The steps for fitting a function by transformation of the variables are as follows:

1. Determine an appropriate nonlinear function, on the basis of theory or a scatter plot of the data. If comparison of the scatter plot with the functions shown in Figure 13.16 indicates several functions that might be appropriate, then the subsequent steps can be completed for each of them and the results compared to determine the most appropriate function.
2. Apply the appropriate transformations, corresponding to the nonlinear function, to the data and estimate the parameters of the linearized function using least squares (see Table 13.5).
3. Determine the proportion of variance accounted for using the linearized (transformed) data and examine the residuals to check for violations of model assumptions.
4. Determine the proportion of variance accounted for using the nonlinear relationship obtained by reversing the transformations. Note again that the parameter estimates obtained using least squares estimation for the transformed data do not, in general, provide least squares (best linear unbiased) estimates of the parameters of the original model when transformed back to obtain estimates of these parameters. Also, the proportion of variance accounted for by the linearized model and the corresponding nonlinear model are not, in general, equal.
5. Determine the appropriateness of the nonlinear function chosen or, if several alternative transformations were estimated, determine which is most appropriate, based on the proportion of variance accounted for and examination of the residuals.

This procedure is illustrated in the following examples.

TABLE 13.5 LINEARIZING TRANSFORMATIONS

Nonlinear Function	Transformation	Linear Form
$y = \beta_0 x^{\beta_1}$ (Fig. 13.16(a), (b))	$y' = \ln y, \qquad x' = \ln x$	$y' = \ln \beta_0 + \beta_1 x'$
$y = \beta_0 x^{\beta_1} + c$ (Fig. 13.16(c), (d))	$y' = \ln(y - c), \qquad x' = \ln x$ $c = \dfrac{y_1 y_2 - y_3}{y_1 + y_2 - 2y_3}$ where $x_3 = \sqrt{x_1 x_2}$ (See Appendix 13.3.)	$y' = \ln \beta_0 + \beta_1 x'$
$y = \beta_0 e^{\beta_1 x}$ (Fig. 13.16(e))	$y' = \ln y$	$y' = \ln \beta_0 + \beta_1 x$
$y = \beta_0 e^{\beta_1 x} + c$ (Fig. 13.16(f))	$y' = \ln (y - c)$ $c = \dfrac{y_1 y_2 - y_3{}^2}{y_1 + y_2 - 2y_3}$ where $x_3 = \dfrac{1}{2}(x_1 + x_2)$ (See Appendix 13.4.)	$y' = \ln \beta_0 + \beta_1 x$
$y = \beta_0 + \beta_1 \ln x$ (Fig. 13.16(g), (h))	$x' = \ln x$	$y = \beta_0 + \beta_1 x'$
$y = k \beta_0{}^{\beta_1{}^x}$ (Fig. 13.16(i))	$y' = \ln(\ln y - \ln k)$ $\ln k = \dfrac{(\ln y_1)(\ln y_2) - (\ln y_3)^2}{(\ln y_1) + (\ln y_2) - 2(\ln y_3)}$ where $x_3 = \dfrac{1}{2}(x_1 + x_2)$ (See Appendix 13.5.)	$y' = \ln(-\ln \beta_0) + (\ln \beta_1)x$
$y = \dfrac{x}{\beta_0 + \beta_1 x}$ (Fig. 13.16(j), (k))	$y' = \dfrac{1}{y} \qquad x' = \dfrac{1}{x}$	$y' = \beta_1 + \beta_0 x'$
$y = \dfrac{x}{\beta_0 + \beta_1 x} + c$ (Fig. 13.16(l), (m))	$y' = \dfrac{x - x_1}{y - y_1}$ where (x_1, y_1) is an observation (point on the curve)	$y' = (\beta_0 + \beta_1 x_1)$ $+ \dfrac{\beta_1}{\beta_0}(\beta_0 + \beta_1 x_1)x$
$y = \dfrac{e^{\beta_0 + \beta_1 x}}{1 + e^{\beta_0 + \beta_1 x}}$ (Fig. 13.16(n))	$y' = \ln\left(\dfrac{y}{1 - y}\right)$	$y' = \beta_0 + \beta_1 x$

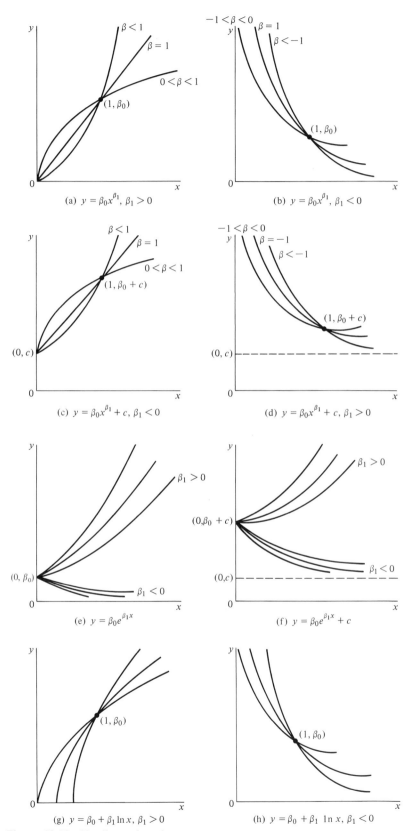

Figure 13.16 Nonlinear functions

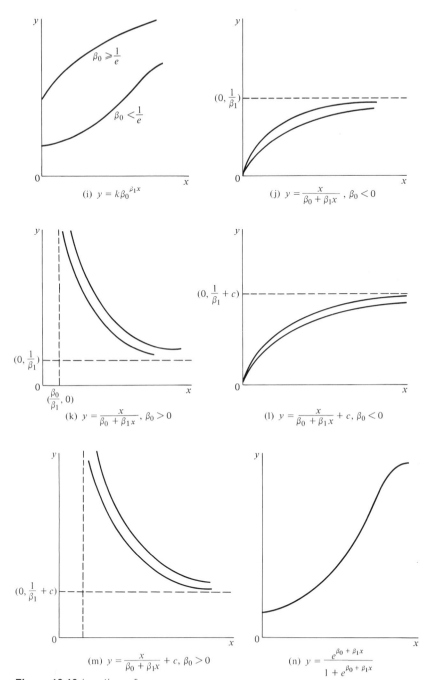

Figure 13.16 (*continued*)

EXAMPLE

Professor Smythe is interested in study patterns over the course of a semester. He asked each of his 30 advisees to keep a record of the number of hours spent studying (in addition to time in classes and labs) for the 16 successive weeks of the semester. The results are given below where X represents week of the semester and Y represents average number of hours the students spent studying.

X	Y
1	18.85
2	20.92
3	22.81
4	24.13
5	25.14
6	25.58
7	26.64
8	27.25
9	28.06
10	28.45
11	28.96
12	29.77
13	29.87
14	30.61
15	30.99
16	31.21

After examining a scatter plot of the data, determine the relationship between X and Y.

On the basis of the scatter plot (Figure 13.17), the relationship between X and Y appears to be nonlinear. The models $Y = \beta_0 X^{\beta_1}$ and $Y = \beta_0 + \beta_1 \ln X$ seem to be possibilities for fitting the data. (Preliminary checking using several points indicate that it is reasonable to assume $c = 0$ for the model $Y = \beta_0 X^{\beta_1} + c$.) The transformed data required for fitting the corresponding linearized models $\ln Y = \ln \beta_0 + \beta_1 \ln X$ and $Y = \beta_0 + \beta_1 \ln X$ are:

Figure 13.17　Scatter plot (example data)

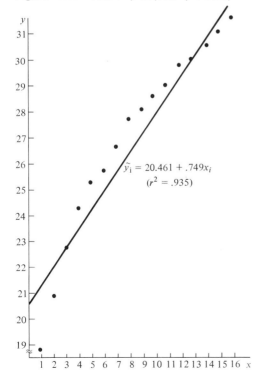

$$\tilde{y}_i = 20.461 + .749 x_i$$
$$(r^2 = .935)$$

$X' = \ln \bar{X}$	$Y' = \ln Y$
0.0000	2.9365
0.6931	3.0407
1.0986	3.1272
1.3863	3.1835
1.6094	3.2245
1.7918	3.2418
1.9459	3.2824
2.0794	3.3051
2.1972	3.3343
2.3026	3.3481
2.3979	3.3659
2.4849	3.3935
2.5649	3.3969
2.6391	3.4213
2.7081	3.4337
2.7726	3.4407

For the original data

$$\sum x_i = 136 \qquad \sum y_i = 429.24$$
$$\sum x_i^2 = 1496 \qquad \sum y_i^2 = 11719.46$$
$$\sum x_i y_i = 3903.20$$

Thus for the model $\bar{y}_i = \beta_0 + \beta_1 x_i$

$$\hat{\beta}_1 = \frac{n \sum x_i y_i - (\sum x_i)(\sum y_i)}{n \sum x_i^2 - (\sum x_i)^2} = \frac{62451.2 - 58376.64}{23936 - 18496} = .749$$

$$\hat{\beta}_0 = \frac{\sum y_i - \hat{\beta}_1 \sum x_i}{n} = \frac{429.24 - 101.864}{16} = 20.461$$

$$r^2 = \frac{[n \sum x_i y_i - (\sum x_i)(\sum y_i)]^2}{[n \sum x_i^2 - (\sum x_i)^2][n \sum y_i^2 - (\sum y_i)^2]} = \frac{(4074.56)^2}{(5440)(3264.39)} = .935$$

and $\bar{y}_i = 20.461 + .749x_i$ with associated coefficient of determination $r^2 = .935$.
    For the transformed data

$$\sum x_i' = 30.6718 \qquad \sum y_i' = 52.4761$$
$$\sum x_i'^2 = 68.1393 \qquad \sum y_i'^2 = 172.4329$$
$$\sum x_i' y_i' = 102.3340$$

For the model $y = \beta_0 x^{\beta_1}$ the linearized form is $\ln y = \ln \beta_0 + \beta_1 \ln x$ or $y' = \beta_0' + \beta_1 x'$. For this model

$$\hat{\beta}_1 = \frac{n \sum x_i' y_i' - (\sum x_i')(\sum y_i')}{n \sum x_i'^2 - (\sum x_i')^2} = \frac{1637.3440 - 1609.5416}{1090.2288 - 940.7654} = 0.186$$

$$\hat{\beta}_0' = \frac{\sum y_i' - \hat{\beta}_1 \sum x_i'}{n} = \frac{52.4761 - 5.7050}{16} = 2.923$$

$$\hat{\beta}_0 = \text{antilog } \hat{\beta}_0' = 18.597$$

$$r^2 = \frac{[n \sum x_i' y_i' - (\sum x_i')(\sum y_i')]^2}{[n \sum x_i'^2 - (\sum x_i')^2][n \sum y_i'^2 - (\sum y_i')^2]} = \frac{(27.8024)^2}{(149.4634)(5.1854)} = .997$$

Thus the estimated linearized model is $\ln y = 2.923 + 0.186 \ln x$ with associated coefficient of determination $r^2 = .997$. The estimated nonlinear model is $\tilde{y}_i = 18.597 x_i^{.186}$ (see Figure 13.18). The values of $\tilde{y}_i$ for this model and the associated errors are:

x_i	y_i	$\tilde{y}_i$	$y_i - \tilde{y}_i$
1	18.85	18.60	.25
2	20.92	21.16	−.24
3	22.81	22.81	.00
4	24.13	24.07	.06
5	25.14	25.09	.05
6	25.58	25.95	−.37
7	26.64	26.71	−.07
8	27.25	27.38	−.13
9	28.06	27.99	.07
10	28.45	28.54	−.09
11	28.96	29.05	−.09
12	29.77	29.52	.25
13	29.87	29.97	−.10
14	30.61	30.38	.23
15	30.99	30.78	.21
16	31.21	31.15	.06

Figure 13.18   Nonlinear regression models (example data)

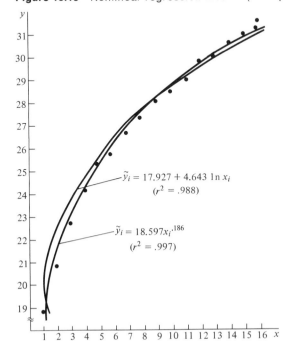

$\tilde{y}_i = 17.927 + 4.643 \ln x_i$
$(r^2 = .988)$

$\tilde{y}_i = 18.597 x_i^{.186}$
$(r^2 = .997)$

Thus the proportion of variance accounted for by the nonlinear model $\tilde{y}_i = 18.597 x_i^{.186}$ is

$$r^2 = 1 - \frac{\sum (y_i - \tilde{y}_i)^2}{\sum (y_i - \bar{y})^2} = 1 - \frac{.4791}{204.025} = .998$$

Although the model $y = \beta_0 x^{\beta_1}$ clearly provides an excellent fit for the data, the model $y = \beta_0 + \beta_1 \ln x$ is estimated for comparison. For this model the linearized form is $y = \beta_0 + \beta_1 x'$ and

$$\hat{\beta}_1 = \frac{n \sum x_i' y_i - (\sum x_i')(\sum y_i)}{n \sum x_i'^2 - (\sum x_i')^2} = \frac{13859.512 - 13165.563}{1090.2288 - 940.7593} = 4.643$$

(note that $\sum x_i' y_i = 866.2195$)

$$\hat{\beta}_0 = \frac{\sum y_i - \hat{\beta}_1 \sum x_i'}{n} = \frac{429.24 - 142.409}{16} = 17.927$$

$$r^2 = \frac{[n \sum x_i' y_i - (\sum x_i')(\sum y_i)]^2}{[n \sum x_i'^2 - (\sum x_i')^2][n \sum y_i^2 - (\sum y_i)^2]} = \frac{(693.949)^2}{(149.4695)(3264.39)} = .988$$

Thus the estimated linearized model is $\tilde{y}_i = 17.927 + 4.642 \ln x_i$ with associated coefficient of determination $r^2 = .987$. The values of $\tilde{y}_i$ for this model and the associated errors are:

x_i'	y_i	$\tilde{y}_i$	$y_i - \tilde{y}_i$
0.0000	18.85	17.93	.92
0.6931	20.92	21.15	−.23
1.0986	22.81	23.03	−.22
1.3863	24.13	24.36	−.23
1.6094	25.14	25.40	−.26
1.7918	25.58	26.25	−.67
1.9459	26.64	26.96	−.32
2.0794	27.25	27.58	−.33
2.1972	28.06	28.13	−.07
2.3026	28.45	28.62	−.17
2.3979	28.96	29.06	−.10
2.4849	29.77	29.46	.31
2.5649	29.87	29.84	.03
2.6391	30.61	30.18	.43
2.7081	30.99	30.50	.49
2.7726	31.21	30.80	.41

The proportion of variance accounted for by the model $\tilde{y}_i = 17.927 + 4.643 \ln x_i$ is

$$r^2 = 1 - \frac{\sum (y_i - \tilde{y}_i)^2}{\sum (y_i - \bar{y})^2} = 1 - \frac{2.4623}{204.025} = .988$$

which is the same value of r^2 obtained before.

On the basis of both proportion of variance accounted for (.997 versus .988) and the pattern of the errors, the model $\tilde{y}_i = 18.597 x_i^{.186}$ seems to be more appropriate for describing the data than the model $\tilde{y}_i = 17.927 + 4.643 \ln x_i$, although both provide excellent fits.

This example illustrates several characteristics of the problem of choosing a model for linearizing data.

1. The scatter plot of the data should be examined and nonlinear models should be considered, even when a linear model provides what would generally be considered a good fit for the data (in this example, $r^2 = .935$). A nonlinear model may provide not only a better fit for the data, but also a more acceptable distribution of the errors.
2. The value of r^2 associated with a linearized model is not, in general, the same as the value of r^2 associated with the corresponding nonlinear model. This occurs because the least squares parameter estimates for the linearized model, when appropriately transformed, are not usually least squares estimates of the parameters of the nonlinear model.
3. Several models may provide better fits for the data than a linear model. In the absence of theoretical reasons, the choice is made on the basis of the proportions of variance accounted for and the error distributions. In choosing a model, one must consider any increase in proportion of variance accounted for and/or improvement in the error distribution in the context of choosing a more complicated model. Clearly, by choosing a sufficiently complicated model, one could fit any set of data exactly, but nothing would be accomplished by doing so.
4. The error distribution in some cases violates the assumptions of the linear regression model essentially because the data are not linear. Thus a transformation that linearizes a model may result in a more appropriate distribution of the errors, although this does not always occur.

EXAMPLE
A physiologist is interested in motor learning skills of young children. She obtained the following data for a random sample of several hundred children on 12 trials of a difficult balancing test. X represents the trial number and Y represents the balancing time (in seconds).

X	Y
1	1.18
2	1.55
3	2.46
4	3.04
5	3.96
6	5.27
7	6.19
8	7.20
9	8.69
10	10.38
11	12.41
12	14.89

After examining a scatter plot of the data, determine the relationship between X and Y.

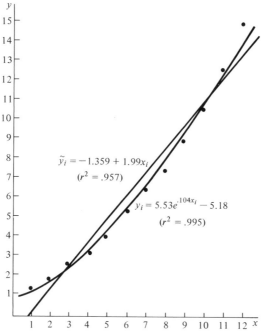

Figure 13.19   Scatter plot and regression models (example data)

The scatter plot indicates a nonlinear relationship between X and Y (Figure 13.19). The model $y = \beta_0 e^{\beta_1 x} + c$ seems to be a possibility for fitting the data. The transformed data required for fitting the corresponding linearized model $\ln y = \ln \beta_0 + \beta_1 x$ are given below. The parameter c is estimated using the points

$$x_1 = 2, \qquad y_1 = 1.55$$
$$x_2 = 11, \qquad y_2 = 12.41$$
$$x_3 = 6.5, \qquad y_3 = 5.7 \qquad \text{(estimated)}$$
$$c = \frac{y_1 y_2 - y_3{}^2}{y_1 + y_2 - 2y_3} = \frac{19.2355 - 32.49}{2.56} = -5.18$$

x	$y' = \ln(y + 5.18)$
1	1.8500
2	1.9066
3	2.0334
4	2.1066
5	2.2127
6	2.3466
7	2.4310
8	2.5161
9	2.6297
10	2.7447
11	2.8673
12	2.9992

For the original data

$$\sum x_i = 78 \qquad \sum y_i = 77.21$$
$$\sum x_i^2 = 650 \qquad \sum y_i^2 = 711.6794$$
$$\sum x_i y_i = 673.37$$

Thus for the model $\tilde{y}_i = \beta_0 + \beta_1 x_i$

$$\hat{\beta}_1 = \frac{n \sum x_i y_i - (\sum x_i)(\sum y_i)}{n \sum x_i^2 - (\sum x_i)^2} = \frac{8080.44 - 6022.38}{7800 - 6084} = 1.199$$

$$\hat{\beta}_0 = \frac{\sum y_i - \hat{\beta}_1 \sum x_i}{n} = \frac{77.21 - 93.522}{12} = -1.359$$

$$r^2 = \frac{[n \sum x_i y_i - (\sum x_i)(\sum y_i)]^2}{[n \sum x_i^2 - (\sum x_i)^2][n \sum y_i^2 - (\sum y_i)^2]} = \frac{(2058.06)^2}{(1716)(2578.7687)} = .957$$

and $\tilde{y}_i = -1.359 + 1.199 x_i$ with associated coefficient of determination $r^2 = .957$. For the transformed data

$$\sum y_i' = 28.6433$$
$$\sum y_i'^2 = 69.9357$$
$$\sum x_i y_i' = 201.1201$$

For the model $y = \beta_0 e^{\beta_1 x} + c$ the linearized form is $\ln(y - c) = \ln \beta_0 + \beta_1 x$ or $y' = \beta_0' + \beta_1 x$. For this model

$$\hat{\beta}_1 = \frac{n \sum x_i y_i' - (\sum x_i)(\sum y_i')}{n \sum x_i^2 - (\sum x_i)^2} = \frac{2413.4412 - 2234.1774}{1716} = .104$$

$$\hat{\beta}_0' = \frac{\sum y_i' - \hat{\beta}_1 \sum x_i}{n} = \frac{28.6433 - 8.112}{12} = 1.711$$

$$\hat{\beta}_0 = \text{antilog } \hat{\beta}_0' = 5.53$$

$$r^2 = \frac{[n \sum x_i y_i' - (\sum x_i)(\sum y_i')]^2}{[n \sum x_i^2 - (\sum x_i)^2][n \sum y_i'^2 - (\sum y_i')^2]} = \frac{(179.2638)^2}{(1716)(18.7898)} = .997$$

Thus the estimated linearized model is $\tilde{y}_i' = 1.711 + .104 x_i$ with associated coefficient of determination $r^2 = .997$. The estimated nonlinear model is $\tilde{y}_i = 5.53 e^{.104 x_i} - 5.18$. The values of $\tilde{y}_i$ for this model and the associated errors are:

x	y_i	$\tilde{y}_i$	$y_i - \tilde{y}_i$
1	1.18	0.96	.22
2	1.55	1.63	−.08
3	2.46	2.38	.08
4	3.04	3.21	−.17
5	3.96	4.13	−.17
6	5.27	5.15	.12

x	y_i	$\bar{y}_i$	$y_i - \bar{y}_i$
7	6.19	6.28	$-.09$
8	7.20	7.54	$-.34$
9	8.69	8.93	$-.24$
10	10.38	10.48	$-.10$
11	12.41	12.19	.22
12	14.89	14.10	.79

Thus the proportion of variance accounted for by the nonlinear model $\bar{y}_i = 5.53e^{.104x_i} - 5.18$ is

$$r^2 = 1 - \frac{.9972}{214.8974} = .995$$

This model clearly provides a better fit for the data than the linear model (Figure 13.19), although still a better fit might be obtained by reestimating the constant c or by using another model. Even though little improvement in r^2 is possible, there is some indication that this model is underestimating Y for larger values of X. However, with only one or two values of X involved, a better fit is difficult to obtain.

In addition to the characteristics of the model choice problem discussed, this example illustrates the choice of the value of a constant that is not estimated from the linearized model. Note that this choice is somewhat arbitrary, with respect to both the points (x_1, y_1) and (x_2, y_2) selected and the estimated value of y_3. For this example the value of y_3 was estimated by rough interpolation of the data. If a rough sketch of the nonlinear model is used as a basis for obtaining the value of y_3, the estimate would be lower, say 5.4. In this case

$$c = \frac{y_1 y_2 - y_3^2}{y_1 + y_2 - 2y_3} = -3.14$$

Using this value of c, the transformed data are:

$y' = \ln(y + 3.14)$	
1.4634	
1.5456	$\sum y_i' = 25.9235$
1.7229	
1.8214	
1.9602	$\sum y_i'^2 = 58.4188$
2.1295	
2.2333	
2.3361	$\sum x_i y_i' = 187.0771$
2.4707	
2.6042	
2.7441	
2.8921	

For the model $y = \beta_0 e^{\beta_1 x} + c$ the linearized form is $\ln y = \ln \beta_0 + \beta_1 x$ or $y' = \beta_0' + \beta_1 x$. For this model

$$\hat{\beta}_1 = \frac{n \sum x_i y_i' - (\sum x_i)(\sum y_i')}{n \sum x_i^2 - (\sum x_i)^2} = \frac{2244.9252 - 2022.033}{1716} = .130$$

$$\hat{\beta}_0' = \frac{\sum y_i' - \hat{\beta}_1 \sum x_i}{n} = \frac{25.9235 - 10.14}{12} = 1.315$$

$$\hat{\beta}_0 = \text{antilog } \beta_0' = 3.725$$

$$r^2 = \frac{[n \sum x_i y_i' - (\sum x_i)(\sum y_i')]^2}{[n \sum x_i^2 - (\sum x_i)^2][n \sum y_i'^2 - (\sum y_i')^2]} = \frac{(222.8922)^2}{(1716)(28.9978)} = .998$$

Thus the estimated linearized model is $\tilde{y}_i' = 1.315 + .130 x_i$ with associated coefficient of determination $r^2 = .998$. The estimated nonlinear model is $y_i = 3.725 e^{.130 x_i} - 3.14$. The values of $\tilde{y}_i$ for this model and the associated errors are:

x	y_i	$\tilde{y}_i$	$y_i - \tilde{y}_i$
1	1.18	1.10	.08
2	1.55	1.69	−.14
3	2.46	2.36	.10
4	3.04	3.13	−.09
5	3.96	3.99	−.03
6	5.27	4.99	.28
7	6.19	6.11	.08
8	7.20	7.40	−.20
9	8.69	8.86	−1.7
10	10.38	10.53	−.15
11	12.41	12.42	−.01
12	14.89	14.59	.30

Thus the proportion of variance accounted for by the nonlinear model $\tilde{y}_i = 3.725 e^{.130 x_i} - 3.14$ is

$$r^2 = 1 - \frac{.3113}{214.8974} = .999$$

This model provides a better fit than the model for which $c = -5.18$ (Figure 13.20). If the data closely follow a curve sketched on their scatter plot, then estimating the constant on the basis of points interpolated from the curve will generally provide at least as good a fit as estimating the constant on the basis of interpolating the data. However, this procedure is not good if there is considerable variability of the data around the sketched curve, making its appropriate form more difficult to determine.

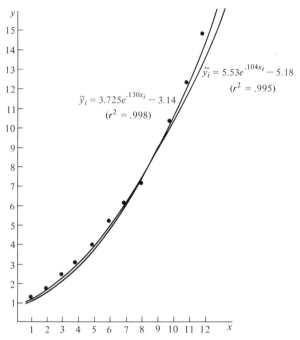

Figure 13.20 Nonlinear regression models (example data)

13.10 SUMMARY

This chapter concerns correlation and regression analysis for bivariate data. Least squares estimation and tests of significance are discussed. The effects of violations of model assumptions are considered and the use of some linearizing transformations is illustrated. Similar analyses are discussed for multivariate observations in Chapter 14.

APPENDIX 13.1
· DERIVATION OF THE NORMAL EQUATIONS

$$\sum_{i=1}^{n} \left[y_i - (\hat{\beta}_0 + \hat{\beta}_1 x_i) \right]^2$$

$$= \sum_{i=1}^{n} \left[y_i^2 - 2y_i(\hat{\beta}_0 + \hat{\beta}_1 x_i) + (\hat{\beta}_0 + \hat{\beta}_1 x_i)^2 \right]$$

$$= \sum_{i=1}^{n} y_i^2 - 2\hat{\beta}_0 \sum_{i=1}^{n} y_i - 2\hat{\beta}_1 \sum_{i=1}^{n} x_i y_i + \sum_{i=1}^{n} \hat{\beta}_0^2$$

$$+ 2\hat{\beta}_0\hat{\beta}_1 \sum_{i=1}^{n} x_i + \hat{\beta}_1^2 \sum_{i=1}^{n} x_i^2$$

Taking partial derivatives, we have

$$\frac{\partial}{\partial \hat{\beta}_0} = -2 \sum_{i=1}^{n} y_i + 2 \sum_{i=1}^{n} \hat{\beta}_0 + 2\hat{\beta}_1 \sum_{i=1}^{n} x_i$$

$$\frac{\partial}{\partial \hat{\beta}_1} = -2 \sum_{i=1}^{n} x_i y_i + 2\hat{\beta}_0 \sum_{i=1}^{n} x_i + 2\hat{\beta}_1 \sum_{i=1}^{n} x_i^2$$

and the normal equations are obtained by setting $\partial/\partial \hat{\beta}_0 = 0$ and $\partial/\partial \hat{\beta}_1 = 0$ (which can be shown to give a minimum):

$$\sum_{i=1}^{n} y_i = \hat{\beta}_0 n + \hat{\beta}_1 \sum_{i=1}^{n} x_i$$

$$\hat{\beta}_0 = \frac{\sum_{i=1}^{n} y_i}{n} - \hat{\beta}_1 \frac{\sum_{i=1}^{n} x_i}{n}$$

$$\sum_{i=1}^{n} x_i y_i = \hat{\beta}_0 \sum_{i=1}^{n} x_i + 2 \sum_{i=1}^{n} x_i^2$$

$$\hat{\beta}_1 \sum_{i=1}^{n} x_i^2 = \sum_{i=1}^{n} x_i y_i - \sum_{i=1}^{n} x_i \left(\frac{\sum_{i=1}^{n} y_i}{n} - \hat{\beta}_1 \frac{\sum_{i=1}^{n} x_i}{n} \right)$$

$$\hat{\beta}_1 \left(\sum_{i=1}^{n} x_i - \frac{\left(\sum_{i=1}^{n} x_i \right)^2}{n} \right) = \sum_{i=1}^{n} x_i y_i - \frac{\left(\sum_{i=1}^{n} x_i \right)\left(\sum_{i=1}^{n} y_i \right)}{n}$$

$$\hat{\beta}_1 = \frac{\sum_{i=1}^{n} x_i y_i - \dfrac{\left(\sum_{i=1}^{n} x_i \right)\left(\sum_{i=1}^{n} y_i \right)}{n}}{\sum_{i=1}^{n} x_i^2 - \dfrac{\left(\sum_{i=1}^{n} x_i \right)^2}{n}}$$

APPENDIX 13.2
DERIVATIONS OF SUMS OF SQUARES
FOR ANALYSIS OF VARIANCE FOR REGRESSION

To show that $\sum_{i=1}^{n} (y_i - \bar{y})^2 = \sum_{i=1}^{n} (y_i - \tilde{y}_i)^2 + \sum_{i=1}^{n} (\tilde{y}_i - \bar{y})^2$

$$\sum_{i=1}^{n} (y_i - \bar{y})^2 = \sum_{i=1}^{n} [(y_i - \tilde{y}_i) + (\tilde{y}_i - \bar{y})]^2$$

$$= \sum_{i=1}^{n} (y_i - \tilde{y}_i)^2 + 2 \sum_{i=1}^{n} (y_i - \tilde{y}_i)(\tilde{y}_i - \bar{y}) + \sum_{i=1}^{n} (\tilde{y}_i - \bar{y})^2$$

$$\sum_{i=1}^{n} (y_i - \tilde{y}_i)(\tilde{y}_i - \bar{y}) = \sum_{i=1}^{n} \tilde{y}_i(y_i - \tilde{y}_i) - \underbrace{\sum_{i=1}^{n} \bar{y}(y_i - \tilde{y}_i)}_{0}$$

$$= \sum_{i=1}^{n} \left[\bar{y} + \hat{\beta}_1(x_i - \bar{x}) \right] \left[(y_i - \bar{y}) - \hat{\beta}_1(x_i - \bar{x}) \right]$$

$$= \underbrace{\bar{y} \sum_{i=1}^{n} (y_i - \bar{y})}_{0} - \underbrace{\hat{\beta}_1 \bar{y} \sum_{i=1}^{n} (x_i - \bar{x})}_{0}$$

$$+ \hat{\beta}_1 \sum_{i=1}^{n} (x_i - \bar{x})(y_i - \bar{y}) - \hat{\beta}_1^2 \sum_{i=1}^{n} (x_i - \bar{x})^2$$

$$= \hat{\beta}_1 \sum_{i=1}^{n} x_i y_i - n\bar{x}\bar{y} - \hat{\beta}_1^2 x_i^2 - n\bar{x}^2$$

But

$$\hat{\beta}_1 = \frac{\sum_{i=1}^{n} x_i y_i - n\bar{x}\bar{y}}{\sum_{i=1}^{n} x_i^2 - n\bar{x}^2}$$

and thus

$$\sum_{i=1}^{n} (y_i - \tilde{y}_i)(\tilde{y}_i - \bar{y}) = \frac{\left[\sum_{i=1}^{n} x_i y_i - n\bar{x}\bar{y} \right]^2}{\sum_{i=1}^{n} x_i^2 - n\bar{x}^2} - \frac{\left[\sum_{i=1}^{n} x_i y_i - n\bar{x}\bar{y} \right]^2}{\sum_{i=1}^{n} x_i^2 - n\bar{x}^2} = 0$$

And

$$\sum_{i=1}^{n} (y_i - \bar{y})^2 = \sum_{i=1}^{n} (y_i - \tilde{y}_i)^2 + \sum_{i=1}^{n} (\tilde{y}_i - \bar{y})^2$$

APPENDIX 13.3
DERIVATION OF FORMULA FOR c WHEN $y = \beta_0 x^{\beta_1} + c$

Since the three points (x_1, y_1), (x_2, y_2), and (x_3, y_3) fall on the curve,

$$y_1 = \beta_0 x_1^{\beta_1} + c \quad y_2 = \beta_0 x_2^{\beta_1} + c \quad \text{and} \quad y_3 = \beta_0 x_3^{\beta_1} + c$$

Since $x_3 = \sqrt{x_1 x_2}$

$$x_3^{\beta_1} = (x_1^{\beta_1} x_2^{\beta_1})^{1/2}$$

and

$$\beta_0 x_3^{\beta_1} = \sqrt{\beta_0 x_1^{\beta_1} \beta_0 x_2^{\beta_1}}$$

But

$$\beta_0 x_3^{\beta_1} = y_3 - c \quad \beta_0 x_2^{\beta_1} = y_2 - c \quad \text{and} \quad \beta_0 x_1^{\beta_1} = y_1 - c$$

and thus

$$y_3 - c = \sqrt{(y_1 - c)(y_2 - c)}$$

$$(y_3 - c)^2 = (y_1 - c)(y_2 - c)$$

$$y_3^2 - 2y_3c + c^2 = y_1y_2 - cy_2 - cy_1 + c^2$$

$$c = \frac{y_1y_2 - y_3^2}{y_1 + y_2 - 2y_3}$$

APPENDIX 13.4
DERIVATION OF FORMULA FOR c WHEN $y = \beta_0 e^{\beta_1 x} + c$

Since the three points (x_1, y_1), (x_2, y_2), and (x_3, y_3) fall on the curve,

$$y_1 = \beta_0 e^{\beta_1 x_1} + c \quad y_2 = \beta_0 e^{\beta_1 x_2} + c \quad \text{and} \quad y_3 = \beta_0 e^{\beta_1 x_3} + c$$

and

$$\log \frac{y_1 - c}{\beta_0} = (\beta_1 \log e)x_1 \quad \log \frac{y_2 - c}{\beta_0} = (\beta_1 \log e)x_2$$

and

$$\log \frac{y_3 - c}{\beta_0} = (\beta_1 \log e)x_3$$

Since

$$x_3 = \frac{(x_1 + x_2)}{2},$$

$$(\beta_1 \log e)x_3 = \frac{1}{2}\left[(\beta_1 \log e)x_1 + (\beta_1 \log e)x_2\right]$$

and thus

$$\log\left(\frac{y_3 - c}{\beta_0}\right) = \frac{1}{2}\left[\log\left(\frac{y_1 - c}{\beta_0}\right) + \log\left(\frac{y_2 - c}{\beta_0}\right)\right]$$

$$= \log \sqrt{\left(\frac{y_1 - c}{\beta_0}\right)\left(\frac{y_2 - c}{\beta_0}\right)}$$

$$\frac{y_3 - c}{\beta_0} = \sqrt{\left(\frac{y_1 - c}{\beta_0}\right)\left(\frac{y_2 - c}{\beta_0}\right)}$$

$$y_3 - c = \sqrt{(y_1 - c)(y_2 - c)}$$

Therefore

$$c = \frac{y_1y_2 - 2y_3^2}{y_1 + y_2 - 2y_3}$$

APPENDIX 13.5
DERIVATION OF FORMULA FOR k WHEN $y = k\beta_0^{\beta_1 x}$

For the model $y = \beta_0 x^{\beta_1} + c$, $\ln(y - c) = \ln \beta_0 + \beta_1 x$ and the value of c is estimated from the points (x_1, y_1), (x_2, y_2), (x_3, y_3), where $x_3 = \frac{1}{2}(x_1 + x_2)$, by the equation

$$c = \frac{y_1 y_2 - y_3^2}{y_1 + y_2 - 2y_3}$$

(See Appendix 13.3.)

For the model $y = k\beta_0^{\beta_1^x}$, $\ln(\ln k - \ln y) = \ln(-\ln \beta_0) + (\ln \beta_1)X$ and the value of $\ln k$ is estimated from the three points $(x_1, \ln y_1)$, $(x_2, \ln y_2)$ and $(x_3, \ln y_3)$, where $x_3 = \frac{1}{2}(x_1 + x_2)$, by the equation

$$\ln k = \frac{(\ln y_1)(\ln y_2) - (\ln y_3)^2}{(\ln y_1) + (\ln y_2) - 2 \ln y_3}$$

(See Appendix 13.3.)

PROBLEMS

13.1 The following data were obtained on a random sample of summer weekdays. X represents temperature and Y represents number of people using a city beach between noon and 2 P.M.
   (a) Compute the linear regression equation of Y on X after making a scatter plot to verify linearity.
   (b) Compute the correlation coefficient and test its significance at the 1 percent level.

x	y
62	10
58	6
79	30
84	48
86	52
76	32
69	20
77	25
78	30
89	50

13.2 The following data were obtained for a random sample of uranium ores. X represents purity grade of ore (expert's grading at site of mine) and Y represents grams of uranium obtained per 1000 lb of ore.
   (a) Compute the least squares regression equation Y on X after verifying linearity by a scatter plot.
   (b) Compute the correlation coefficient and test its significance at the 5 percent level.

x	y
85	2.3
65	1.2
73	1.5
90	1.9
82	1.8
80	2.0
68	1.3
88	2.1

13.3 The following computations were made for data obtained from a random sample of 20 male students. X represents height and Y represents weight. Compute the correlation coefficient and test its significance at the 1 percent level.

$$\sum_{i=1}^{20} x_i = 840 \qquad \sum_{i=1}^{20} y_i = 2030$$

$$\sum_{i=1}^{20} x_i^2 = 58{,}900 \qquad \sum_{i=1}^{20} y_i^2 = 350{,}420$$

$$\sum_{i=1}^{20} x_i y_i = 143{,}100$$

13.4 The following computations were made from data obtained from a random sample of 10 businessmen. X represents age and Y represents thousands of dollars in stocks. Assuming linearity,
   (a) Compute the least squares regression equation of Y on X
   (b) Compute the correlation coefficient and test its significance at the 5 percent level

$$\sum_{i=1}^{10} x_i = 350 \qquad \sum_{i=1}^{10} y_i = 480$$

$$\sum_{i=1}^{10} x_i^2 = 19{,}900 \qquad \sum_{i=1}^{10} y_i^2 = 35{,}400$$

$$\sum_{i=1}^{10} x_i y_i = 22{,}200$$

13.5 The following computations were made for data obtained from a random sample of 50 companies. X represents number of departments and Y represents volume of business (per year in millions of dollars). Assuming linearity,
   (a) Compute the least squares regression equation of Y on X.
   (b) Use the regression equation obtained in (a) to predict the value of Y if $X = 5$ and obtain 95 percent confidence limits for this prediction.
   (c) Compute the correlation coefficient of Y and X.
   (d) Compute $s_{y \cdot x}$.

$$\sum_{i=1}^{50} x_i = 350 \qquad \sum_{i=1}^{50} y_i = 200$$

$$\sum_{i=1}^{50} x_i^2 = 2650 \qquad \sum_{i=1}^{50} y_i^2 = 928$$

$$\sum_{i=1}^{50} x_i y_i = 1500$$

13.6 Suppose a random sample of 67 observations is drawn from a bivariate population for which the correlation coefficient is .90. What is the probability that the correlation coefficient in the sample is more than .80?

13.7 Suppose that the correlation coefficient for a random sample of 67 observations is .80. Find 95 percent confidence limits for the population correlation coefficient.

13.8 A random sample of size 52 is taken from a population in which the correlation coefficient is .59. Find the probability that the correlation coefficient for the sample is .40 or less.

13.9 From a population with correlation coefficient .65, a large number of random samples of equal size are taken. If 15 percent of these samples have a correlation coefficient of .75 or more, what is the size of the samples?

13.10 From a population with correlation coefficient .80, an experimenter selects 500 random samples, all of equal size. If 115 of these samples show a correlation coefficient greater than .85, what is the sample size?

13.11 In a random sample of 203 cases from population A, a correlation coefficient of .66 is obtained. In a random sample of 153 cases from population B, a correlation coefficient of .70 is obtained. Test $H_0: \rho_A = \rho_B$ at the 5 percent level of significance.

13.12 In each of the following there is some inconsistency—real data would not yield these figures unless some mistake had been made. Determine what is wrong in each case.
(a) $\bar{y} = .96x + 13.3$ and $\bar{x} = 1.42y - 12.5$
(b) $\sum(x_i - \bar{x})(y_i - \bar{y}) = 1543, \sum(x_i - \bar{x})^2 = 1136, \sum(y_i - \bar{y})^2 = 1560$
(c) $r = .65, b_{yx} = -.24, b_{xy} = -.72$
(d) $r = .60, b_{yx} = .80, b_{xy} = .90$
(e) $r = .60, s_y = 7.2, s_{y \cdot x} = 12$
(f) $r = .20, b_{yx} = .80, b_{xy} = .50$
(g) $\bar{y} = -.9x + 5, \bar{x} = -.6y + 4, r = .43$
(h) $r = .5, s_x = 4, s_y = 10, b_{yx} = .25$
(i) $r = .5, s_x = 6, s_y = 3, \bar{y} = x, \bar{x} = .25y$

13.13 Two characteristics are known to have a population correlation coefficient of .585. Of a very large number of random samples of the same size, 7.5 percent have a correlation coefficient of .5 or less. Determine the size of the samples.

13.14 A consumer's research organization is interested in the relationship between the assessed value of retail men's clothing stores and their annual sales volume. Assessed value, X, and annual sales, Y, were obtained for a random sample of 16 men's clothing stores and a scatter plot indicated that the relationship was linear. The following least squares regression lines were obtained:

$$\bar{y} = 3.2 + 2.25x$$

$$\bar{x} = 4.5 + .25y$$

and

$$s_y = 3, \qquad s_x = 1$$

(a) Compute r_{xy} and s_y.
(b) Test ρ and β_{yx} for significance (5 percent level).
(c) Find 95 percent confidence intervals for ρ and β_{yx}.

13.15 A group of 150 students (ages 5 to 14) were given a mathematical aptitude test; scores are given below. A scatter plot shows considerable departure from linearity. Compute the appropriate measure of association and test its significance at the 1 percent level. If the correlation coefficient is .763, is the apparent nonlinearity significant (1 percent level)?

AGE	NUMBER OF STUDENTS	AVERAGE SCORE
14	10	11.0
13	15	14.0
12	12	14.5
11	19	16.0
10	18	18.1
9	21	20.8
8	18	25.1
7	15	31.3
6	13	40.5
5	9	49.8

(Helpful facts: $s_y^2 = 157.1262$ and $\bar{y} = 23.0$)

13.16 A record was kept in the XYZ school system of the absences (number of days per year) of a random sample of 150 students (grades 1 to 12). The data are summarized below; a scatter plot shows considerable departure from linearity. Compute the appropriate measure of association and test its significance at the 1 percent level; test the hypothesis of linearity (1 percent level of significance) if $r^2 = .396$.

GRADE (x_j)	NUMBER OF STUDENTS (n_j)	AVERAGE DAYS ABSENT $(\bar{y}_j)$	$n_j(\bar{y}_j - \bar{y})^2$
1	10	16.8	462.4
2	15	15.0	375.0
3	14	13.4	161.84
4	12	11.1	14.52
5	11	12.2	53.24
6	15	10.7	7.35
7	14	9.2	8.96
8	10	8.4	25.6
9	12	7.0	108.0
10	15	6.1	228.15
11	13	5.1	312.13
12	9	4.1	313.29
	$\sum = 150$		$\sum = 2070.48$

$$\left(\text{Helpful facts: } \bar{y} = \sum_{j=1}^{12} \frac{n_j \bar{y}_j}{150} = 10 \text{ and } s_y^2 = \frac{4000}{149}\right)$$

13.17 A test determining concentration of a certain chemical in the blood was administered to 112 patients in 12 different age groups. A scatter plot indicates considerable departure from linearity. Using the computational information below,
(a) Compute the appropriate measure of association and test its significance (1 percent level)
(b) Test for significance of departure from linearity (1 percent level)

$$\sum_{i=1}^{12} n_j(\bar{y}_j - \bar{y})^2 = 8880 \qquad \sum_{j=1}^{12} n_j = 112 \qquad s_y^2 = 100 \qquad r_{xy}^2 = .60$$

where $\bar{y}_j$ is the average for the n_j members of the jth age group, $j = 1, 2, \ldots, 12$.

13.18 A doctor who is investigating the effect of amount of hormone injection on weight of carcinogenic tumors in white mice has obtained data for a random sample of 20 white mice. For each animal he has the amount (cubic centimeters) of hormone injected (represented by X) and the weight (milligrams) of the resulting tumor (represented by Y). These data are summarized as follows:

$$\sum x = 100 \qquad \sum y = 400$$
$$\sum x^2 = 600 \qquad \sum y^2 = 9000$$
$$\sum xy = 2080$$

(a) Obtain a linear least squares equation for predicting weight of tumor from amount of hormone injected.
(b) Obtain 95 percent limits of prediction for weight of the tumor if the amount of hormone injected is 10 cc.

13.19 A market research organization is interested in the relationship between advertising expenditures, x, and total sales volume y, for grocery stores. Total sales volume was obtained for a random sample of 126 grocery stores in six advertising expenditures categories and the following computations were made:

$$\sum_{j=1}^{6} n_j(\bar{y}_j - \bar{y})^2 = 1250 \qquad s_y^{\,2} = 12.5 \qquad s_x^{\,2} = 1.5$$

The least squares regression line $\bar{y} = 5.6 + 2.5x$ has been obtained. The data show some departure from linearity on a scatter plot.
(a) Compute the appropriate measure of nonlinear association and test its significance at the 1 percent level.
(b) Test the hypothesis of linearity at the 1 percent level of significance.

13.20 A flour processing company has discovered that its machines are giving excess fill. In planning its servicing schedule for the machines, the company is interested in the relationship between the time since the last servicing and the amount of excess fill. The quality control engineer obtained a random sample of ten bags of flour from the stockroom, measured the excess fill and determined which machine had been used (from a code on the bag) and the time since its last servicing. The data are:

EXCESS WEIGHT (GRAMS)	TIME SINCE SERVICING (DAYS)
15	2
23	6
20	5
17	4
19	3
15	2
15	1
10	0
16	8
30	9

(a) Make a scatter plot of the data.
(b) Compute the appropriate least squares regression line and add it to the graph in (a).
(c) Use the least squares regression line obtained in (b) to predict the excess weight given by a machine last serviced 5 days previously.
(d) Compute the standard error of estimate and obtain .95 limits of prediction for the excess weight given by a machine last serviced 5 days previously.
(e) Obtain .95 limits of prediction for the *average* excess weight given by a machine last serviced 5 days previously.
(f) Compute the correlation coefficient and test at the .01 level of significance whether there is a relationship between the amount of excess fill and the time since the last servicing of the machine.

13.21 Dr. Cardiac evaluated (on a continuous scale) the severity of heart damage following stroke, X, for 125 patients and classified them in 10 categories on this basis. He then administered to them a drug supposed to reduce tendency toward clotting. Response to this drug, Y, was measured continuously on the basis of a laboratory test procedure. The following computations were made where T_j is the total of the n_j observations of Y in category $x_j, j = 1, 2, \ldots, 10$:

$$\sum_{i=1}^{10} \frac{T_j^{\,2}}{n_j} = 789 \qquad T = 75 \qquad s_y^{\,2} = 8 \qquad s_x^{\,2} = 3$$

Regression: $\bar{y} = .28 + 1.2x$
(a) Compute the appropriate measure of nonlinear correlation and test its significance (1 percent level).
(b) Test linearity of regression (1 percent level.)

13.22 For a random sample of 323 semiskilled workers, the correlation between years of formal education and salary is .30; for a random sample of 83 professional people, the correlation between years of formal education and salary is .53.

(a) Test the hypothesis that the two population correlation coefficients are equal (1 percent level of significance).

(b) Determine a 95 percent confidence interval for the correlation between years of formal education and salary for professional people.

13.23 The American Petro-Chemical Company measured (for a total of 175 observations) the yield of polythene, Y, as a function of pressure in the reactor, X, for 15 different (randomly selected) pressure settings. The following computations were made, where T_j is the total of the n_j observations of Y in category $x_j, j = 1, 2, \ldots, 15$.

$$\sum_{j=1}^{15} \frac{T_j^2}{n_j} = 933 \qquad \sum_{j=1}^{15} T_j = 105 \qquad s_y^2 = 10 \qquad s_x^2 = 15$$

Regression: $\bar{y} = 2.3 + .18x$

(a) Compute the appropriate measure of nonlinear correlation and test its significance (1 percent level).

(b) Test linearity of regression (1 percent level).

13.24 For a random sample of 10 communities, an index of consumer prices (denoted by Y) and population (denoted by X) were recorded and the following computations were made:

$$\sum x = 15 \qquad \sum x^2 = 30$$
$$\sum y = 50 \qquad \sum y^2 = 280$$
$$\sum xy = 63$$

(a) Obtain the least squares regression equation for Y on X.

(b) Compute the correlation between X and Y.

(c) State an appropriate hypothesis and test the significance of the linear relationship between X and Y.

13.25 Assume that the following computations have been performed for a sample of 50 observations (Y represents annual product and X represents number of employees of the firm).

$$\sum x = 250 \qquad \sum x^2 = 2500$$
$$\sum y = 100 \qquad \sum y^2 = 650$$
$$\sum xy = 1000$$

(a) Obtain the least squares regression equation of Y on X.

(b) Compute the standard error for $\hat{\beta}$ and test the significance of the coefficient $\hat{\beta}$.

(c) Determine the correlation coefficient for X and Y.

13.26 In 1971 the men's clothing industry felt a sharp change in consumer tastes. Tailored men's clothing suffered while "high fashion" fluorished. Measurements of correlation between amounts spent for tailored and "high fashion" goods were obtained from 59 eastern and 43 west coast retailers, producing coefficients of $-.55$ and $-.47$, respectively. Test to see whether the difference between east and west coast preferences is significant at the 5 percent level.

13.27 When the operations of two copper mines were compared, it was found that there was a relationship between the number of heavy equipment operators and the number of minority group employees. Cross-section samples of 53 and 88, respectively, produced correlation coefficients of .75 for the first mine and .86 for the second. Test to see whether the difference between two samples can be attributed to chance (5 percent level of significance).

13.28 A consumers research group recently conducted a survey of the automobile industry to investigate the relationship between implementation of new body styles and current sales. A random sample of 93 dealers from corporation A produced a correlation coefficient of .73 while a random sample of 118 dealers from corporation B produced a correlation coefficient of .97. Test the hypothesis that the two coefficient of .97. Test the hypothesis that the two coefficients are equal at the 5 percent level of significance.

13.29 For a random sample of 30 observations of a bivariate population the following have been calculated:

$$\sum x = 45 \qquad \sum y = 190$$
$$\sum x^2 = 327 \qquad \sum y^2 = 2120$$
$$\sum xy = 755$$

(a) Assuming linearity, compute the regression equations for X on Y and Y on X.
(b) Using the coefficients thus obtained, calculate the correlation coefficient.
(c) Obtain a 95 percent confidence interval for the correlation coefficient.

13.30 For each of the following sets of observations, sketch in the approximate least squares linear regression line and determine by observation whether the residuals indicate the presence of heteroskedasticity and/or autocorrelation.

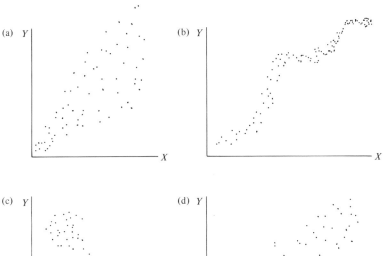

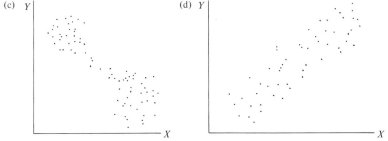

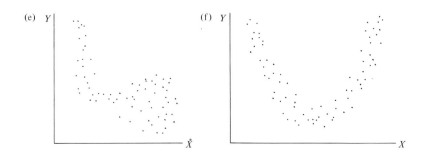

13.31 Write each of the following linearized models in the corresponding nonlinear form. For (e) and (f) sketch the nonlinear equation, assuming positive parameters.

(a) $\ln Y - \ln c + bX$

(b) $\ln(Y - a) = \ln b + c \ln X$

(c) $\dfrac{X}{Y - c} = a + bX$

(d) $\ln(\ln Y - \ln c) = \ln(-\ln k + \ln a)X$

(e) $\ln(Y - k) = \ln b + aX$

(f) $\dfrac{1}{Y} = \dfrac{c}{X} + k$

(g) $\dfrac{X}{Y} = c + kX$

(h) $\ln(Y - k) = \ln b + c \ln X$

13.32 Consider the classical linear model

$$y_i = \beta_0 + \beta_1 X_i + \varepsilon_i$$

with the following assumptions: (i) linearity, (ii) ε_i normally distributed, (iii) homogeneity of variance of ε_i, (iv) ε_i and ε_j uncorrelated for all i and j.

(a) Suppose that least squares estimation provides the regression equation

$$\tilde{y}_i = 138.72 - 2.50x_i$$

with the following residuals $y_i - \tilde{y}_i = e_i$:

$y_i - \tilde{y}_i$
0
0
0
0
4
4
6
6
5
3
0
−3
−2
−2
−2
−2
−3
−5
−3
−3
−2
−1

For each of the assumptions (i) to (iv) state whether the validity of the assumption can be evaluated using *only* the information given; if the validity of an assumption cannot be evaluated, state the reason; if the validity of an assumption can be evaluated, state whether the assumption seems to be valid.

(b) State and test an appropriate hypothesis for evaluating validity of the assumption of uncorrelated errors.

(c) Estimate the first-order autocorrelation for the errors.

(d) Suppose that $\sum (y_i - \bar{y})^2 = 1320$; determine r^2 and test its significance.

13.33 Technical Instruments Corporation has been in operation for eight and one half years and has recently experienced rapid growth. Age of the company in years (represented by X) and annual sales in tens of thousands of current dollars (represented by Y) are:

X	Y
1.5	6.069
2.0	5.719
2.5	32.90
3.0	55.6
3.5	84.91
4.0	189.1
4.5	160.7
5.0	196.9
5.5	179.2
6.0	207.4
6.5	428.9
7.0	655.5
7.5	636.3
8.0	855.0
8.5	1457.0

Make a scatter plot of the data and estimate the appropriate relationship between X and Y.

13.34 The XYZ School Board has interviewed large numbers of parents and teachers of students ages 8 to 18. On the basis of these interviews, average yearly cost of supplies (paper, notebooks, pencils, pens, etc.) has been estimated for students of different ages. The results are given below; X represents age of the student (because of the school board's strict enforcement of the age requirement for starting school and liberal promotion policies, ages are readily translated into grades) and Y represents estimated average yearly cost of supplies.

X	Y
8	34.5
9	54.1
10	68.7
11	79.7
12	95.6
13	102.3
14	115.3
15	122.0
16	133.7
17	138.7
18	142.0

Make a scater plot of the data and estimate the appropriate relationship between X and Y.

13.35 The prospectors' Union keeps detailed records of the sales of purified uranium and has compiled the following data; X represents milligram of impurities per gram and Y represents price per gram.

X	Y
4.922	467
6.670	435
8.295	422
9.750	400
12.420	370
14.000	357
15.080	345
17.955	317
19.140	303
20.400	294
21.960	278

Make a scatter plot of the data and estimate the appropriate relationship between X and Y.

13.36 A particular production process involves the use of two chemicals, one of which is very expensive but also very effective in causing the precipitation on which the process

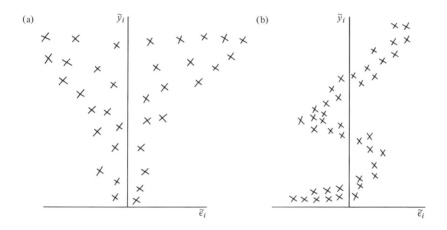

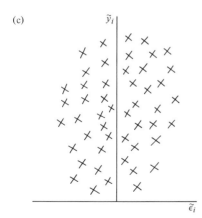

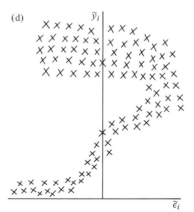

depends. The following data have been obtained for various runs where X represents ratio of the expensive to the less expensive chemical and Y represents grams precipitated under standard conditions.

X	Y
1.5	11.09
2.0	18.04
2.5	53.65
3.0	61.96
3.5	242.1
4.0	420.5
4.5	940.0
5.0	1869.0
5.5	4106.0
6.0	8546.0
6.5	12,400.0
7.0	27,210.0
7.5	69,000.0
8.0	144,800.0
8.5	383,400.0

Make a scatter plot of the data and estimate the appropriate relationship between X and Y.

13.37 For each of the sets of standardized residuals plotted below, state which assumptions of the classical regression model appear(s) to be violated.

13.38 Estimated corporate worth of the Astro-Sonics Corporation at the end of the fiscal year for each of the first 13 years of its existence is given below. X represents year and Y represents corporate worth (in hundreds of thousands of dollars).

X	Y
1	1.634
2	3.563
3	5.338
4	6.296
5	7.521
6	8.366
7	9.099
8	9.352
9	9.586
10	9.690
11	9.777
12	9.843
13	9.957

Make a scatter plot of the data and estimate the appropriate relationship between X and Y.

13.39 Least squares estimation of the model

$$y_i = \beta_0 + \beta_1 x_{1i} + \varepsilon_i$$

resulted in the following predicted values of y_i for the corresponding observed values. State and test an appropriate hypothesis concerning first order autocorrelation after examining the residuals.

y	$\hat{y}$
35	32
26	28
42	40
29	32
27	24
28	31
33	33
31	31
27	27
33	36
35	33
39	43
43	39
34	34
28	27
26	29
35	32

13.40 The Industrial Manufacturers' Association has collected data for 10 representative manufacturers concerning estimated corporate worth and average number of employees for each top-level executive. The results are given below; X represents estimated corporate worth and Y represents average number of employees for each top-level executive.

X	Y
0.8	8.0
1.9	12.1
4.4	15.2
10.0	18.5
21.4	21.7
48.4	25.3
92.5	28.3
218.7	31.9
437.3	35.2
980.0	38.2

Make a scatter plot of the data and estimate the appropriate relationship between X and Y.

13.41 Average price of the stock of Computeronics, Inc., is given below for its first 6 years on the New York Stock Exchange. X represents year on the exchange and Y represents average price.

X	Y
1	34.3
2	56.2
3	71.9
4	83.4
5	92.0
6	100.0

On the basis of a scatter plot of the data, determine the appropriate relationship between X and Y and estimate the parameters.

13.42 The purity of sheet polythene (measured by a standard method) is related to the time it remains in the reactor, if temperature and pressure are held constant. The following data were obtained from one reactor; X represents number of seconds and Y represents purity.

X	Y
1.5	0.890
2.0	0.974
2.5	1.175
3.0	1.096
3.5	1.349
4.0	1.347
4.5	1.417
5.0	1.440
5.5	1.492
6.0	1.519
6.5	1.523
7.0	1.531
7.5	1.538
8.0	1.555
8.5	1.560

On the basis of a scatter plot of the data, determine the appropriate relationship between X and Y and estimate the parameters.

13.43 Professor Practice collected problem assignments from classes for a number of years; students were given 10 points for each assignment completed. Number of points earned on problem assignments and average points earned on the final examination follow. X represents points on problem assignments and Y represents average points on the final examination.

X	Y
0	27
10	39
20	51
30	60
40	64
50	73
60	83
70	90
80	93
90	99
100	105
110	110
120	111
130	113
140	117
150	120
160	125
170	130
180	133
190	133
200	134
210	138
220	140
230	145
240	146

On the basis of a scatter plot of the data, determine the appropriate relationship between X and Y and estimate the parameters.

13.44 For each of the following sets of data, make a scatter plot and estimate the appropriate relationship between X and Y.

(a) X	Y		(b) X	Y
3	83.1		3	22.0
4	80.0		4	36.8
5	77.5		5	61.9
6	75.6		6	99.1
7	73.8		7	167.6
8	72.4		8	277.1
9	71.1		9	453.3
10	70.1		10	740.2
11	69.0		11	1247.1
12	67.9		12	2018.4

(c) X	Y		(d) X	Y
3	.090		1	35.3
4	.211		3	56.0
5	.469		5	71.7
6	.313		7	83.5
7	.834		9	89.7
8	1.223		11	93.9
9	1.517		13	96.0
10	1.878		15	98.9
11	5.447		17	99.5
12	12.198		19	98.4

(e) X	Y		(f) X	Y
3	44.2		3	49.0
4	79.7		4	93.6
5	127.0		5	154.2
6	177.7		6	223.0
7	248.0		7	328.5
8	324.8		8	475.4
9	407.9		9	619.6
10	498.8		10	785.8
11	616.7		11	993.5
12	720.5		12	1301.2
13	843.0		13	1588.3
14	991.5		14	1836.8
15	1116.7		15	2173.0
16	1292.8		16	2585.5
17	1432.4		17	2950.6
18	1605.0		18	3557.7

MULTIPLE REGRESSION ANALYSIS

14.1 INTRODUCTION

Multiple regression analysis can be viewed as an extension or generalization of bivariate regression analysis. In *multiple regression analysis* there are two or more regressor variables and the dependent variable is predicted as a linear function of these regressor variables. In most applications the dependent variable can be assumed to be related to several other variables. For example, price of a given security is a function of its earnings per share during the past year, its profit margins during recent months, various general economic conditions, and so forth. Profits of a company are a function of sales volume, advertising expenses, production expenses, and so forth. Thus multiple regression is usually a more realistic model than bivariate regression. Estimation of the parameters of a multiple regression equation and interpretation of the results are discussed in this chapter.

There are several problems associated with multiple regression analysis that do not occur in bivariate regression analysis. The most important of these problems are:

1. A multiple regression model and its parameter estimates can be expressed conveniently only in terms of matrices, for the general case. For practical purposes, a computer is required to obtain parameter estimates and to perform tests of significance and other analyses for multiple regression. The simplest case of multiple linear regression, in which there are two regressor variables, can be analyzed without matrix algebra and computer assistance and is discussed in detail to illustrate concepts and problems. The general case is discussed briefly in terms of computer printouts.
2. A multiple regression analysis frequently involves selection of the regressor variables to be included in the regression equation. This is a somewhat complicated problem for which several procedures have been proposed. These procedures are summarized, but not discussed in detail.

3. A multiple regression equation is considerably more complicated to interpret than a bivariate regression equation, partly because of the limitation of graphical representation to at most three dimensions. In addition, interpretation of a multiple regression equation is complicated by the fact that the regressor variables are usually correlated among themselves and thus their contributions to the prediction of the dependent variable are not independent. This problem, referred to as multicollinearity, is discussed briefly.

14.2 THE GENERAL CASE: k REGRESSOR VARIABLES

The model for multiple linear regression assumes that a dependent variable, Y, is linearly related to several regressor variables $X_1, X_2, \ldots, X_k$. The model is thus assumed to be of the form

$$Y_i = \beta_0 + \beta_1 X_{1i} + \beta_2 X_{2i} + \cdots + \beta_k X_{ki} + \varepsilon_i, \qquad i = 1, 2, \ldots, n$$

where, as in the bivariate case, the errors ε_i are assumed to be independently normally distributed with common variance σ^2. Note that each X_j, $j = 1, 2, \ldots, k$, has a value for each Y_i, $i = 1, 2, \ldots, n$ and that $k < n$.

The least squares estimates of $\beta_0, \beta_1, \ldots, \beta_k$, denoted by $\hat{\beta}_0, \hat{\beta}_1, \ldots, \hat{\beta}_k$, are obtained by minimizing the sum of squared errors

$$\sum_{i=1}^{n} (y_i - \hat{\beta}_0 - \hat{\beta}_1 x_{1i} - \hat{\beta}_2 x_{2i} - \cdots - \hat{\beta}_k x_{ki})^2$$

where the sets of observations on $Y, X_1, X_2, \ldots, X_k$ for each of the n items or individuals are represented, respectively, by

$$y_1, x_{11}, x_{21}, \ldots, x_{k1}$$

$$y_2, x_{12}, x_{22}, \ldots, x_{k2}$$

$$\cdot \quad \cdot \quad \cdot \quad \quad \cdot$$

$$\cdot \quad \cdot \quad \cdot \quad \quad \cdot$$

$$\cdot \quad \cdot \quad \cdot \quad \quad \cdot$$

$$y_n, x_{1n}, x_{2n}, \ldots, x_{kn}$$

The least squares estimates $\hat{\beta}_0, \hat{\beta}_1, \ldots, \hat{\beta}_k$ are in fact the solution of the following set of equations that result from minimizing the sum of squared errors:*

---

* The sum of squared errors is differentiated partially with respect to $\hat{\beta}_i$, $i = 0, 1, 2, \ldots, k$, and these derivatives are set equal to zero to obtain the normal equations that are solved to obtain least squares parameter estimates.

$$n\hat{\beta}_0 \qquad + \hat{\beta}_1 \sum x_{1i} \quad + \hat{\beta}_2 \sum x_{2i} \quad + \cdots + \hat{\beta}_k \sum x_{ki} \quad = \sum y_i$$

$$\hat{\beta}_0 \sum x_{1i} + \hat{\beta}_1 \sum x_{1i}^2 \quad + \hat{\beta}_2 \sum x_{1i}x_{2i} + \cdots + \hat{\beta}_k \sum x_{1i}x_{ki} = \sum x_{1i}y_i$$

$$\hat{\beta}_0 \sum x_{2i} + \hat{\beta}_1 \sum x_{2i}x_{1i} + \hat{\beta}_2 \sum x_{2i}^2 \quad + \cdots + \hat{\beta}_k \sum x_{2i}x_{ki} = \sum x_{2i}y_i$$

$$\vdots \qquad\qquad \vdots \qquad\qquad \vdots \qquad\qquad \vdots \qquad\qquad \vdots$$

$$\hat{\beta}_0 \sum x_{ki} + \hat{\beta}_1 \sum x_{ki}x_{1i} + \hat{\beta}_2 \sum x_{ki}x_{2i} + \cdots + \hat{\beta}_k \sum x_{ki}^2 \quad = \sum x_{ki}y_i$$

Note that the second equation is the first equation multiplied by x_{1i} before summation, the third equation is the first equation multiplied by x_{2i} before summation, and so forth. These equations and their solutions can be stated more concisely in terms of matrices.

For k even as large as 4 or 5, solving these equations without using a computer is tedious and the results are likely to be inaccurate due to rounding errors. This is particularly true for problems that require not only the computation and evaluation of the regression equation, but also the selection of the regressor variables to be included in the equation. For this reason and because multiple regression programs are readily available at most computing centers, the general multivariate case is discussed in a later section in the context of computer printouts.

In order to illustrate and clarify some of the concepts and procedures involved in multiple regression analysis, we discuss the special case $k = 2$ next.

14.3 THE SPECIAL CASE OF TWO REGRESSOR VARIABLES

The linear regression model with two regressor variables is given by

$$\tilde{y}_i = \beta_0 + \beta_1 x_{1i} + \beta_2 x_{2i} + \varepsilon_i$$

with the usual assumptions concerning ε_i. The method of least squares provides estimates $\hat{\beta}_0, \hat{\beta}_1$, and $\hat{\beta}_2$ which minimize the sum of squared deviations from prediction

$$\sum_{i=1}^{n} (y_i - \tilde{y}_i)^2 = \sum_{i=1}^{n} [y_i - (\hat{\beta}_0 + \hat{\beta}_1 x_{1i} + \hat{\beta}_2 x_{2i})]^2$$

The three equations that must be solved to obtain $\hat{\beta}_0, \hat{\beta}_1$, and $\hat{\beta}_2$ are

$$\hat{\beta}_0 + \hat{\beta}_1 \sum x_{1i} + \hat{\beta}_2 \sum x_{2i} = \sum y_i$$

$$\hat{\beta}_0 \sum x_{1i} + \hat{\beta}_1 \sum x_{1i}^2 + \hat{\beta}_2 \sum x_{1i}x_{2i} = \sum x_{1i}y_i$$

$$\hat{\beta}_0 \sum x_{2i} + \hat{\beta}_1 \sum x_{1i}x_{2i} + \hat{\beta}_2 \sum x_{2i}^2 = \sum x_{2i}y_i$$

For any given set of data the required sums of squares and cross products can be calculated and $\hat{\beta}_0$, $\hat{\beta}_1$, and $\hat{\beta}_2$ can be determined by solving the three equations obtained by substituting these calculated values into the equations above. The three equations can also be solved algebraically, most conveniently in matrix notation. For the special case of two regressor variables the least squares estimates $\hat{\beta}_0, \hat{\beta}_1$, and $\hat{\beta}_2$ can be written in terms of correlation coefficients and standard deviations as follows:

$$\hat{\beta}_1 = \frac{r_{y1} - r_{y2}r_{12}}{1 - r_{12}^2} \cdot \frac{s_y}{s_1}$$

$$\hat{\beta}_2 = \frac{r_{y2} - r_{y1}r_{12}}{1 - r_{12}^2} \cdot \frac{s_y}{s_2}$$

$$\hat{\beta}_0 = \frac{\sum\limits_{i=1}^{n} y_i - \hat{\beta}_1 \sum\limits_{i=1}^{n} x_{1i} - \hat{\beta}_2 \sum\limits_{i=1}^{n} x_{2i}}{n} = \bar{y} - \hat{\beta}_1\bar{x}_1 - \hat{\beta}_2\bar{x}_2$$

where r_{y1}, r_{y2}, and r_{12} are the correlation coefficients for y and x_1, y and x_2, and x_1 and x_2, respectively, and s_y, s_1, and s_2 are the standard deviations of the observations of y_i, x_{1i}, and x_{2i}, respectively.

Because of rounding errors, these equations generally do not provide very accurate parameter estimates. However, the equations are useful for considering relationships among the estimates of β_0, β_1, and β_2 and various correlation coefficients and are discussed for this purpose. Note, for example, that $\hat{\beta}_1$ is an increasing function of r_{y1} and s_y and a decreasing function of r_{y2} and s_1; $\hat{\beta}_2$ is an increasing function of r_{y2} and s_y and a decreasing function of r_{y1} and s_2.

It is important to understand clearly that $\hat{\beta}_0, \hat{\beta}_1$, and $\hat{\beta}_2$ are estimated jointly. The least squares estimates $\hat{\beta}_0, \hat{\beta}_1$, and $\hat{\beta}_2$ are such that, used together in a regression equation, they minimize the sum of squared errors of prediction. In general, these coefficients are *not* the same as would be obtained for individual bivariate linear regressions. The relationships just noted emphasize the non-independences of the multiple regression coefficients. Note that this nonin-dependence, discussed in some detail later, occurs because x_1 and x_2 are not independent. For the special case of independence, $r_{12} = 0$, the values of $\hat{\beta}_1$ and $\hat{\beta}_2$ are the same when they are estimated jointly as when they are estimated in individual bivariate regressions.

It is sometimes convenient to express the variables in a multiple regression equation in standardized or coded form as follows:

$$\frac{\tilde{y} - \bar{y}}{s_y} = \hat{\beta}_1^* \frac{x_1 - \bar{x}_1}{s_1} + \hat{\beta}_2^* \frac{x_2 - \bar{x}_2}{s_2}$$

Note that $\hat{\beta}_0^* = 0$, $\hat{\beta}_1^* = \hat{\beta}_1(s_1/s_y)$, and $\hat{\beta}_2^* = \hat{\beta}_2(s_2/s_y)$. Standardization of the variables in a regression analysis facilitates comparison of the regression coefficients and reduces rounding errors. Since standardization is accomplished by a linear transformation of the variable, the proportion of variance accounted for by the regression is unchanged.

The standard errors of the estimated standardized regression coefficients are given by

$$s_{\hat{\beta}_1^*} = s_{\hat{\beta}_2^*} = \left[\frac{1 + 2r_{y1}r_{y2}r_{12} - r_{y1}^2 - r_{y2}^2 - r_{12}^2}{(n-3)(1-r_{12}^2)^2} \right]^{1/2} = \left[\frac{1 - R_{y \cdot 12}^2}{(n-3)(1-r_{12}^2)} \right]^{1/2}$$

and the standard errors of the estimated (nonstandardized) regression coefficients are given by

$$s_{\hat{\beta}_1} = s_{\hat{\beta}_1*} \cdot \frac{s_y}{s_1} \quad \text{and} \quad s_{\hat{\beta}_2} = s_{\hat{\beta}_2*} \cdot \frac{s_y}{s_2}$$

MULTIPLE CORRELATION

The *coefficient of multiple correlation* or the *multiple correlation coefficient* is defined as

$$r_{y \cdot 12} = \pm \sqrt{1 - \frac{\sum_{i=1}^{n} (y_i - \tilde{y}_i)^2}{\sum_{i=1}^{n} (y_i - \bar{y})^2}} = \pm \sqrt{1 - \frac{\text{unexplained variance}}{\text{total variance}}}$$

$$= \pm \sqrt{\frac{\text{explained variance}}{\text{total variance}}}$$

where $\tilde{y}_i = \hat{\beta}_0 + \hat{\beta}_1 x_{1i} + \hat{\beta}_2 x_{2i}$ and $r_{y \cdot 12}$ denotes the multiple correlation of y with x_1 and x_2. The value of $r_{y \cdot 12}^2$ is the proportion of the variance in y that is attributable (jointly) to variance in x_1 and x_2. It is customary to denote $r_{y \cdot 12}^2$ by $R_{y \cdot 12}^2$ or, if there can be no confusion concerning which regression equation is involved, by R^2.

The standardized regression coefficients $\hat{\beta}_1*$ and $\hat{\beta}_2*$ are related to $r_{y \cdot 12}$ by the equation

$$r_{y \cdot 12} = \sqrt{r_{y1}\hat{\beta}_1* + r_{y2}\hat{\beta}_2*}$$

Since

$$\hat{\beta}_1* = \hat{\beta}_1 \frac{s_1}{s_y} \quad \text{and} \quad \hat{\beta}_2* = \hat{\beta}_2 \frac{s_2}{s_y}$$

$$r_{y \cdot 12} = \sqrt{\frac{r_{y1}\hat{\beta}_1 s_1 + r_{y2}\hat{\beta}_2 s_2}{s_y}}$$

or, substituting for $\hat{\beta}_1$ and $\hat{\beta}_2$, we get

$$r_{y \cdot 12} = \sqrt{\frac{r_{y1}^2 + r_{y2}^2 - 2r_{y1}r_{y2}r_{12}}{1 - r_{12}^2}}$$

The standard error of estimate, an estimate of σ, is given by

$$s = s_y \sqrt{\frac{1 - r_{y1}^2 - r_{y2}^2 - r_{12}^2 + 2r_{y1}r_{y2}r_{12}}{1 - r_{12}^2}}$$

and is sometimes denoted by $s_{y \cdot 12}$.

EXAMPLE
On the basis of the following computational information, estimate the parameters of the model $\tilde{y}_i = \beta_0 + \beta_1 x_{1i} + \beta_2 x_{2i} + \varepsilon_i$ and evaluate the regression equation.

$$\sum_{i=1}^{50} y_i = 50 \qquad \sum_{i=1}^{50} x_{1i} = 50 \qquad \sum_{i=1}^{50} x_{2i} = 100$$

$$\sum_{i=1}^{50} y_i^2 = 100 \qquad \sum_{i=1}^{50} x_{1i}^2 = 75 \qquad \sum_{i=1}^{50} x_{2i}^2 = 250$$

$$\sum_{i=1}^{50} x_{1i}y_i = 80 \qquad \sum_{i=1}^{50} x_{2i}y_i = 120 \qquad \sum_{i=1}^{50} x_{1i}x_{2i} = 125$$

The least squares estimates $\hat{\beta}_0$, $\hat{\beta}_1$, and $\hat{\beta}_2$ can be obtained by solving the following set of normal equations:

$$50\hat{\beta}_0 + 50\hat{\beta}_1 + 100\hat{\beta}_2 = 50$$
$$50\hat{\beta}_0 + 75\hat{\beta}_1 + 125\hat{\beta}_2 = 80$$
$$100\hat{\beta}_0 + 125\hat{\beta}_1 + 250\hat{\beta}_2 = 120$$

Eliminating $\hat{\beta}_0$ from the first and second equations, we get

$$25\hat{\beta}_1 + 25\hat{\beta}_2 = 30$$
$$25\hat{\beta}_1 + 50\hat{\beta}_2 = 20$$

and

$$\hat{\beta}_2 = -.4$$

By substitution,

$$\hat{\beta}_1 = 1.6$$
$$\hat{\beta}_0 = .2$$

Thus the estimated regression equation is

$$\tilde{y}_i = .2 + 1.6x_{1i} - .4x_{2i}$$

The pairwise correlation coefficients can be obtained as follows:

$$r_{y1}^2 = \frac{[\sum (y_i - \bar{y})(x_{1i} - \bar{x}_1)]^2}{\sum (y_i - \bar{y})^2 \sum (x_{1i} - \bar{x}_1)^2} = \frac{(30)^2}{(50)(25)} = .72$$

$$r_{y1} \approx .849$$

$$r_{y2}^2 = \frac{[\sum (y_i - \bar{y})(x_{2i} - \bar{x}_2)]^2}{\sum (y_i - \bar{y})^2 \sum (x_{2i} - \bar{x}_2)^2} = \frac{(20)^2}{(50)(50)} = .16$$

$$r_{y2} = .4$$

$$r_{12}^2 = \frac{[\sum (x_{1i} - \bar{x}_1)(x_{2i} - \bar{x}_2)]^2}{\sum (x_{1i} - \bar{x}_1)^2 \sum (x_{2i} - \bar{x}_2)^2} = \frac{(25)^2}{(25)(50)} = .5$$

$$r_{12} \approx .707$$

The least squares parameter estimates can also be obtained using the formulas involving correlation coefficients and standard deviations as follows:

$$\hat{\beta}_1 = \frac{r_{y1} - r_{y2}r_{12}}{1 - r_{12}^2} \cdot \frac{s_y}{s_1} = \frac{.849 - (.4)(.707)}{1 - .5}(\sqrt{2}) = 1.60$$

$$\hat{\beta}_2 = \frac{r_{y2} - r_{y1}r_{12}}{1 - r_{12}^2} \cdot \frac{s_y}{s_1} = \frac{.4 - (.849)(.707)}{1 - .5}(1) = -.40$$

$$\hat{\beta}_0 = \bar{y} - \hat{\beta}_1\bar{x}_1 - \hat{\beta}_2\bar{x}_2 = 1 - 1.60 + .80 = .20$$

Note that these estimates are the same as those obtained above by solving the normal equations.

The coefficient of determination is

$$R_{y \cdot 12}^2 = \frac{r_{y1}^2 + r_{y2}^2 - 2r_{y1}r_{y2}r_{12}}{1 - r_{12}^2} = \frac{.72 + .16 - 2(.849)(.4)(.707)}{.5} = .80$$

and the standard errors of $\hat{\beta}_1$ and $\hat{\beta}_2$ are:

$$s_{\hat{\beta}_1} = \left[\frac{1 - R_{y \cdot 12}^2}{(1 - r_{12}^2)(n - 3)} \right]^{1/2} \cdot \frac{s_y}{s_1} = \left[\frac{1 - .8}{(.5)(47)} \right]^{1/2} (2)^{1/2} = .130$$

and

$$s_{\hat{\beta}_2} = \left[\frac{1 - R_{y \cdot 12}^2}{(1 - r_{12}^2)(n - 3)} \right]^{1/2} \cdot \frac{s_y}{s_2} = \left[\frac{1 - .8}{(.5)(47)} \right]^{1/2} (1) = .092$$

Note that $R_{y \cdot 12}^2$ can also be obtained using the formula

$$R_{y \cdot 12}^2 = r_{y1}\hat{\beta}_1{}^* + r_{y2}\hat{\beta}_2{}^* = (.849)(1.132) + (-.4)(.4005) = .80$$

which is the same value obtained above.

These estimation results can be summarized as follows:

$$\tilde{y}_i = .2 + 1.6x_{1i} - .4x_{2i} \qquad R^2 = .80$$

$$(.130) \qquad (.092)$$

where the numbers in parentheses are the standard errors of the corresponding parameter estimates.

PARTIAL CORRELATION

In some cases, interpretation of the relationship between two variables, as indicated by their correlation coefficient, can be complicated by their dependence on one or more other variables. The *coefficient of partial correlation* or the *partial correlation coefficient* between two variables measures their relationship after the effects of one or more other variables have been taken into account or "partialed out." For example,

$$r_{y1 \cdot 2} = \frac{r_{y1} - r_{y2}r_{12}}{\sqrt{(1 - r_{y2}^2)(1 - r_{12}^2)}}$$

denotes the partial correlation coefficient of y and x_1 with respect to x_2, or the correlation of y and x_1 partialed on x_2.

If the least squares regressions of both y on x_2 and x_1 on x_2 are computed, then $r_{y1 \cdot 2}$ is the correlation coefficient of the corresponding errors of prediction; that is,

$$r_{y1 \cdot 2} = r_{y - \tilde{y}, x_1 - \tilde{x}_1} = \frac{\sum_{i=1}^{n} (y_i - \tilde{y}_i)(x_{1i} - \tilde{x}_{1i})}{\sqrt{\sum_{i=1}^{n} (y_i - \tilde{y}_i)^2 \sum_{i=1}^{n} (x_{1i} - \tilde{x}_{1i})^2}}$$

where

$$\tilde{y} = \hat{\beta}_0 + \hat{\beta}_1 x_2$$
$$\tilde{x}_1 = \hat{\beta}_0' + \hat{\beta}_1' x_2$$

and $\hat{\beta}_0$ and $\hat{\beta}_1$ are the least squares coefficients for the regression of y on x_2 and $\hat{\beta}_0'$ and $\hat{\beta}_1'$ are the least squares coefficients for the regression of x_1 on x_2. Although this approach is clearly not computationally practical, it provides a mathematical basis for the statement that $r_{y1 \cdot 2}$ is the correlation of y and x_1, after the effects of x_2 have been taken into account.

Additional subscripts following the dot indicate additional variables that are partialed out, that is, variables that are included in the regression equations from which the deviations to be correlated are taken. For example,

$$r_{y1 \cdot 23} = r_{y - \tilde{y}, x_1 - \tilde{x}_1} = \frac{\sum_{i=1}^{n} (y_i - \tilde{y}_i)(x_{1i} - \tilde{x}_{1i})}{\sqrt{\sum_{i=1}^{n} (y_i - \tilde{y}_i)^2 \sum_{i=1}^{n} (x_{1i} - \tilde{x}_{1i})^2}}$$

where

$$\tilde{y} = \hat{\beta}_0 + \hat{\beta}_2 x_2 + \hat{\beta}_3 x_3$$
$$\tilde{x}_1 = \hat{\beta}_0' + \hat{\beta}_2' x_2 + \hat{\beta}_3' x_3$$

are the least squares equations for predicting y and x_1, respectively, from x_2 and x_3.

In partial correlation the subscripts preceding the dot are referred to as *primary subscripts*, those following the dot are referred to as *secondary subscripts*. The number of secondary subscripts is the *order* of the partial correlation coefficient; thus $r_{y1 \cdot 2}$ is a first-order partial correlation coefficient, $r_{y1 \cdot 23}$ is a second-order partial correlation coefficient, and so forth; r_{y1} is a zero-order correlation coefficient.

Any partial correlation coefficient can be obtained from partial correlation coefficients of the next lowest order. As noted,

$$r_{y1 \cdot 2} = \frac{r_{y1} - r_{y2} r_{12}}{\sqrt{(1 - r_{y2}^2)(1 - r_{12}^2)}}$$

If additional secondary subscripts are added to every r in the aforementioned formula, then the formula for a higher order partial correlation coefficient is obtained. For example,

$$r_{y1 \cdot 23} = \frac{r_{y1 \cdot 3} - r_{y2 \cdot 3} r_{12 \cdot 3}}{\sqrt{(1 - r_{y2 \cdot 3}^2)(1 - r_{12 \cdot 3}^2)}}$$

(Note that the order of the secondary subscripts does not affect the value of a partial correlation coefficient; i.e., $r_{y1 \cdot 23} = r_{y1 \cdot 32}$). Thus it is possible to obtain partial correlation coefficients of any order by working up from zero-order coefficients. However, except for partials of low order, this method is not only laborious but involves considerable rounding error. Also, it is possible to obtain a partial correlation coefficient as the geometric mean of the corresponding $\hat{\beta}$'s or $\hat{\beta}^*$'s. For example,

$$r_{y1 \cdot 2}^2 = \hat{\beta}_{y1 \cdot 2}\hat{\beta}_{1y \cdot 2} = \hat{\beta}_{y1 \cdot 2}^*\hat{\beta}_{1y \cdot 2}^*$$

where $\hat{\beta}_{y1 \cdot 2}$ is the regression coefficient of x_1 in the least squares equation for predicting y from x_1 and x_2 and $\hat{\beta}_{1y \cdot 2}$ is the regression coefficient of y in the least squares equation for predicting x_1 from y and x_2; $\hat{\beta}_{y1 \cdot 2}^*$ and $\hat{\beta}_{1y \cdot 2}^*$ are the corresponding standardized regression coefficients. This method also involves considerable rounding error and, in practice, partial correlation coefficients are usually obtained using matrix operations performed on a computer. However, the formula does emphasize the fact that the regression parameter estimates in a multiple regression equation are partial regression coefficients, that is, regression coefficients estimated when one or more other regressor variables are also included in the equation.

NOTE
The tests of significance previously discussed for zero-order correlation coefficients are appropriate for partial correlation coefficients, if the number of degrees of freedom is decreased by one for each secondary subscript.

RELATION OF PARTIAL TO ZERO-ORDER AND MULTIPLE CORRELATION COEFFICIENTS

A multiple correlation coefficient is as large as or larger in absolute value than any of the partial correlation coefficients having the same primary subscript. For example, $r_{y \cdot 1234}^2$ is as large as or larger than any of the squared partial correlation coefficients $r_{y1 \cdot 234}^2, r_{y2 \cdot 134}^2, r_{y3 \cdot 124}^2$ or $r_{y4 \cdot 123}^2$. Thus the set of partial correlation coefficients that have as a primary subscript the primary subscript of the multiple correlation coefficient provide a lower bound for the multiple correlation coefficient. This is equivalent to saying that addition of a variable to a regression cannot decrease the proportion of variance accounted for by the regression. Even if a variable does not contribute to the prediction of the dependent variable, its addition cannot decrease the predictability associated with the simpler model.

Note that there is no necessary relationship between the multiple correlation coefficient $r_{y \cdot 1234}$ and partial correlation coefficients not having y as a primary subscript, for example, $r_{12 \cdot 34}$ or $r_{24 \cdot 13}$.

The relations between zero-order and partial correlation coefficients are not so simple as for partial and multiple correlation coefficients. In practice, it frequently happens that a zero-order correlation coefficient is larger than a partial correlation coefficient having the same primary subscripts, but this is not always the case. For example, in most samples, number of years married and net worth tend to be highly positively correlated; however, if age is partialed out, the correlation decreases, possibly even becoming negative. Suppose, for example,

$$x_1 = \text{years married} \qquad r_{12} = .80$$

$$x_2 = \text{net worth} \qquad r_{13} = .70$$

$$x_3 = \text{age} \qquad r_{23} = .90$$

then

$$r_{12 \cdot 3} = \frac{r_{12} - r_{13}r_{23}}{\sqrt{(1 - r_{13}^2)(1 - r_{23}^2)}} = \frac{.8 - (.7)(.9)}{\sqrt{(1 - .49)(1 - .81)}} = .55$$

and

$$r_{12 \cdot 3} < r_{12}$$

In such cases the zero-order coefficient is spuriously inflated through correlation of both primary variables with a third variable. That is, the correlation between years married and net worth is decreased if the effect of age is taken into account.

Occasionally a zero-order correlation coefficient is larger than a partial correlation coefficient having the same primary subscripts. For example, education and income tend to be positively correlated and if age is partialed out, then the correlation increases. Suppose for example,

$$x_1 = \text{education} \qquad r_{12} = .70$$

$$x_2 = \text{income} \qquad r_{13} = -.40$$

$$x_3 = \text{age} \qquad r_{23} = .30$$

then

$$r_{12 \cdot 3} = \frac{r_{12} - r_{13}r_{23}}{\sqrt{(1 - r_{13}^2)(1 - r_{23}^2)}} = \frac{.7 - (.4)(.3)}{\sqrt{(1 - .16)(1 - .09)}} = .93$$

and

$$r_{12 \cdot 3} > r_{12}$$

In such cases the zero-order coefficient is deflated or supressed by the nature of the correlation of both variables with a third variable. That is, the correlation between education and income is increased if the effects of age are taken into account.

The three correlation coefficients r_{12}, r_{13}, and r_{23} must be related in such a way that none of the partial coefficients exceeds 1 in numerical value. This requirement is equivalent to the inequality

$$1 + 2r_{12}r_{13}r_{23} \geq r_{12}^2 + r_{13}^2 + r_{23}^2$$

where equality holds if $r_{12} = r_{13} = r_{23} = 1$, or, equivalently,

$$1 - r_{12}^2 - r_{13}^2 - r_{23}^2 + 2r_{12}r_{13}r_{23} \geq 0$$

(note that this inequality is satisfied in the two examples above).

Clearly the interpretation of correlation coefficients is not at all simple and the investigator must consider carefully the nature of the relationship in which he is interested. In some cases a zero-order correlation coefficient is appropriate; in other cases a partial correlation coefficient is more meaningful—it depends entirely on the purpose for which the data were obtained and the interpretation to be given to the correlation coefficient.

For the example data, $r_{y1} = .849$, $r_{y2} = .4$, and $r_{12} = .707$; thus

$$r_{y1 \cdot 2} = \frac{r_{y1} - r_{y2}r_{12}}{\sqrt{(1 - r_{y2}^2)(1 - r_{12}^2)}} = \frac{.849 - (.4)(.707)}{\sqrt{(.84)(.5)}} = .874$$

$$r_{y2 \cdot 1} = \frac{r_{y2} - r_{y1}r_{12}}{\sqrt{(1 - r_{y1}^2)(1 - r_{12}^2)}} = \frac{.4 - (.849)(.707)}{\sqrt{(.28)(.5)}} = -.535$$

Although r_{y1} and $r_{y1 \cdot 2}$ do not differ appreciably, r_{y2} and $r_{y2 \cdot 1}$ differ not only in value but in sign. This results from the fact that r_{y1} is substantially larger than r_{y2} and r_{12} is relatively large. Note that the signs of the regression coefficients in a multiple regression always have the same sign as the corresponding partial correlation coefficients.

EVALUATING THE REGRESSION EQUATION

The regression equation $\tilde{y}_i = \hat{\beta}_0 + \hat{\beta}_1 x_{1i} + \hat{\beta}_2 x_{2i}$ can be evaluated with respect to the individual regression coefficients and with respect to the overall fit of the regression equation. It can be shown that

$$\frac{\beta_i - \hat{\beta}_i}{s_{\hat{\beta}_i}} \sim t_{n-3} \qquad i = 1, 2, 3$$

The hypothesis $\beta_i = 0$ (or some other constant) can be tested on the basis of this t statistic. Again note that $\hat{\beta}_0$, $\hat{\beta}_1$, and $\hat{\beta}_2$ are estimated jointly and are not independent. Thus a test of the hypothesis $H_0 : \beta_i = 0$ is actually a test of the significance of the partial regression coefficient β_i. In addition, the significance level is affected when more than one significance test is performed. This is perhaps not serious when there are two regressor variables, but it can be a problem when a larger number of regressor variables are included in the equation.

The overall fit of the regression equation is evaluated on the basis of the associated coefficient of determination, $R_{y \cdot 12}^2$. It can be shown that

$$\frac{\dfrac{R_{y \cdot 12}^2}{2}}{\dfrac{1 - R_{y \cdot 12}^2}{n - 3}} \sim F_{2, n-3}$$

The hypothesis $\rho_{y \cdot 12} = 0$ can be tested on the basis of this F statistic. This is a test of the significance of the overall regression equation, that is, a test of the null hypothesis that the overall regression equation does not account for a significant proportion of the variation in y.

For the case of two regressor variables the F statistic can also be written as

$$F = \frac{t_1^2 + t_2^2 + 2t_1 t_2 r_{12}}{2(1 - r_{12}^2)}$$

where $t_1 = \hat{\beta}_1 / s_{\hat{\beta}_1}$ and $t_2 = \hat{\beta}_2 / s_{\hat{\beta}_2}$. This equation again indicates the importance of the degree of correlation between X_1 and X_2. Note that if $r_{12} = 0$, then $F = \frac{1}{2}(t_1{}^2 + t_2{}^2)$. If $r_{12} \neq 0$, then the value of F depends on r_{12} as well as on $t_1{}^2$ and $t_2{}^2$.

The goodness of fit of the regression equation can be evaluated in more detail by considering the conditional contribution of each regressor variable. Suppose the following least squares equations are estimated:

$$\tilde{y}_i = \hat{\beta}_0 + \hat{\beta}_1 x_{1i} \qquad\qquad R_{y1}{}^2$$

$$\tilde{y}_i = \hat{\beta}_0' + \hat{\beta}_2 x_{2i} \qquad\qquad R_{y2}{}^2$$

$$\tilde{y}_i = \hat{\beta}_0'' + \hat{\beta}_1' x_{1i} + \hat{\beta}_2' x_{2i} \qquad R_{y \cdot 12}^2$$

where R_{y1}, R_{y2}, and $R_{y \cdot 12}$ denote the associated coefficients of determination. Note that unless $r_{12} = 0$, $\hat{\beta}_1 \neq \hat{\beta}_1'$ and $\hat{\beta}_2 \neq \hat{\beta}_2'$, since $\hat{\beta}_1'$ and $\hat{\beta}_2'$ are conditional regression coefficients. The proportion of variance accounted for by adding x_2 given that x_1 is already included in the regression equation is given by

$$R_{y \cdot 12}^2 - R_{y1}{}^2$$

Similarly, the proportion of variance accounted for by adding x_1 given that x_2 is already included in the regression equation is given by

$$R_{y \cdot 12}^2 - R_{y2}{}^2$$

The statistical significance of the additional variable x_2 can be tested using the F statistic

$$F_{1, n-3} = \frac{R_{y \cdot 12}^2 - R_{y1}{}^2}{\dfrac{1 - R_{y \cdot 12}^2}{n - 3}}$$

and similarly for the addition of x_1 given that x_2 is already included in the regression equation.

For the example given on p. 449,

$$\tilde{y}_i = .2 + 1.6 x_{1i} - .4 x_{2i} \qquad R^2 = .80$$

$$(.130) \qquad (.092)$$

Significance tests of the individual coefficients can be performed using t statistics.

For $H_0 : \beta_1 = 0$

$$t_1 = \frac{\hat{\beta}_1}{s_{\hat{\beta}_1}} = \frac{1.60}{.130} = 12.308$$

For $H_0 : \beta_2 = 0$

$$t_2 = \frac{\hat{\beta}_2}{s_{\hat{\beta}_2}} = \frac{-.40}{.092} = -4.348$$

Since the critical value of t with 47 degrees of freedom is approximately 2.7 for

$\alpha = .01$, both hypotheses are rejected; that is, both regression coefficients are significant.

The overall regression equation can be evaluated using the F statistic

$$F = \frac{\dfrac{R^2_{y \cdot 12}}{2}}{\dfrac{1 - R^2_{y \cdot 12}}{n - 3}} = \frac{\dfrac{.80}{2}}{\dfrac{.2}{47}} = 94$$

The critical value of $F_{2, 47}$ is approximately 5.1 for $\alpha = .01$, so the hypothesis $\rho_{y \cdot 12} = 0$ is rejected; that is, the regression equation accounts for a significant proportion of variance in Y.

Note that

$$F = \frac{t_1^2 + t_2^2 + 2t_1 t_2 r_{12}}{2(1 - r_{12}^2)} = \frac{151.479 + 18.904 - 75.670}{2(.5)} \approx 94.713$$

which differs from the above result only due to rounding.

The contributions of x_1 and x_2 to the regression equation can also be evaluated by considering the regression equations

$$\tilde{y}_i = -.2 + 1.2 x_1 \qquad r^2 = .72$$

and

$$\tilde{y}_i = .2 + .4 x_2 \qquad r^2 = .16$$

Note that the coefficients of x_1 and x_2 are not the same as in the multiple regression equation and, in particular, the coefficient of x_2 has a different sign. Again note that the coefficients are the same only if $r_{12} = 0$.

The significance of the contribution of x_2 to the regression equation, given that x_1 is included, can be tested using the F statistic

$$F = \frac{R^2_{y \cdot 12} - R_{y1}^2}{\dfrac{1 - R^2_{y \cdot 12}}{n - 3}} = \frac{.80 - .72}{\dfrac{.20}{47}} = 18.8$$

The critical value of $F_{1, 47}$ for $\alpha = .01$ is approximately 7.2, so the conditional contribution of x_2 to the regression is significant.

14.4 SOME GENERALIZATIONS FOR $k > 2$

The concepts discussed for the case $k = 2$ can be generalized for $k > 2$, although for practical purposes matrix notation is required to state the results conveniently and a computer is required to perform analyses. For the general model

$$\tilde{y}_i = \beta_0 + \beta_1 x_{1i} + \beta_2 x_{2i} + \cdots + \beta_k x_{ki} + \varepsilon_i$$

the $(k + 1)$ least squares equations can be obtained by minimizing $\sum_{i=1}^{n} (y_i - \tilde{y}_i)^2$, as previously discussed. The individual regression coefficients can be evaluated by tests of significance based on the t-statistics

$$t_{n-k-1} = \frac{\beta_i - \hat{\beta}_i}{s_{\hat{\beta}_i}}$$

Note that the degrees of freedom for the t-statistic decreases by 1 for each regressor variable added to the equation. Note also that, again, the least squares estimates are obtained jointly and are not independent and the significance level is inaccurate if multiple significance tests are performed. For example, if the confidence of each of four independent statements is .90, the confidence associated with the joint statement is $.9 \times .9 \times .9 \times .9 = .6561$.*

The coefficient of determination associated with a multiple regression equation is given by

$$R^2_{y \cdot 12 \ldots k} = 1 - \frac{\sum\limits_{i=1}^{n} (y_i - \tilde{y}_i)^2}{\sum\limits_{i=1}^{n} (y_i - \bar{y})^2}$$

where $\tilde{y}_i = \hat{\beta}_0 + \hat{\beta}_1 x_{1i} + \hat{\beta}_2 x_{2i} + \cdots + \hat{\beta}_k x_{ki}$. As for the special case $k = 2$, the coefficient of determination indicates the proportion of variance accounted for by the regression equation including k regressor variables. Significance of $R^2_{y \cdot 12 \ldots k}$ does not imply that all—or even any—of the individual regression coefficients are significant. In particular, significance of $R^2_{y \cdot 12 \ldots k}$ does not imply that each of the k regressor variables is contributing significantly to the regression. The problem of selecting the regressor variables to be included in a regression equation is discussed later.

The values of $\hat{\beta}_i$ and $s_{\hat{\beta}_i}$ for $i = 0, \ldots, k$ and the value of $R^2_{y \cdot 12 \ldots k}$ are included in standard multiple regression computer printouts, as are the values of the appropriate t and F statistics. However, as k increases, these results may become increasingly difficult to interpret because of nonindependence of the regressor variables. This problem is considered briefly in the following section.

MULTICOLLINEARITY

Reference has been made several times in preceding sections to the fact that the regressor variables generally are not uncorrelated with each other, that their coefficients are estimated jointly not independently, and that their individual contributions must be analyzed in evaluating the regression equation. The usual interpretation of a multiple regression equation depends implicitly on the assumption that the regressor variables are not strongly correlated with each other. A *regression coefficient* is usually interpreted as a measure of the change in the dependent variable when the value of the corresponding regressor variable is changed and the values of all other regressor variables are unchanged. Although it is possible conceptually to change the value of one regressor variable while the values of the other regressor variables remain constant, this is not always possible in practice. For example, in economics even when the values

---

* This is a conservative estimate of the joint confidence, since it assumes independence of the individual statements with respect to their confidence coefficients.

of a variable can be changed, they usually cannot be changed without also changing the values of other variables. When the regressor variables do not vary independently, interpretation of the regression equation is difficult conceptually and, in addition, there are statistical problems associated with its evaluation.

When there is no linear relationship among the regressor variables, that is, when $r_{i_j} = 0$ for $i, j = 1, 2, \ldots, k$ and $i \neq j$, the variables are said to be *orthogonal*. Nonindependence or nonorthogonality of the regressor variables is referred to as the problem of collinear data or *multicollinearity*. Clearly, multicollinearity is a matter of degree. In most cases the regressor variables are not orthogonal, although frequently the degree of multicollinearity is not sufficient to affect the analyses seriously. However, in some situations the regressor variables are so strongly interrelated that results of the regression analysis cannot be interpreted unambiguously.

When the regressor variables are strongly intercorrelated, several difficulties occur in interpreting and evaluating the regression equation. The most serious of these problems are: (1) the importance and effects of individual regressor variables cannot be estimated; (2) the estimates of the regression coefficients are very sensitive to slight changes in the data and to addition and deletion of regressor variables in the equation; and (3) the estimated regression coefficients have large sampling errors that affect both inferences concerning the regression model and forecasts based on the model.

The existence of multicollinearity is sometimes difficult to detect. Multicollinearity is not an error in the model that can be detected by examining the residuals, but is essentially a matter of deficient data and may therefore be extremely difficult or impossible to correct. When better data are not available, there are some methods which can be used to improve the interpretation of the available data, as discussed briefly below.

Detection of Multicollinearity   Multicollinearity is associated with instability of the estimated regression coefficients. As noted, multicollinearity is indicated if there are large changes in the estimated regression coefficients when a variable is added or deleted or a data point is altered slightly or omitted. In addition, if analysis of the residuals indicates that the model is correctly specified, multicollinearity may be present if (1) the signs of the regression coefficients are not as expected or (2) the regression coefficients associated with regressor variables expected to be important have large standard errors.

Also note that the estimated regression coefficients are in fact partial coefficients; the sign of each regression coefficient is the same as the sign of the corresponding partial correlation coefficient. Thus, for example, the sign of $\hat{\beta}_i$ is the same as the sign of

$$r_{yi \cdot 1\, 2, \ldots, i-1, i+1, \ldots, k}$$

which is the correlation of y and x_i partialed on $x_1, \ldots, x_{i-1}, x_{i+1}, \ldots, x_k$. If the regressor variables are intercorrelated, then this partial correlation coefficient may not have the same sign as r_{yi}.

Most standard computer printouts for multiple regression analysis include a table of pairwise correlations of the variables. If one or more of these correlations is large, then multicollinearity is indicated. However, a linear relationship may involve several regressor variables; in this case, serious multicollinearity may be present although it is not indicated by examination of the pairwise correlations. A more complete check for multicollinearity would involve examination of the value of R^2 for the multiple regression of each regressor variable on all other regressor variables.

Several other methods of detecting multicollinearity have been proposed, including principal components analysis and ridge regression analysis. These analyses, which may also be used to aid in interpreting multicollinear data, are beyond the scope of the present discussion.

SELECTION OF REGRESSOR VARIABLES

There are some situations in which theoretical or other considerations determine the regressor variables to be included in an equation. However, in many instances selection of the regressor variables is an important aspect of the analysis. In these cases, data are obtained for a number of variables, say $x_1, x_2, \ldots, x_p$, and the problem is to select a subset of $k < p$ of these variables to be included in the regression equation. Usually the value of k is not specified in advance. The process of variable selection can be viewed as an extensive analysis of the correlation structure of the regressor variables and their individual and joint relationships with the dependent variable. Usually there is no "best" set of regressor variables; there may be several subsets of the regressor variables that are adequate. A good selection procedure should identify these several subsets rather than obtain a "best" subset.

A number of selection procedures and criteria have been proposed and several of those most frequently used are now discussed.

Evaluating All Possible Equations A very direct method, which is useful for both collinear and noncollinear data, involves estimating all possible subset regression equations. For a set of p regressor variables, this involves examining p equations with one regressor variable, $\binom{p}{2}$ equations with two regressor variables, and so forth—a total of 2^p equations. The most promising equations are determined using one of several criteria based on mean squared error. These equations are then analyzed carefully by examining the residuals for violations of the assumptions. The advantage of this procedure is that it provides the maximum possible amount of information as a basis for choosing the most appropriate regression model. However, when the number of regressor variables is large, evaluation of all possible subset regressions is not feasible in practice.

Stepwise Procedures Several variable selection procedures that do not involve evaluating all possible regression equations have been proposed. These procedures, referred to as stepwise procedures, add or delete variables one at a time and involve evaluation of at most $(k + 1)$ equations, where k is the number of variables selected. Stepwise procedures are of two general types, forward selection and backward elimination.

The forward selection procedure starts with only a constant term. The first variable entered is the regressor variable having the highest correlation with the dependent variable Y. Additional regressor variables are included one at a time on the basis of the highest partial correlation coefficient with the dependent variable given the regressor variables already included in the equation. The procedure is terminated when the variable entering the equation has an insignificant regression coefficient or all the regressor variables are included in the equation.

The backward elimination procedure starts with an equation including all the regressor variables and deletes one variable at a time. Variables are dropped on the basis of their contribution to the reduction of the error sum of squares. The procedure is terminated when all regressor variables included in the equation contribute significantly to the reduction in the error sum of squares.

A particular version of forward selection, referred to as the *stepwise method*, provides for deletion of a variable, as in backward elimination. Thus a variable entered at an earlier stage of selection may be eliminated at a later stage.

Stepwise procedures have the advantage that they involve much less analysis than evaluating all possible subset regressions. However, stepwise procedures should not be used mechanically to determine the "best" regression equation. The results of stepwise procedures may be particularly difficult to interpret when data are collinear, since in these cases forward and backward methods frequently produce different results. When the data are collinear, other methods such as principal components or ridge regression are recommended as a basis for selecting variables (References in Appendix A).

VIOLATIONS OF ASSUMPTIONS

Violations of the assumptions concerning the error distributions of the multiple regression model are detected by examining the residuals, as for bivariate regression. Statistical tests for normality and homogeneity of variance that require multiple observations of y for a given set of values of $X_1, \ldots, X_k$ generally are not possible for multiple regression analysis. The Durbin–Watson test can be used with appropriate degrees of freedom.

Nonlinearity The linearizing transformations discussed for bivariate regression can be generalized for multiple regression. However, the situation for multiple regression is complicated by the fact that scatter plots of the dependent variable Y versus each of the regressor variables $X_1, X_2, \ldots, X_k$ may indicate different forms of nonlinearity. Clearly, it is more difficult to determine the appropriate linearizing transformation in the multiple regression case and many types of multivariate nonlinear functions cannot be linearized.

The following nonlinear functions are most frequently used for multivariate data:

(a) $$y = \beta_0 x_1^{\beta_1} x_2^{\beta_2} \cdots x_k^{\beta_k} \cdot \varepsilon$$

linearizing transformation: $\beta_0' = \ln \beta_0$

$$y' = \ln y$$

$$x_i' = \ln x_i \qquad i = 1, \ldots, k$$

(b)
$$y = e^{\beta_1 x_1 + \beta_2 x_2 + \cdots + \beta_k x_k} \cdot \varepsilon$$

linearizing transformation: $y' = \ln y$

(c)
$$y = \frac{1}{\beta_0 + \beta_1 x_1 + \beta_2 x_2 + \cdots + \beta_k x_k + \varepsilon}$$

linearizing transformation: $y' = \dfrac{1}{y}$

(d)
$$y = \frac{1}{1 + e^{\beta_0 + \beta_1 x_1 + \beta_2 x_2 + \cdots + \beta_k x_1 + \varepsilon}}$$

linearizing transformation: $y' = \ln\left(\dfrac{1}{y} - 1\right)$

These models are straightforward generalizations of the corresponding bivariate models.

14.5 SUMMARY

The concepts and some of the statistical tests and inferences associated with multiple regression analysis are discussed in this chapter. Since the general case involving k regressor variables requires matrix notation and computer analysis, the concepts and procedures are illustrated for the simplest case $k = 2$. Problems of interpretation and procedures for selection of variables are discussed briefly for the general case.

PROBLEMS

14.1 Given the following computational information:

$$\bar{y} = 12 \qquad s_y = 10 \qquad r_{12} = .6$$
$$\bar{x}_1 = 10 \qquad s_1 = 5 \qquad r_{y1} = .2$$
$$\bar{x}_2 = 3 \qquad s_2 = 2 \qquad r_{y2} = .4$$

(a) Estimate the regression of y on x_1 and x_2.
(b) Estimate y if $x_1 = 20$, $x_2 = 1$.
(c) Compute $s_{y \cdot 12}^2$, the estimate of σ^2.
(d) Compute $R_{y \cdot 12}^2$ and $r_{y1 \cdot 2}^2$ and explain how they differ in interpretation.

14.2 Given the following computational information (for 25 observations):

$$\sum x_1 = 35 \qquad \sum x_1^2 = 265 \qquad \sum x_1 x_2 = 350$$
$$\sum x_2 = 50 \qquad \sum x_2^2 = 700 \qquad \sum x_1 y = 120$$
$$\sum y = 30 \qquad \sum y^2 = 132 \qquad \sum x_2 y = 150$$

(a) Compute $r_{y1}, r_{y2}, r_{12}, \hat{\beta}_0, \hat{\beta}_1, \hat{\beta}_2, \hat{\beta}_1^*, \hat{\beta}_2^*$, and $R_{y \cdot 12}^2$.
(b) Verify that $1 + 2r_{12} r_{y1} r_{y2} > r_{12}^2 + r_{y1}^2 + r_{y2}^2$.

14.3 During the summer, daily records were kept at Sandy Beach of the number of cups of coffee sold, the number of people who used the beach (there is a turnstile that admits one person every time a quarter is put in it), and the average temperature during the hours the beach was open. The correlation between number of cups of coffee sold and number of people using the beach is .4; the correlation between number of cups of coffee sold and average temperature is $-.3$; the correlation between number of people using the beach and average temperature is .6. Compute the relevant partial correlations and interpret them.

14.4 For a group of married men ages 18 to 25, data are obtained concerning number of children, number of years of formal education, and number of years married. Number of children and number of years of education have a correlation coefficient of $-.5$; number of children and number of years married have a correlation coefficient of .8; number of years of education and number of years married have a correlation coefficient of $-.7$. Compute the relevant partial correlations and interpret them.

14.5 For a random sample of communities, data concerning population, number of practicing physicians, and yearly deaths from cancer (average for the last five years) were obtained. The correlation between population and number of doctors is .8; the correlation between number of doctors and yearly deaths from cancer is .5; the correlation between yearly deaths from cancer and population is .7. Compute the relevant partial correlations and interpret them.

14.6 A group of people, ages 10 to 25, were given a general information test and their ages, test scores, and years of formal education were recorded. Age and years of formal education have a correlation of .7; age and test score have a correlation of .6; test score and years of formal education have a correlation of .9. Compute the relevant partial correlations and interpret them.

14.7 According to Statistical Research Corporation, in a recent study of 250 randomly selected communities, the correlation between crime rate and population was .80, the correlation between population and church membership was .75 and the correlation between crime rate and church membership was .50. Compute the three partial correlation coefficients and interpret the one which seems most meaningful.

14.8 The Shiny-Brite Company sells toothpaste nationwide but varies its advertising expenditures in various communities. The company's research department is investigating the relationship of sales, Y, to its own advertising budget, X_1, and the combined advertising budgets of its competitors, X_2. Data obtained for a random sample of 100 communities are summarized below. (Y denotes yearly sales in millions of dollars, X_1 denotes advertising budget in thousands of dollars, and X_2 denotes advertising budget of competitors in thousands of dollars.)

$$r_{y1} = .3 \qquad s_y = 21 \qquad \bar{y} = 105$$
$$r_{y2} = .4 \qquad s_1 = 2 \qquad \bar{x}_1 = 14$$
$$r_{12} = -.5 \qquad s_2 = 11 \qquad \bar{x}_2 = 25$$

(a) Determine the least squares regression line of y on x_1 and x_2 and use it to estimate y if $x_1 = 16$, $x_2 = 20$.

(b) Determine the multiple correlation coefficient $R_{y.12}$ and the partial correlation coefficient $r_{y1.2}$ and explain the difference in their interpretations.

(c) Determine the standard error of estimate $s_{y.12}$.

14.9 Professor Linear computed a least squares regression of y on x_1 and x_2 for 24 observations and obtained the equation

$$\tilde{y} = 1.5 + 2.6x_1 + 0.48x_2$$
$$(1.3) \qquad (.96)$$

(numbers in parentheses are standard deviations). The correlation between x_1 and x_2 is 0.5.

(a) What is the value of F corresponding to the regression equation?

(b) What is the value of R^2 corresponding to the regression equation?

14.10 Suppose that the following equations were obtained as part of an analysis of a sample of 83 observations on time series data. (Numbers in parentheses are standard deviations.)

$$\bar{y} = 6.008 + 0.809x_1 + 9.115x_3 + 0.172x_4$$
$$(0.293)\quad(2.005)\quad(0.053)$$

$$R^2 = .685$$

$$D.W. = 2.26$$

$$\bar{y} = 4.682 + 0.657x_1 + 2.315x_2 + 8.009x_3 + 0.037x_4$$
$$(0.653)\quad(0.809)\quad(2.113)\quad(0.033)$$

$$R^2 = .691$$

$$D.W. = 2.31$$

(a) For each equation, test the significance of each regression coefficient, the significance of the overall regression, and the autocorrelation of the residuals. (State the hypothesis for each significance test.)

(b) On the basis of the tests performed in (a), discuss choice of the most appropriate regression equation for the data.

(c) What can be inferred about the nature of the multicollinearity of the variables in these equations?

14.11 Suppose that the following equations were obtained as part of a regression analysis of a sample of 50 observations. (Numbers in parentheses are standard deviations of the estimated regression coefficients.)

$$y = 2.684 + 0.956x_1$$
$$(0.239)$$

$$y = -1.982 + 0.236x_1 - 0.520x_2$$
$$(0.118)\quad(0.104)$$

(a) Determine R^2 for the first equation.

(b) If $F = 62.5$ and $r_{y2} = -.6$ for the second equation, determine r_{12} and $r_{y1 \cdot 2}$.

(c) Determine R^2 for the second equation.

14.12 For a sample of 53 observations the correlation between expenditures and income is 0.8, the correlation between income and education is 0.5, and the correlation between expenditures and education is 0.3.

(a) Compute the partial correlation between expenditures and education given income and test its significance.

(b) Compute the multiple correlation of expenditures with income and education.

14.13 A dependent variable Y has been regressed on five independent variables for a sample of 15 time series observations. The observed values of Y and the values predicted by the regression equation are as follows:

y	$\bar{y}$
14	12
13	15
22	18
18	21
26	24
25	27
31	30
30	33
39	36
38	39
46	42
41	45
51	48
50	51
56	54

(a) Obtain the residuals.

(b) State and test the appropriate hypothesis concerning serial correlation.

14.14 Assume that a set of predicted values of y were obtained from a least squares regression of the form

$$\hat{y} = \hat{\beta}_0 + \hat{\beta}_1 x_1 + \hat{\beta}_2 x_2 + \hat{\beta}_3 x_3$$

Using the values of y given below and the corresponding y estimates,

(a) Obtain the residuals and determine by observation whether autocorrelation seems to be present and, if so, of what form (positive or negative)

(b) State the appropriate hypothesis and test for autocorrelation using the Durbin–Watson statistic

y	$\tilde{y}$
35	32
26	28
42	40
29	32
27	24
28	31
33	33
31	31
27	27
33	36
35	33
29	33
43	39
34	34
28	27
26	29
35	32

14.15 For a multivariate population of one dependent and four independent variables a regression equation was obtained from 20 time series observations. The observed and the predicted values are:

y	$\tilde{y}$
3	5
5	6
4	3
7	6
10	9
9	7
12	13
15	16
17	16
14	16
18	15
21	20
20	20
22	23
24	25
23	22
22	21
20	19
19	21
17	19

(a) Compute the residuals.

(b) State the appropriate hypothesis and test for serial correlation.

14.16 Given the following information:

$$\bar{y} = 13 \qquad s_y = 9 \qquad r_{12} = .3$$

$$\bar{x}_1 = 7 \qquad s_1 = 12 \qquad r_{y1} = .5$$

$$\bar{x}_2 = 9 \qquad s_2 = 15 \qquad r_{y2} = .6$$

(a) Compute the regression of y on x_1 and x_2.

(b) Estimate y if $x_1 = 15$ and $x_2 = -20$.

(c) Compute $s_{y \cdot 12}$.

(d) Compute $R^2_{y \cdot 21}$ and $r^2_{y2 \cdot 1}$ and explain their different interpretations.

14.17 Observations from a multivariate population are:

y	x_1	x_2
0	1	1
-8	-1	3
-3	0	1
-10	-2	2
3	3	4
1	2	3
12	3	-5
1	1	0
-2	-1	-3
0	2	4

Obtain the least squares estimates for the regression coefficients by setting up and solving the appropriate normal equations and determine the least squares regression equation of y on x_1 and x_2.

14.18 For the data of Problem 14.17, calculate the least squares regression line of y on x_1 and x_2 using the formulas for two predictors.

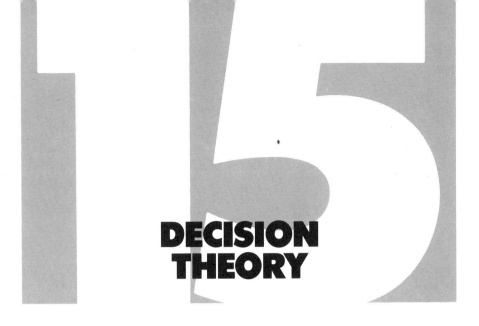

DECISION THEORY

15.1 INTRODUCTION

Decision theory is the study of quantitative techniques for formulating and analyzing decision problems in order to provide a framework of methods for making decisions in a systematic and rigorous fashion. Decision theory makes the decision process more explicit and presumes to improve it, by requiring that alternatives and their associated gains or losses be precisely stated. Because most interesting decision problems involve uncertainty, even the most precise mathematical framework does not mean that, in retrospect, it will necessarily be the optimal decision. But decision theory does provide rules for determining the best decision, with respect to some specified criterion, given the knowledge available.

A decision theory problem has three essential parts or aspects:

1. Alternative choices or courses of action available to the decisionmaker
2. Consequences or outcomes of these choices or actions that depend, in general, not only on the choice itself but also on the configuration of the environment, sometimes referred to as states of nature, in which the choice is made
3. Goals or objectives that the decision maker wishes to achieve and which are quantifiable, that is, measurable so that numbers can be attached to them

A decision problem thus involves a choice among alternative actions, whose outcomes depend on factors not under the control of the decision maker, in order to achieve some goal or objective. Decision problems are classified for analysis on the basis of the knowledge the decision maker has about the likelihood of occurrence of the possible states of nature. On this basis, decisions are classified as being made under (1) certainty, (2) risk, (3) uncertainty, or (4) conflict.

For decisions under certainty there is only one state of nature and the decision maker's choice of action uniquely determines the outcome; under risk, the probability distribution of the possible states of nature is known, either

	States of nature		
Choices	N_1	N_2	N_3
A_1	P_{11}	P_{12}	P_{13}
A_2	P_{21}	P_{22}	P_{23}
A_3	P_{31}	P_{32}	P_{33}

Figure 15.1   Payoff matrix

objectively or subjectively; under uncertainty, the probability distribution of the states of nature is assumed unknown; and under conflict, the states of nature represent competitive strategies available to a rational opponent, whose goal is in conflict with that of the decision maker. The choices of action open to the decision maker then represent available competitive strategies.

A *payoff matrix* is used to represent the value of the consequences or outcome of each choice available to the decision maker for each possible state of nature. The value of an outcome must be expressed in numerical, although not necessarily monetary, terms and is referred to as a *payoff*. In a payoff matrix the choices or alternative actions are represented by the rows and the states of nature are represented by the columns. The entries in the matrix are the payoffs to the decision maker; positive values are gains and negative values are losses. In Figure 15.1, the A's represent choices or alternatives, the N's represent states of nature, and the p's represent payoffs.

15.2 DECISIONS UNDER CERTAINTY

Decisions under certainty represent to the statistician the least interesting of the four types of decisions because they require less statistical effort, although in practice they may be quite important—such as deterministic linear programming. Many decisions by firms have subparts that can be viewed as decisions under certainty, but most in their entirety are in either the risk or uncertainty category. For example, a firm may be considering the purchase of a new machine to replace an existing machine. The possible alternative new machines available can be priced on the market with certainty. If minimizing purchase price of a new machine were the only consideration, then this information would be sufficient for making the decision under certainty. Such a simple situation might be realistic, for example, if the decision concerns choosing the supplier from whom to purchase a large number of standardized parts for the machine. However, the optimal decision concerning which machine to buy, if any, usually involves considerations other than the purchase price, some of which cannot be assessed with certainty. For example, the quality of the parts produced by and the maintenance costs of a new machine cannot be known in advance with certainty. If the machine's purchase depends partly on an expectation of increased sales, then this increase cannot be known with certainty. The proper discount rate to use in evaluating the capital expenditure on a present worth basis is also known only approximately. Thus, in practice, decisions under certainty seldom occur, except as subparts of a decision made under either risk or uncertainty. A decision under certainty requires only that the payoffs for the various alternatives be obtained. (In practice, this is not always easy.) Once the payoffs are obtained, the alternative corresponding to the maximum payoff is then chosen.

EXAMPLE
Suppose an investor is considering investing in one of a series of possible short-and long-term bonds and notes and her *sole* criterion for such investment is the yield to maturity to be obtained on the investment.

INVESTMENT	YIELD (%)
Short-term treasury notes	$6\frac{1}{2}$
Long-term treasury bonds	$6\frac{1}{2}$
Savings bonds—Series E	$4\frac{1}{2}$
AAA Electric Utility bonds	7
BBB Railroad bonds	$7\frac{1}{2}$

In such a situation, with the possibilities given above, the investor would choose the BBB railroad bonds. Of course, one might argue with her that the criterion of yield alone is not good because it ignores time to maturity and default risk. But, given the criterion, the decision is made under conditions of certainty.

15.3 DECISIONS UNDER RISK

For decisions under risk the probability distribution of the states of nature is assumed known. In practice, the probability distribution of the states of nature is frequently subjective; as discussed in Chapter 4, subjectivity is a matter of degree and is always present to some extent. In a risk situation the expected value of the payoff for each choice is calculated, and the choice with the highest expected payoff is chosen. The argument for using expected value as the criterion for decisions under risk is that if a decision is repeated a large number of times, the average returns to the decision maker are maximized if expected value is used as a criterion for choosing among alternative acts. There are some disadvantages in using the expected value criterion when a decision can be made only once, not over and over again.

The expected value of a random variable, discrete or continuous, is defined in Chapter 5 as the mean of its distribution. For purposes of the payoff matrix, the possible states of nature are discrete, although actually the variable defining the state of nature may be continuous. In the case of risk the probability of each state of nature occurring is assumed known. Therefore the expected value of the payoff for a given choice is the sum of the products of the possibilities of the states of nature occurring times their respective payoffs for the given choice.

One of the classic prototype problems for decisions under risk is that of a store that stocks a perishable item periodically; the demand for this item follows some known probability distribution based on relative frequency information. If the item is not sold, it must be disposed of at a loss. The question is: How many of the items should the store stock? The following example illustrates this.

EXAMPLE

On Monday of every week a grocer stocks a perishable commodity, which if not sold by Friday must be thrown away. The commodity is purchased at $1.00 each and costs $.10 each to pack when sold. It is sold for $2.00 each. Overhead in the store is $.15 per dollar of sales. Based on purchases of and requests for the commodity over a rather long period of time, the grocer has compiled the following probability distribution of demand for the commodity.

DEMAND, x	PROBABILITY $P(x)$
0	0.05
1	0.15
2	0.25
3	0.30
4	0.25
5+	0.00

Two things should be noted at once in filling in the payoff matrix for this problem and any similar problems:

1. The matrix must be filled in with net values.
2. The irrelevant costs must be excluded from consideration in any kind of an incremental problem. For example, the overhead is an allocated cost, and thus irrelevant for the decision.

The payoff matrix can be set up as follows:

Acts, stock	States of nature; demand				
	0	1	2	3	4
0	0	0	0	0	0
1	-1.00	0.90	0.90	0.90	0.90
2	-2.00	-0.10	1.80	1.80	1.80
3	-3.00	-1.10	0.80	2.70	2.70
4	-4.00	-2.10	-0.20	1.70	3.60

The calculation of a payoff must include the cost of items, the cost of packing if sold, and the revenue derived from sale. For example, to calculate the payoff for demand 2, stock 3:

$$\begin{array}{ll} \text{Revenue from demand 2: } 2(\$2.00) = \$4.00 \\ \text{Cost of 3: } 3(\$1.00) = \$3.00 \\ \text{Cost of packing 2: } 2(\$0.10) = 0.20 & \underline{3.20} \\ \text{Payoff} & 0.80 \end{array}$$

It is implicit in the payoffs calculated for this matrix that no cost is assumed for demands larger than the number stocked, which would probably be realistic for this sort of situation, but might not be in others. If the commodity was, for example, electric motors for an electrical wholesaler and if customers met with many out of stock situations for their demands, then they might stop dealing with the wholesaler. The costs associated with this sort of possibility are difficult to assess. Since, in the example, the probabilities are given for each possible demand, the expected value of each strategy can be calculated.

$$E(0) = 0(.05) + 0(.15) + 0(.25) + 0(.30) + 0(.25) = 0$$

$$E(1) = (1.00)(.05) + (.90)(.15) + (.90)(.25) + (.90)(.30) + (.90)(.25) = 0.805$$

$$E(2) = (2.00)(.05) + (.10)(.15) + (1.80)(.25) + (1.80)(.30) + (1.80)(.25) = 1.325$$

$$E(3) = (3.00)(.05) + (1.10)(.15) + (.80)(.25) + (2.70)(.30) + (2.70)(.25) = 1.370$$

$$E(4) = (4.00)(.05) + (2.10)(.15) + (.20)(.25) + (1.70)(.30) + (3.60)(.25) = 0.855$$

The strategy with the highest expected value is 3; therefore the best strategy is to stock 3. This means that over a long period if the demand distribution remains stable, the grocer will make on the average $1.37 per week by stocking 3 items of the commodity per week.

EXAMPLE
Due to limited time and financial resources, a candidate can campaign in either District A or District B, but not both. District A is four times as important to win as District B. The candidate's advisors feel that if he campaigns in a district he is certain to carry it. If he campaigns in District A, he is certain to lose District B. If he campaigns in District B, he will carry District A only if Election Day is rainy (so that there is a light voter turnout). The candidate thinks the probability that Election Day will be rainy is $\frac{1}{6}$. In which district should he campaign in order to maximize the expected votes? What probability of a rainy Election Day would make campaigning in the two districts equally advantageous in terms of expected votes?

		ELECTION DAY	
		RAINY	NOT RAINY
Campaign	District A	4	4
	District B	5	1

NOTE:
Any multiple of the above payoff table would result in an equivalent analysis; here the value of B is 1 and the value of A is 4.

$$E(\text{District } A) = \left(\frac{1}{6}\right)(4) + \left(\frac{5}{6}\right)(4) = 4$$

$$E(\text{District } B) = \left(\frac{1}{6}\right)(5) + \left(\frac{5}{6}\right)(1) = \frac{2}{3}$$

so the candidate should campaign in District A. In order for $E(\text{District } A) = E(\text{District } B)$,

$$4p + 4(1 - p) = 5p + (1 - p)$$

$$3 = 4p$$

$$p = \frac{3}{4}$$

Thus if the probability of a rainy election day is $\frac{3}{4}$, then E(District A) = E(District B) and campaigning in either district is equally advantageous in terms of expected votes.

EXAMPLE

A general can defend installation A, B, or C. Installation A is twice as valuable as B and B is twice as valuable as C. The enemy will attack one of the installations, but if it is defended, the installation will suffer no loss. Installation A is geographically removed from B and C and there is no chance of moving forces to or from A after an attack has begun; however, B and C are close enough so that forces may be moved from one to the other in time to save half the installation attacked. The general's intelligence officer believes that the probability of attack on A is $\frac{1}{8}$; on B, $\frac{1}{2}$, and on C, $\frac{3}{8}$. Which installation should the general defend in order to maximize his expected remaining installations?

		ATTACK		
		A	B	C
	A	14	10	12
Defend	B	6	14	13
	C	6	12	14

Note that any multiple of the above payoff table would result in an equivalent analysis, as would use of negative values to indicate losses; here the value of A is 8, B is 4, and C is 2.

$$\text{E(Defend } A) = \left(\frac{1}{8}\right)(14) + \left(\frac{1}{2}\right)(10) + \left(\frac{3}{8}\right)(12) = \frac{90}{8} = 11\frac{1}{4}$$

$$\text{E(Defend } B) = \left(\frac{1}{8}\right)(6) + \left(\frac{1}{2}\right)(14) + \left(\frac{3}{8}\right)(13) = \frac{101}{8} = 12\frac{5}{8}$$

$$\text{E(Defend } C) = \left(\frac{1}{8}\right)(6) + \left(\frac{1}{2}\right)(12) + \left(\frac{3}{8}\right)(14) = \frac{96}{8} = 12$$

So the general should defend installation B.

UTILITY APPROACH

There is an alternative decision criterion for risk type situations that may be maximized rather than expected value. This criterion is expected utility. Since the concept of utility appeared in the economic literature of the nineteenth century, a great deal has been written about it in the context of individual consumer behavior and welfare economics. Utility is sometimes defined as that quality which when possessed by something, causes it to render satisfaction. It is defined in terms of preferences—one alternative has more utility for an individual than another if he or she prefers or chooses the first alternative when both are available.

Originally, utility was considered to be measurable and additive. Thus if an individual obtained a certain utility (satisfaction) from one unit of a commodity, then he or she should obtain twice as much from two units of the commodity. This concept is known as *cardinal utility*. During the first third of the twentieth century, cardinal utility fell into disrepute and was replaced by the concept of *ordinal utility*. This view of utility holds that it is possible to rank alternatives, commodities, or items on a preference basis but not possible to assign numerical values to them in the usual sense of an interval scale. That is, if people were faced with various alternatives then they could indicate their order of preference for the alternatives but could not say how much they preferred one alternative to another.

About the time of World War II, John Von Neumann and Oscar Morgenstern proposed a type of utility considered to be numerical but not additive. This type of utility falls between the traditional cardinal utility of the nineteenth century and the ordinal utility of the twentieth, although it is often referred to as a cardinal utility. The Von Neumann-Morgenstern utility (as it is called) is widely used in decision theory as an alternative to expected value.

The Von Neumann–Morgenstern determination of utility for some item is numerical but is on an arbitrary scale. Suppose the problem is to determine the utility of some prize, call it *Z*. The first question is, "for whom." That utility is an individual matter is important. Therefore the utility of prize *Z* must be considered for a specified person. It is unlikely that the utility of a particular item would be the same for two individuals, even on the same scale. So the utility of prize *Z* will be calculated for a hypothetical person, say Mr. Zud. In fact, Mr. Zud's cooperation is needed in order to calculate the utility of prize *Z* for him.

First, it is necessary to set up the arbitrary utility scale along which the utility of *Z* will be placed. Mr. Zud is asked to imagine something more desirable and something less desirable than the item or items whose utility (for Mr. Zud) is being assessed. In other words, these two somethings lie outside the range of preferences of the current decision. The more desirable something will be called *D*, with an assigned utility of 1000. The less desirable something will be called *U*, with an assigned utility of 1. Thus the scale of 1 to 1000 is established; notice that negative values are excluded from the scale. This is only for convenience, since, as has been mentioned, the scale is arbitrary.

Mr. Zud is then confronted with what is called the *standard lottery ticket*, which offers only two possible outcomes, *D* and *U*. Mr. Zud is asked if he had the free choice of either the standard lottery ticket or prize *Z* what probability (*P*) of the occurrence of *D* and probability (1 − *P*) of the occurrence of *U* would make him indifferent between the ticket and prize *Z*. When these probabilities are obtained from Mr. Zud, the prize *Z* is said to be the *certainty equivalent* of the gamble represented by the standard lottery ticket. It should be obvious that the probabilities *P* and 1 − *P* must exist. If *P* = 1 and 1 − *P* = 0, then the standard lottery ticket would be chosen; if, on the other hand, *P* = 0 and 1 − *P* = 1, then the prize would be chosen. Therefore some value of *P* between 0 and 1 must make Mr. Zud indifferent between the ticket and the prize. If the utility of the standard lottery ticket is calculated using the probabilities *P* and 1 − *P*, then it is equivalent to the utility of the prize, because those probabilities make the prize and the ticket of equal worth in the eyes of the decision maker, Mr. Zud.

The calculation of the utility of the standard lottery ticket is as follows:

$$U_s = (P)(U_D) + (I - P)(U_U)$$

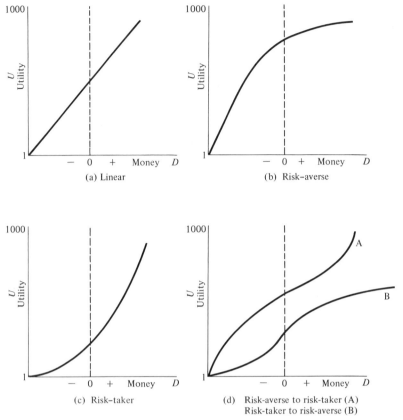

Figure 15.2 Utility functions for money

Suppose Mr. Zud gives the probabilities $P = 0.4$ and $1 - P = 0.6$ for the standard lottery ticket to be indifferent to prize Z. Then

$$U_s = (.4)(1000) + (.6)(1) = 400.6 = U_X$$

Stating Z as a sum of money usually makes it easier to consider it as a certainty equivalent. Then by changing Z to different sums and questioning Mr. Zud about the probabilities in each case, one could construct Mr. Zud's utility function for money based on the scale 1 to 1000.

There are essentially four forms that the utility function for money may take, as shown in Figure 15.2. The possibilities D and U, just discussed, represent an extremely large amount of money and an extremely large money loss, respectively. In case (a) the utility function is linear, the implicit assumption in the use of expected value as a decision criterion. In case (b) the utility function is concave from below, the case of the risk averter. This utility function for money represents the risk averter because utility increases less rapidly than money amount. In case (c) the utility function is convex from below, the case of the risk taker (or gambler). This utility function for money represents the risk taker because utility increases more rapidly than money amount. Case (d) illustrates the situation in which the individual is a risk taker over part of the range and a risk averter over part of the range. This situation exists, for example, when the same person buys fire insurance on his or her house (risk avoidance) and also gambles (risk taking). At some intermediate range, where the decision maker changes from one behavior to another with regard to risk, the utility is approximately linear with respect to money.

In using expected value (sometimes referred to as expected monetary value) in decision making, we are implicitly assuming 15.2(a) as the relevant utility function with regard to money, that is, utility is directly proportional to payoff. Most persons are thought to have a utility function for money similar to 15.2(b), that is, diminishing marginal utility for money. Yet, as was just pointed out, 15.2(b) may not be generally applicable because many persons will gamble small sums even where the odds are very much against them and will purchase fire insurance on their house where the probability of loss of the house from fire is very small. In other words, although the expected value in each of these decisions is negative, numerous individuals make them every day. Persons who make such decisions clearly do not have the implicit utility function for money (15.2(a)) assumed when expected monetary value is used as the decision criterion.

In situations involving risk decisions, it can be argued that the decision maker should make that decision which maximizes his or her expected utility rather than expected monetary value. There is a classic example that has been used for several hundred years to illustrate this argument and also to show that, with respect to money, the individual's utility function must be concave from below for sufficient amounts of money. This example, which follows, is known as the St. Petersburg Paradox (so called because J. Bernoulli published the paradox in a journal in St. Petersburg, Russia).

EXAMPLE
Suppose someone offered to pay you $1.00, $2.00, $4.00, $8.00, and so forth, doubling the amount for every head you were able to obtain *in a row* in tossing a fair coin. However, payments stop and the game ends as soon as a tail is obtained on a toss. The question is, "How much would you be willing to pay as a lump sum in advance to play this game if you used expected value as your decision criterion?" The answer is that you should be willing to pay an infinite amount because the expected value of the game is infinite. This is shown as follows:

		Cumulative
Expected value of first toss:	$(.5)(\$1.00) = \$0.50$	0.50
Expected value of second toss:	$(.25)(\$2.00) = 0.50$	1.00
Expected value of third toss:	$(.125)(\$4.00) = 0.50$	1.50
Expected value of nth toss:	$\left(\dfrac{1}{2}\right)^n (2)^{n-1} = 0.50$	$n(.50)$
Expected value of game		∞

Bernoulli called this a paradox because, of course, no one would pay a large amount, not anywhere near approaching infinity, to play the game. The explanation lies in the negative utility associated with the potential large loss. From this follows the argument that expected utility rather than expected value should be used as the decision criterion. Suppose a businessman, for instance, is faced with an investment decision which if successful promises a very large profit and whose expected value is quite a large positive number. It is is unsuccessful, however, the result will be bankruptcy for the business. Despite the positive expected value, the businessman might reasonably refuse to undertake the investment because the possibility of bankruptcy has an extremely large disutility associated with it.

REGRET CRITERION

When a decision is made under risk it may turn out to be nonoptimal after the fact—even though it was the best possible decision in terms of the information available before the fact and therefore was the optimal decision to make. This is not a criticism of the analysis but, rather, an inherent aspect of decisions made under risk. Because decisions made under risk cannot be guaranteed to be optimal after the fact, the losses that may thus be incurred are of particular interest. *The opportunity loss of a decision* is the difference between the payoff actually realized under the decision and the payoff that would have been realized if the decision had been the best one possible for the state of nature that actually occurred. Strictly speaking, the opportunity loss cannot be known until after the decision is made. However, before the fact it is possible to view the opportunity loss of a decision as the difference between the expected payoff resulting from making the decision under risk and the expected payoff resulting from making the decision under certainty.

Frequently, it may not be reasonable in practice to make a decision solely on the basis of expected payoff or profit, because of the possible (opportunity) losses that could be incurred. In such cases, computing the expected opportunity loss of a contemplated decision may be useful. If the expected opportunity loss is intolerably large, then the person making the decision may be able to find some way of reducing it, for example, by delaying the decision until more information can be obtained.

The corresponding opportunity loss table can be obtained from a payoff table by subtracting each entry of the payoff table from the maximum entry in its column. Note that the maximum entry in each column represents the greatest possible profit for the state of nature corresponding to that column. Thus the opportunity loss for any decision is the difference between this maximum possible profit and the profit actually obtained under the decision. We see that opportunity loss is zero for the best possible act for any state of nature and is positive for all other acts—opportunity loss, by definition, cannot be negative.

In many cases, costs and losses are analyzed as negative profits. In some problems that are concerned primarily with costs it may be more logical to write the payoff table in terms of costs rather than profits and then to minimize expected cost (rather than maximize expected negative profit). If the payoff table is in terms of costs rather than profit, then the corresponding opportunity loss table is obtained by subtracting the minimum entry in each column from the other entries in that column of the payoff table. The minimum entry in each column represents the least possible cost for the event corresponding to that column. Thus the opportunity loss for any decision is the difference between the cost actually incurred under the decision and this minimum possible cost. As when the payoff table is in terms of profits, the opportunity loss is zero for the best possible act for any event and is positive for all other acts.

EXAMPLE

A grocer must place his weekend order for pears. Each crate costs him $1.50 and is sold (by the crate) for $2.50. The pears are very ripe and those not sold during weekend shopping will be a total loss by Monday. The grocer feels sure he can sell at least 5 crates and space limitations prevent stocking more than 10 crates. The payoff table is

CRATES STOCKED	CRATES COULD SELL					
	5	6	7	8	9	10
5	$5.00	$5.00	$5.00	$5.00	$5.00	$5.00
6	3.50	6.00	6.00	6.00	6.00	6.00
7	2.00	4.50	7.00	7.00	7.00	7.00
8	0.50	3.00	5.50	8.00	8.00	8.00
9	−1.00	1.50	4.00	6.50	9.00	9.00
10	−2.50	0	2.50	5.00	7.50	10.00

The opportunity loss table, or regret matrix, for the grocer is

CRATES STOCKED	CRATES COULD SELL					
	5	6	7	8	9	10
5	$0	$1.00	$2.00	$3.00	$4.00	$5.00
6	1.50	0	1.00	2.00	3.00	4.00
7	3.00	1.50	0	1.00	2.00	3.00
8	4.50	3.00	1.50	0	1.00	2.00
9	6.00	4.50	3.00	1.50	0	1.00
10	7.50	6.00	4.50	1.00	1.50	0

The expected oportunity loss of stocking 7 crates (the act that maximizes expected profit) is

$$E(\text{opportunity loss of } 7) = \left(\frac{1}{10}\right)(3) + \left(\frac{1}{5}\right)(1.50)$$

$$+ \left(\frac{1}{4}\right)(0) + \left(\frac{1}{5}\right)(1) + \left(\frac{1}{6}\right)(2) + \left(\frac{1}{12}\right)(3)$$

$$= \$1.38$$

Alternately, expected opportunity loss can be computed as expected profit under certainty minus expected profit under uncertainty

$$E(\text{profit under certainty}) = \left(\frac{1}{10}\right)(5) + \left(\frac{1}{5}\right)(6)$$

$$+ \left(\frac{1}{4}\right)(7) + \left(\frac{1}{5}\right)(8) + \left(\frac{1}{6}\right)(9) + \left(\frac{1}{12}\right)(10)$$

$$= \$7.38$$

$$E(\text{profit for } 7) = \$6.00$$

$$E(\text{opportunity loss for } 7) = \$7.38 - \$6.00 = \$1.38, \text{ as above.}$$

Note that, computationally, the difference in the two methods consists in whether the subtraction between rows is done before or after multiplication by the corresponding probabilities. The meaning of expected opportunity loss can be understood intuitively by supposing that the grocer stocked 7 crates for a number of weeks, say 100, and that during this time the various values of demand occurred with relative frequencies equal to the probabilities given in the problem. Then the grocer's average weekly profit was $1.38 less than it would have been if he had known each week's demand accurately in advance.

EXAMPLE

Mrs. Magoo owns some stock in a British Steel Company. She can sell it currently for $100 profit. If she holds the stock and the Conservative party wins the upcoming election her profit will be $300, but if the Labor party wins she will lose $200. She is also considering increasing her stock (by a certain fixed amount); with these large holdings, she would profit $600 in the event of a Conservative party victory and would lose $400 in the event of a Labor party victory. According to the latest poll (in which she has great faith), the probability of a victory for the Conservatives is .45 and for Labor is .55. In order to maximize her expected profit, should Mrs. Magoo sell, hold or increase her stock? What is the expected opportunity loss of this decision?

| | ELECTION VICTORY | |
ACTION ON STOCK	CONSERVATIVE	LABOR
Sell	100	100
Hold	300	− 200
Buy	600	− 400

$$E(\text{sell}) = (.45)(100) + (.55)(100) = \$100$$

$$E(\text{hold}) = (.45)(300) + (.55)(-200) = \$25$$

$$E(\text{buy}) = (.45)(600) + (.55)(-400) = \$50$$

so in order to maximize expected profit, she should sell her stock.

$$
\begin{aligned}
E(\text{opportunity loss of selling}) &= E(\text{profit under certainty}) \\
&\quad - E(\text{profit from selling}) \\
&= [(.45)(600) + (.55)(100)] - 100 \\
&= 325 - 100 \\
&= \$225
\end{aligned}
$$

15.4 DECISIONS UNDER UNCERTAINTY

When all the facts relevant to a decision are known accurately (when a decision is made under certainty), careful analysis is sufficient to prevent the decision from turning out, after the fact, to have been wrong. However, when the relevant facts are not all known (when a decision is made under risk or uncertainty), the decision is necessarily a gamble and cannot be guaranteed to be optimal, after the fact, by careful thinking (or by any other means). In such circumstances a correct decision consists of choosing the best possible alternative, in the sense of a bet, whether or not the choice turns out, after the fact, to have been optimal.

Many business decisions and, in fact, most decisions in life must be made under uncertainty. For example, inventory problems involve uncertain demand, investment problems involve uncertain market conditions of various types, marketing problems involve uncertain consumer reaction, life insurance problems involve uncertain death rates, vacation plans involve uncertain weather, career planning involves uncertain availability of jobs, and so forth.

Therefore, when decisions are made under uncertainty the possible states of nature are known to the decision maker but their probabilities of occurrence are unknown. In these situations the decision maker does not have sufficient experience and information to state even a subjective probability distribution. The most difficult problem in such cases is the choice of a decision criterion; there is none that is generally accepted as satisfactory. In lieu of a unique "correct" decision criterion, there are three approaches frequently used in dealing with uncertainty in terms of selecting a decision criterion:

1. Principle of insufficient reason (Laplace criterion)
2. Maximin or minimax
3. Hurwicz criterion

Note that for those decision makers who accept the Bayesian approach to probability (covered in Chapter 16), there is almost never (some would say never) a siatuation in which nothing is known about the possibilities of the various states of nature occurring. For such persons, decision making under uncertainty as described previously does not exist. They would argue that uncertainty blends into risk, and their division into two dichotomous situations is arbitrary. Bayesians maintain that there are simply degrees of confidence that one can have with regard to probabilities assigned and that sometimes one has better information than at other times. Deferring further discussion of this until the next chapter, we will take up the three suggested criteria for decision making under (strict) uncertainty.

PRINCIPLE OF INSUFFICIENT
REASON OR LAPLACE CRITERION

The principle of insufficient reason, sometimes called the Laplace criterion, assumes equally likely occurrence of the possible states of nature. Thus the same probability is assigned to each of the states of nature and the expected value of each strategy is calculated. The strategy having the largest expected value or expected utility is chosen. This criterion has been criticized on the basis that lack of information about a probability distribution in no way implies logically that the distribution is uniform. In spite of this criticism, the Laplace criterion is frequently used.

EXAMPLE

Suppose the Tiny TV Company is considering selling a color television set under its own name, something it has never done before. The sets would be purchased in a large lot from a major manufacturer, the Tiny name would be attached, and the sets would then be priced at a discount from the major's retail price. Tiny's management sees two possible results: either the sets will sell well to the public and Tiny will make about $200,000 profit or they will not sell well, in which case Tiny must dispose of the sets at a $100,000 loss. The probabilities of the sets selling or not selling are unknown to Tiny's management. With equally likely probabilities of the two possibilities, the payoff table and expected values are:

	STATES OF NATURE	
ACTS	SETS WILL SELL	SETS WILL NOT SELL
Introduce color	+200,000	−100,000
Do not introduce color	0	0

$$E(A_1) = \left(\frac{1}{2}\right)(200,000) + \left(\frac{1}{2}\right)(-100,000) = +50,000$$

$$E(A_2) = \left(\frac{1}{2}\right)(0) + \left(\frac{1}{2}\right)(0) = 0$$

On the basis of expected monetary value, Tiny should introduce the color sets. However, if the color sets are introduced and do not sell, then the $100,000 loss may mean Tiny is wiped out. In this case the negative utility associated with being put out of business may be considerably greater than the positive utility of gaining $200,000. If so, expected utility would be maximized by not introducing the color sets.

MAXIMIN OR MINIMAX

The maximin or minimax criterion is essentially the criterion of the pessimist. It implies that the decision maker assumes the worst and then maximizes the minimum gain or minimizes the maximum loss. (A loss can be viewed as a negative gain, so maximin and minimax are equivalent.)

EXAMPLE

Sam Speculator has been given three hot tips from his friend the stock tout, Boiler R. Operator. All three stocks are over the counter stocks that are "sure things" to double in a year. Sam can get a 1000 share clock of any or all three stocks, Kansas Blue Sky, Mississippi Bubble, Teapot Hula Hoop. The stocks are selling for $2, $3, and $4 a share, respectively. Sam, being unable to turn a deaf ear to a hot tip, decides immediately that he will buy at least one of the stocks. He has a streak of caution, however, because he has picked up a few "cats and dogs" in the past that have become worthless, even though they promised to do very well. Sam assumes that there are only two possibilities for each of the stocks—to double or to go to zero. Sam also rules out the possibility that all three stocks will go down; he optimistically believes that at least one must be good. Sam sets up the following matrix to help decide whether to buy one, two, or all three stocks, and if less than all three, which one or ones:

States of nature

Acts	$K\uparrow$ $M\uparrow$ $T\uparrow$	$K\uparrow$ $M\downarrow$ $T\downarrow$	$K\uparrow$ $M\uparrow$ $T\downarrow$	$K\uparrow$ $M\downarrow$ $T\uparrow$	$K\downarrow$ $M\downarrow$ $T\uparrow$	$K\downarrow$ $M\uparrow$ $T\uparrow$	$K\downarrow$ $M\uparrow$ $T\downarrow$	Maximum loss
Buy all 3	+9000	−5000	+1000	+3000	−1000	+5000	−3000	−5000
Buy K	+2000	+2000	+2000	+2000	−2000	−2000	−2000	−2000
Buy M	+3000	−3000	+3000	−3000	−3000	+3000	+3000	−3000
Buy T	+4000	−4000	−4000	+4000	+4000	+4000	−4000	−4000
Buy KM	+5000	−1000	+5000	−1000	−5000	+1000	+1000	−5000
Buy KT	+6000	−2000	−2000	+6000	−2000	−2000	−6000	−6000
Buy MT	+7000	−7000	−1000	+1000	+1000	+7000	−1000	−7000

If Sam uses the minimax criterion, he will select the strategy that will minimize his maximum loss. In this case it would be S_2, buy K. (In the above matrix, the letters K, M, T represent the first letter of the stock, and the arrows up (↑) indicate doubling and the arrows down (↓) indicate becoming worthless.)

A variant of the minimax or maximin strategy is the criterion of the optimist. Here the best is assumed, and the decision maker selects the strategy that maximizes his maximum gain or minimizes his minimum loss; this is referred to as the maximax or minimin strategy. Using maximax in the previous example, Sam would buy all three stocks, since his maximum gain ($9000) would occur as a possibility under that strategy.

HURWICZ CRITERION

In order to use the Hurwicz criterion, the decision maker must select an index of optimism, α, between 0 and 1. For each choice the maximum of the possible outcomes is multiplied by α, the minimum of the possible outcomes is multiplied by $1 - \alpha$, and these two products are added. The choice for which this sum is the largest is selected. For the Hurwicz criterion, the α and $1 - \alpha$ do not represent probabilities; they represent an index of relative optimism and pessimism. The Hurwicz criterion calculation is

$$H = \alpha(\text{max}) + (1 - \alpha)(\text{min})$$

If α is taken to be 0, the Hurwicz criterion is equivalent to the minimax criterion; if α is taken to be 1, the Hurwicz criterion is equivalent to the maximax criterion.

The determination of α for any individual is clearly difficult. One method (the value of which is open to some question) is as follows: In the payoff matrix below, that value of x is determined

States of nature

	N_1	N_2
A_1	0	1
A_2	x	x

which causes the decision maker to be indifferent between A_1 and A_2. This will occur when the Hurwicz criterion for each is the same, that is,

$$H(A_1) = H(A_2)$$
$$\alpha(1) + (1 - \alpha)(0) = x(\alpha) + x(1 - \alpha)$$
$$\alpha = x$$

If the decision maker can express his indifference between A_1 and A_2 by giving a value to x between 0 and 1, he has determined his α and $(1 - \alpha)$. This procedure, of course, is not unlike the determination of the Von Neumann–Morgenstern utility discussed in Section 15.3.

EXAMPLE

Take the previous example of Sam Speculator. Suppose Sam is feeling lucky the day he gets the "hot tips," and assigns $\alpha = 0.7$. Then the Hurwicz value for each set is calculated as

$$A_1 = (.7)(9000) + (.3)(-5000) = 4800$$
$$A_2 = (.7)(2000) + (.3)(-2000) = 800$$
$$A_3 = (.7)(3000) + (.3)(-3000) = 1200$$
$$A_4 = (.7)(4000) + (.3)(-4000) = 3600$$
$$A_5 = (.7)(5000) + (.3)(-5000) = 2000$$
$$A_6 = (.7)(6000) + (.3)(-6000) = 2400$$
$$A_7 = (.7)(7000) + (.3)(-7000) = 2700$$

Using the Hurwicz criterion, Sam would choose strategy A_1, that is, buy all three stocks.

15.5 DECISIONS UNDER CONFLICT

The main characteristic of decisions under conflict is that the decision maker is faced with one or more rational opponents. The choices of acts open are possible competitive strategies available to the decision maker, while the states of nature then become competitive strategies available to the opponent or opponents. The decision maker's objective and the opponent's objective are in conflict. Each is assumed to know the other's objective as well as his or her own, and this knowledge influences the choice of strategies' As far as the likelihood of occurrence of the various strategies open to the opponent, the decision maker can assume that the opponent is rational and thus will attempt to obtain as much as possible for himself or herself at the expense of the decision maker. The decision maker is, of course, attempting the same thing with respect to the opponent.

The analysis of decisions under conflict is called *game theory*. This methodology was originally developed in an attempt to solve an economic problem—the oligopoly problem.* It did not solve the oligopoly problem, but found application elsewhere, especially in the analysis of business and warfare situations. Any situation that fits the above description of one or more persons who are engaged in a process of opposition and who are trying to make some gain from the other (or others) may be called a game. The use of the word "game" derives from the fact that most recreational games such as ticktacktoe, checkers, backgammon, chess, poker, bridge, and other card games can be analyzed as games of strategy. As usually formulated, gambling games such as dice and roulette are not games of strategy—a person playing one of these games is "playing against the odds," not against a rational opponent, and the problem thus involves maximizing expected profits or minimizing expected losses.

---

* It was developed by Von Neumann and Morgenstern in the same context in which they developed their concept of utility, mentioned earlier in this chapter, *The Theory of Games and Economic Behavior*, (Princeton, N.J.: Princeton University Press, 1944).

There are several fundamental distinctions by which games are classified for solution. The most important are: (1) the number of persons, (2) the nature of the payoff, and (3) the number of available strategies.

Number of Persons Games are classified according to the number of distinct sets of interests or objectives present in the game. From the game theory point of view, the number of persons in the game is not necessarily the same as the number of people playing the game. If two or more players form a coalition in which they agree to pool their winnings or losses, game theory analysis treats them as a single person.

One-person games, for example, solitaire and various gambling games, are solved as problems involving maximizing expected profit. There is only one person involved who selects the course of action that results in the highest profit on the average. As noted, such situations are not really games of strategy, since the player does not have a responsive rational opponent. However, under certain circumstances, one-person games may be regarded as a special kind of two-person game in which Nature is the other player. For example, this may be a useful viewpoint if the player does not know enough about Nature's habits to select the course of action that will yield the most on the average. In such cases game theory is applicable.

Most of the work done thus far in game theory deals with two-person games, and the following discussion is confined to the two-person case. For the others there is a lack of a satisfactory decision criterion due to the possibility of side payoffs or losses (nonzero sum aspect mentioned below), and the possibility of temporary coalitions formed among otherwise opponents when there are more than two persons. Note, however, the many situations that are not strictly two-person games may be analyzed as though they were. For example, the interests in a game of cards can be considered as "his" or "her" and "everybody else's."

The Payoff Another important characteristic for classifying games is the payoff—what happens at the end of the game? The distinction in this respect is between zero-sum games and nonzero-sum games. If the sum of the payoffs to all players of a game is zero, counting winnings as positive and losses as negative, then the game is *zero-sum*; otherwise, it is *nonzero-sum*. Thus in a zero-sum game anything won by one player is lost by another player. The importance of this distinction lies in the fact that a zero-sum game is a closed system and a nonzero-sum game is not. Almost all parlor games are zero-sum and many other situations can be analyzed as zero-sum games. Most of the work in game theory has concerned zero-sum games and the discussion in subsequent sections will be confined to such games. (Note that a non-zero-sum game may be made zero-sum by adding a fictitious player, say Nature, but this necessitates a more difficult analysis, especially if the original game was a two-person game.)

Strategies In game theory a strategy for a particular player is a plan that specifies the action for every possible action of his or her opponent; a strategy is a complete plan for playing the game, without any connotation of skillfulness on the part of the player. In a game completely amenable to analysis, it is possible (conceptually, if not actually) to foresee all eventualities and thus to catalog all possible strategies. Games are classified according to the number of strategies available to each player: if player 1 has m possible strategies and player 2 has n possible strategies (and they are the only players), then the game is $m \times n$ (that is, m rows by n columns). The important distinction for classifying games on the basis of strategies is between finite games and infinite games. If the greatest number of strategies available to any player is finite, then the game is *finite*; if at least one player has an infinite number of available strategies, then the game is *infinite*. The theory of infinite games is very difficult and will not be discussed. For the analysis of finite games, it is convenient to distinguish three cases: those in which the player having the least number of strategies has two, three, or more than three. The discussions in following sections will generally concern finite, zero-sum, two-person games.

The Game Matrix   A problem is usually set up for game theory analysis in the form of a game matrix. A *game matrix* (or *payoff matrix*) is a rectangular array of the payoffs where the rows represent the strategies of one player and the columns represent the strategies of the other player (thus an $m \times n$ game is represented by an $m \times n$ game matrix.) It is conventional to write the payoffs from the point of view of the player whose strategies are associated with the rows of the matrix—the payoffs for the other player (in a zero-sum game) are then given by the negative of this matrix.

Game theory thus assumes that the strategies available to each player can be enumerated and that the corresponding payoffs can be expressed in meaningful (though not necessarily monetary) units. This information is sufficient for solution of the game—for determining which choice of strategies each player should make—assuming that each player wishes to maximize his or her minimum winnings or minimize his or her maximum losses.* Note that this is not the only possible criterion for solving a game matrix and that its use leads to a conservative theory, since the opponent is assumed to be skillful and to use his or her best strategy.

Saddle Points   If a game matrix contains an entry that is simultaneously a maximum of row minima and a minimum of column maxima, this minimax entry is said to be a *saddle point* of the game and the game is said to be *strictly determined*. In this case, according to the criterion of game theory, the *optimal strategies* for the respective players are represented by the row and column whose intersection is the saddle point. The *value*, v, of the game is the value of the saddle point; a game is said to be *fair* if its value is zero. (See Figure 15.3.)

The first step in the solution of a matrix game is to check for the existence of a saddle point. If one is found, the game is solved; if not, further analysis is necessary. Checking for a saddle point is usually done by writing the row minimum beside each row and the column maximum at the bottom of each column and then determining the maximum of the minima and the minimum of the

---

* The minimax theorem, the key result of the theory of games, states that such a minimax solution exists for every finite zero-sum two-person game.

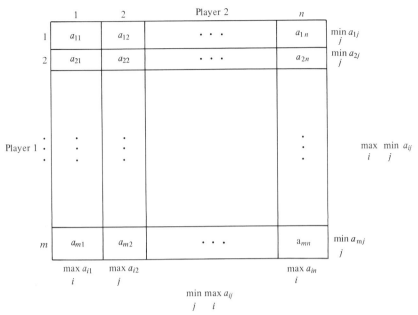

Figure 15.3 Saddle point search in game matrix

maxima. (A saddle point can also be determined by checking for an entry that is simultaneously the minimum of the row in which it occurs and the maximum of the column in which it occurs.)

EXAMPLE
Check the following games for saddle points

(a)   *Minimum across rows*

−2	−3	0	−1	−3
14	①	6	5	1
6	−6	13	6	−6
−7	−9	18	−12	−12

Maximum down columns   14   1 18   6

There is a saddle point at the intersection of the second row (player 1's optimal strategy) and the second column (player 2's optimal strategy); the value of the game is 1.

(b)   *Minimum across rows*

15	13	6	3	−20	3	−20
20	−12	7	5	−10	0	−12
−19	9	8	−6	−17	2	−17
12	3	0	4	−8	−6	−8

Maximum down columns   20   13 8   5   −8   3

There is a saddle point at the intersection of the fourth row (player 1's optimal strategy) and the fifth column (player 2's optimal strategy); the value of the game is −8.

(c)   *Minimum across rows*

-5	6	15	-5
4	3	0	0
-6	-7	18	-7	*max min = 6*
2	-3	0	-3
7	14	6	6

Maximum down columns   7   14   18

There is no saddle point.

(d)   *Minimum across rows*

3	-5	6	17	8	12	-4	0	6	5	-8
8	7	-1	2	0	3	-4	-5	16	0	9	$\boxed{-5}$ *max min = -5*

Maximum down columns 8   7   6 17 8 12 $\boxed{-4}$ 0 16 5   9

min max = -4

There is no saddle point.

Two-Person Two-Strategy Games   The basic concepts of game theory analysis are illustrated in the following sections for two-person two-strategy games, which are games where at least one of the two players has only two strategies. By convention, the strategies for player 1 are listed and indexed in a column along the left edge of the game matrix and the strategies for player 2 are listed and indexed in a row along the top edge. The payoffs are to player 1; a positive number indicates a payoff from player 1 to player 2.

2 × 2 Games   The most easily analyzed two-strategy games are 2×2, which are games for which each player has only two possible strategies. Solution of 2×2 subgames is also frequently necessary as a step in the solution of larger ($2 \times n$ or $m \times 2$) two-strategy games.

EXAMPLE
The game

Player 2
                1   2

Player 1 1 | 0 | 6 | ⓪
            2 | -5 | -1 | -5

                ⓪   6

is strictly determined and fair. Player 1's optimal strategy is 1; player 2's optimal strategy is 1.

EXAMPLE
The game

Player 2

                1   2

Player 1 1 | 10 | 3 | ③
            2 | 9 | -8 | -8

                10   ③

is strictly determined, but not fair (its value is 3). Player 1's optimal strategy is 1; player 2's optimal strategy is 2.

EXAMPLE
The game

Player 2

$$
\begin{array}{c c}
& 1\ \ 2 \\
\textit{Player 1}\ \ \begin{array}{c} 1 \\ 2 \end{array} &
\begin{array}{|c|c|}
\hline
0 & 4 \\
\hline
5 & 0 \\
\hline
\end{array}
\begin{array}{c} 0 \\ 0 \end{array} \\
& 5\ \ 4
\end{array}
$$

is not strictly determined.
Theorem: The matrix game

$$
G = \begin{array}{|c|c|}
\hline
a & b \\
\hline
c & d \\
\hline
\end{array}
$$

is nonstrictly determined if and only if one of the following two conditions
is satisfied:
 (i) $a < b$, $a < c$, $d < b$, and $d < c$
 (ii) $a > b$, $a > c$, $d > b$, and $d > c$
(These equations mean that the two entries on one diagonal of the matrix
must each be greater than each of the two entries on the other diagonal.)

In a nonstrictly determined game there is no clearly optimal strategy for
either player to use consistently. Furthermore, consistent use of any particular
strategy by either player can be used to the advantage of the other player. Thus
there is an important difference between strictly determined and nonstrictly
determined games: in a strictly determined game an optimal strategy for each
player and no "security measures" are necessary; in a nonstrictly determined
game optimal play involves preventing the opponent from knowing what
strategy one plans to use on a given play. This is accomplished by selecting
the strategy to be used for each play at random, according to probabilities
computed from the game matrix. Such a strategy which consists of a probability
mixture of more than one (pure) strategy is called a *mixed strategy*.

Thus the solution of a 2×2 nonstrictly determined game consists of a
pair of probabilities p_1 and $p_2 = 1 - p_1$ with which player 1 selects (at random)
his or her strategies 1 and 2, respectively, and a pair of probabilities q_1 and $q_2 =
1 - q_1$ with which player 2 selects (at random) his or her strategies 1 and 2,
respectively. These probabilities provide the optimal mixed strategies by which
each player can maximize his or her minimum expected profits or minimize
his or her maximum expected losses, respectively.

The value of p_1 can be obtained by equating the two possible expected
payoffs for player 1 and solving for p_1, since for any other p_1 one or the other
of the two expected payoffs is less and thus the minimax criterion is violated.
Similarly, q_1 can be obtained by equating the two possible expected payoffs
for player 2 and solving.

If

$$
G = \begin{array}{|c|c|}
\hline
a & b \\
\hline
c & d \\
\hline
\end{array}
$$

represents the game matrix, then the expected payoff to player 1 is $ap_1 +
c(1 - p_1)$ if player 2 uses strategy 1. The expected payoff to player 1 is $bp_1 +
d(1 - p_1)$ if player 2 uses strategy 2. Thus, equating the expected payoffs, we

have

$$ap_1 + c(1 - p_1) = bp_1 + d(1 - p_1)$$

$$p_1(a - b - c + d) = d - c$$

$$p_1 = \frac{d - c}{a - b - c + d}$$

$$p_2 = 1 - p_1 = \frac{a - b}{a - b - c + d}$$

Similarly, equating the (negative) expected payoffs to player 2, we have

$$aq_1 + b(1 - q_1) = cq_1 + d(1 - q_1)$$

$$q_1(a - b - c + d) = d - b$$

$$q_1 = \frac{d - b}{a - b - c + d}$$

$$q_2 = 1 - q_1 = \frac{a - c}{a - b - c + d}$$

The value of a game has the same meaning for strictly and nonstrictly determined games: the value of a game is the quantity that a player can expect to win by good play against good play by the opponent. On the average, a player cannot win more than the value of a game unless his, or her opponent plays poorly, nor can the player win less than the value of a game unless he or she plays poorly.

The value of the game

$$
G = \begin{array}{c} & \begin{array}{cc} q_1 & q_2 \end{array} \\ \begin{array}{c} p_1 \\ p_2 \end{array} & \begin{array}{|c|c|} \hline a & b \\ \hline c & d \\ \hline \end{array} \end{array}
$$

(to player 1) is

$$V = ap_1 + c(1 - p_1) = bp_1 + d(1 - p_1) = -[aq_1 + b(1 - q_1)]$$

$$= -[cq_1 + d(1 - q_1)]$$

$$V = \frac{ad - bc}{a - b - c + d}$$

EXAMPLE
The game

$$
\begin{array}{|c|c|} \hline 3 & 1 \\ \hline 0 & 2 \\ \hline \end{array}
$$

is nonstrictly determined. The optimal mixed strategies are:

$$p_1 = \frac{2}{4} = \frac{1}{2}$$

$$p_2 = \frac{2}{4} = \frac{1}{2}$$

$$q_1 = \frac{1}{4}$$

$$q_2 = \frac{3}{4}$$

$$V = \frac{6}{4} = \frac{3}{2} \qquad \text{(that is, the game is biased in favor of player 1)}$$

EXAMPLE
The game

−6	2
3	−2

is nonstrictly determined. The optimal mixed strategies are:

$$p_1 = \frac{-5}{-13} = \frac{5}{13}$$

$$p_2 = \frac{-8}{-13} = \frac{8}{13}$$

$$q_1 = \frac{-4}{-13} = \frac{4}{13}$$

$$q_2 = \frac{-9}{-13} = \frac{9}{13}$$

$$V = \frac{6}{-13} = \frac{-6}{13} \qquad \text{(the game is biased in favor of player 2)}$$

EXAMPLE
The game

4	−9
2	5

is nonstrictly determined. The optimal mixed strategies are:

$$p_1 = \frac{3}{16}$$

$$p_2 = \frac{13}{16}$$

$$q_1 = \frac{14}{16} = \frac{7}{8}$$

$$q_2 = \frac{2}{16} = \frac{1}{8}$$

$$V = \frac{38}{16} = \frac{19}{8} \qquad \text{(the game is biased in favor of player 1)}$$

EXAMPLE
The game

5	−35
−15	15

is nonstrictly determined. The optimal mixed strategies are:

$$p_1 = \frac{30}{70} = \frac{3}{7}$$

$$p_2 = \frac{40}{70} = \frac{4}{7}$$

$$q_1 = \frac{50}{70} = \frac{5}{7}$$

$$q_2 = \frac{20}{70} = \frac{2}{7}$$

$$V = \frac{-450}{70} = \frac{-45}{7}$$ (the game is biased in favor of player 2)

The optimal strategies of a game are not affected by adding a constant to all payoffs or by multiplying all payoffs by a positive constant. The value of the game is affected by the same transformation as that applied to the payoffs of the game matrix.

EXAMPLE
For each of the games

$$G_1 = \begin{array}{|c|c|} \hline 4 & -3 \\ \hline 0 & 2 \\ \hline \end{array} \qquad G_2 = \begin{array}{|c|c|} \hline 7 & 0 \\ \hline 3 & 5 \\ \hline \end{array} \qquad G_3 = \begin{array}{|c|c|} \hline 8 & -6 \\ \hline 0 & 4 \\ \hline \end{array}$$

$$p_1 = \frac{2}{9}, \qquad p_2 = \frac{7}{9}, \qquad q_1 = \frac{5}{9}, \qquad q_2 = \frac{4}{9}$$

NOTE:

$$G_2 = G_1 + 3 \quad and \quad V(G_1) = \frac{8}{9}$$

$$V(G_2) = 3\frac{8}{9} = \frac{8}{9} + (3)$$

$$G_3 = 2G_1 \quad and \quad V(G_3) = \frac{16}{9} = 2\left(\frac{8}{9}\right)$$

2 × n-GAMES AND m × 2-GAMES

2 × n-games and m × 2-games are games in which one player has two strategies and the other player has more than two strategies; their solution can be reduced to the solution of a 2 × 2 subgame.

As in the solution of 2 × 2 games, the first step is to check for a saddle point. If one exists, the optimal (pure) strategies and the value of the game are thus determined.

EXAMPLE
The game

$$\begin{array}{|c|c|} \hline -6 & -6 \\ \hline -2 & 2 \\ \hline 14 & 5 \\ \hline 6 & 1 \\ \hline -8 & 3 \\ \hline \end{array}$$

has a saddle point at the intersection of the third row (player 1's optimal strategy) and the second column (player 2's optimal strategy); the value of the game is 5.

EXAMPLE
The game

5	1	0	2
14	−6	−3	−8

has a saddle point at the intersection of the first row (player 1's optimal strategy) and the third column (player 2's optimal strategy); the value of the game is zero.

In an $m \times n$ matrix game, row i is said to *majorize* or to *dominate* row h if every entry in row i is as large as or larger than the corresponding entry in row h. Similarly, column j is said to *minorize* or to be *dominated* by column k if every entry in column j is as small as or smaller than the corresponding entry in column k. Any dominated row or dominating column can be omitted from the matrix game without affecting its solution, since such strategies are clearly not optimal. Thus if a $2 \times n$ or $m \times 2$ game does not have a saddle point (i.e., is not strictly determined) all majorized rows and minorized columns should be eliminated as the next step in the solution.

EXAMPLE
The game

has no saddle point. However, row 3 dominates row 2, column 2 is dominated by column 1, and column 3 is dominated by column 4. Thus the game is reduced for calculation to

3	1
2	9

EXAMPLE
The game

8	−1	0	−2
0	6	−1	−8
2	0	−1	−2
3	11	5	10

has no saddle point. However, row 4 dominates rows 2 and 3 and column 2 is dominated by column 4. Thus the game is reduced, for calculation, to

8	0	−2
3	5	10

The solution of a nonstrictly determined $2 \times n$-game consists of probabilities p_1 and $p_2 = 1 - p_1$ with which player 1 selects (at random) his or her strategies 1 and 2, respectively, and probabilities $q_1, q_2, \ldots, q_n$ (where

$\sum_{i=1}^{n} q_i = 1$) with which player 2 selects (at random) his or her strategies $1, 2, \ldots,$ n, respectively. Similarly, the solution of a nonstrictly determined $m \times 2$ game consists of probabilities $p_1, p_2, \ldots, p_m$ (where $\sum_{i=1}^{m} p_i = 1$) for player 1 and probabilities q_1 and $q_2 = 1 - q_1$ for player 2. After dominance has been used to reduce a $2 \times n$ or $m \times 2$ game for calculation, all possible 2×2 games derived from the matrix of this reduced game can be solved. The value of the original game is the value of one of these derived 2×2 games, and the optimal strategies of the original game are also those of that derived 2×2 game (extended for one player by the addition of zeros). Which of the derived games provides the solution of the original game can be determined either by trial and error or graphically.

The trial and error procedure consists of solving the derived 2×2 games until a subgame is found for which the two-strategy player does at least as well (usually better) against all of his or her opponent's other strategies as the player does against the pair appearing in the 2×2 subgame. When such a game is determined, its solution provides the solution of the original game.

EXAMPLE
In the first example above, only one 2×2 game remained

$$\begin{array}{|c|c|} \hline 3 & 1 \\ \hline 2 & 9 \\ \hline \end{array}$$

Its solution is $p_1 = \frac{7}{9}, p_2 = \frac{2}{9}, q_1 = \frac{8}{9}, q_2 = \frac{1}{9}$ and the solution of the 3×4 game is thus $p_1 = \frac{7}{9}, p_2 = 0, p_3 = \frac{2}{9}; q_1 = \frac{8}{9}, q_2 = q_3 = 0, q_4 = \frac{1}{9}$. For both games, $V = \frac{25}{9}$.

EXAMPLE
In the second example above, a 2×3 game remained

$$\begin{array}{|c|c|c|} \hline 8 & 0 & -2 \\ \hline 3 & 5 & 10 \\ \hline \end{array}$$

and thus there are three 2×2 games for (possible) solution. The first of these 2×2 games

$$\begin{array}{|c|c|} \hline 8 & 0 \\ \hline 3 & 5 \\ \hline \end{array}$$

has the solution $p_1 = \frac{1}{5}, p_2 = \frac{4}{5}; q_1 = \frac{1}{2}, q_2 = \frac{1}{2}$. Its value is 4. Against the other remaining strategy of player 2 (column 3 of the reduced games), $p_1 = \frac{1}{5}, p_2 = \frac{4}{5}$ has value

$$\left(\frac{1}{5}\right)(-2) + \left(\frac{4}{5}\right)(10) = \frac{38}{5}$$

which is greater than 4, so the solution of this 2×2 game (extended) is the solution of the original game: $p_1 = \frac{1}{5}, p_2 = 0, p_3 = \frac{4}{5}, p_4 = 0; q_1 = \frac{1}{2}, q_2 = 0,$ $q_3 = \frac{1}{2}, q_4 = 0.$

NOTE:
The 2×2 game

$$\begin{array}{|c|c|} \hline 8 & -2 \\ \hline 3 & 10 \\ \hline \end{array}$$

has the solution $p_1 = \frac{7}{17}$, $p_2 = \frac{10}{17}$; $q_1 = \frac{12}{17}$, $q_2 = \frac{5}{17}$; the value of the game is $\frac{86}{17}$. Against the other remaining strategy of player 2 (column 2 of the reduced game), $p_1 = \frac{7}{17}$, $p_2 = \frac{10}{17}$ has value $\frac{60}{17}$ which is not greater than $\frac{86}{17}$, so the solution of this game is not the solution of the original game.

The 2×2 game

has a saddle point; the solution is $p_1 = 0$, $p_2 = 1$; $q_1 = 1$, $q_2 = 0$; the value of the game is 5. Against the other remaining strategy of player 2 (column 1 of the reduced game), $p_1 = 0$, $p_2 = 1$ has value 3 which is *not greater* than 5, so the solution of this game is *not* the solution of the original game.

Graphically, the 2×2 game whose solution is the solution of a $2 \times n$ game is determined as follows: Plot the payoffs of the n strategies of player 2 on separate vertical axes and connect the pairs of points by straight lines; locate the highest maximum point on the lower boundary of the figure. The lines that intersect at this point identify the strategies player 2 should use in his or her optimal strategy. Note that it may be possible to reduce the game by dominance before doing graphical analysis.

EXAMPLE
In the game

$$
\begin{array}{|c|c|c|c|}
\hline
5 & 4 & 2 & 1 \\
\hline
7 & 6 & 3 & 6 \\
\hline
12 & 8 & 10 & 9 \\
\hline
6 & 18 & -9 & 14 \\
\hline
\end{array}
$$

row 3 dominates rows 1 and 2 and column 3 is dominated by column 1, thus the reduced game is

$$
\begin{array}{|c|c|c|}
\hline
8 & 10 & 9 \\
\hline
18 & -9 & 14 \\
\hline
\end{array}
$$

which can be reduced further by graphical analysis.

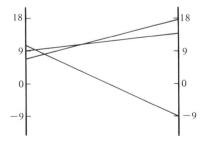

The 2×2 game to be solved is

$$
\begin{array}{|c|c|}
\hline
8 & 10 \\
\hline
18 & -9 \\
\hline
\end{array}
$$

and its solution is $p_1 = \frac{27}{29}$, $p_2 = \frac{2}{29}$; $q_1 = \frac{19}{29}$, $q_2 = \frac{10}{29}$. Thus the solution of the original game is $p_1 = p_2 = 0$, $p_3 = \frac{27}{29}$, $p_4 = \frac{2}{29}$; $q_1 = 0$, $q_2 = \frac{19}{29}$, $q_3 = \frac{10}{29}$, $q_4 = 0$. The value is -4.

Similarly, for an $m \times 2$ game, the payoffs of the m strategies of player 1 are plotted and the lines that intersect at the lowest point on the line segments that form the upper boundary of the figure identify the strategies player 1 should use in his or her optimal strategy.

EXAMPLE
In the game

10	8	11	-2
14	6	-5	5
9	7	5	-4
5	4	-3	3

Row 1 dominates row 3 and column 4 is dominated by columns 1 and 2. Thus the reduced game is

11	-2
-5	5
-3	3

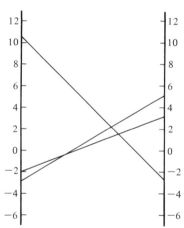

which can be reduced further by graphical analysis. The 2×2 game to be solved is

11	-2
-5	5

and its solution is $p_1 = \frac{10}{23}$, $p_2 = \frac{13}{23}$; $q_1 = \frac{7}{23}$, $q_2 = \frac{16}{23}$. Thus the solution of the original game is $p_1 = \frac{10}{23}$, $p_2 = \frac{13}{23}$, $p_3 = p_4 = 0$; $q_1 = q_2 = 0$, $q_3 = \frac{7}{23}$, $q_4 = \frac{16}{23}$. The value is $\frac{45}{23}$.

EXAMPLE
Mrs. Business Tycoon has a choice of buying one of three companies: A, B, C. The outcome of her purchase depends on whether a particular conglomerate firm executes a merger, divests itself of a subsidiary or maintains the status quo. In case of a merger, Company A results in a gain of 20, Company B in

a loss of 25, and Company C in a gain of 12. In case of divesting, Company A results in a loss of 5, Company B in a loss of 10, and Company C in a loss of 12. If the status quo is maintained, Company A results in a gain of 5, Company B in a gain of 30, and Company C in a loss of 4 (amounts are in thousands of dollars).

(a) In order to maximize her minimum expected gains, what company should Mrs. Business Tycoon buy?

(b) If divesting were impossible, but the gains and losses from the other possibilities were unchanged, what company should she buy to maximize her minimum expected gains.

(a)	MERGER	DIVEST	STATUS QUO
A	20	-5	5
B	-25	-10	30
C	12	-12	-4

There is a saddle point at row 1 and column 2; so buy Company A.

(b)	MERGER	STATUS QUO
A	20	5
B	-25	30
C	-12	-4

$$20p - 25(1 - p) = 5p + 30(1 - p)$$

$$45p - 25 = 30 - 25p$$

$$70p = 55$$

$$p = \frac{11}{14} \qquad 1 - p = \frac{3}{14}$$

So buy Company A with probability $\frac{11}{14}$ and Company B with probability $\frac{3}{14}$.

There are various devices available for solving games larger than those having two strategies for one player. However, unless such a game has a saddle point or can be reduced considerably by dominance, solution usually involves rather tedious trial and error. The development of linear programming has led to the solution of games by electronic computers. Linear programming is concerned with the problem of maximizing or minimizing a linear form (or linear function) whose variables are restricted to values satisfying a system of constraints that are a set of linear equations and/or linear inequalities. A matrix game can be expressed as a problem of this type. The simplex method, a systematic iterative procedure readily adapted for use on computers, is generally considered to be the most effective procedure for solving linear programming (and thus game theory) problems. Deterministic and probabilistic linear programming are topics outside the scope of this book.

PROBLEMS

15.1 A gambler bets $1.00 on a game of dice. Two dice are thrown: if the result is 3, 4, 9, 10, or 11, the gambler wins $1.00; if the result is 2, the gambler wins $2.00; if the result is 12, the gambler wins $3.00; otherwise he loses. What is the gambler's expectation?

15.2 The two teams playing the World Series are evenly matched (that is, they each have probability $\frac{1}{2}$ of winning any particular game). What is the expected number of games in the World Series?

15.3 An urn contains 2 black balls and 3 red balls. Player A draws a ball first, then (without replacement) player B draws a ball, and so on alternately. The first player to draw a red ball wins $10.00. What is player A's expectation?

15.4 An urn contains 6 red balls and 4 white balls. Two balls are drawn at random without replacement; a red ball is worth $2 and a white ball is worth $5. What is the expected value of a draw?

15.5 A large batch of light bulbs contains an unknown proportion of defectives. The producer can either reject the batch or offer it for sale with a double-your-money-back guarantee. What proportion defective can the producer tolerate and still prefer (in the minimax sense) to sell rather than reject the bulbs?

15.6 For the week of the World Series a hotel manager can either lease his hotel for a convention, make reservations as they are requested, or refuse reservations. His profit if he leases is $10,100. If he accepts reservations at the normal price until the pennant winners are determined and then increases the price, his profit is $15,000 if his home team wins the pennant and $5,000 if his home team loses. If he refuses reservations at the normal price, hoping to increase the price if his home team wins the pennant, his profit will be $20,000 if the team wins the pennant but only $1,000 if it loses. What should the manager do to maximize his minimum expected profit if he thinks his home team has probability $\frac{5}{8}$ of winning the pennant? What is the expected opportunity loss of this decision?

15.7 A man has leased a refreshment stand at an amusement park for the summer and must order supplies in advance for the entire summer in order to qualify for a sizable discount. If he serves hot coffee, his profit will be $600 if the summer is relatively cool, but he will lose $150 if the summer is relatively hot. If he serves cold pop, he will lose $100 if the summer is cool and profit $300 if it is hot. If he serves both, he will profit $200 if the summer is cool and $100 if it is hot. If he thinks the probability of a relatively cool summer is $\frac{1}{4}$, what should he do in order to maximize his minimum expected profit? What is the expected opportunity loss of this decision?

15.8 A uranium prospector must decide whether to continue his search. He finds it difficult to assess the possible outcomes in monetary terms, but in terms of relative satisfaction he will lose 400 if he stops and there is uranium, he will gain 200 if he stops and there is no uranium, he will gain 1000 if he continues and finds uranium, and he will lose 300 if he continues and does not find uranium. How probable must he think it is that uranium is present in order to continue looking (based on the minimax criterion)?

15.9 It has turned unseasonably cold in mid-July and the proprietor of a boat rental business must decide whether to close the business for the summer (store the boats, dismiss the help, etc.), close it temporarily (dismiss the help but leave the boats out), or enter into a contract with a man who will operate the business and share profits or losses with the proprietor. If he closes, his profit for the remainder of the summer would, of course, be zero. If he closes temporarily and the weather warms within two weeks, he will reopen and profit $5000; if the weather does not warm within two weeks but does warm by Labor Day, he will reopen for that weekend and profit $1000; if the weather stays cold and he does not reopen it will cost him $100 to have the boats put into storage at the end of the season. If he signs the contract with a manager, he will profit $3000 if they reopen immediately, $2000 if they reopen only for Labor Day and will lose $1000 if the weather stays cold. If he thinks the probability it will warm within two weeks is $\frac{1}{4}$, the probability it will warm for Labor Day is $\frac{1}{4}$, and the probability it will stay cold is $\frac{1}{2}$, what should he do to maximize his minimum expected profit? What is the expected opportunity loss of this decision?

15.10 An investor has $100,000 either to place in stocks or to deposit in a savings and loan at 4 percent. If business conditions are relatively good, the stocks will earn 16 percent; if business conditions are relatively poor, the stocks will earn only 2 percent. What probability must be attached to relatively good business conditions in order to maximize the minimum expected profits by investing in stocks?

15.11 Solve the following matrix games using the minimax criterion.

(a)

8	10	13	16	9
10	12	6	15	10
16	18	9	13	25
4	9	18	20	6

(b)

5	8	7
−1	−3	10
2	12	−6

(c)

10	8	6	2
15	12	2	4
−4	6	−3	1
12	−2	8	−6
16	13	7	12

(d)

14	3	5	8	4
−3	−1	6	−2	5
2	12	10	13	16

15.12 Mrs. A and Mrs. B each have a small garden in which they grow produce to sell to their neighbors. Each of them grows either tomatoes, flowers, or strawberries in any particular year but they are very secretive about their plans. If Mrs. A grows tomatoes, her profit is $100 if Mrs. B also grows tomatoes, $150 if Mrs. B grows strawberries, and $200 if Mrs. B grows flowers. If Mrs. A grows strawberries, her profit is $180, $125, $200 if Mrs. B grows tomatoes, strawberries, flowers, respectively. If Mrs. A grows flowers, her profit is $140, $125, $100 if Mrs. B grows tomatoes, strawberries, flowers, respectively. In order to maximize their minimum expected profit, what should Mrs. A and Mrs. B grow?

15.13 A plant manager must set up his reactors to produce a certain type of polythene using either process 1, 2, 3, 4, or 5. Unfortunately, the chemical raw material varies in nitrogen content and may contain 3, 4, 5, or 6 percent nitrogen. The nitrogen content affects the relative efficiency of the 5 processes. With 3 percent nitrogen, the 5 processes have an output of 50, 45, 60, 50, and 30 tons, respectively. With 4 percent nitrogen, the outputs are 60, 70, 75, 90, 60 tons, respectively; with 5 percent nitrogen, the outputs are 30, 55, 60, 45, 70 tons, respectively; with 6 percent nitrogen, the outputs are 45, 80, 80, 65, 85 tons, respectively. Testing for nitrogen content is too expensive to be practical. What process should the plant manager use in order to maximize the minimum output?

15.14 The Defense Department plans to award a contract for a new missile range at one of two locations, A or B. A real estate speculator intends to invest $5000 in land—all at location A, all at location B, or half at each location. If she buys at location A, the land will be worth $10,000 if the missile range is built there, but $3000 if it is built at location B. If she buys at location B, the land will be worth $4000 if the missile range is built at location A and $8000 if it is built at location B. If she buys at both locations, the land will be worth $6000 if the missile range is built at location A and $5000 if it is built at location B. In order to maximize her minimum expected profits, what should be speculator do?

15.15 Player 1 can choose (on each play of a game) either of the numbers 2, 5, 6, and player 2 can choose either of the numbers 1, 3, 4. If the sum of the two numbers is even, player 1 wins that amount; if the sum is odd, player 1 pays player 2 $5. What should each player do in order to maximize his minimum expected winnings? On the average, which player will win how much in the long run?

15.16 An automobile manufacturer has 5 proposed designs for next year's new cars. Which of these is likely to sell best depends largely on whether business conditions are excellent, good, fair, or poor. If business conditions are excellent, his profit (millions of dollars) will be 100, 150, 50, 125, and 90, respectively; if business conditions are good, his profit will be 80, 55, 55, 60, and 70, respectively; if business conditions are fair, his profit will be 150, 100, 100, 100, and 125, respectively; if business conditions are poor, his profit will be 50, 80, 25, 80, and 75, respectively. What design should he choose to maximize his minimum expected profit?

15.17 The president of Technoproducts, Inc. is planning his budget for the coming year. He can put supplementary funds into research and development, sales, or management. Technoproducts is negotiating for two large contracts and the outcome of these negotiations will determine the efficacy of the possible budget allocations as follows:

Number of Contracts

	0	1	2
R & D	-100	200	500
Sales	1000	-300	-800
Management	-200	400	500

(a) The president of Technoproducts thinks that the probability of obtaining each contract is $\frac{1}{2}$ and that these events are independent. On this basis, determine the probabilities of obtaining neither contract, one (either) contract, both contracts.

(b) On the basis of the probabilities obtained in (a), what should the president do to maximize expected value?

(c) What would the president be willing to pay for perfect information?

15.18 Consider the following payoff matrix where $C_1, C_2,$ and C_3 represent possible choices or decisions and N_1, N_2, N_3, N_4 and N_5 represent possible states of nature.

	N_1	N_2	N_3	N_4	N_5
C_1	0	100	150	-80	-10
C_2	450	0	-10	-300	30
C_3	-100	50	400	0	100

Determine the optimal decision on the basis of:
(a) Hurwicz criterion, alpha = .7
(b) Maximin
(c) Maximax
(d) Minimax regret
(e) Minimax expectation

15.19 Using the minimax criterion, find the optimal strategies and the value for each of the following two-person zero-sum matrix games.

(a)

18	3	-20	-6
7	-2	-8	5
9	7	-10	-6
3	0	-5	10
10	8	15	12
2	6	-13	5

(b)

-3	0	5	-6
2	5	10	0
-5	-1	3	-8
4	9	-2	3

(c)

0	−2	3	5	−6	−1	9	−8
1	0	4	3	−5	−4	5	−10

15.20 For each of the payoff matrices, obtain the minimax solution for each player and the expected value of the game.

(a)

	B_1	B_2	B_3	B_4
A_1	0	2	−6	2
A_2	−6	9	3	−2
A_3	10	−8	4	1
A_4	4	14	10	3

(b)

	B_1	B_2	B_3
A_1	3	0	4
A_1	4	5	−3

15.21 Team A and team B are going to play a football game and Mr. Jones has been offered 4 different bets: the first is a $10 bet on team A, in which Mr. Jones wins $10 if team A wins or ties and loses $10 if team B wins; the second is a $10 bet on team B, in which Mr. Jones win $10 is team B wins and $5 if team B ties and loses $10 if team A wins; the third is a $20 bet on team A, in which Mr. Jones wins $20 if team A wins and $10 if team A ties and loses $20 if team B wins; the fourth is a $20 bet on team B, in which Mr. Jones wins $20 if team B wins and loses $20 if team A wins and neither wins nor loses in case of a tie. If Mr. Jones is going to accept one and only one bet and he thinks the probability team A will win is $\frac{2}{8}$, the probability team B will win is $\frac{5}{8}$ and the probability of a tie is $\frac{1}{8}$, what bet should he accept to maximize his expected winnings? What is the expected opportunity loss of this decision?

15.22 A contractor is going to build a large number of houses for a housing development. Four types of houses have been discussed: colonial, ranch, split-level, and contemporary; the development committee will choose two of these styles for the contractor to construct. The contractor has the opportunity to buy materials in carload lots, thus saving considerable money, but he must order in advance of the committee's decision and can only order one type of materials. If the committee chooses colonial and ranch, the contractor will make an (extra) profit, in thousands of dollars, of 125, 120, 60, and 50 if he ordered colonial, ranch, split-level, and contemporary materials, respectively. If the committee chooses colonial and split-level, he will make 90, 40, 80, 75, respectively; if the committee chooses colonial and contemporary, he will make 150, 30, 75, 100 respectively; if the committee chooses ranch and split-level, he will make 70, 70, 75, 65, respectively; if the committee chooses ranch and contemporary, he will make 90, 80, 80, 120, respectively; if the committee chooses split-level and contemporary, he will make 80, 40, 130, and 80, respectively. How should the contractor order to maximize his minimum expected extra profit?

15.23 A student must decide how to study for a final exam in history. He studies differently for true-false, multiple choice, and essay exams and he doesn't know which type this exam will be. The student thinks that if he studies for a true-false test, he will score 85 on a true-false test, 80 on a multiple choice test, and 75 on an essay test. If he studies for a multiple choice test, he will score 85 on a true-false test, 90 on a multiple choice test, and 85 on an essay test. If he studies for an essay test, he will score 80 on a true-false test, 90 on a multiple choice test and 90 on an essay test.
(a) In order to maximize his minimum expected score, for what type of test should the student study?
(b) If the student thinks the probabilities of a true-false, multiple choice and essay test are, respectively, $\frac{1}{5}$, $\frac{2}{5}$, and $\frac{2}{5}$, for what type of test should he study in order to maximize his expected score?

15.24 Joe Political and Ben Softsell are campaigning for their party's presidential nomination. There are three states (A, B, and C) in which they think campaigning (at this late stage) could change the voting at the party's convention. State A has 40 electoral votes, state B has 30 electoral votes and state C has 80 electoral votes. Joe and Ben each must decide whether to campaign thoroughly in one of the three states (and, if so, in which one) or, alternatively, to divide effort between states A and B. State C is large and is geographically remote from states A and B, so campaigning there precludes campaigning elsewhere. Joe is "favored" in states A and B and Ben is "favored" in state C. A candidate who is "favored" in a state will carry that state unless he does not campaign in the state and his opponent does (with either divided or concentrated effort) or he campaigns in the state with only divided effort and his opponent campaigns with concentrated effort—in these cases, his opponent carries the state. What should each candidate do to maximize his minimum expected electoral vote from these three states?

15.25 Mr. Reno has decided to bet $5 on a basketball game between Anglewood and Barleyville. Mr. Las and Mrs. Vegas have each offered Mr. Reno a bet. Mr. Las wants to bet on Barleyville and is willing to pay Mr. Reno $10 if Anglewood wins and collect $5 if Barleyville wins. Mrs. Vegas wants to bet on Anglewood and is willing to pay Mr. Reno $4 if Barleyville wins and collect $5 if Anglewood wins.
(a) If he wants to maximize his minimum expected winnings, what should Mr. Reno do?
(b) If he wants to maximize his expected winnings, with what probability must Mr. Reno feel Barleyville will win before choosing Mrs. Vegas's bet?

15.26 Transisteronics produces a computer whose performance depends on (among other things) the proper functioning of a small electronic component which is time-consuming to install or replace. Each component costs $10. If a component is defective, another is supplied free of charge; however, Transisteronics incurs the cost of actually replacing the component in the computer. This replacement cost is $10. Alternatively, Transisteronics can pretest each component before installing it in a computer. This pretest costs $5 and, if a component is defective in such a pretest, an additional cost of $1 is incurred due to delay while another component is obtained. It is also possible for Transisteronics to buy components pretested by the supplier (and thus never defective) for $17.
(a) If $\frac{1}{5}$ of the components supplied are defective, what should Transisteronics do to minimize expected cost?
(b) If the proportion of defectives is unknown, what should Transisteronics do to minimize maximum expected cost?

16.1 INTRODUCTION

The basic philosophical difference between classical and Bayesian statistics is concerned with the appropriateness of the use of prior information in inference. Classical statisticians contend that inference, to be defensible, must be based only on the observation or measurement of appropriate data and must not be "biased" by the investigator's prior information or beliefs. Classical statisticians have the attitude that the sample should speak for itself and judgment or prior information should not enter into the decision once the data are obtained. Thus, for example, in classical inference a hypothesis is stated and a procedure is specified so that for every possible set of values of the sample observations, the corresponding decision is known in advance. Thus the decision procedure is specified prior to drawing the sample observations and it is considered incorrect, and even unfair, to look at the data and then specify the decision procedure. Furthermore, the decision is definite—the hypothesis is either accepted or rejected.*

    Bayesian statisticians, on the other hand, contend that the investigator's prior information and beliefs are themselves relevant data and should be considered, along with other more "objective" data, in making inferences. This difference regarding the use of prior information is related, at least in part, to a disagreement concerning the definition of the concept of probability. In Bayesian inference the investigator is assumed to have a prior (subjective) probability distribution that represents his beliefs about the population parameter(s) under consideration—for example, instead of concerning himself with the hypothesis that the mean of a population is, say 50, the Bayesian statistician has a prior probability distribution representing his beliefs about the value of this parameter; his belief about the mean of the population might be represented by a normal distribution with a mean of 50 and a standard deviation of, say 5.

---

    * There are some classical sequential procedures in which one of the permissible decisions at any stage is to obtain additional data before making a final decision. However, the eventual decision is definite, not probabilistic.

The Bayesian statistician also obtains a random sample of observations from the population in question; he uses these data not to accept or reject a hypothesis, but to revise the prior probabilities representing his belief about the value of the parameter. This new revised probability distribution represents his beliefs after incorporating the information provided by the sample with the prior probabilities held before obtaining the sample.

The axioms of probability theory and the algebraic rules for manipulating probabilities that follow from them are accepted by both classical and Bayesian statisticians—even Bayes' rule is unquestioned with respect to its algebraic validity. The controversy is concerned entirely with the definition and interpretation of probabilities, not with their algebraic manipulation.

In a sense, Bayesian statisticians redefine the classical concept of probability as long-run relative frequency in terms of probability as degree of confirmation or belief. The relative frequency concept of probability applies, strictly speaking, only to events that can be repeated indefinitely many times, at least in theory. Although the relative frequency concept of probability is sometimes referred to as "objective" probability, it is clearly subjective in practice, in the sense that a limit is never observed empirically and must be inferred from a finite (though perhaps very large) set of data or from a model, both of which involve some subjectivity. The concept of probability as a degree of confirmation or belief is sometimes referred to as "subjective" probability or "personalistic" probability; however, in many cases, such probabilities are based as much on objective data and as little on subjective judgment as are "objective" probabilities.

The prior information or beliefs used in Bayesian analysis are frequently stated in terms of subjective probabilities and many Bayesian decision problems involve the revision of prior subjective probability distributions. Thus acceptance of the validity and meaningfulness of subjective probability is necessary (although not sufficient) for acceptance of the validity and usefulness of Bayesian analysis.

Actually the classical and Bayesian points of view are not nearly so diametrically opposed as statisticians on both sides of the argument seem to believe. Unfortunately, relatively little has been written concerning the similarities between the two approaches, except for the case when prior information is very weak and, in a sense, the Bayesian approach reduces to the classical approach.

Certainly classical and Bayesian inference differ, both in philosophical approach and in the application of specific statistical analyses. However, at least in some respects, the differences may be more apparent than real.

Many sophisticated methods for incorporating prior information have been developed by Bayesian statisticians and are being applied with increasing frequency in practice. But the argument can be made that classical statisticians also incorporate prior information in their inference procedures although not explicitly, thus, perhaps without realizing it, classical statisticians may be somewhat Bayesian in some of their procedures and attitudes. This is illustrated in the context of statistical hypothesis testing in Section 16.3.

Bayesian analysis is frequently applied in decision theory, even by statisticians who do not use Bayesian methods in other contexts. Bayesian decision analysis is discussed next. Subsequent sections consider Bayesian methods of estimation and testing hypotheses.

16.2 BAYESIAN DECISION ANALYSIS

A number of criteria for determining the best decision under uncertainty are discussed in Chapter 15. In some practical problems it is feasible to obtain further data or information on the basis of which the (objective or subjective) probabilities concerning the states of nature can be revised. Revision of the probabilities is based on Bayes' theorem, and decision analysis involving such revision is referred to as Bayesian decision analysis. Bayes' theorem can be written as follows (refer to Section 4.5):

$$P(\text{State of Nature}|\text{Data}) = \frac{P(\text{Data}|\text{State of Nature})P(\text{State of Nature})}{P(\text{Data})}$$

where the set of probabilities $P(\text{State of Nature})$ are the *prior probabilities* and the set of probabilities $P(\text{State of Nature}|\text{Data})$ are the *posterior probabilities*. Each of these sets of probabilities must sum to 1. Note, however, that the set of conditional probabilities $P(\text{Data}|\text{State of Nature})$ need not add to 1.

One of the most frequently stated objections to Bayesian inference is that the prior distribution is almost always subjective and therefore Bayesian inference is subject to error. Bayesians answer this criticism in three ways:

1. Classical inference is also subjective (with respect to specification of the null and alternative hypotheses and the choice of α and β), and all statistical inference is inherently subject to error.
2. Prior distribution becomes less and less important as sample information accumulates, as illustrated later in this section.
3. Subjectivity is a matter of degree, and there is some subjectivity even in the statement of probabilities based on relative frequencies. Many statements now considered "objective" facts were once considered highly subjective matters of belief, for example, that the world is round. Most scientific theories become "accepted" as a result of repeated revision of the subjective probabilities concerning them.

Suppose a bond trader says that he bought a bond from those listed as New York Stock Exchange bonds, and he doubled his money in a month. The question is whether he made his profit as a result of stock price movements (assume he did if the bond was one convertible into common stock), or if he made his profit on some other basis. It is known that NYSE bonds can be divided into utility and industrial bonds, and that 80 percent of the total are industrials and 20 percent are utilities. Of the total convertible bonds listed, 75 percent are industrials and 25 percent are utilities. Of the total nonconvertible bonds listed, 60 percent are industrials and 40 percent are utilities.

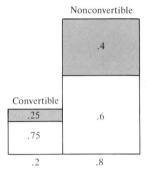

Figure 16.1   Proportion of utility and industrial bonds

Without any further information, the probability of the bond purchased at random being a convertible is .2. But now suppose the trader gave out the information that the bond is an industrial bond. Now what is the probability that the bond is a convertible? The situation can be represented as in Figure 16.1.

The calculations of the probabilities can be laid out as in Table 16.1. Since the additional information given (that the bond is an industrial) corresponds to sample information, it is designated S.

Note that the joint probability is calculated by the multiplication rule:

$$P(S|X) = P(X)P(S|X)$$

and the posterior probability is calculated using Bayes' Rule:

$$P(X|S) = \frac{P(X)P(S|X)}{P(S)}$$

The unconditional probability of S (.63) is often called a marginal probability, because it is found in the margin of the table.

In the above illustration the prior probabilities and the conditional probabilities of the sample are based on relative frequencies and thus are "objective." In the following example these probabilities are "subjective."

TABLE 16.1   PRIOR AND POSTERIOR PROBABILITY DISTRIBUTIONS

Event (X)	Prior Probability $P(X)$	Conditional Probability of Sample $P(S|X)$	Joint Probability of Sample and Event $P(S,X)$		Posterior Probability $P(X|S)$
C	.2	.75	.15	.15/.63 =	.237
NC	.8	.60	.48	.48/.63 =	.763
			.63		1.000

EXAMPLE

Consider the example of Chapter 15 (Section 15.3) concerning Mrs. Magoo and the upcoming British election. Recall that Mrs. Magoo assesses the probability of a Conservative victory as .45 and the probability of a Labor victory as .55 (her prior probabilities). Suppose that Mrs. Magoo asks a reporter friend to predict the outcome of the upcoming election, and the reporter predicts a victory for the Conservative party. On the basis of past experience with the reporter, Mrs. Magoo thinks she would predict a Conservative victory if one were in fact going to occur with probability .9, and she would predict a Conservative victory if in fact the Labor party were going to win with probability .3. On the basis of this information, how should Mrs. Magoo revise her probabilities concerning the outcome of the election? That is, what are her posterior probabilities after talking to her friend?

$$P(C|\text{Predict } C) = \frac{P(\text{Predict } C|C)P(C)}{P(\text{Predict } C)}$$

$$P(L|\text{Predict } C) = \frac{P(\text{Predict } C|L)P(L)}{P(\text{Predict } C)}$$

Note that $P(\text{Predict } C) = P(\text{Predict } C|C)P(C) + P(\text{Predict } C|L)(P(L)$
$$= (.9)(.45) + (.3)(.55) = .57$$

thus

$$P(C|\text{Predict } C) = \frac{(.9)(.45)}{.57} = \frac{.405}{.570} = .71$$

and

$$P(L|\text{Predict } C) = \frac{(.3)(.55)}{(.57)} = \frac{.165}{.570} = .29$$

If her reporter friend had predicted a Labor victory, Mrs. Magoo should revise her probabilities as follows:

$$P(C|\text{Predict } L) = \frac{P(\text{Predict } L|C)P(C)}{P(\text{Predict } L)}$$

$$P(L|\text{Predict } L) = \frac{P(\text{Predict } L|L)P(L)}{P(\text{Predict } L)}$$

Note that $P(\text{Predict } L) = P(\text{Predict } L|L)P(L) + P(\text{Predict } L|C)P(C)$
$$= (.7)(.55) + (.1)(.45) = .43$$

thus

$$P(C|\text{Predict } L) = \frac{(.1)(.45)}{(.43)} = .11$$

and

$$P(L|\text{Predict } L) = \frac{(.7)(.55)}{(.43)} = .89.$$

The revision of Mrs. Magoo's prior probabilities on the basis of her reporter friend's predictions is in the direction common sense would dictate. Mrs. Magoo's prior probabilities are nearly equal, $P(C) = .45$ and $P(L) = .55$. A prediction of Conservative victory by her friend, in whose predictions Mrs. Magoo has considerable confidence, changes these probabilities to $P(C|C) = .71$ and

$P(L|C) = .29$, while a prediction of Labor victory changes them to $P(C|L) = .11$ and $P(L|L) = .89$. These posterior probabilities show clearly the combination of prior probabilities and sample information on which they are based.

The revision of prior probabilities using Bayes' theorem can be represented by a tree diagram as follows:

(Prior) Probability of event	Conditional probability of sample	Joint probability sample, event	Marginal probability of sample	Posterior probability of event

The revision of probabilities using Bayes' rule is often set up as in Table 16.1, which relates the prior and posterior probability distributions. Defining S_C = Predict C, and S_L = Predict L, the following table applies to the previous example:

Event (X)	Prior Probability $P(X)$	Conditional Probability of Sample $P(S|X)$	Joint Probability of Sample and Event $P(S,X)$	Posterior Probability $P(X|S)$
C	.45	$P(S_C|C) = .9$	$P(S_C, C) = .405$	$P(C|S_C) = .71$
L	.55	$P(S_C|L) = .3$	$P(S_C, L) = \underline{.165}$	$P(L|S_C) = .29$
			$P(S_C) = .57$	
C	.45	$P(S_L|L) = .7$	$P(S_L, L) = .385$	$P(L|S_L) = .89$
L	.55	$P(S_L|C) = .1$	$P(S_L, C) = \underline{.045}$	$P(C|S_L) = .11$
			$P(S_L) = .43$	

Notice that only the first two revised probabilities in the above table are needed, because the sample information consists of Mrs. Magoo's friend predicting a Conservative victory. The latter two revised probabilities for sample information of a Labor victory prediction are calculated for illustration.

Using the revised (posterior) probabilities for the payoff matrix given on p. 505, we get

Election victory

		.71 Conservative	.26 Labor
Action on stock	Sell	100	100
	Hold	300	− 200
	Buy	600	− 400

$$E(\text{Sell}) = .71(100) + .29(100) = 100$$

$$E(\text{Hold}) = .71(300) + .29(-200) = \frac{5900}{38} = 155.26$$

$$E(\text{Buy}) = .71(600) + .29(-400) = 310.53$$

Therefore, in order to maximize expected profit, Mrs. Magoo should buy more stock. Note that this is not the same decision as that based on the original probabilities. In the solution in Section 15.3, sell had the highest expected value.

Bayesian decision analysis consists essentially of revising the prior (objective or subjective) probabilities concerning the states of nature in view of additional data or information and basing the decision on these revised probabilities. In the preceding example the conditional probabilities of the reporter's predictions given that certain election outcomes were going to occur were of a rather subjective nature. If the reporter had made predictions for a large number of elections, these conditional probabilities could, of course, have been based on (objective) relative frequencies. In many problems relatively objective determination of these probabilities is possible, as in the following example.

EXAMPLE
The Blue Bolt Company has had a series of mix-ups on its production line, and as a result there is a case containing 100,000 screws in the shipping room ready for shipment but unmarked as to size, production run, or other identification. The company produces three sizes of screws having lengths whose means and standard deviations are, respectively, 5 and 1.5, 6 and 2.5, and 8 and 4 in. The company must decide what to do with the case of screws; there are costs involved in sending the case if it is not the size ordered and costs involved in discarding the case. These costs (including loss of good will and partial replacements costs when relevant) are given by the following matrix:

	$\frac{1}{3}$ $\mu = 5$	$\frac{1}{3}$ $\mu = 6$	$\frac{1}{3}$ $\mu = 8$	Expected value
Ship as $\mu = 5$	0	− 300	− 500	− 800/3 ←
Ship as $\mu = 6$	− 1600	0	− 400	− 2000/3
Ship as $\mu = 8$	− 2000	− 1800	0	− 3800/3
Discard	− 1000	− 1200	− 1500	− 3700/3

(a) Assuming that the screws are equally likely to be each of the three sizes, what should the Blue Bolt Company do to maximize its expected value? (b) Assuming that the lengths of each size of bolts are normally distributed what should the company do if a random sample of 100 bolts has a mean length of 6.8 in.? (Consider 6.8 as including the interval $6.8 \pm .1$ to solve the problem of continuity;—this can be justified on the basis of errors of measurement).

(a) The company would ship the screws to fill an order for screws having an average length of 5 in., since this decision has the minimum expected loss (given at the side of the above table).

(b) Given the sample of size 100 with $\bar{x} = 6.8$ in., the probability of this sample coming from the three possible distributions would be calculated as follows

$$n = 100$$

$$z = \frac{\bar{x} - \mu}{\sigma/\sqrt{n}}$$

$$z = \frac{\bar{x} - \mu}{\sigma/\sqrt{n}} = \frac{6.7 - 5}{.15} = \frac{1.7}{.15} = 12.6; \ P(\bar{x} = 6.8 \pm .1 \,|\, \mu = 5, \sigma = 1.5) = 0$$

$$\left.\begin{array}{l} z = \dfrac{6.9 - 6}{.25} = \dfrac{.9}{.25} = 3.6 \\[2mm] z = \dfrac{6.7 - 6}{.25} = \dfrac{.7}{.25} = 2.8 \end{array}\right\} P(2.8 \leq z \leq 3.6 \,|\, \mu = 6, \sigma = 2.5) = .0026 - .0011 = .0015$$

$$\left.\begin{array}{l} z = \dfrac{6.9 - 8}{.4} = \dfrac{-1.1}{.4} = -2.75 \\[2mm] z = \dfrac{6.7 - 8}{.4} = \dfrac{-0.9}{.4} = -2.25 \end{array}\right\} \begin{array}{l} P(-2.75 \leq z \leq -2.25 \,|\, \mu = 8, \sigma = 4) = 0.122 - .0030 \\ \qquad\qquad\qquad\qquad\qquad\quad = .0092 \end{array}$$

$$P(5 \,|\, 6.8) = \frac{P(6.8 \,|\, 5)P(5)}{P(6.8)} = \frac{0(\frac{1}{3})}{0 + (.0015)(\frac{1}{3}) + (.0092)(1.3)} = 0$$

$$P(6 \,|\, 6.8) = \frac{P(6.8 \,|\, 6)P(6)}{P(6.8)} = \frac{.0005}{.00357} = .14$$

$$P(8 \,|\, 6.8) = \frac{P(6.8 \,|\, 8)P(8)}{P(6.8)} = \frac{.00307}{.00357} = .86$$

The following table summarizes the prior and posterior probability distributions for this example:

EVENT (X)	PRIOR PROBABILITY $P(X)$	CONDITIONAL PROBABILITY OF SAMPLE $P(S\,|\,X)$	JOINT PROBABILITY OF SAMPLE AND EVENT $P(S,X)$	POSTERIOR PROBABILITY $P(X\,|\,S)$
$\mu = 5$ $\sigma = 1.5$	1/3	0	0	0
$\mu = 6$ $\sigma = 2.5$	1/3	.0015	.0005	.14
$\mu = 8$ $\sigma = 4$	1/3	.0092	$\dfrac{.00307}{.00357}$	$\dfrac{.86}{1.00}$

Considering the decision payoff matrix in view of these revised (posterior) probabilities, we have

	0 $\mu = 6$	.14 $\mu = 6$	.86 $\mu = 8$	Expected value
Ship as $\mu = 5$	0	-300	-500	-472
Ship as $\mu = 6$	-1600	0	-400	-344
Ship as $\mu = 8$	-2000	-1800	0	-252
Discard	-1000	-1200	-1500	-1458

So the company would ship the screws to fill an order for screws having an average length of 8 in.

16.3 TESTING SCIENTIFIC AND STATISTICAL HYPOTHESES

A statistical hypothesis is a statement about the distribution of a random variable; most statistical hypotheses concern the value of a parameter of a distribution of known or assumed form, although some hypotheses concern the form of the distribution itself. A test of a statistical hypothesis is a procedure for deciding whether to accept or reject the hypothesis on the basis of the outcome of a random experiment—that is, on the basis of the observed value of a random variable.

The use of a random experiment to test a statistical hypothesis is based on an extension of the method of testing scientific hypotheses by nonrandom experiments. For centuries, hypotheses in the physical sciences have been tested by experiments. Galileo reportedly dropped a large and a small cannon ball from the leaning tower of Pisa in order to test the hypothesis that two objects of different weights fall at the same speed. His hypothesis was supported by the fact that the cannon balls did in fact strike the ground at nearly the same instant. More recently and more spectacularly, Einstein's hypothesis that energy equals mass times the square of the speed of light was supported by the explosion of the atomic bomb.

The logic of testing scientific hypotheses by nonrandom experiments is as follows: According to the hypothesis, a specific result should occur or an observable quantity should have a specified value. If the occurrence predicted by the hypothesis is observed when the experiment is performed, then the hypothesis is supported. It is not proved, however, since there may be other hypotheses that make the same prediction. This procedure is based on the assumption that the experiment is performed precisely as required to test the hypothesis and that there is no error in measurement or observation—that is, it is based on the assumption that there is a "critical experiment" for the hypothesis.

Even under ideal circumstances with regard to conduct of experiments and precision of measurement, a theory or hypothesis can never be proved. As data consistent with a theory or hypothesis accumulate and contradictory data are not observed, the theory or hypothesis is supported and may even eventually be regarded as proved for practical purposes. However, one counterexample or one set of contradictory data is sufficient to disprove a theory or hypothesis and this may occur after a theory or hypothesis is generally considered to be established. No doubt the best-known example of a theory that was virtually universally accepted as true before being disproved is Newton's theory of gravitation, which was accepted for many years before Einstein disproved it.

In the case of a statistical hypothesis the situation regarding acceptance and rejection is not so clear-cut. A statistical hypothesis states that a random variable is distributed in a particular way or that a parameter of its distribution has a specified value. The observed quantity is itself a random variable and generally the hypothesis assigns a nonzero probability to each of its possible values. Since there is no value of the random variable that could not be observed under the stated hypothesis, no observation can lead to rejection of the hypothesis with certainty; that is, there is no critical experiment. However, if certain values of the random variable are very unlikely to occur if the hypothesis is true, then their observation causes the hypothesis to be seriously doubted or rejected.

The distinction between statistical and nonstatistical or scientific hypotheses is not so sharp as it might appear, since in fact all observations involve some uncertainty or error of measurement. For example, a more precise model for Galileo's experiment would consider the difference in time of impact of the two cannon balls to be a random variable to which Galileo's hypothesis assigns an expected value of zero. This formulation emphasizes the fact that although a theory or hypothesis is deterministic by nature, its test may be statistical because of errors in experimentation or measurement.

In statistical hypothesis testing the stated hypothesis is usually referred to as the null hypothesis, since it is often an assertion that a particular variable, treatment, process, or factor has no effect on whatever is being measured or that there is no difference between the effects of two variables, treatments, processes, or factors. Usually the investigator also states an alternative hypothesis which is accepted if the null hypothesis is rejected. In fact, decision theory considers hypothesis testing from the point of view of choosing between two hypotheses (representing two states of nature) rather than from the point of view of testing a null hypothesis against an alternative hypothesis. Frequently, a hypothesis is stated and tested in order to decide between two possible actions, one appropriate if the null hypothesis is true and the other appropriate if it is false. When nonstatistical or scientific hypotheses are tested, the investigator may have an alternative in mind but frequently he does not.

Thus the usual procedure and approach in testing statistical hypotheses can be summarized as follows. An experiment is performed or a random sample is drawn and the value of a random variable is observed. Suppose certain values of the random variable are very unlikely to occur if the null hypothesis is true but are very likely to occur if the alternative hypothesis is true. Then if the random variable is observed to have a sufficiently "unlikely" value, the null hypothesis is rejected and the alternative hypothesis is accepted.

PRIOR DECISIONS IN TESTING HYPOTHESES

The investigator must make two basic decisions in testing a hypothesis. First, he must state precisely the null hypothesis and the alternative hypothesis. Second, he must state the criterion for testing the hypothesis in terms of the observed values of the random variable for which the null hypothesis is to be accepted and the values for which it is to be rejected and the alternative hypothesis accepted. The criterion for testing the hypothesis affects the decision; for a given observation the decision made concerning the hypothesis depends not only on the data observed, but also on several prior decisions made somewhat arbitrarily by the investigator.

Stating a criterion for testing the null hypothesis H_0 against an alternative hypothesis H_A is, in effect, partitioning the sample space (the possible sample values), of the observed random variable into two sets, a region of acceptance for H_0 and a region of rejection for H_0.

The statement of the null and alternative hypotheses determines the general location of the rejection region by placing it in the lower tail, in the upper tail, or in both tails of the sampling distribution of the test statistic. In addition, the investigator must specify the size of the rejection region, α, the probability of rejecting the null hypothesis when it is true (discussed in Chapter 8). The investigator's choice of α depends, at least in part, on his prior confidence or belief in the null hypothesis. If he is quite firmly convinced of the validity of the null hypothesis, he requires very strong evidence against it before rejecting the hypothesis. If, on the other hand, his belief in the null hypothesis is rather weak, he requires less convincing evidence against it before rejecting the hypothesis.

The choice of α is complicated by the fact that the null hypothesis may in fact be false, in which case an error is committed if the hypothesis is accepted. The probability of this error of false acceptance is denoted by β, which can be precisely specified only for a specific alternative hypothesis. (See Sections 8.2 and 8.3).

Unfortunately, for any fixed sample size, if the probability of false rejection of the null hypothesis is decreased, the probability of false acceptance of the null hypothesis is increased. Thus the investigator must base his choice of α and β on his judgment of the relative seriousness of the two types of error.

Suppose, for example, that a company is interested in whether a certain advertising project would increase net profit by a certain amount, say θ_0. Consider the problem first from the point of view of a company executive who believes the cost of the project is excessive and must be shown evidence to the contrary in order to change his view. He would set up the hypothesis testing procedure so that only relatively conclusive evidence favoring the advertising project would change his view of the situation. (See Figure 16.2.)

Figure 16.2   Decision rule for $H_0 : \theta \leq \theta_0$
$H_A : \theta > \theta_0$

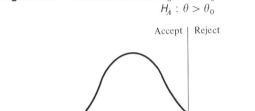

Decision: $\theta \leqslant \theta_0$ $\theta > \theta_0$

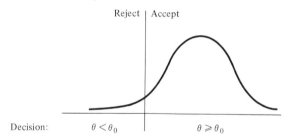

Figure 16.3   Decision rule for $H_0 : \theta \geq \theta_0$
$$H_A : \theta < \theta_0$$

Now consider the problem from the point of view of the advertising manager who believes that the project would be effective and must be shown evidence to the contrary to change his view. He would set up the hypothesis testing procedure so that only relatively conclusive evidence against the advertising project would change his view of the situation. (See Figure 16.3.)

This dependence of the decision on specification of the hypothesis and thus on prior beliefs is summarized in Figure 16.4 for the advertising situation discussed. For sample values corresponding to a test statistic $\tilde{\theta} < A$, both the company executive and the advertising manager decide that the project would not increase net profit by as much as θ_0. Note that this represents a change of opinion on the part of the advertising manager as a result of strong contrary empirical evidence. For sample values corresponding to a test statistic $\bar{\theta} > B$, both the company executive and the advertising manager decide that the project would increase net profit by at least θ_0. This represents a change of opinion on the part of the company executive as a result of strong contrary empirical evidence. For sample values corresponding to a test statistic $\tilde{\theta}$ whose value is

Figure 16.4   Dependence of decision on prior beliefs

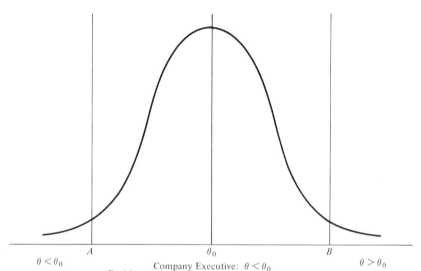

between A and B, both the company executive and the advertising manager maintain their prior beliefs—the company executive believing that the project would not be successful and the advertising manager believing that it would be successful. In this case the empirical evidence is not sufficiently strong to cause either to change his opinion.

Assume that this very simple two-decision analysis is appropriate. If the decision must be made between the null hypothesis and the alternative hypothesis and there is no option of reserving judgment when the data are not "significant," the decision will differ according to the prior beliefs of the decision maker. Except when the empirical evidence is strongly contradictory, the decisionmaker maintains belief in his prior hypothesis, whatever that hypothesis is.

This seems reasonable, assuming that the investigator has a reasonable basis for formulating his hypothesis in the first place. It is also quite subjective and contradicts the contention that in classical hypothesis testing the decision depends only on empirical evidence, while in Bayesian hypothesis testing the decision depends also on the prior information or opinion of the investigator. Clearly, even in classical analysis the decision is influenced by the investigator's statement of the problem (prior belief)—unless the empirical evidence is compelling, the investigator maintains his prior hypothesis. Similarly, in Bayesian analysis prior information is more important when empirical evidence is inconclusive and virtually of no importance when it is weak relative to empirical evidence.

In this context the primary differences between classical and Bayesian analysis concern the explicit degree of emphasis on the prior information or beliefs of the investigator and the degree of refinement in their use. Bayesian analysis assigns probabilities to various prior beliefs or hypotheses of the investigator. Classical analysis says nothing specifically about prior beliefs of the investigator, although the null hypothesis is certainly in some sense such a prior belief and the investigator's confidence in it is indicated by his choice of α— by the probability with which he rejects the hypothesis when it is true. The difference between classical and Bayesian analysis for the problem of testing hypotheses is thus apparently a matter of the explicitness and specificity of the use of prior beliefs in inference, not a matter of whether such beliefs influenced inference. The situation is similar for other problems of statistical inference.

In classical statistical inference a null hypothesis and an alternative hypothesis are stated, and hypothesis testing consists essentially of deciding whether the null or the alternative hypothesis is true. As discussed in the preceding section, prior information can influence the choice of null and alternative hypotheses and the values of α and β, but the decision consists of concluding that one hypothesis is true and the other hypothesis is false. This all-or-none type of decision does not correspond with most observed behavior. A scientist rarely completely discards an idea or hypothesis in which he had great confidence because an experimental result seems to refute it. He is much more likely to modify his (probabilistic) belief in its validity or, perhaps, modify the hypothesis itself. In practical problems it may be necessary at a certain time to choose one particular hypothesis and the corresponding course of action. However, this choice usually reflects the behavior thought to be optimal with respect to the probabilities associated with various hypotheses and the corresponding loss

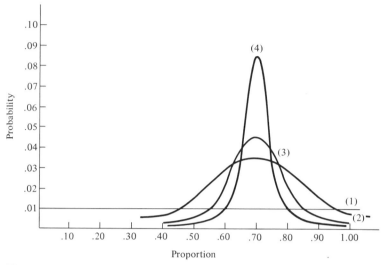

Figure 16.5 Posterior distributions for uniform prior

function, but no certainty that the corresponding hypothesis is correct. Examples of this type of decision analysis are given in Section 16.5.

There are many interesting and important problems for which the possible hypotheses cannot be stated as a discrete set without artificial and usually arbitrary categorization. For example, suppose an investment broker is interested in the proportion of the adult population of the United States that expect an inflation of at least 4 percent during the next year. Clearly, any discrete set of probabilities or intervals representing the possible range of the proportion (from zero to 1) is an artificial representation of a continuous variable. Logically, Bayesian inference proceeds in exactly the same manner for continuous distributions as for discrete distributions. Computationally, the problem in general is more complicated for continuous distributions and in some cases is best handled by a computer. Although exact formulas are not available for most types of prior distributions, the approximate analysis for various prior distributions and sample results can easily be represented graphically. Suppose the broker has a flat (uniform) prior distribution (a) as shown in Figure 16.5. If he obtains a random sample of 12 people of whom 9 expect at least at 6 percent inflation, his revised (posterior) probability distribution is represented by (b); if, say, 45 people of a random sample of 60 expect at least 6 percent inflation, his posterior probability distribution is represented by (c); similarly, if, say, 90 people of a random sample of 120 expect at least 6 percent inflation, the posterior distribution is represented by (d).

Note that for this example, prior information is very weak and the results of the Bayesian analysis are essentially identical to the results of classical analysis. Now suppose that the broker has a prior distribution peaked at .50, on the basis that about half the people believe any proposition not manifestly ridiculous. Figure 16.6 shows the posterior distributions for this peaked prior distribution and the same sample results as in Figure 16.5. We see that as sample information increases, the prior distribution becomes less important.

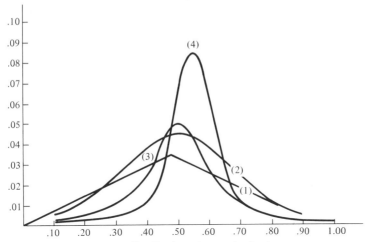

Figure 16.6 Posterior distributions for peaked prior

If normality is assumed, the formulas for the posterior distribution are simple. Suppose a random sample of n observations is drawn from a normal distribution having unknown mean μ and known variance σ^2. The investigator's prior distribution for μ is normal with mean μ_0 and variance σ^2. Then the posterior distribution of μ is normal with mean and variance given by

$$\mu_1 = \frac{\left(\dfrac{1}{\sigma_0{}^2}\right)\mu_0 + \left(\dfrac{n}{\sigma^2}\right)\bar{x}}{\dfrac{1}{\sigma_0{}^2} + \dfrac{n}{\sigma^2}}$$

$$\sigma_1{}^2 = \frac{1}{\dfrac{1}{\sigma_0{}^2} + \dfrac{1}{\sigma^2}}$$

where $\bar{x}$ and σ^2/n are the sample mean and variance. If σ^2 is also unknown the sample variance can be estimated by s^2/n.

The posterior mean is a weighted average of the prior mean μ_0 and the sample mean $\bar{x}$; the weights are proportional to the reciprocals of the corresponding variances.

If the prior information is relatively weak, then the prior distribution is "diffuse" and its variance is relatively large; in this case the sample information has relatively little influence in determining the mean of the posterior distribution. Similarly, if the prior information is relatively strong, then the prior distribution is "peaked" and its variance is relatively small; in this case the sample information has considerable influence in determining the mean of the posterior distribution.

The posterior variance of μ is equal to the reciprocal of the sum of the reciprocal variances; thus it is smaller than either the prior variance or the sample variance.

The influence of prior information on the posterior distribution can be summarized as follows:

If $\sigma_0^2 \to \infty$, $\mu_1 \to \bar{x}$ and $\sigma_1^2 \to \sigma_x^2$ (classical inference)

If $\sigma_0^2 \to 0$, $\mu_1 \to \mu_0$ and $\sigma_1^2 \to \sigma_0^2$ (Bayesian inference)

Clearly, the relative influence of prior sample information on the posterior distribution depends on the relative "strength" of these sources of information; in this context, information is measured by the reciprocal variance of the relevant distribution.

EXAMPLE
Supermarkets, Inc., has 5000 stores. The marketing manager estimates that average monthly sales per store of a proposed new product would be normally distributed with a mean of $500 and a standard deviation of $25. A random sample of 100 stores is obtained and the new product is introduced. Sales records indicate average sales of $400 with a standard deviation of $150. On the basis of this information, what is the manager's posterior distribution of sales of the new products?

$$\mu_0 = 500$$

$$\sigma_0 = 25$$

$$\bar{x} = 400$$

$$s_x = 150$$

$$n = 100$$

$$\mu_1 = \frac{\left(\dfrac{1}{\sigma_0^2}\right)\mu_0 + \left(\dfrac{n}{\sigma^2}\right)\bar{x}}{\dfrac{1}{\sigma_0^2} + \dfrac{n}{\sigma^2}} = \frac{\left(\dfrac{1}{625}\right)500 + \left(\dfrac{100}{22500}\right)400}{\dfrac{1}{625} + \dfrac{100}{22500}} = 426.47$$

$$\sigma_1^2 = \frac{1}{\frac{1}{625} + \frac{100}{22500}} = 165.44$$

$$\sigma_1 = 12.86$$

The posterior distribution is normal with mean 426.47 and standard deviation 12.86.

In summary, Bayesian inference depends less and less on prior information or beliefs as the sample or empirical evidence becomes stronger. At least qualitatively, this is the way scientific theories are established. At first, widely divergent hypotheses or theories may be considered equally or nearly equally probable (e.g., hypotheses concerning forms of life on the moon prior to man's landing on its surface). However, as empirical information is accumulated (as by man's landing on the moon's surface), these divergent hypotheses are no longer equally tenable; in fact the possible range of reasonable disagreement narrows as evidence accumulates. Thus Bayesian analysis can be viewed as a restatement of the scientific method in precise statistical terms.

PROBLEMS

16.1 There are three boxes, A, B, and C, each of which contains 100 balls. Box A contains 20 red balls, box B contains 40 red balls, box C contains 80 red balls. One of the boxes is to be selected at random, but since the boxes look identical it is impossible to tell which has been chosen. A random sample of 5 balls is taken from the box chosen above—the balls are drawn one at a time, with replacement after each draw. Suppose of the 5 balls, 2 are red.
   (a) What would now be designated the probability of the box being box A, box B, or box C?
   (b) Set up the problem in a table and calculate the conditional, joint, marginal, and posterior probabilities.

16.2 Mr. Smallpants is a good tennis player, but not so good at golf. He can beat Mr. Bigdome 60 percent of the time at tennis, but only 30 percent of the time at golf. During the summer they play either tennis or golf every day; and during the summer it rains every afternoon in New Ireland, where they live. If they play in the morning it would be golf, in the afternoon it would be tennis (enclosed court). They try to play in the morning, but if either gentleman gets drunk the night before, he has a hangover and cannot play the next morning. Each gets drunk three nights of the week at random and independently of each other. One evening in the Planters' Club you overhear two chaps talking, and one says, "I understand Smallpants lost today." If this statement is true, what is the probability you would now assign to
   (a) Mr. Smallpants being drunk last evening.
   (b) Mr. Bigdome being drunk last evening.

16.3 Three engineers who know the operating properties of a new machine guess that it will require an average of 18, 20, and 21 min, respectively, to complete a certain process. From experience with similar machines the time is known to have a standard deviation of 3 min. It is assumed that one of the engineers is correct and that *a priori* each is equally likely to be correct. In a random sample of 25 runs the average time to complete the process exceeded 20 min. Obtain the Bayesian estimate of the true average time to complete the process.

16.4 It is a well-known meteorological fact that it rains on an average of every other day along the Banana River. Mrs. Merritt lives along the Banana River, and one day she remarked that it nearly always was a clear day following an evening with a brilliant red sunset. After some pressure to be more specific, Mrs. Merritt says she thinks that the probability of rain on a day after a brilliant red sunset is .1. Being analytically minded, Mrs. Merritt during the subsequent year kept careful record of the weather on days following evenings of a brilliant red sunset. Of 100 such evenings, rain fell on 20 and did not fall on 80 following days. Mrs. Merritt is a confirmed Bayesian in her statistical analysis. On the basis of the record that she kept, what would she now say was the probability of rain on a day following an evening of a brilliant red sunset? (If this problem cannot be solved, identify specifically why not.)

16.5 Suppose a production process turning out gadgets can be thought of as a binomial process, each gadget is either defective (*D*) or nondefective (*N*). The quality control engineer has a discrete prior probability distribution on the percent defective being produced by the process. He believes from experience that the percent defective can be 2, 10, 20, or 30 percent with probabilities: $P(2\%D) = .30$, $P(10\%D) = .30$, $P(20\%D) = .20$, $P(30\%D) = .20$. Assuming the production process to be turning out a very large number of gadgets, the engineer takes a random sample of 20 from the production line and finds 5 defective.
   (a) Set up the table and calculate the posterior probabilities.
   (b) After performing (a), the engineer takes another sample of 20 and finds 3 defective; repeat (a).

16.6 The Gizmo Manufacturing Company, through a production error, has produced far too many of its novelty Number 10. The company does not know whether to attempt to sell the excess, scrap them, or store them. Which would be the best policy depends on the demand for the novelty over the next year. The three basic states of demand are high, low, and steady. There are both out-of-pocket and opportunity costs associated with each policy, so that the improper policy will depress the profit on this item. Profit would appear as follows under the various possibilities:

	DEMAND		
	HIGH	LOW	STEADY
Sell	10,000	2,000	4,000
Scrap	3,000	8,000	5,000
Store	6,000	4,000	10,000

(a) In advance of any sampling or surveying, Gizmo feels that each of the three states of demand is equally likely. What should the firm do to maximize expected value?

(b) The firm performs some market surveys on the demand for the novelty and from the information gathered concludes that the demand will most likely be low, high, steady in that order. Gizmo feels that the probability of predicting demand high, if it were to be high, is .5, if it were to be steady is .3, and if it were to be low is .2. Gizmo feels that the probability of predicting demand low, if it were to be high is .1, if it were to be steady is .3, if it were to be low is .6. Gizmo feels that the probability of predicting demand steady, if it were to be high is .2, if it were to be steady is .6, if it were to be low is .2. On the basis of this added information, what should the firm do to maximize expected value?

16.7 The Wow Chemical Company makes plastic sheets. One type it makes is designed to substitute for canvas tarpaulins, so that it is rather thick and tough. On one batch the dial setting records were lost so that the thickness is unknown. Because of variation in the process, the three possible thicknesses have different standard deviation; the mean thicknesses and their standard deviations are as follows: $\frac{1}{4}$ and $\frac{3}{32}$ in., $\frac{3}{16}$ and $\frac{2}{32}$ in., $\frac{1}{8}$ and $\frac{1}{32}$ in. If Wow can sell the batch at its correct thickness, a profit will be realized; if the batch is sold at the wrong thickness, a loss will be realized. The following table summarizes the situation, (discard is given as a fourth possibility):

	$\mu = \frac{1}{4}$	$\mu = \frac{3}{16}$	$\mu = \frac{1}{8}$
Sell as $\frac{1}{4}$	10,000	$-5,000$	$-8,000$
Sell as $\frac{3}{16}$	$-2,000$	10,000	$-4,000$
Sell as $\frac{1}{8}$	$-4,000$	$-3,000$	10,000
Discard	$-8,000$	$-8,000$	$-8,000$

(a) What should Wow do?

(b) Assuming that the thicknesses are normally distributed, what should Wow do if a random sample of 64 sheets have a mean thickness of $\frac{7}{32}$?

16.8 Freak-Out Clothes has 81 outlets for its mod fashions. It wants to introduce a 1909 model men's bathing suit. It believes the average monthly sales per store of this item would be normally distributed with mean $360 and a standard deviation of $100. A random selection of 20 stores is chosen and the new-old fashion is introduced. Records after several months indicate mean sales of $500 with a standard deviation of $200. On the basis of this information, what would be the posterior distribution of sales of the item?

16.9 Fordor Motor Company must decide which of three types of pollution control devices, A, B, or C to put into production for next year's models. Unfortunately, the government standards for pollution control are not yet firm, and any of three specifications, S_1, S_2, and S_3, may become the standard. Fordor must begin production planning of one of the devices soon, although the government's decision will not be firm for some time. If Fordor misjudges the eventual standard, there will be expense in adjusting production. The payoff function is given by the following table:

Standard

		S_1	S_2	S_3
Pollution	A	150	-50	-100
Control	B	-100	40	100
Device	C	-50	100	60

(a) From the newspapers, Fordor's management thinks that the probabilities of standards of S_1, S_2, and S_3 are, respectively, $\frac{1}{2}, \frac{1}{4}, \frac{1}{4}$. What decision should Fordor make?

(b) In order to obtain more information about the government's decision, Fordor hires a consultant who predicts that standard S_3 will be adopted. If Fordor's confidence in the consultant is represented by the following matrix of conditional probabilities, what decision should Fordor make?

	S_1	S_2	S_3
Predict S_1	.6	.2	.1
Predict S_2	.3	.4	.1
Predict S_3	.1	.4	.8

16.10 The average number of widgets sold per day by Warco, Inc., is assumed by the management to be normally distributed with a mean of 100 and a standard deviation of 15 (in thousands).

(a) Based on this assumption, how many widgets should Warco have in stock each day in order to be unable to satisfy demand only 25 percent of the time?

(b) One of Warco's consultants is doubtful about management's assumption concerning the distribution of daily sales. He takes a random sample of 25 days from the records for the last 3 years, during which Warco has had a stable sales record, and finds that the sample has a mean of 100 and a standard deviation of 100. On the basis of both management's assumption and his sample information, what proportion of the time would the consultant expect Warco to be unable to satisfy demand using the stock policy determined in (a)?

16.11 Desert Land Development Corporation must decide immediately whether to buy property I, II, or III. The board of directors has decided to buy one and only one of these properties. All three properties are in an area that will be affected by the outcome of a rezoning hearing which will not be completed for several months. The board thinks that the probability that complete rezoning will be permitted is $\frac{1}{2}$; if it is permitted, properties I, II, and III will be worth 50, 75, and 50 thousand dollars respectively. The board thinks that the probability that partial rezoning will be permitted is $\frac{1}{4}$; in this case, properties I, II and III will be worth 200, 100, and 25 thousand dollars, respectively. The board thinks that the probability that no rezoning will be permitted is $\frac{1}{4}$; in this case, properties I, II, and III will be worth 50, 150, and 50 thousand dollars, respectively.

(a) Determine the optimal decision on the basis of expected value.

(b) The board of directors wishes to obtain additional information before deciding which property to buy. They ask a well-known expert his opinion concerning the outcome of the hearing for rezoning. The expert predicts that the hearing will result in permission for complete rezoning. After examination of the past record of the expert in making similar decisions, the board concludes that the expert predicts that complete rezoning will be allowed when it is in fact refused with probability .3; he predicts that complete rezoning will be allowed when partial rezoning is in fact allowed with probability .8; and he predicts that complete rezoning will be allowed when in fact complete rezoning is allowed with probability .6. Using this additional information, determine the optimal decision on the basis of expected value.

16.12 Farmer John, Inc., is considering the introduction of a vegetable protein (no meat) TV dinner. The cost of introducing this new product would be $840,000 and each dinner sold would provide $.20 profit for Farmer John. Management wishes to introduce the dinner only if costs would be recovered in one year. Farmer John distributes its products to 1000 stores. Management estimates that the stores would sell an average of 375 dinners per month. Management is somewhat uncertain about this sales prediction but believes that with probability .68 average yearly sales for the stores would be between 4020 and 4980.

(a) Should Farmer John introduce the new dinner?

(b) Farmer John decides to test market the new dinner for a random sample of 25 stores. For this sample, average monthly sales for the stores average 400 with a standard deviation of 30. In view of this information, should Farmer John introduce the dinner?

NONPARAMETRIC METHODS

17.1 INTRODUCTION

The statistical tests discussed in this chapter are examples of a class of procedures referred to as *nonparametric methods*. The term nonparametric methods is generally interpreted as including a variety of procedures that are appropriate for data that do not satisfy all the requirements of the classical methods discussed, for example, in previous chapters. In particular, methods appropriately referred to as nonparametric and methods more appropriately referred to as distribution-free are generally included in the term nonparametric methods.

Strictly speaking, a *nonparametric test* does not state a hypothesis about the specific value of a parameter, while a *distribution-free test* makes no assumption about the precise form of the sampled population. Most of the statistical methods already discussed are neither nonparametric nor distribution-free, since they involve both hypotheses about the values of parameters and assumptions concerning the distributions of the populations from which samples are obtained. The terms nonparametric and distribution-free are not mutually exclusive; several tests are both distribution-free and nonparametric, for example, the median test considered in this chapter.

Distribution-free, rather than nonparametric, is more descriptive of the property that makes a test desirable. The distribution from which the test is free is that of the sampled population and this freedom is usually relative; for example, in a distribution-free test the population is sometimes assumed to be continuously or even symmetrically distributed, but the precise form of its distribution is not specified. The distribution of the test statistic must be known in all cases, but for distribution-free tests this distribution is based on properties of the sample, not on properties of the population.

An important concept in nonparametric statistics is that of a *randomization distribution*, which is essentially a probability distribution of a statistic obtained by considering all of its possible sample outcomes and determining their probabilities under the null hypothesis. Many of the tests discussed in this chapter are based on the randomization distribution of the relevant test statistic.

Nonparametric tests are frequently based on ordinal or nominal measurement, such as rank, position in a sequence, or frequency of occurrence, rather than on continuous measurement. For example, nonparametric tests are concerned with medians rather than means, interquartile ranges rather than variances, sign rather than size of observations, and so forth.

Nonparametric tests have several advantages over classical tests. In general, their derivation and application are simpler and, for samples of small or moderate size, they are usually less time consuming to apply than classical tests. The most important advantage of nonparametric tests is that they can be applied in many situations for which the assumptions of classical tests are not valid. The fact that, in general, more powerful tests can be obtained if more restrictive assumptions are made is the disadvantage of nonparametric tests. This problem is now discussed in the context of relative efficiency.

RELATIVE EFFICIENCY OF NONPARAMETRIC TESTS

In the sense of statistical efficiency, the relative efficiency of nonparametric and classical tests depends on the nature of the assumptions that can be made about the population. Nonparametric tests are often more efficient than their most efficient classical counterparts when both tests are applied under conditions meeting all the assumptions of the nonparametric test, but failing to meet some of the assumptions of the classical tests. When both tests are applied under conditions meeting all the assumptions of both tests, the nonparametric test is usually somewhat less efficient for small samples and increasingly less efficient as sample size increases.

Efficiency is an indication of the power of a particular test compared with the power of another test; this second (standard) test is usually the most powerful test available for the conditions under which the comparison is made. The *relative efficiency* of test A with respect to test B can be defined as b/a, where a is the number of observations required by test A in order to have the same power that test B has when based on b observations; both tests are for the same null hypothesis against the same alternative hypothesis at the same significance level (both being either one-tailed or two-tailed tests). Relative efficiency of a test thus depends on and is defined in terms of the comparison test, the size of the sample, the alternative hypothesis, the significance level, and the location of the rejection region.

The relative efficiency of each of the nonparametric tests discussed in this chapter has been investigated and found to be satisfactory under a variety of conditions. The specific results of these studies are reported in texts on nonparametric methods (see Appendix A).

When both types of tests are available for a particular problem, the choice between a nonparametric test and its classical counterpart should be based on the assumptions that can be made concerning the distribution of the sampled population. Almost any violation of the assumptions on which a parametric test is based alters the distribution of the test statistic and changes the probabilities of type I and type II errors.

A test is said to be *valid* at level α if the probability of type I error does not exceed α for any distribution. A test is said to be *robust* against violation of a particular assumption if its probabilities of type I and type II errors are not appreciably affected by the violation.

Various results concerning robustness of particular tests with respect to specific violations of assumptions, particularly the assumption of normality, have been obtained. Unfortunately, the conditions under which a test tends to be robust are highly idiosyncratic. In general, robustness is inversely related to the degree of violation of the assumption, but both concepts are difficult to quantify. Also, unfortunately, a test's robustness with respect to violation of a given assumption is often strongly dependent on factors not involved in the statement of the assumption itself. For example, size and location of the rejection region, number of samples and absolute and relative sizes of the samples, and relative variances of the sampled populations are factors not involved in stating assumptions or in describing their violation. However, the effects on type I and type II errors of a given violation of the assumptions of a test are frequently greatly influenced by these seemingly irrelevant factors.

Even when all the relevant population and sampling information is available, it may be impossible in complex situations to predict mathematically the direction of nonrobustness and thus to state whether the true significance level is greater or less than the nominal significance level. Generally, when the conditions under which a test is to be applied are known or even strongly suspected to violate the assumptions of the appropriate classical test, it is advisable to use a nonparametric test.

In the following sections some of the most frequently applicable nonparametric tests are discussed and compared with their classical counterparts. The tests are grouped according to the hypotheses that they are designed to test; these hypotheses concern central tendency, randomness, correlation, and distribution. It should be noted that the tests included are only a few of the more commonly used nonparametric tests; many additional nonparametric tests have been devised for specific types of situations.

17.2 TESTS FOR HYPOTHESES CONCERNING CENTRAL TENDENCY

Nonparametric tests for hypotheses concerning central tendency usually consider the median as the measure of central tendency. The statistical tests in this section are for hypotheses concerning the value of the median of one population or the equality of the medians of two or more populations.

ONE SAMPLE

Two tests for hypotheses concerning the median of one population are discussed below: the *sign test* for the hypothesis that the median of a continuously distributed population has a specified value and the *signed rank test* for the hypothesis that the median of a symmetric population has a specified value.

The Sign Test   Suppose that a population is known to be continuously distributed and its median is hypothesized to have the value M_0. If this hypothesis is true, half the population lies above M_0 and half lies below M_0. The value of a randomly drawn observation x can be classified as either $x > M_0$ or $x < M_0$ or, equivalently, as $x - M_0 > 0$ or $x - M_0 < 0$. Under the hypothesis, if there are n observations, the number of observations, r, in either category, say $x > M_0$, has the binomial distribution

$$P(r) = \binom{n}{r}(.5)^n$$

The corresponding cumulative probability distribution is

$$P(r \le r^*) = \sum_{r=0}^{r^*} \binom{n}{r}(.5)^n$$

and the hypothesis can be tested using tables of the cumulative binomial distribution. (Table B.2)

CASE

The sign test can be summarized as follows:

$$H_0 : M = M_0$$

$$H_A : \text{(a)} \quad M > M_0$$
$$\text{(b)} \quad M < M_0 \quad \text{possible alternative hypotheses}$$
$$\text{(c)} \quad M \ne M_0$$

Assumptions:
The population is continuous and the observations are independently and randomly distributed.

Sample Statistic:
$r =$ number of observations above (or below) M_0.

Sampling Distribution:
Binomial, $\pi = \frac{1}{2}$.

Critical Region:
(a) $r \ge r_{1-\alpha}$
(b) $n - r \ge r_{1-\alpha}$
(c) $\max[r, n - r] \ge r_{1-\alpha/2}$

where r is the number of observations $> M_0$
and n is the total number of observations.

For a normal distribution, the median M and the mean μ are equal, so the sign test and student's t-test are comparable in the sense that they test the same hypothesis $H_0 : \mu = \mu_0$.

EXAMPLE
A spokesman for the State Insurance Commission maintains that the median annual interest rate for short-term nonsecured loans in New York City obtained for the purpose of purchasing automobiles is 20 percent. A legislative committee thought this to be an underestimate and obtained a random sample of 25 such loans whose annual interest rates were as given below. On the basis of these data, can the legislative committee refute the commission's claim (5 percent level of significance)?

$$H_0 : M = .20$$

$$H_A : M > .20$$

x	SIGN OF $x - M_0$
32.2	+
18.6	−
40.1	+
45.0	+
27.6	+
33.4	+
16.3	−
10.2	−
43.9	+
46.2	+
28.5	+
30.0	+
59.6	+
28.9	+
17.8	−
32.7	+
29.3	+
19.3	−
25.2	+
12.5	−
42.8	+
19.4	−
18.1	−
39.6	+
37.5	+

Of the 25 observations, there are 17 for which $x - M_0 > 0$ and 8 for which $x - M_0 < 0$. Referring to Table B.2, when $n = 25$, the largest value of r for which $P(r \text{ or fewer} - s) \leq .05$ is 7. Therefore $P(r \leq 8) > .05$, where r is the number of observations for which $x - M_0 < 0$. Thus H_0 is accepted and, on the basis of these data, the claim that the median interest rate is 20 percent cannot be refuted.

The Wilcoxon Signed-Rank Test   Suppose that a population is known to be symmetric and its median is hypothesized to have the value M_0. If the hypothesis is true, $X - M_0$ is symmetrically distributed about zero. The signed rank test is based on a statistic obtained by ranking the absolute values of the differences between the observations and the hypothesized median M_0, attaching the sign of each difference to its corresponding rank and summing the positive ranks. Suppose $x_1, x_2, \ldots, x_n$ are the values of the observations in a random sample of size n; let S_i be the sign of the difference $x_i - M_0$ and let R_i be the rank (in increasing order) of $|x_i - M_0|$ when the n absolute values of the differences between the observations and the hypothesized population median are ordered.

If the hypothesis is true, S_i is as likely to be positive as to be negative for every value of $|x_i - M_0|$ and thus for every value of R_i. Therefore, under the hypothesis $m = m_0$, each of the 2^n possible assignments of algebraic signs to the n ranks R_i is equally likely. For each of these 2^n sets of n signed ranks, denoted $S_i R_i$, the test statistic

$$W_+ = \sum_{S_i +} S_i R_i$$

(where the sum is over positive S_i) can be calculated and its null (randomization) distribution can be obtained. (An equivalent test statistic can be obtained by summing over negative values of $S_i R_i$.) Fairly extensive tables of the null distribution of W_+ are available; Table B.19 gives critical values of W_+ for several values of α and a considerable range of n.

CASE

The signed-rank test can be summarized as follows:

$$H_0: M = M_0$$

$$H_A: \begin{array}{l} \text{(a)} \ M > M_0 \\ \text{(b)} \ M < M_0 \\ \text{(c)} \ M \neq M_0 \end{array} \right\} \quad \text{possible alternative hypotheses}$$

Assumptions:

The population is continuous and symmetric and the observations are independently and randomly distributed.

Sample Statistic:

$$W_+ = \sum_{S_i +} S_i R_i$$

Sampling Distribution:

Null distribution of W_+ over the 2^n equally likely sets of signed ranks.

Critical Region:

   (a) $W_+ \geq W_{+, 1-\alpha}$
   (b) $|W_-| \geq W_{+, 1-\alpha}$
   (c) $\max(W_+, |W_-|) \geq W_{+, 1-\alpha/2}$

$$\text{where } W_- = \sum_{S-} S_i R_i.$$

Note that $W_+ + |W_-| = \dfrac{n(n+1)}{2}$.

The null distribution of W_+ (or of $|W_-|$) has a mean of $n(n+1)/4$ and a variance of $n(n+1)(2n+1)/24$ and is asymptotically normal. When n is too large for the exact distribution tables, an approximate test can be obtained using the standard normal distribution, since

$$z = \frac{W_+ - \left[\dfrac{n(n+1)}{4} \right]}{\sqrt{\dfrac{n(n+1)(2n+1)}{24}}}$$

is approximately standard normal.

EXAMPLE

Suppose that the assumptions for the Wilcoxon signed rank test are met and that this test is applied to the data for the preceding example.

$$H_0: M = 20$$

$$H_A: M > 20$$

$x - 20$	SIGNED RANK
-0.6	-1
-0.7	-2
-1.4	-3
-1.9	-4
-2.2	-5
-3.7	-6
5.2	7
-7.5	-8
7.6	9
8.5	10
8.9	11
9.3	12
-9.8	-13
10.0	14
12.2	15
12.7	16
13.4	17
17.5	18
19.6	19
20.1	20
22.8	21
23.9	22
25.0	23
26.2	24
39.6	25

$$\sum_{S_i -} S_i R_i = -3 - 6 - 13 - 5 - 2 - 8 - 1 - 4$$

$$= -42$$

$$|W_-| = 42$$

The lower tail critical value for $\alpha = .05$ is 100 (Table B.19). So accept the hypothesis $M = 20$ and conclude that the claim cannot be rejected.

TWO RELATED SAMPLES

Both the sign test and the signed rank test can be modified to test the hypothesis that the median of a population of difference scores is zero and thus to test the hypothesis that the medians of two populations differ by a specified amount.

The Sign Test   Suppose that a continuously distributed variable X is hypothesized to be as likely to exceed as to be exceeded by another continuously distributed variable Y; that is, for a randomly drawn pair of observations on the two variables it is hypothesized that $P(X - Y) > 0 = P(X - Y) < 0$. This is equivalent to hypothesizing that the median of a continuously distributed population of difference scores is zero, and thus this hypothesis can be tested using the sign test just described. If n pairs of observations are drawn randomly from the two populations, then the number r of differences $x_i - y_i$ that are positive (or negative) is binomially distributed with parameter $\pi = .5$ if the hypothesis $P(X - Y) > 0 = P(X - Y) < 0$ is true.

CASE

The sign test for two related samples can be summarized as follows:

$$H_0 : P(X > Y) = P(X < Y) = .5$$

$$H_A : \text{(a)}\ \ P(X > Y) > P(X < Y)$$
$$\text{(b)}\ \ P(X > Y) < P(X < Y) \bigg\} \quad \text{possible alternative hypotheses}$$
$$\text{(c)}\ \ P(X > Y) \neq P(X < Y)$$

NOTE:
By adding C to each y observation before subtracting from its paired x observation, we can test the null hypothesis that the median difference is C.

Assumptions:

The population of difference scores is continuous and the observations are independently and randomly distributed.

Sample Statistic:

Number of positive (or negative) differences.

Sampling Distribution:

Binomial, $\pi = \dfrac{1}{2}$

Critical Region:

(a) $r \geq r_{1-\alpha}$
(b) $n - r \geq r_{1-\alpha}$
(c) $\max[r, n - r] \geq r_{1-\alpha/2}$

where n is the total number of observations and
r is the number of observations for which $x_i - y_i > 0$

Note that if the median of the population of difference scores is zero, this does not imply that the medians of the X and Y distributions are equal, unless the X and Y distributions are known to be either identical or symmetric. Similarly, if the medians of the X and Y distributions are equal, this does not imply that the median of $X - Y$ is zero, unless the X and Y distributions are known to be either identical or symmetric.

EXAMPLE

Twelve matched pairs of grocery stores were used in an advertising research experiment designed to compare the effects on sales of having a pretty girl give samples of sausage to customers in the store (method A) and of selling the sausage for 5c less per pound and advertising this fact on a large poster in the store (method B). The results, in terms of per cent change in scales, are given below. Test the hypothesis that the median difference between the changes in sales is zero (5 percent level).

$$H_0 : P(X > Y) = P(X < Y)$$

$$H_A : P(X > Y) \neq P(X < Y)$$

STORE PAIR	X(METHOD A)	Y(METHOD B)	$X - Y$
1	5.0	3.0	2.0
2	2.8	2.3	0.5
3	3.2	3.1	0.1
4	2.7	2.5	0.2
5	1.9	-0.6	2.5
6	4.3	5.6	-1.3
7	2.5	3.9	-1.4
8	-1.8	1.0	-2.8
9	1.5	1.1	0.4
10	-0.8	-0.5	0.3
11	2.7	4.8	-2.1
12	1.6	3.8	-2.2

Using Table B.2, reject H_0 if the number of observations for which $(X > Y)$ or the number of observations for which $(X < Y)$ is fewer than 2. In the sample there are 7 observations for which $X > Y$ and 5 observations for which $X < Y$. So accept H_0 and conclude that the median of the distribution of differences is zero.

The Wilcoxon Signed-Rank Test Suppose x_i and y_i are the values of the ith matched pair of observations from the symmetric populations X and Y and that the median M of the distribution of differences $x_i - y_i$ is hypothesized to be zero. If the hypothesis is true, $X - Y$ is distributed symmetrically about zero and the signed-rank test previously described for one sample can be used with S_i denoting the algebraic sign of the difference $x_i - y_i$. Note that the statistic $W_+ = \sum_{S_i+} S_i R_i$ may fall in the rejection region because the symmetric distributions X and Y differ in location or because one or both distributions are asymmetric and differ in some respect. Thus if it is known that X or Y or both are asymmetric, the signed-rank test is a test of the hypothesis that the distributions of X and Y are identical. If it is known that the X and Y distributions are either (a) both symmetrical or (b) identical in shape and variance, then the hypothesis that the median of the X distribution exceeds that of the Y distribution can be tested by determining W_+ for the differences $X_i - Y_i$. (Similarly, the hypothesis that the median of the X distribution exceeds that of the Y distribution by C units can be tested by determining W_+ for the differences $X_i - Y_i - C$.)

CASE

The Wilcoxon signed-rank test can be summarized as follows:

$H_0: M = 0$

$H_A:$ (a) $M > 0$
     (b) $M < 0$
     (c) $M \neq 0$

Assumptions:

The populations X and Y are symmetric and the pairs of observations are independently and randomly distributed.

Sample Statistic:

$$W_+ = \sum_{S_i+} S_I R_I$$

Sampling Distribution:
Null distribution of W_+ over the 2^n sets of signed ranks.

Critical Region:
(a) $W_+ \geq W_{+,1-\alpha}$
(b) $|W_-| \geq W_{+,1-\alpha}$
    where $W_- = \sum_{S^-} S_i R_i$
(c) $\max(W_+, |W_-|) \geq W_{+,1-\alpha/2}$

As noted above, for large n,

$$z = \frac{W_+ - [n(n+1)/4]}{\sqrt{n(n+1)(2n+1)/24}}$$

is approximately normally distributed with mean zero and variance 1.

EXAMPLE
Suppose the assumptions of the Wilcoxon signed-rank test are met and the test is applied to the data of the preceding example

$H_0: M = 0$

$H_A: M \neq 0$

$x - y$	SIGNED RANK
2.0	$+8$
0.5	$+5$
0.1	$+1$
0.2	$+2$
2.5	$+11$
-1.3	-6
-1.4	-7
-2.8	-12
0.4	$+4$
0.3	$+3$
-2.1	-9
-2.2	-10

$$W_+ = 34, \ W_- = 44 \qquad \min(W_+, |W_-|) \text{ is } 34.$$

From Table B.19, for $n = 12$, the critical value is between 17 and 18. So accept the hypothesis $M = 0$ and conclude that there is no evidence that the changes in sales differ.

TWO INDEPENDENT SAMPLES

Two tests for hypotheses concerning the medians of two independent populations are: the *median test* for the hypothesis that equal proportions of two populations lie below a particular sample value and the *Wilcoxon rank-sum test* for the hypothesis that two populations are identical (this test is particularly sensitive to differences in location).

The Median Test   Suppose that populations X and Y are both continuously distributed and that the median of a combined sample of n_1 observations from X and n_2 observations from Y has the value $\hat{M}$. Then the hypothesis that $\hat{M}$ is the same, but unknown, quantile in the two populations can be tested, and if the hypothesis is true and $n = n_1 + n_2$ is large, $\hat{M}$ will tend to be close to the population median.

It is convenient to tabulate the sample observations as follows:

	$< \hat{M}$	$> \hat{M}$	
X sample			n_1
Y sample			n_2
	$\dfrac{n_1 + n_2}{2}$	$\dfrac{n_1 + n_2}{2}$	

Note that if the number of observations $n = n_1 + n_2$ is odd, the single observation equal to $\hat{M}$ is omitted from the table; thus the number of observations in the table is always even. Under the hypothesis, the distribution of the observations in a table of this type is hypergeometric. If the following notation is used for the cells of the table

a	$A - a$	A
b	$B - b$	B
$a + b$	$(A + B) - (a + b)$	$A + B$

where $A \geq B$

$$\frac{a}{A} \geq \frac{b}{B}$$

then Table B,20 can be entered with A, B, a and the one-tailed significance level α to obtain the (largest) value of b such that a/A is just significantly larger than b/B. The exact one-tailed cumulative probability for this value of b is also given. For large n, the χ^2 approximation is appropriate.

CASE

The median test can be summarized as follows:

H_0: $P(X < \hat{M}) = P(Y < \hat{M})$ and $P(X > \hat{M}) = P(Y > \hat{M})$

H_A: (a) $P(X < \hat{M}) > P(Y < \hat{M})$ and $P(X > \hat{M}) < P(Y > \hat{M})$
(b) $P(X < \hat{M}) < P(Y < \hat{M})$ and $P(X > \hat{M}) > P(Y > \hat{M})$
(c) $P(X < \hat{M}) \neq P(Y < \hat{M})$ and $P(X > \hat{M}) \neq P(Y > \hat{M})$

possible alternative hypotheses

When H_0 is accepted, $\hat{M}$ can be regarded as an estimated median provided n is large.

Assumptions:

The populations are continuously distributed and the observations are independently and randomly distributed.

Sample Statistic:

Number of X's $< M$ and number of Y's $< M$

Sampling Distribution:

$$\text{Hypergeometric } P(a, b) = \frac{\dbinom{A}{a}\dbinom{B}{b}}{\dbinom{A + b}{a + b}}$$

Critical Region:
   (a) $b \le b_\alpha$
   (b) $b \le b_\alpha$
   (c) $b \le b_{\alpha/2}$

where $a/A > b/B$ and the data must be checked for direction in cases (a) and (b).

EXAMPLE
Electric light bulbs are produced by two machines in the Edison Factory. Over a
period of time, bulbs from each machine were randomly sampled and tested to
determine the number of hours they would burn. The results (in hours) given below
were obtained. Is there evidence (5 percent level) that the machines are not equally
likely to produce bulbs whose lives are above and below the sample median?

$$H_0: P(X < \hat{M}) = P(Y < \hat{M}) \quad \text{and} \quad P(X > \hat{M}) = P(Y > \hat{M})$$

$$H_A: P(X < \hat{M}) \ne P(Y < \hat{M}) \quad \text{and} \quad P(X > \hat{M}) \ne P(Y > \hat{M})$$

MACHINE X	MACHINE Y
175.6	163.8
182.3	149.5
192.7	172.9
179.5	163.4
169.4	150.8
165.3	151.7
172.8	146.3
189.4	152.6
193.4	153.5
	161.0

COMBINED SAMPLES	
Y	14.63
Y	149.5
Y	150.8
Y	151.7
Y	152.6
Y	153.5
Y	161.0
Y	163.4
Y	163.8
X	165.3 ← M
X	169.4
X	172.8
Y	172.9
X	175.6
X	179.5
X	182.3
X	189.4
X	192.7
X	193.4

$$< \hat{M} \quad > \hat{M}$$

	$< \hat{M}$	$> \hat{M}$
X	0	8
Y	9	1

$$\frac{a}{A} = \frac{9}{10}$$

$$A = 10 \qquad B = 8 \qquad a = 9 \qquad b = 0$$

$$b_{.05} = 3, \qquad \text{from Table B.20.}$$

So reject H_0 and conclude that the machines are equally likely to produce bulbs whose lives are above and below the sample median.

The Wilcoxon Rank-Sum Test   Suppose that n observations from an X population and m observations from a Y population (where $m \geq n$) are obtained randomly and independently. The hypothesis to be tested is that the two populations are identical. If the hypothesis is true, the $m + n$ observations are, in effect, drawn from the same population; thus each of the $\binom{m+n}{n}$ possible partitions of the combined sample of $m + n$ observations into the subsets of m and n observations is as likely under the null hypothesis as the obtained partition. For each of these $\binom{m+n}{n}$ partitions there corresponds a value of the statistic

$$W_n = \sum_{i=1}^{n} R_i$$

where R_i is the rank of x_i in the combined sample of $m + n$ observations with ranks increasing in size as observations increase. Fairly extensive tables of the null distribution of W_n are available (Table B.21). Tables of the Mann–Whitney U statistic, which is a linear transformation of W_n, are also available.

CASE

The Wilcoxon rank-sum test can be summarized as follows:

H_0: X and Y identically distributed

H_A: X and Y not identically distributed

Assumptions:

The populations are continuous and the observations are independently and randomly obtained.

Sample Statistic:

$$W_n = \sum_{i=1}^{n} R_i$$

Sampling Distribution:

Null distribution of W_n over the $\binom{m+n}{n}$ equally likely partitions of the $m + n$ observations

Critical Region:

$$W_n \leq W_{n, \alpha/2}$$

Although nonidentical populations having equal means are not only theoretically possible but are easy to construct, they occur rather infrequently in practice.

Thus rejection of the hypothesis of identical populations can generally be taken as evidence that the populations differ with respect to location (although not necessarily only with respect to location). If the alternative hypothesis is that one population is located below the other population, then the corresponding one-sided test is appropriate. Similarly, if the populations are assumed to be identical except perhaps for location, the hypothesis that one population is larger than the other by a specified amount can be tested using the appropriate one-sided test.

As m and n increase (with n/m constant), the distribution of W_n approaches normality and

$$z = \frac{W_n - \dfrac{[n(n + m + 1)]}{2}}{\sqrt{\dfrac{nm(N + M + 1)}{12}}}$$

has approximately the standard normal distribution. This approximation can be corrected for continuity by subtracting $\frac{1}{2}$ from the absolute value of the numerator; the approximation is reasonably accurate if $n = m \geq 25$ and $\alpha \geq .025$.

Although the rank-sum test is especially sensitive to differences in location, it is not necessarily insensitive to differences in shape between populations having the same location, since such differences also usually result in the $\binom{m+n}{n}$ partitions being unequally likely. Thus, unless two populations are known to be identical except perhaps with respect to location, a value of W_n that falls in the rejection region can be interpreted only as indicating that the two populations are not identical, not that the difference is necessarily with respect to location.

EXAMPLE
Suppose the assumptions for the Wilcoxon rank-sum test are met and the test is applied to the data from the preceding example.

$$H_0: X \text{ and } Y \text{ identically distributed}$$

$$H_A: X \text{ and } Y \text{ not identically distributed}$$

1	Y	146.3
2	Y	149.5
3	Y	150.8
4	Y	151.7
5	Y	152.6
6	Y	153.5
7	Y	161.0
8	Y	163.4
9	Y	163.8
10	X	165.3
11	X	169.4
12	X	172.8
13	Y	172.9
14	X	175.6
15	X	179.5
16	X	182.3
17	X	189.4
18	X	192.7
19	X	193.4

$$\sum_{i=1}^{9} R_i = 10 + 11 + 12 + 14 + 15 + 16 + 17 + 18 + 19$$

$$= 132$$

$$m = 10 \qquad n = 9$$

Using Table B.21, for $\alpha = .01$, the lower value is 61. So reject H_0 and conclude that X and Y are not identically distributed and may differ in location.

17.3 TESTS FOR CORRELATION

There are several nonparametric tests for correlation. Two of these tests are: the Spearman test for rank-order correlation and the Kendall test for correlation.

Spearman Test for Rank-Order Correlation Suppose that measurements for variables X and Y are obtained on each of n randomly selected units. The hypothesis is that X and Y are independent and hence uncorrelated. Suppose there are no ties among the X measurements nor among the Y measurements and that each of the X's is replaced by its rank in order of increasing size and each of the Y's is replaced by its rank in order of increasing size. If the units are arranged in order of increasing X ranks, then if the hypothesis is true, each of the $n!$ possible corresponding permutations of the Y ranks is equally likely. For each of these $n!$ possible patterns of association between the X ranks and the Y ranks there is a value of Spearman's rank-difference correlation coefficient

$$r_s = 1 - \frac{6 \sum\limits_{i=1}^{n} d_i^2}{n(n^2 - 1)} = 1 - \frac{6D}{n(n^2 - 1)}$$

where d_i is the difference between the X rank and the Y rank for the ith unit. The frequency distribution of these $n!$ values of r_s is the null distribution of r_s and can be used as the basis for a test of the hypothesis that X and Y are independent. Equivalently, since r_s is a monotonic function of D for any fixed n, D can be used as the test statistic and referred to the null distribution of D (Table B.22).

CASE
The Spearman test can be summarized as follows:

H_0: X and Y are statistically independent

H_A: (a) X and Y are positively (directly) related ⎫ possible
    (b) X and Y are negatively (inversely) related ⎬ alternative
    (c) X and Y are not independent ⎭ hypotheses

Assumptions:
Random sample from a continuous bivariate population

Sample Statistic:
$$r_s = 1 - \frac{6 \sum d_i^2}{n(n^2 - 1)} = 1 - \frac{6D}{n(n^2 - 1)}$$

Sampling Distribution:
Randomization distribution of r_s over the $n!$ possible assignments of ranks

Critical Region:
    (a) $r_s \geq r_{s,1-\alpha}$
    (b) $r_s \leq r_{s,\alpha}$
    (c) $r_s \geq r_{s,1-\alpha/2}$ or $r_s \leq r_{s,\alpha/2}$

Spearman's rank difference correlation coefficient is equivalent to Pearson's product-moment correlation coefficient (Chapter 13) computed for the ranks, rather than the original observations. If the hypothesis is true, the distributions of r_s and of D are symmetric; r_s ranges from -1 to $+1$ and D ranges from 0 to $n(n-1)/3$. Exact tables of critical values of r_s and D (Table B.22) are available for small sample sizes. The distributions of r_s and D are asymptotically normal, and for large samples the normal approximation can be used:

$$z = r_s\sqrt{n-1}$$

is approximately standard normal or, equivalently,

$$z = \frac{\sum\limits_{i=1}^{n} d_i^2 - [(n^3-n)/6]}{\dfrac{[(n^3-n)/6]}{\sqrt{n-1}}}$$

is approximately standard normal. Similarly, if the population correlation is zero,

$$t = r_s\sqrt{\frac{n-2}{1-r_s^2}}$$

is approximately distributed as Student's t with $n-2$ degrees of freedom. This approximation apparently is more accurate than the normal approximation given above.

EXAMPLE
An investor is interested in the possible relationship between the median price of a stock and its percentage variation (measured as 100 times the ratio of the high price minus the low price divided by the low price) during the past year. He obtained the following data for a random sample of ten stocks listed on the New York Stock Exchange. Do these data indicate that median price and percentage variation are negatively correlated?

$$H_0: r_s = 0$$

$$H_A: r_s < 0$$

STOCK	MEDIAN PRICE	RANK	PERCENTAGE VARIATION	RANK
A	$55\frac{7}{8}$	3	41	9
B	68	4	29	5
C	$42\frac{1}{8}$	2	34	7
D	$30\frac{3}{8}$	1	52	10
E	$94\frac{5}{8}$	7	27	4
F	10	8	18	3
G	$142\frac{1}{8}$	10	12	2
H	$87\frac{1}{4}$	6	32	6
I	$72\frac{3}{4}$	5	36	8
J	$120\frac{3}{8}$	9	9	1

X-RANK	Y-RANK	d_i
1	10	81
2	7	25
3	9	36
4	5	1
5	8	9
6	6	0
7	4	9
8	3	25
9	1	64
10	2	64
		$\sum d_i^2 = 314$

$$r_s = 1 - \frac{6(314)}{10(99)} = 1 - \frac{314}{165} = -0.90$$

From Table B.22, for $n = 10$, $D_{.95} = 258$. So reject H_0 and conclude that median price and percentage variation are negatively correlated.

Kendall Test for Correlation   Suppose that measurements for variables X and Y are obtained on each of n randomly selected units. The hypothesis that X and Y are independent can be tested by arranging the n units in increasing order with respect to the X variable and testing the resulting order, with respect to the Y variable, for randomness. If the two variables are independent, the Y sequence is equally likely to be any of the $n!$ possible permutations of the n values of Y. However, if the variables are linearly (or even monotonically) related, the Y observations tend to form an increasing or decreasing sequence, and any statistic that reflects this increase or decrease can be used to test for independence.

Suppose that the n observations are arranged in increasing order of the value of the X observation, and let I be the number of times a Y observation is followed in the sequence by a smaller Y observation. The statistic I is referred to as the number of inversions, since it is the number of pairs of Y values in the sequence whose rank order is the reverse of the ascending order. Let T be the number of times a Y observation is followed in the sequence by a larger Y observation. The statistic used for the Kendall test is $S = T - I$. Except when there are tied observations, the statistics S, I, and T are mathematically equivalent and the value of one of them determines the values of the other two. The maximum possible value of I is the number of pairs of Y values $\binom{n}{2}$ and, in all cases, $I + T = \binom{n}{2}$. Thus $T = \binom{n}{2} - I$, $S = \binom{n}{2} - 2I$, and all three statistics are functions of I. The value of I is usually more easily computed if the Y values are replaced by their ranks.

The null distribution of S is the relative frequency distribution of S over the $n!$ possible permutations of the Y's. This distribution is symmetric about zero and its range is $-n(n-1)/2$ to $n(n-1)/2$; tables are available for small values of n (Table B.23). Under the null hypothesis, the distribution of S is asymtotically normal; a test based on the normal approximation can be used for values of n

larger than those for which the distribution is tabled:

$$z = \frac{S}{\sqrt{\dfrac{(n - 1)(2n + 5)}{18}}}$$

is approximately standard normal. This approximation can be improved by a correction for continuity that consists of reducing by 1 the absolute value of the numerator S of this ratio.

The Kendall coefficient of rank correlation is defined as

$$t = \frac{S}{\dfrac{n}{2}(n - 1)}$$

and is on the range -1 to $+1$.

CASE

The Kendall test can be summarized as follows:

$$H_0 : X \text{ and } Y \text{ are independent}$$

H_A: (a) X and Y are positively correlated ⎫ possible
    (b) X and Y are negatively correlated ⎬ alternative
    (c) X and Y are not independent ⎭ hypotheses

Assumptions:
    Random sample from a continuous bivariate population

Sample Statistic:
    $S = T - I = \binom{n}{2} - 2I$, where I is the number of times y is followed by a smaller value of y.

Sampling Distribution:
    Randomization distribution of S over the $n!$ possible permutations of the y's.

Critical Region:
    (a) $S \geq S_{1-\alpha}$
    (b) $S \leq S_{\alpha}$
    (c) $S \geq S_{1-\alpha/2}$ or $S \leq S_{\alpha/2}$

EXAMPLE
Considering the data for the preceding example, we have

$$H_0 : S = 0$$
$$H_A : S < 0$$

X-RANK	Y-RANK
1	10
2	7
3	9
4	5
5	8
6	6
7	4
8	3
9	1
10	2

$$I = 9 + 6 + 7 + 4 + 5 + 4 + 3 + 2 = 40$$

$$T = 2 + 2 + 1 = 5$$

$$I + T = \frac{(10)(9)}{2} = 45$$

$$S = T - I = -35$$

For $n = 10$, $S_{.05} = -21$ (Table B.23). So reject H_0 and conclude that median price and percentage variation are negatively correlated.

The Spearman and Kendall tests can be used under the same conditions to test the same hypothesis. These two tests are not mathematically equivalent, but their respective test statistics are highly correlated for samples from a bivariate normal population when the product-moment correlation ρ is not too large.

The Spearman test is computationally easier than the Kendall test when n is moderately large. The Spearman test is based on paired rank differences and thus is preferable when it is desirable to weight large discrepancies more heavily than small ones. However, in most other respects the Kendall test seems preferable—its test statistic is mathematically much more tractable than the Spearman statistic and, for a given n, its distribution is usually smoother and better approximated by a normal distribution than the distribution of the Spearman statistic.

17.4 TESTS FOR RANDOMNESS

Several nonparametric tests already discussed, including the Spearman, Kendall, Sign, and Median tests, can be used as tests for randomness against the alternative of trend. These tests are sensitive to monotonic trend but not to cyclical trend. Tests for cyclical trend are not discussed but can be found in references given in Appendix A. The runs test for randomness versus sequential dependence is also discussed in this section.

Daniel Test (based on Spearman rank-order correlation) If the X observation is the time at which the unit was drawn from the population and the Y observation is a measurement of some characteristic of the unit, the

Spearman test for rank-order correlation is a test for the correlation of Y with time and is thus a test for randomness of Y over time.

EXAMPLE

An index of average price for a group of stocks of companies in the computer industry was computed for ten successive quarters. The data are given below. Do they indicate an upward trend in this index? (5 percent level)

$$H_0: r_s = 0$$

$$H_A: r_s > 0$$

QUARTER	INDEX	RANK
1	100.0	2
2	98.5	1
3	101.2	3
4	103.4	5
5	105.7	6
6	102.5	4
7	106.1	7
8	108.3	8
9	109.9	9
10	110.6	10

$$r_s = 1 - \frac{6 \sum_{i=1}^{n} d_i^2}{n(n^2 - 1)} = 1 - \frac{6(8)}{10(99)} = 1 - .048 = 0.952$$

So reject H_0 (see Table B.22) and conclude that there is an upward trend in the index over time.

Kendall's test for rank-order correlation can be used as a test for randomness versus trend in the same way that the Spearman test is used, that is, by considering the correlation of the variable with time.

Median Test Cox and Stuart have suggested using the median test for randomness versus trend in certain cases. If the first half of a series of observations obtained sequentially is taken to be one sample and the second half is taken to be the other sample, the median test can be used to test for randomness versus trend of any type that affects the proportion of the population lying below $\hat{M}$. One such type of trend is slippage in location without change in shape. Note that this type of trend affects the proportions of the population lying above and below any quantile, so the closeness with which $\hat{M}$ approximates the population median is not a problem.

EXAMPLE

Suppose the test is applied to the data of the preceding example:

$$H_0: P(X < \hat{M}_x | \text{sample 1}) = P(X > \hat{M}_x | \text{sample 2})$$

$$H_A: P(X < \hat{M}_x | \text{sample 1}) > P(X > \hat{M}_x | \text{sample 2}).$$

98.5	2
100.0	1
101.2	3
102.5	6
103.4	4
105.7 $\leftarrow \hat{M}_x$	5
106.1	7
108.3	8
109.9	9
110.6	10

	$x < \hat{M}_x$	$x > \hat{M}_x$	
Sample 1	4	1	5
Sample 2	1	4	5

$$A = 5 \qquad B = 5 \qquad a = 4 \qquad b = 1$$

For a two-tailed test for $\alpha = .05$, the critical value of b is 0 (Table B.20). So reject the null hypothesis and conclude that there is a trend in the index.

The sign test can also be used to test for randomness versus trend.

Runs Test   Consider a sequence of n elements of which n_1 are of one type, say a, and n_2 are of another type, say b; the pattern in which the a's and b's are arranged is random if the probabilities π and $1 - \pi$ of a and b, respectively, are constant throughout the sequence. If, however, these probabilities are not constant—in particular, if there is sequential dependency among elements of the same type—then there may be an unusually small number of runs (unbroken sequences of like elements); that is, there may be runs of considerable length.

Thus the total number of runs, the length of the longest run, and other run statistics can be used to test for randomness of a sequence against the alternative of sequential dependency. Runs tests are extremely simple to apply. Although in some cases a runs test is inefficient, even in those cases, if data are sufficient for adequate power, a runs test may be useful because of its simplicity.

The runs test discussed in the following paragraphs is based on the statistic U, defined as the total number of runs of any length for both types of elements. If a sequence consists of n_1 elements of type a and n_2 elements of type b, then each of the $\binom{n_1 + n_2}{n_1}$ distinguishable arrangements is equally likely to occur if the probabilities of the two types of elements are constant throughout the sequence. Usually, if the hypothesis of randomness is rejected, it is because there are too few runs. Note that if the probabilities change monotonically throughout a sequence, then elements of one type are more likely to occur at one end of the sequence than at the other, thus decreasing the number of runs. The hypothesis of randomness may also be rejected because there are too many runs. This occurs when the suspected nonrandomness is such that the probability of occurrence of an element of one type is decreased by its occurrence at an immediately preceding point in the sequence, thus increasing the number of runs.

Tables of the exact cumulative probability distribution of U are available for small n_1 and n_2 (Table B.24). The distribution of U is asymptotically normal as $n_1, n_2 \to \infty$ and n_1/n_2 is constant. The statistic U has mean

$$\bar{U} = \frac{2n_1 n_2}{n_1 + n_2} - 1$$

and variance

$$\sigma_U{}^2 = \frac{2n_1 n_2(2n_1 n_2 - n_1 n_2)}{(n_1 + n_2)^2(n_1 + n_2 - 1)}$$

Thus for large n_1 and n_2

$$z = \frac{U - \dfrac{2n_1 n_2}{n_1 + n_2} - 1}{\sqrt{\dfrac{2n_1 n_2(2n_1 n_2 - n_1 - n_2)}{(n_1 + n_2)^2(n_1 + n_2 - 1)}}}$$

is approximately normally distributed with mean zero and variance 1. To correct for continuity, we decrease the absolute value of the numerator by $\frac{1}{2}$. If r_1 denotes the number of runs of one type of element and r_2 denotes the number of runs of the other type of element, then $U = r_1 + r_2$ and $r_1 - r_2$ is zero or 1; thus r_1, r_2, and U are essentially equivalent statistics. The distributions of r_1 and r_2 are hypergeometric and tables of the hypergeometric distribution may thus be used for testing the hypothesis, rather than tables of the distribution of U.

The runs test for randomness is appropriate when the two types of elements arise from a natural dichotomy. If the variable whose randomness is to be tested can assume more than two values, on at least an ordinal scale, then nonparametric tests based on ranks are more efficient.

CASE

The runs test can be summarized as follows:

H_0: sequence random

H_A: (a) too many runs
      (b) too few runs
      (c) sequence not random (too many or too few runs)

Assumptions:

Random sampling (unless sampling is part of the process tested for randomness), unambiguous dichotomization of the elements into two classes

Sample Statistic:

U = total number of runs

Sampling Distribution:

Randomization distribution of U over the $\binom{n_1 + n_2}{n_1}$ possible arrangements of elements

Critical Region:
(a) $U \geq U_{1-\alpha}$
(b) $U \leq U_{\alpha}$
(c) $U \geq U_{1-\alpha/2}$ or $U \leq U_{\alpha/2}$

EXAMPLE
The quality control engineer for Big Bolts Inc. has recorded the results of an inspection of 50 bolts obtained sequentially from one machine. The sequence is given below; P indicates passed inspection and R indicates rejected. He wishes to determine whether the distribution of defectives is random (5 percent level):

$$PPPPP \mid R \mid PPP \mid R \mid PPPPPPPPP \mid R \mid P \mid R \mid PP$$

$$PP \mid R \mid PPPPPPPPPP \mid R \mid PPPP \mid R \mid PPPP \mid R \mid$$

$$H_0 : \text{sequence random}$$

$$H_A : \text{sequence not random}$$

r_1 (passed inspection) $= 8$

r_2 (rejected) $= 8$

$U = 16$

This problem exceeds tabled values, so the normal approximation is used:

$$Z = \frac{U - \dfrac{2n_1 n_2}{n_1 + n_2} - 1}{\sqrt{\dfrac{2n_1 n_2 (2n_1 n_2 - n_1 - n_2)}{(n_1 + n_2)^2 (n_1 + n_2 - 1)}}}$$

$$= \frac{16 - \dfrac{2(8)(42)}{50} - 1}{\sqrt{\dfrac{2(8)(42)(672 - 8 - 42)}{(50)^2(49)}}}$$

$$= \frac{(15 - 13.44)(50)(7)}{\sqrt{(672)(622)}}$$

$$= \frac{546}{8\sqrt{(21)(311)}} = \frac{273}{4(80.81)} = .84$$

$$P(Z > .84) > .05$$

So accept H_0 and conclude that the sequence is random.

17.5 TEST CONCERNING DISTRIBUTIONS

The tests discussed in this section are for hypotheses that concern the distribution or distributions from which samples are drawn. In some cases these hypotheses are of interest for theoretical reasons; in other cases the purpose is to test the validity of distribution assumptions required for other tests of hypotheses.

ONE-SAMPLE TESTS

The Kolmogorov–Smirnov test of the hypothesis that a sampled distribution is of a specified form is discussed next. This test is based on a comparison of theoretical and observed cumulative distribution functions.

The Kolmogorov–Smirnov (K–S) Test   If a random sample is obtained from a population, the cumulative distribution function of the sample observations presumably would not differ substantially from the cumulative distribution function of the population. On the other hand, a sample cumulative distribution function presumably would differ from the cumulative distribution function of a population different from that sampled. The Kolmogorov–Smirnov test is based on a statistic that indicates the deviation of a sample cumulative distribution function from a specified population cumulative distribution function.

Suppose $F(X)$ is the cumulative distribution function of a continuously distributed variable X and $S_n(X)$ is the empirical cumulative distribution function of a random sample of n observations from the X population. If $F(X)$ and $S_n(X)$ are plotted on the same graph, $F(X)$ is a monotonically increasing curve and $S_n(X)$ is a step function that increases in steps of $1/n$ or some multiple of $1/n$; $S_n(X) - F(X)$ is the difference in the ordinate values of the two cumulative distribution functions at a common abscissa value. The distributions of the statistics

$$K^+ = \max[S_n(X) - F(X)]$$

$$K^- = \min[S_n(X) - F(X)] = -\max[F(X) - S_n(X)] = -K^+$$

$$K \ \ = \max[|S_n(X) - F(X)|]$$

have been obtained and shown to be independent of $F(X)$. If the hypothesis to be tested is $F(X) = F_0(X)$, where $F_0(X)$ is the completely specified cumulative distribution function hypothesized to be the distribution of X, then $F_0(X)$ is used in the above formulas. The statistic K^+ is used for one-sided alternatives $F(X) > F_0(X)$ for some values of X, K^- is used for one-sided alternatives $F(X) < F_0(X)$ for some values of X, and K is used for general alternatives $F(X) \neq F_0(X)$. Extensive tables of the distributions of K^+ and K are available (See Table B.25).

The Kolmogorov–Smirnov test is not designed for any specific types or classes of alternatives; it is a general test of goodness of fit. Thus the most appropriate classical test with which to compare it is the chi-square test of goodness of fit. The Kolmogorov–Smirnov test has been found to be more powerful than the chi-square test for the cases examined.

CASE

The Kolmogorov–Smirnov (K–S) test can be summarized as follows:

$$H_0: F(X) = F_0(X)$$

$$H_A: \text{(a)} \ F(X) > F_0(X) \text{ for some } X$$
$$\text{(b)} \ F(X) < F_0(X) \text{ for some } X$$
$$\text{(c)} \ F(X) \neq F_0(X) \text{ for some } X$$

Assumptions:

Random sample of observations from a continuous population having a completely specified distribution

Sample Statistic:
(a) $K^+ = \max[S_n(X) - F(X)]$
(b) $K^- = \min[S_n(X) - F(X)] = -\max[F(X) - S_n(X)] = -K^+$
(c) $K = \max[|S_n(X) - F(X)|]$

Sampling Distribution:

Null distributions of K^+ or K when $F(X) = F_0(X)$; these distributions have been shown to be independent of the form of $F_0(X)$.

Critical Region:
(a) $K^+ > K^+_{1-\alpha}$
(b) $K^- < K_\alpha{}^-$ or, equivalently, $K^- < -K_\alpha{}^+$
(c) $K > K_{1-\alpha}$

Note that a test based on K is *not* a two-tailed version of a test based on K^+; however, if $\alpha \leq .05$, $K^+_{1-\alpha} \approx K_{1-2\alpha}$

EXAMPLE

An investigator hypothesizes that the population distribution of the length of a life of a certain type of transistor is normal with a mean of 125 hr and a variance of 25 hr. The lengths of life for a random sample of 10 transistors are measured and found to be 130, 122, 115, 127, 125, 129, 114, 112, 109, 128, respectively. Do these data support the hypothesis (5 percent level)?

$$H_0: F(X) \text{ is normal, } \mu = 125, \sigma^2 = 25$$

$$H_A: F(X) \text{ is not normal, } \mu = 125, \sigma^2 = 25$$

X	$S_n(X)$	$Z = \dfrac{X - 125}{5}$	$F_0(X)$	$S_n(X) - F_0(X)$
109	0.1	−3.2	.0007	.9993
112	0.2	−2.6	.005	.195
114	0.3	−2.2	.014	.286
115	0.4	−2.0	.023	.377
122	0.5	−0.6	.274	.226
125	0.6	0	.500	.100
127	0.7	0.4	.655	.045
128	0.8	0.6	.726	.074
129	0.9	0.8	.788	.112
130	1.0	1.0	.841	.159

$$K = \max[|S_n(X) - F_0(X)|] = .9993$$

For $n = 10$, $K_{.95} = .40925$ (Table B.25). So reject H_0 and conclude that the distribution is not $N(125, 25)$.

TWO-SAMPLE TEST

The Smirnov test of the hypothesis that two populations are identical against the general alternative that they are not identical is discussed below. This test is a generalization of the Kolmogorov–Smirnov one-sample test and is based on the sample cumulative frequency functions.

The Smirnov Test   Suppose that a random sample of n observations from the distribution of X and a random sample of m observations from the distribution of Y are obtained and that the $n + m$ observations are then arranged in increasing order of magnitude. Denote the ith smallest observation in the combined sample by Z_i; let r_i be the number of Z's $\leq Z_i$ that are X's and let s_i be the number of Z's $\leq Z_i$ that are Y's; calculate

$$d_i = \frac{r_i}{n} - \frac{s_i}{m} = \frac{k}{n} \qquad i = 1, 2, \ldots, n + m$$

The values of $D^+ = \max d_i$ and $D = \max |d_i|$ can be used as statistics for testing the hypothesis that the X and Y populations are identical against a one-sided or two-sided alternative, respectively.

If the distributions of X and Y are in fact identical, then the observations can be regarded as a sample of $n + m$ observations from a single population and each of the $\binom{n+m}{n}$ partitions of the $n + m$ observations into n observations of X and m observations Y is equally likely. The null distribution of D(or D^+) is the relative frequency distribution of D(or D^+) over these $\binom{n+m}{n}$ partitions. The D and D^+ statistics have a graphical interpretation; if $S_n(X)$ and $T_m(Y)$ denote the empirical distribution functions of the X and Y samples, respectively, then

$$D^+ = \max d_i = \max[S_n(Z) - T_m(Z)]$$
$$D = \max |d_i| = \max[|S_n(Z) - T_m(Z)|]$$

where $S_n(Z) - T_m(Z)$ is the difference in ordinates between $S_n(X)$ and $T_m(Y)$ when $X = Y$.

Tables of the cumulative distributions of D and D^+ are available for small n and m (Table B.27), and tables of limiting distributions of functions of the statistics can be used for large n and m.

CASE

The Smirnov test can be summarized as follows:

$$H_0: F(X) = F(Y)$$

$$
\begin{aligned}
H_A: &\text{(a)} \;\; F(X) > F(Y) \\
&\text{(b)} \;\; F(X) < F(Y) \\
&\text{(c)} \;\; F(X) \neq F(Y)
\end{aligned}
$$

Assumptions:

Random sampling from continuous populations (or from infinite populations with no tied observations)

Sample Statistic:
  (a) $D^+ = \max d_i$
  (b) $D^- = \min d_i = -\max d_i = -D^+$
  (c) $D = \max |d_i|$
  Note that a test based on D is *not* a two-tailed version of a test based on D^+; however,
  if $\alpha \leq .05$, $D^+_{1-\alpha} \approx D_{1-2\alpha}$.

Sampling Distribution:

  Null distribution of D^+ or D over the $\binom{n+m}{n}$ equally likely partitions of X and Y

Critical Region:
  (a) $D^+ \geq D^+_{1-\alpha}$
  (b) $D^- < D_\alpha^-$ or, equivalently, $D^- < -D_\alpha^+$
  (c) $D \geq D_{1-\alpha}$

EXAMPLE

The quality control engineer for Poly Plastics Corp. obtained a random sample of 12 sheets of polythene from each of two machines X and Y and measured the clarity (using a standard measurement procedure) for each of the 24 samples. The data are recorded below. Are the clarity distributions identical for the two machines (5 percent level)?

$$H_0: F(X) = F(Y)$$

$$H_A: F(X) \neq F(Y)$$

X OBSERVATIONS	Y OBSERVATIONS
3.70	3.82
3.25	4.55
2.56	3.35
3.09	4.27
2.97	3.63
2.66	4.10
2.38	3.89
3.19	2.90
2.62	4.37
2.21	4.79
3.51	4.92
4.06	2.86

The calculations for the Smirnov test are:

	JOINT RANKING	i	r_i	s_i	$di = \dfrac{r_i - a_i}{12}$
X	2.21	1	1	0	$\frac{1}{12}$
X	2.38	2	2	0	$\frac{2}{12}$
X	2.56	3	3	0	$\frac{3}{12}$
X	2.62	4	4	0	$\frac{4}{12}$
X	2.66	5	5	0	$\frac{5}{12}$
Y	2.86	6	5	1	$\frac{4}{12}$
Y	2.90	7	5	2	$\frac{3}{12}$

	JOINT RANKING	i	r_i	s_i	$di = \dfrac{r_i - a_i}{12}$
X	2.97	8	6	2	$\frac{4}{12}$
X	3.09	9	7	2	$\frac{5}{12}$
X	3.19	10	8	2	$\frac{6}{12}$
X	3.25	11	9	2	$\frac{7}{12}$
Y	3.35	12	9	3	$\frac{6}{11}$
X	3.51	13	10	3	$\frac{7}{12}$
Y	3.63	14	10	4	$\frac{6}{12}$
X	3.70	15	11	4	$\frac{7}{12}$
Y	3.82	16	11	5	$\frac{6}{12}$
Y	3.89	17	11	6	$\frac{5}{12}$
X	4.06	18	12	6	$\frac{6}{12}$
Y	4.10	19	12	7	$\frac{5}{12}$
Y	4.27	20	12	8	$\frac{4}{12}$
Y	4.37	21	12	9	$\frac{3}{12}$
Y	4.55	22	12	10	$\frac{2}{12}$
Y	4.79	23	12	11	$\frac{1}{12}$
Y	4.92	24	12	12	0

$$D^+ = \frac{7}{12} \quad \text{or} \quad k = 7$$

From Table B.26, for $n = m = 12$, the critical (smallest) value of k for which $P(D^+ \geq k/n) \leq .025$ is 7. So reject H_0 and conclude that $F(X) \neq F(Y)$.

The Kolmogorov–Smirnov test for one sample may be regarded as a special case of the Smirnov test for two samples when one sample size is infinite, thus making it the null distribution. The Smirnov tests are the best known of the maximum deviation tests for identical populations, although a number of similar tests have been proposed. For example, The Cramer–Von Mises test is based on the test statistic

$$\frac{nm}{(n + m)^2} \sum d_i^2$$

and a similar test is based on the test statistic

$$\frac{nm}{(n + m)^2} \sum (d_i - \bar{d})^2$$

These tests have also been generalized for more than two populations.

TABLE 17.1 SUMMARY OF NONPARAMETRIC TESTS

	Test	Hypothesis	Population Assumption
	Central Tendency		
One sample	Sign test	$M = M_0$	continuous
	Wilcoxon signed rank test	$M = M_0$	continuous symmetric
Two related samples	Sign test	$P(X > Y) = P(X < Y)$	continuous
	Wilcoxon signed rank test	$M = 0$	continuous symmetric
Two independent samples	Median test	$P(X < \hat{M}) = P(Y < \hat{M})$ $P(X > \hat{M}) = P(Y > \hat{M})$	continuous
	Wilcoxon rank-sum test	X and Y identically distributed	continuous
	Correlation		
Two independent samples	Spearman test	X and Y independent	continuous
	Kendall test	X and Y independent	continuous
	Randomness		
One sample	Daniel test	$r_s = 0$	continuous
	Median test	$P(X < \hat{M}_x$ sample 1) $= P(X > \hat{M}_x$ sample 2)	continuous
	Runs test	sequence random	dichotomous
	Distribution		
One sample	Kolmogorov–Smirnov	$F(X) = F_0(X)$	continuous specified
Two sample	Smirnov	X and Y identically Distribution	continuous

SUMMARY TABLE

The nonparametric tests examined in this chapter are summarized in Table 17.1. These and additional tests are discussed in texts listed in Appendix A.

PROBLEMS

17.1 An air conditioning company is concerned with the distribution of monthly average temperatures in Megolopolis. The marketing department maintains that the average temperature in May is normally distributed with a mean of $82.5°$F and a variance of $6.25°$F. Do the following mean temperatures for a random sample of ten years support the marketing department's statement (5 percent level of significance)?

78.6
82.9
91.8
73.7
75.4
90.6
89.5
93.0
76.9
87.8

17.2 Consumer Research Associates are conducting research on the discretionary spending habits of husbands and wives. A random sample of 15 husbands and wives
were asked (separately) how much money they felt free to spend without consulting
their spouse and the following data (in dollars) were recorded.

HUSBAND	WIFE
500	320
250	310
350	100
400	230
100	210
300	260
350	300
450	300
100	70
150	50
200	80
150	190
100	20
300	100
75	100

Do these data indicate a difference in the reported discretionary spending of husbands
and wives (5 percent level of significance)? Use both the sign test and the signed
rank test and state carefully the hypothesis and assumptions for each test.

17.3 Compute the correlation (Spearman and Kendall) between number of years in
business and number of accounts for the following advertising agencies and test
the hypothesis that the rank order correlation is zero, against the alternative hypothesis that it is positive (5 percent level of significance).

AGENCY	NUMBER OF YEARS IN BUSINESS	NUMBER OF ACCOUNTS
A	10	125
B	15	100
C	12	120
D	5	60
E	6	70
F	2	50

17.4 Suppose that the following residuals were obtained from a least squares regression. Using the runs test, state and test the appropriate hypothesis concerning serial dependency. (Note that the normal approximation is appropriate.)

+5.682	−0.025
+3.124	−0.293
+4.897	+1.113
+0.089	+2.682
+0.682	+4.890
+2.178	+5.090
+3.009	+3.272
−0.097	+3.669
−2.223	+2.898
−1.007	+1.009
−1.118	+0.907
−2.003	+1.212
−2.958	+1.973
−3.804	+0.909
−3.617	−0.997
−6.446	−0.884
−5.888	−1.982
−4.101	−2.772
−0.925	−4.693
+2.169	−3.335
+3.083	−5.005
+4.997	−3.526
+5.002	−2.229
+1.678	−4.907
+1.541	−2.920

17.5 Patterns of consumption were determined for a random sample of 15 respondents from City A and 20 respondents from City B. The ages of the respondents are given below.

CITY A	CITY B
25	32
19	45
37	24
22	26
35	43
48	38
29	39
22	30
27	20
23	21
32	43
33	35
55	37
47	42
52	26
	53
	49
	31
	34
	45

Do the data indicate that the respondents from Cities A and B are comparable with respect to age (5 percent level of significance)? Use both the median test and the rank-sum test and state carefully the hypothesis and assumptions for each test.

17.6 A promotion brochure for hospitalization insurance claims that the median hospital stay is currently 5 days. The supervisor of a large general hospital suspects that the median duration of stay in that hospital is shorter than 5 days. He obtained a random sample of 30 records for the past month. The stays recorded (in days) and ordered were as follows:

.5	2.75	4.75
.75	3	5
.75	3.25	5.5
1	3.5	6
1.25	3.5	8.25
1.5	3.75	9.5
2	4	10
2.25	4.25	10.25
2.25	4.5	16
2.5	4.75	25

Note that the hospital has three check-out times: 8 P.M. (counted as .25 day), noon (counted as .5 day) and 4 P.M. (counted as .75 day). Do these data support the supervisor's suspicion (5 percent level of significance)? Use both the sign test and the signed rank test and state carefully the assumptions for each test.

17.7 A lumber company is concerned with whether there is a decrease in the percentage of its sales made to private individuals (rather than to construction companies.) Do the following data (in percents) for the past 20 weeks indicate a downward trend in percentage of sales to private individuals (5 percent level of significance)? Use the Daniel test, the Kendall test, and the median test.

MONTH	PRIVATE SALES
1	20.2
2	21.3
3	18.4
4	19.6
5	17.5
6	17.8
7	16.9
8	17.0
9	23.2
10	24.8
11	19.9
12	20.7
13	21.3
14	24.6
15	23.9
16	23.7
17	23.8
18	22.9
19	21.6
20	17.7

17.8 Precision Research Associates employ two key punch operators. Jobs are assigned at random to the operators; all data are recorded on standardized forms for punching. When the cards are verified, the number of punching errors is recorded. The following data are the numbers of errors made by the operators on random samples of jobs. Are the error distributions for the two operators identical (5 percent level of significance)?

OPERATOR A	OPERATOR B
2	5
10	4
14	10
0	13
12	19
6	7
13	16
9	8
8	17
18	12
10	20
1	9
11	3
16	15
17	6
7	14

17.9 An advertising research company is concerned that its computer expenses are increasing over time. Do the following data (in hundreds of dollars) for the past 15 months indicate an upward trend in computer expenses (5 percent level of significance)? Use the Daniel test, the Kendall test, and the median test.

MONTH	EXPENSES
1	20.2
2	19.1
3	26.3
4	28.9
5	27.5
6	23.6
7	30.4
8	32.5
9	35.6
10	36.0
11	37.9
12	35.2
13	36.9
14	40.5
15	42.7

17.10 The following residuals were obtained from a regression of an economic variable on time. Using the runs test, state and test the appropriate hypotheses concerning serial independence (5 percent level of significance).

+ 5.5	− 4.8
+ 6.2	− 5.0
+ 8.3	− 5.9
+ 4.2	− 4.7
+ 1.1	+ 2.8
− 3.0	+ 2.1
− 4.6	+ 4.9
− 3.2	+ 3.9
+ 1.0	+ 3.6
+ 3.8	+ 5.2
+ 4.5	+ 4.6
− 5.9	+ 1.3
− 4.3	− 2.9
− 7.6	− 2.7
− 6.0	− 3.4
− 5.2	− 3.0
− 6.5	− 3.2
− 6.2	− 4.1

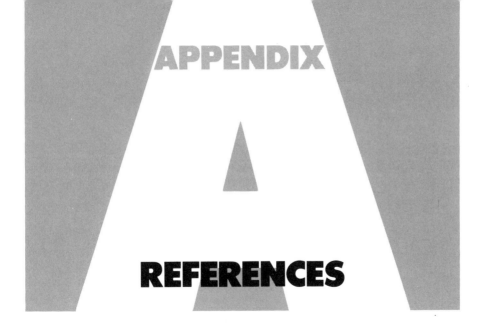

APPENDIX A

REFERENCES

Chatterjee, Samprit, and Bertram Price. *Regression Analysis by Example*. New York: Wiley, 1977.

Cochran, William G., and Gertrude M. Cox. *Experiment Designs* (2nd ed.). New York: Wiley, 1957.

Fox, Karl A. *Intermediate Economic Statistics*, New York: Wiley, 1968.

Gibbons, Jean D. *Nonparametric Methods for Quantitative Analysis*. New York: Holt, Rinehart and Winston, 1976.

Guenther, Clayton A. *Concepts of Probability*. New York: McGraw-Hill, 1968.

Guenther, William C. *Analysis of Variance*, Englewood Cliffs, N.J.: Prentice-Hall, 1964.

Harris, Richard J. *A Primer of Multivariate Statistics*. New York: Academic Press, 1975.

Hicks, Charles R. *Fundamental Concepts in the Design of Experiments*. New York: Holt, Rinehart and Winston, 1964.

Hollander, Myles, and Douglas A. Wolfe. *Nonparametric Statistical Methods*. New York: Wiley, 1973.

Johnston, J. *Econometric Methods* (2nd ed.). New York: McGraw-Hill, 1972.

Kish, Leslie, *Survey Sampling*, New York: Wiley, 1965.

Kleinbaum, David G., and Lawrence L. Kupper. *Applied Regression Analysis and Other Multivariate Methods*. North Scituate, Mass.: Duxbury Press, 1978.

Kmenta, Jan. *Elements of Econometrics*. New York: Macmillan, 1971.

Levin, Richard I., and Charles A. Kirkpatrick. *Quantitative Approaches to Management* (3rd ed.). New York: McGraw-Hill, 1975.

Mendenhall, William, Lyman Ott, and Richard L. Schaeffer. *Elementary Survey Sampling*. Duxbury Press, 1971.

Montgomery, Douglas C. *Design and Analysis of Experiments*. New York: Wiley. 1976.

Mood, Alexander M., Franklin A. Graybill, and Duane C. Boes. *Introduction to the Theory of Statistics*. New York: McGraw-Hill, 1974.

Ott, Lyman. *An Introduction to Statistical Methods and Data Analysis*. North Scituate: Duxbury Press, 1977.

Pindyck, R. S., and D. L. Rubinfeld. *Econometric Models and Economic Forecasts*. New York: McGraw-Hill, 1976.

Press, S. James. *Applied Multivariate Analysis*. New York: Holt, Rinehart and Winston, 1972.

Wesolowsky, George O. *Multiple Rregression and Analysis of Variance*. New York: Wiley, 1976.

TABLE B.1 BINOMIAL DISTRIBUTION (INDIVIDUAL TERMS)

						π					
n x	.05	.10	.15	.20	.25	.30	.35	.40	.45	.50	
1 0	.9500	.9000	.8500	.8000	.7500	.7000	.6500	.6000	.5500	.5000	
1	.0500	.1000	.1500	.2000	.2500	.3000	.3500	.4000	.4500	.5000	
2 0	.9025	.8100	.7225	.6400	.5625	.4900	.4225	.3600	.3025	.2500	
1	.0950	.1800	.2550	.3200	.3750	.4200	.4550	.4800	.4950	.5000	
2	.0025	.0100	.0225	.0400	.0625	.0900	.1225	.1600	.2025	.2500	
3 0	.8574	.7290	.6141	.5120	.4219	.3430	.2746	.2160	.1664	.1250	
1	.1354	.2430	.3251	.3840	.4219	.4410	.4436	.4320	.4084	.3750	
2	.0071	.0270	.0574	.0960	.1406	.1890	.2389	.2880	.3341	.3750	
3	.0001	.0010	.0034	.0080	.0156	.0270	.0429	.0640	.0911	.1250	
4 0	.8145	.6561	.5220	.4096	.3164	.2401	.1785	.1296	.0915	.0625	
1	.1715	.2916	.3685	.4096	.4219	.4116	.3845	.3456	.2995	.2500	
2	.0135	.0486	.0975	.1536	.2109	.2646	.3105	.3456	.3675	.3750	
3	.0005	.0036	.0115	.0256	.0469	.0756	.1115	.1536	.2005	.2500	
4	.0000	.0001	.0005	.0016	.0039	.0081	.0150	.0256	.0410	.0625	
5 0	.7738	.5905	.4437	.3277	.2373	.1681	.1160	.0778	.0503	.0312	
1	.2036	.3280	.3915	.4096	.3955	.3602	.3124	.2592	.2059	.1562	
2	.0214	.0729	.1382	.2048	.2637	.3087	.3364	.3456	.3369	.3125	
3	.0011	.0081	.0244	.0512	.0879	.1323	.1811	.2304	.2757	.3125	
4	.0000	.0004	.0022	.0064	.0146	.0284	.0488	.0768	.1128	.1562	
5	.0000	.0000	.0001	.0003	.0010	.0024	.0053	.0102	.0185	.0312	
6 0	.7351	.5314	.3771	.2621	.1780	.1176	.0754	.0467	.0277	.0156	
1	.2321	.3543	.3993	.3932	.3560	.3025	.2437	.1866	.1359	.0938	
2	.0305	.0984	.1762	.2458	.2966	.3241	.3280	.3110	.2780	.2344	
3	.0021	.0146	.0415	.0819	.1318	.1852	.2355	.2765	.3032	.3125	
4	.0001	.0012	.0055	.0154	.0330	.0595	.0951	.1382	.1861	.2344	
5	.0000	.0001	.0004	.0015	.0044	.0102	.0205	.0369	.0609	.0938	
6	.0000	.0000	.0000	.0001	.0002	.0007	.0018	.0041	.0083	.0156	

TABLE B.1   BINOMIAL DISTRIBUTION (INDIVIDUAL TERMS) (*continued*)

						π					
n x	.05	.10	.15	.20	.25	.30	.35	.40	.45	.50
7 0	.6983	.4783	.3206	.2097	.1335	.0824	.0490	.0280	.0152	.0078
1	.2573	.3720	.3960	.3670	.3115	.2471	.1848	.1306	.0872	.0547
2	.0406	.1240	.2097	.2753	.3115	.3177	.2985	.2613	.2140	.1641
3	.0036	.0230	.0617	.1147	.1730	.2269	.2679	.2903	.2918	.2734
4	.0002	.0026	.0109	.0287	.0577	.0972	.1442	.1935	.2388	.2734
5	.0000	.0002	.0012	.0043	.0115	.0250	.0466	.0774	.1172	.1641
6	.0000	.0000	.0001	.0004	.0013	.0036	.0084	.0172	.0320	.0547
7	.0000	.0000	.0000	.0000	.0001	.0002	.0006	.0016	.0037	.0078
8 0	.6634	.4305	.2725	.1678	.1001	.0576	.0319	.0168	.0084	.0039
1	.2793	.3826	.3847	.3355	.2670	.1977	.1373	.0896	.0548	.0312
2	.0515	.1488	.2376	.2936	.3115	.2965	.2587	.2090	.1569	.1094
3	.0054	.0331	.0839	.1468	.2076	.2541	.2786	.2787	.2568	.2188
4	.0004	.0046	.0185	.0459	.0865	.1361	.1875	.2322	.2627	.2734
5	.0000	.0004	.0026	.0092	.0231	.0467	.0808	.1239	.1719	.2188
6	.0000	.0000	.0002	.0011	.0038	.0100	.0217	.0413	.0703	.1094
7	.0000	.0000	.0000	.0001	.0004	.0012	.0033	.0079	.0164	.0312
8	.0000	.0000	.0000	.0000	.0000	.0001	.0002	.0007	.0017	.0039
9 0	.6302	.3874	.2316	.1342	.0751	.0404	.0207	.0101	.0046	.0020
1	.2985	.3874	.3679	.3020	.2253	.1556	.1004	.0605	.0339	.0176
2	.0629	.1722	.2597	.3020	.3003	.2668	.2162	.1612	.1110	,0703
3	.0077	.0446	.1069	.1762	.2336	.2668	.2716	.2508	.2119	.1641
4	.0006	.0074	.0283	.0661	.1168	.1715	.2194	.2508	.2600	.2461
5	.0000	.0008	.0050	.0165	.0389	.0735	.1181	.1672	.2128	.2461
6	.0000	.0001	.0006	.0028	.0087	.0210	.0424	.0743	.1160	.1641
7	.0000	.0000	.0000	.0003	.0012	.0039	.0098	.0212	.0407	.0703
8	.0000	.0000	.0000	.0000	.0001	.0004	.0013	.0035	.0083	.0176
9	.0000	.0000	.0000	.0000	.0000	.0000	.0001	.0003	.0008	.0020
10 0	.5987	.3487	.1969	.1074	.0563	.0282	.0135	.0060	.0025	.0010
1	.3151	.3874	.3474	.2684	.1877	.1211	.0725	.0403	.0207	.0098
2	.0746	.1937	.2759	.3020	.2816	.2335	.1757	.1209	.0763	.0439
3	.0105	.0574	.1298	.2013	.2503	.2668	.2522	.2150	.1665	.1172
4	.0010	.0112	.0401	.0881	.1460	.2001	.2377	.2508	.2384	.2051
5	.0001	.0015	.0085	.0264	.0584	.1029	.1536	.2007	.2340	.2461
6	.0000	.0001	.0012	.0055	.0162	.0368	.0689	.1115	.1596	.2051
7	.0000	.0000	.0001	.0008	.0031	.0090	.0212	.0425	.0746	.1172
8	.0000	.0000	.0000	.0001	.0004	.0014	.0043	.0106	.0229	.0439
9	.0000	.0000	.0000	.0000	.0000	.0001	.0005	.0016	.0042	.0098
10	.0000	.0000	.0000	.0000	.0000	.0000	.0000	.0001	.0003	.0010

TABLE B.1   BINOMIAL DISTRIBUTION (INDIVIDUAL TERMS)(*continued*)

						π					
n x		.05	.10	.15	.20	.25	.30	.35	.40	.45	.50
11	0	.5688	.3138	.1673	.0859	.0422	.0198	.0088	.0036	.0014	.0005
	1	.3293	.3835	.3248	.2362	.1549	.0932	.0518	.0266	.0125	.0054
	2	.0867	.2131	.2866	.2953	.2581	.1998	.1395	.0887	.0513	.0269
	3	.0137	.0710	.1517	.2215	.2581	.2568	.2254	.1774	.1259	.0806
	4	.0014	.0158	.0536	.1107	.1721	.2201	.2428	.2365	.2060	.1611
	5	.0001	.0025	.0132	.0388	.0803	.1321	.1830	.2207	.2360	.2256
	6	.0000	.0003	.0023	.0097	.0268	.0566	.0985	.1471	.1931	.2256
	7	.0000	.0000	.0003	.0017	.0064	.0173	.0379	.0701	.1128	.1611
	8	.0000	.0000	.0000	.0002	.0011	.0037	.0102	.0234	.0462	.0806
	9	.0000	.0000	.0000	.0000	.0001	.0005	.0018	.0052	.0126	.0269
	10	.0000	.0000	.0000	.0000	.0000	.0000	.0002	.0007	.0021	.0054
	11	.0000	.0000	.0000	.0000	.0000	.0000	.0000	.0000	.0002	.0005
12	0	.5404	.2824	.1422	.0687	.0317	.0138	.0057	.0022	.0008	.0002
	1	.3413	.3766	.3012	.2062	.1267	.0712	.0368	.0174	.0075	.0029
	2	.0988	.2301	.2924	.2835	.2323	.1678	.1088	.0639	.0339	.0161
	3	.0173	.0852	.1720	.2362	.2581	.2397	.1954	.1419	.0923	.0537
	4	.0021	.0213	.0683	.1329	.1936	.2311	.2367	.2128	.1700	.1208
	5	.0002	.0038	.0193	.0532	.1032	.1585	.2039	.2270	.2225	.1934
	6	.0000	.0005	.0040	.0155	.0401	.0792	.1281	.1766	.2124	.2256
	7	.0000	.0000	.0006	.0033	.0115	.0291	.0591	.1009	.1489	.1934
	8	.0000	.0000	.0001	.0005	.0024	.0078	.0199	.0420	.0762	.1208
	9	.0000	.0000	.0000	.0001	.0004	.0015	.0048	.0125	.0277	.0537
	10	.0000	.0000	.0000	.0000	.0000	.0002	.0008	.0025	.0068	.0161
	11	.0000	.0000	.0000	.0000	.0000	.0000	.0001	.0003	.0010	.0029
	12	.0000	.0000	.0000	.0000	.0000	.0000	.0000	.0000	.0001	.0002
13	0	.5133	.2542	.1209	.0550	.0238	.0097	.0037	.0013	.0004	.0001
	1	.3512	.3672	.2774	.1787	.1020	.0540	.0259	.0113	.0045	.0016
	2	.1109	.2448	.2937	.2680	.2059	.1388	.0836	.0453	.0220	.0095
	3	.0214	.0997	.1900	.2457	.2517	.2181	.1651	.1107	.0660	.0349
	4	.0028	.0277	.0838	.1535	.2097	.2337	.2222	.1845	.1350	.0873
	5	.0003	.0055	.0266	.0691	.1258	.1803	.2154	.2214	.1989	.1571
	6	.0000	.0008	.0063	.0230	.0559	.1030	.1546	.1968	.2169	.2095
	7	.0000	.0001	.0011	.0058	.0186	.0442	.0833	.1312	.1775	.2095
	8	.0000	.0000	.0001	.0011	.0047	.0142	.0336	.0656	.1089	.1571
	9	.0000	.0000	.0000	.0001	.0009	.0034	.0101	.0243	.0495	.0873
	10	.0000	.0000	.0000	.0000	.0001	.0006	.0022	.0065	.0162	.0349
	11	.0000	.0000	.0000	.0000	.0000	.0001	.0003	.0012	.0036	.0095
	12	.0000	.0000	.0000	.0000	.0000	.0000	.0000	.0001	.0005	.0016
	13	.0000	.0000	.0000	.0000	.0000	.0000	.0000	.0000	.0000	.0001

TABLE B.2   CUMULATIVE BINOMIAL DISTRIBUTION

						π					
$n\ x'$	.05	.10	.15	.20	.25	.30	.35	.40	.45	.50
2 1	.0975	.1900	.2775	.3600	.4375	.5100	.5775	.6400	.6975	.7500
2	.0025	.0100	.0225	.0400	.0625	.0900	.1225	.1600	.2025	.2500
3 1	.1426	.2710	.3859	.4880	.5781	.6570	.7254	.7840	.8336	.8750
2	.0072	.0280	.0608	.1040	.1562	.2160	.2818	.3520	.4252	.5000
3	.0001	.0010	.0034	.0080	.0156	.0270	.0429	.0640	.0911	.1250
4 1	.1855	.3439	.4780	.5904	.6836	.7599	.8215	.8704	.9085	.9375
2	.0140	.0523	.1095	.1808	.2617	.3483	.4370	.5248	.6090	.6875
3	.0005	.0037	.0120	.0272	.0508	.0837	.1265	.1792	.2415	.3125
4	.0000	.0001	.0005	.0016	.0039	.0081	.0150	.0256	.0410	.0625
5 1	.2262	.4095	.5563	.6723	.7627	.8319	.8840	.9222	.9497	.9688
2	.0226	.0815	.1648	.2627	.3672	.4718	.5716	.6630	.7438	.8125
3	.0012	.0086	.0266	.0579	.1035	.1631	.2352	.3174	.4069	.5000
4	.0000	.0005	.0022	.0067	.0156	.0308	.0540	.0870	.1312	.1875
5	.0000	.0000	.0001	.0003	.0010	.0024	.0053	.0102	.0185	.0312
6 1	.2649	.4686	.6229	.7379	.8220	.8824	.9246	.9533	.9723	.9844
2	.0328	.1143	.2235	.3446	.4661	.5798	.6809	.7667	.8364	.8906
3	.0022	.0158	.0473	.0989	.1694	.2257	.3529	.4557	.5585	.6562
4	.0001	.0013	.0059	.0170	.0376	.0705	.1174	.1792	.2553	.3438
5	.0000	.0001	.0004	.0016	.0046	.0109	.0223	.0410	.0692	.1094
6	.0000	.0000	.0000	.0001	.0002	.0007	.0018	.0041	.0083	.0156
7 1	.3017	.5217	.6794	.7903	.8665	.9176	.9510	.9720	.9848	.9922
2	.0444	.1497	.2834	.4233	.5551	.6706	.7662	.8414	.8976	.9375
3	.0038	.0257	.0738	.1480	.2436	.3529	.4677	.5801	.6836	.7734
4	.0002	.0027	.0121	.0333	.0706	.1260	.1998	.2898	.3917	.5000
5	.0000	.0002	.0012	.0047	.0129	.0288	.0556	.0963	.1529	.2266
6	.0000	.0000	.0001	.0004	.0013	.0038	.0090	.0188	.0357	.0625
7	.0000	.0000	.0000	.0000	.0001	.0002	.0006	.0016	.0037	.0078
8 1	.3366	.5695	.7275	.8322	.8999	.9424	.9681	.9832	.9916	.9961
2	.0572	.1869	.3428	.4967	.6329	.7447	.8309	.8936	.9368	.9648
3	.0058	.0381	.1052	.2031	.3215	.4482	.5722	.6846	.7799	.8555
4	.0004	.0050	.0214	.0563	.1138	.1941	.2936	.4059	.5230	.6367
5	.0000	.0004	.0029	.0104	.0273	.0580	.1061	.1737	.2604	.3633
6	.0000	.0000	.0002	.0012	.0042	.0113	.0253	.0498	.0885	.1445
7	.0000	.0000	.0000	.0001	.0004	.0013	.0036	.0085	.0181	.0352
8	.0000	.0000	.0000	.0000	.0000	.0001	.0002	.0007	.0017	.0039
9 1	.3698	.6126	.7684	.8658	.9249	.9596	.9793	.9899	.9954	.9980
2	.0712	.2252	.4005	.5638	.6997	.8040	.8789	.9295	.9615	.9805
3	.0084	.0530	.1409	.2618	.3993	.5372	.6627	.7682	.8505	.9102
4	.0006	.0083	.0339	.0856	.1657	.2703	.3911	.5174	.6386	.7461
5	.0000	.0009	.0056	.0196	.0489	.0988	.1717	.2666	.3786	.5000
6	.0000	.0001	.0006	.0031	.0100	.0253	.0536	.0994	.1658	.2539
7	.0000	.0000	.0000	.0003	.0013	.0043	.0112	.0250	.0498	.0898
8	.0000	.0000	.0000	.0000	.0001	.0004	.0014	.0038	.0091	.0195
9	.0000	.0000	.0000	.0000	.0000	.0000	.0001	.0003	.0008	.0020

TABLE B.2   CUMULATIVE BINOMIAL DISTRIBUTION (*continued*)

n x′	.05	.10	.15	.20	.25	.30	.35	.40	.45	.50
10 1	.4013	.6513	.8031	.8926	.9437	.9718	.9865	.9940	.9975	.9990
2	.0861	.2639	.4457	.6242	.7560	.8507	.9140	.9536	.9767	.9893
3	.0115	.0702	.1798	.3222	.4744	.6172	.7384	.8327	.9004	.9453
4	.0010	.0128	.0500	.1209	.2241	.3504	.4862	.6177	.7340	.8281
5	.0001	.0016	.0099	.0328	.0781	.1503	.2485	.3669	.4956	.6230
6	.0000	.0001	.0014	.0064	.0197	.0473	.0949	.1662	.2616	.3770
7	.0000	.0000	.0001	.0009	.0035	.0106	.0260	.0548	.1020	.1719
8	.0000	.0000	.0000	.0001	.0004	.0016	.0048	.0123	.0274	.0547
9	.0000	.0000	.0000	.0000	.0000	.0001	.0005	.0017	.0045	.0107
10	.0000	.0000	.0000	.0000	.0000	.0000	.0000	.0001	.0003	.0010
11 1	.4312	.6862	.8327	.9141	.9578	.9802	.9912	.9964	.9986	.9995
2	.1019	.3026	.5078	.6779	.8029	.8870	.9394	.9698	.9861	.9941
3	.0152	.0896	.2212	.3826	.5448	.6873	.7999	.8811	.9348	.9673
4	.0016	.0185	.0694	.1611	.2867	.4304	.5744	.7037	.8089	.8867
5	.0001	.0028	.0159	.0504	.1146	.2103	.3317	.4672	.6029	.7256
6	.0000	.0003	.0027	.0117	.0343	.0782	.1487	.2465	.3669	.5000
7	.0000	.0000	.0003	.0020	.0076	.0216	.0501	.0994	.1738	.2744
8	.0000	.0000	.0000	.0002	.0012	.0043	.0122	.0293	.0610	.1133
9	.0000	.0000	.0000	.0000	.0001	.0006	.0020	.0059	.0148	.0327
10	.0000	.0000	.0000	.0000	.0000	.0000	.0002	.0007	.0022	.0059
11	.0000	.0000	.0000	.0000	.0000	.0000	.0000	.0000	.0002	.0005
12 1	.4596	.7176	.8578	.9313	.9683	.9862	.9943	.9978	.9992	.9998
2	.1184	.3410	.5565	.7251	.8416	.9150	.9576	.9804	.9917	.9968
3	.0196	.1109	.2642	.4417	.6093	.7472	.8487	.9166	.9579	.9807
4	.0022	.0256	.0922	.2054	.3512	.5075	.6533	.7747	.8655	.9270
5	.0002	.0043	.0239	.0726	.1576	.2763	.4167	.5618	.6956	.8062
6	.0000	.0005	.0046	.0194	.0544	.1178	.2127	.3348	.4731	.6128
7	.0000	.0001	.0007	.0039	.0143	.0386	.0846	.1582	.2607	.3872
8	.0000	.0000	.0001	.0006	.0028	.0095	.0255	.0573	.1117	.1938
9	.0000	.0000	.0000	.0001	.0004	.0017	.0056	.0153	.0356	.0730
10	.0000	.0000	.0000	.0000	.0000	.0002	.0008	.0028	.0079	.0193
11	.0000	.0000	.0000	.0000	.0000	.0000	.0001	.0003	.0011	.0032
12	.0000	.0000	.0000	.0000	.0000	.0000	.0000	.0000	.0001	.0002
13 1	.4867	.7458	.8791	.9450	.9762	.9903	.9963	.9987	.9996	.9999
2	.1354	.3787	.6017	.7664	.8733	.9363	.9704	.9874	.9951	.9983
3	.0245	.1339	.2704	.4983	.6674	.7975	.8868	.9421	.9731	.9888
4	.0031	.0342	.0967	.2527	.4157	.5794	.7217	.8314	.9071	.9539
5	.0003	.0065	.0260	.0991	.2060	.3457	.4995	.6470	.7721	.8666
6	.0000	.0009	.0053	.0300	.0802	.1654	.2841	.4256	.5732	.7095
7	.0000	.0001	.0013	.0070	.0243	.0624	.1295	.2288	.3563	.5000
8	.0000	.0000	.0002	.0012	.0056	.0182	.0462	.0977	.1788	.2905
9	.0000	.0000	.0000	.0002	.0010	.0040	.0126	.0321	.0698	.1334
10	.0000	.0000	.0000	.0000	.0001	.0007	.0025	.0078	.0203	.0461
11	.0000	.0000	.0000	.0000	.0000	.0001	.0003	.0013	.0041	.0112
12	.0000	.0000	.0000	.0000	.0000	.0000	.0000	.0001	.0005	.0017
13	.0000	.0000	.0000	.0000	.0000	.0000	.0000	.0000	.0000	.0001

TABLE B.3   BINOMIAL COEFFICIENTS

n \ $\binom{n}{x}$	$\binom{n}{0}$	$\binom{n}{1}$	$\binom{n}{2}$	$\binom{n}{3}$	$\binom{n}{4}$	$\binom{n}{5}$	$\binom{n}{6}$	$\binom{n}{7}$	$\binom{n}{8}$	$\binom{n}{9}$	$\binom{n}{10}$
0	1										
1	1	1									
2	1	2	1								
3	1	3	3	1							
4	1	4	6	4	1						
5	1	5	10	10	5	1					
6	1	6	15	20	15	6	1				
7	1	7	21	35	35	21	7	1			
8	1	8	28	56	70	56	28	8	1		
9	1	9	36	84	126	126	84	36	9	1	
10	1	10	45	120	210	252	210	120	45	10	1
11	1	11	55	165	330	462	462	330	165	55	11
12	1	12	66	220	495	792	924	792	495	220	66
13	1	13	78	286	715	1287	1716	1716	1287	715	286
14	1	14	91	364	1001	2002	3003	3432	3003	2002	1001
15	1	15	105	455	1365	3003	5005	6435	6435	5005	3003
16	1	16	120	560	1820	4368	8008	11440	12870	11440	8008
17	1	17	136	680	2380	6188	12376	19448	24310	24310	19448
18	1	18	153	816	3060	8568	18564	31824	43758	48620	43758
19	1	19	171	969	3876	11628	27132	50388	75582	92378	92378
20	1	20	190	1140	4845	15504	38760	77520	125970	167960	184756

TABLE B.4 FACTORIALS AND THEIR LOGARITHMS

No.	Factorial	Logarithm	No.	Factorial	Logarithm	No.	Factorial	Logarithm
1	1	0.000 0000	26	4.03291	26.605 6190	51	1.55112	66.190 6450
2	2	0.301 0300	27	1.08889	28.036 9828	52	8.06582	67.906 6484
3	6	0.778 1513	28	3.04888	29.484 1408	53	4.27488	69.630 9243
4	24	1.380 2112	29	8.84176	30.946 5388	54	2.30844	71.363 3180
5	120	2.079 1812	30	2.65253	32.423 6601	55	1.26964	73.103 6807
6	720	2.857 3325	31	8.22284	33.915 0218	56	7.10999	74.851 8687
7	5040	3.702 4305	32	2.63131	35.420 1717	57	4.05269	76.607 7436
8	40320	4.605 5205	33	8.68332	36.938 6857	58	2.35056	78.371 1716
9	362880	5.559 7630	34	2.95233	38.470 1646	59	1.38683	80.142 0236
10	3.62880	6.559 7630	35	1.03331	40.014 2326	60	8.32099	81.920 1748
11	3.99168	7.601 1557	36	3.71993	41.570 5351	61	5.07580	83.705 5047
12	4.79002	8.680 3370	37	1.37638	43.138 7369	62	3.14700	85.497 8964
13	6.22702	9.794 2803	38	5.23023	44.718 5205	63	1.98261	87.297 2369
14	8.71783	10.940 4084	39	2.03979	46.309 5851	64	1.26887	89.103 4169
15	1.30767	12.116 4996	40	8.15915	47.911 6451	65	8.24765	90.916 3303
16	2.09228	13.320 6196	41	3.34525	49.524 4289	66	5.44345	92.735 8742
17	3.55687	14.551 0685	42	1.40501	51.147 6782	67	3.64711	94.561 9490
18	6.40237	15.806 3410	43	6.04153	52.781 1467	68	2.48004	96.394 4579
19	1.21645	17.085 0946	44	2.65827	54.424 5993	69	1.71122	98.233 3070
20	2.43290	18.386 1246	45	1.19622	56.077 8119	70	1.19786	100.078 4050
21	5.10909	19.708 3439	46	5.50262	57.740 5697	71	8.50479	101.929 6634
22	1.12400	21.050 7666	47	2.58623	59.412 6676	72	6.12345	103.786 9959
23	2.58520	22.412 4944	48	1.24139	61.093 9088	73	4.47012	105.650 3187
24	6.20448	23.792 7057	49	6.08282	62.784 1049	74	3.30789	107.519 5505
25	1.55112	25.190 6457	50	3.04141	64.483 0749	75	2.48091	109.394 6117

The power of 10 by which to multiply the factorial is given by the whole number of logarithm.

TABLE B.4 FACTORIALS AND THEIR LOGARITHMS (continued)

No.	Factorial	Logarithm	No.	Factorial	Logarithm	No.	Factorial	Logarithm
76	1.88549	111.275 4253	101	9.42595	159.974 3250	126	2.37217	211.375 1464
77	1.45183	113.161 9160	102	9.61447	161.982 9252	127	3.01266	213.478 9501
78	1.13243	115.054 0106	103	9.90290	163.995 7624	128	3.85620	215.586 1601
79	8.94618	116.951 6377	104	1.02990	166.012 7958	129	4.97450	217.696 7498
80	7.15695	118.854 7277	105	1.08140	168.033 9851	130	6.46686	219.810 6932
81	5.79713	120.763 2127	106	1.14628	170.059 2909	131	8.47158	221.927 9645
82	4.75364	122.677 0266	107	1.22652	172.088 6747	132	1.11825	224.048 5384
83	3.94552	124.596 1047	108	1.32464	174.122 0985	133	1.48727	226.172 3900
84	3.31424	126.520 3840	109	1.44386	176.159 5250	134	1.99294	228.299 4948
85	2.81710	128.449 8029	110	1.58825	178.200 9176	135	2.69047	230.429 8286
86	2.42271	130.384 3013	111	1.76295	180.246 2406	136	3.65904	232.563 3675
87	2.10776	132.323 8206	112	1.97451	182.295 4586	137	5.01289	234.700 0881
88	1.85483	134.268 3033	113	2.23119	184.348 5371	138	6.91779	236.839 9672
89	1.65080	136.217 6933	114	2.54356	186.405 4419	139	9.61572	238.982 9820
90	1.48572	138.171 9358	115	2.92509	188.466 1398	140	1.34620	241.129 1100
91	1.35200	140.130 9772	116	3.39311	190.530 5978	141	1.89814	243.278 3291
92	1.24384	142.094 7650	117	3.96994	192.598 7836	142	2.69536	245.430 6174
93	1.15677	144.063 2480	118	4.68453	194.670 6656	143	3.85437	247.585 9535
94	1.08737	146.036 3758	119	5.57459	196.746 2126	144	5.55029	249.744 3160
95	1.03300	148.014 0994	120	6.68950	198.825 3938	145	8.04793	251.905 6840
96	9.91678	149.996 3707	121	8.09430	200.908 1792	146	1.17500	254.070 0368
97	9.61928	151.983 1424	122	9.87504	202.994 5390	147	1.72725	256.237 3542
98	9.42689	153.974 3685	123	1.21463	205.084 4442	148	2.55632	258.407 6159
99	9.33262	155.970 0037	124	1.50614	207.177 8658	149	3.80892	260.580 8022
100	9.33262	157.970 0037	125	1.88268	209.274 7759	150	5.71338	262.756 8934

The power of 10 by which to multiply the factorial is given by the whole number of logarithm.

x	λ 0.1	0.2	0.3	0.4	0.5	0.6	0.7	0.8	0.9	1.0
0	.9048	.8187	.7408	.6703	.6065	.5488	.4966	.4493	.4066	.3679
1	.0905	.1637	.2222	.2681	.3033	.3293	.3476	.3595	.3659	.3679
2	.0045	.0164	.0333	.0536	.0758	.0988	.1217	.1438	.1647	.1839
3	.0002	.0011	.0033	.0072	.0126	.0198	.0284	.0383	.0494	.0613
4	.0000	.0001	.0002	.0007	.0016	.0030	.0050	.0077	.0111	.0153
5	.0000	.0000	.0000	.0001	.0002	.0004	.0007	.0012	.0020	.0031
6	.0000	.0000	.0000	.0000	.0000	.0000	.0001	.0002	.0003	.0005
7	.0000	.0000	.0000	.0000	.0000	.0000	.0000	.0000	.0000	.0001

x	λ 1.1	1.2	1.3	1.4	1.5	1.6	1.7	1.8	1.9	2.0
0	.3329	.3012	.2725	.2466	.2231	.2019	.1827	.1653	.1496	.1353
1	.3662	.3614	.3543	.3452	.3347	.3230	.3106	.2975	.2842	.2707
2	.2014	.2169	.2303	.2417	.2510	.2584	.2640	.2678	.2700	.2707
3	.0738	.0867	.0998	.1128	.1255	.1378	.1496	.1607	.1710	.1804
4	.0203	.0260	.0324	.0395	.0471	.0551	.0636	.0723	.0812	.0902
5	.0045	.0062	.0084	.0111	.0141	.0176	.0216	.0260	.0309	.0361
6	.0008	.0012	.0018	.0026	.0035	.0047	.0061	.0078	.0098	.0120
7	.0001	.0002	.0003	.0005	.0008	.0011	.0015	.0020	.0027	.0034
8	.0000	.0000	.0001	.0001	.0001	.0002	.0003	.0005	.0006	.0009
9	.0000	.0000	.0000	.0000	.0000	.0000	.0001	.0001	.0001	.0002

x	λ 2.1	2.2	2.3	2.4	2.5	2.6	2.7	2.8	2.9	3.0
0	.1225	.1108	.1003	.0907	.0821	.0743	.0672	.0608	.0550	.0498
1	.2572	.2438	.2306	.2177	.2052	.1931	.1815	.1703	.1596	.1494
2	.2700	.2681	.2652	.2613	.2565	.2510	.2450	.2384	.2314	.2240
3	.1890	.1966	.2033	.2090	.2138	.2176	.2205	.2225	.2237	.2240
4	.0992	.1082	.1169	.1254	.1336	.1414	.1488	.1557	.1622	.1680
5	.0417	.0476	.0538	.0602	.0668	.0735	.0804	.0872	.0940	.1008
6	.0146	.0174	.0206	.0241	.0278	.0319	.0362	.0407	.0455	.0504
7	.0044	.0055	.0068	.0083	.0099	.0118	.0139	.0163	.0188	.0216
8	.0011	.0015	.0019	.0025	.0031	.0038	.0047	.0057	.0068	.0081
9	.0003	.0004	.0005	.0007	.0009	.0011	.0014	.0018	.0022	.0027
10	.0001	.0001	.0001	.0002	.0002	.0003	.0004	.0005	.0006	.0008
11	.0000	.0000	.0000	.0000	.0000	.0001	.0001	.0001	.0002	.0002
12	.0000	.0000	.0000	.0000	.0000	.0000	.0000	.0000	.0000	.0001

x	λ 3.1	3.2	3.3	3.4	3.5	3.6	3.7	3.8	3.9	4.0
0	.0450	.0408	.0369	.0334	.0302	.0273	.0247	.0224	.0202	.0183
1	.1397	.1304	.1217	.1135	.1057	.0984	.0915	.0850	.0789	.0733
2	.2165	.2087	.2008	.1929	.1850	.1771	.1692	.1615	.1539	.1465
3	.2237	.2226	.2209	.2186	.2158	.2125	.2087	.2046	.2001	.1954
4	.1734	.1781	.1823	.1858	.1888	.1912	.1931	.1944	.1951	.1954
5	.1075	.1140	.1203	.1264	.1322	.1377	.1429	.1477	.1522	.1563
6	.0555	.0608	.0662	.0716	.0771	.0826	.0881	.0936	.0989	.1042
7	.0246	.0278	.0312	.0348	.0385	.0425	.0466	.0508	.0551	.0595
8	.0095	.0111	.0129	.0148	.0169	.0191	.0215	.0241	.0269	.0298
9	.0033	.0040	.0047	.0056	.0066	.0076	.0089	.0102	.0116	.0132
10	.0010	.0013	.0016	.0019	.0023	.0028	.0033	.0039	.0045	.0053
11	.0003	.0004	.0005	.0006	.0007	.0009	.0011	.0013	.0016	.0019
12	.0001	.0001	.0001	.0002	.0002	.0003	.0003	.0004	.0005	.0006
13	.0000	.0000	.0000	.0000	.0001	.0001	.0001	.0001	.0002	.0002
14	.0000	.0000	.0000	.0000	.0000	.0000	.0000	.0000	.0000	.0001

TABLE B.6 CUMULATIVE POISSON DISTRIBUTION

					λ					
x'	0.1	0.2	0.3	0.4	0.5	0.6	0.7	0.8	0.9	1.0
0	1.0000	1.0000	1.0000	1.0000	1.0000	1.0000	1.0000	1.0000	1.0000	1.0000
1	.0952	.1813	.2592	.3297	.3935	.4512	.5034	.5507	.5934	.6321
2	.0047	.0175	.0369	.0616	.0902	.1219	.1558	.1912	.2275	.2642
3	.0002	.0011	.0036	.0079	.0144	.0231	.0341	.0474	.0629	.0803
4	.0000	.0001	.0003	.0008	.0018	.0034	.0058	.0091	.0135	.0190
5	.0000	.0000	.0000	.0001	.0002	.0004	.0008	.0014	.0023	.0037
6	.0000	.0000	.0000	.0000	.0000	.0000	.0001	.0002	.0003	.0006
7	.0000	.0000	.0000	.0000	.0000	.0000	.0000	.0000	.0000	.0001

					λ					
x'	1.1	1.2	1.3	1.4	1.5	1.6	1.7	1.8	1.9	2.0
0	1.0000	1.0000	1.0000	1.0000	1.0000	1.0000	1.0000	1.0000	1.0000	1.0000
1	.6671	.6988	.7275	.7534	.7769	.7981	.8173	.8347	.8504	.8647
2	.3010	.3374	.3732	.4082	.4422	.4751	.5068	.5372	.5663	.5940
3	.0996	.1205	.1429	.1665	.1912	.2166	.2428	.2694	.2963	.3233
4	.0257	.0338	.0431	.0537	.0656	.0788	.0932	.1087	.1253	.1429
5	.0054	.0077	.0107	.0143	.0186	.0237	.0296	.0364	.0441	.0527
6	.0010	.0015	.0022	.0032	.0045	.0060	.0080	.0104	.0132	.0166
7	.0001	.0003	.0004	.0006	.0009	.0013	.0019	.0026	.0034	.0045
8	.0000	.0000	.0001	.0001	.0002	.0003	.0004	.0006	.0008	.0011
9	.0000	.0000	.0000	.0000	.0000	.0000	.0001	.0001	.0002	.0002

					λ					
x'	2.1	2.2	2.3	2.4	2.5	2.6	2.7	2.8	2.9	3.0
0	1.0000	1.0000	1.0000	1.0000	1.0000	1.0000	1.0000	1.0000	1.0000	1.0000
1	.8775	.8892	.8997	.9093	.9179	.9257	.9328	.9392	.9450	.9502
2	.6204	.6454	.6691	.6916	.7127	.7326	.7513	.7689	.7854	.8009
3	.3504	.3773	.4040	.4303	.4562	.4816	.5064	.5305	.5540	.5768
4	.1614	.1806	.2007	.2213	.2424	.2640	.2859	.3081	.3304	.3528
5	.0621	.0725	.0838	.0959	.1088	.1226	.1371	.1523	.1682	.1847
6	.0204	.0249	.0300	.0357	.0420	.0490	.0567	.0651	.0742	.0839
7	.0059	.0075	.0094	.0116	.0142	.0172	.0206	.0244	.0287	.0335
8	.0015	.0020	.0026	.0033	.0042	.0053	.0066	.0081	.0099	.0119
9	.0003	.0005	.0006	.0009	.0011	.0015	.0019	.0024	.0031	.0038
10	.0001	.0001	.0001	.0002	.0003	.0004	.0005	.0007	.0009	.0011
11	.0000	.0000	.0000	.0000	.0001	.0001	.0001	.0002	.0002	.0003
12	.0000	.0000	.0000	.0000	.0000	.0000	.0000	.0000	.0001	.0001

					λ					
x'	3.1	3.2	3.3	3.4	3.5	3.6	3.7	3.8	3.9	4.0
0	1.0000	1.0000	1.0000	1.0000	1.0000	1.0000	1.0000	1.0000	1.0000	1.0000
1	.9550	.9592	.9631	.9666	.9698	.9727	.9753	.9776	.9798	.9817
2	.8153	.8288	.8414	.8532	.8641	.8743	.8838	.8926	.9008	.9084
3	.5988	.6201	.6406	.6603	.6792	.6973	.7146	.7311	.7469	.7619
4	.3752	.3975	.4197	.4416	.4634	.4848	.5058	.5265	.5468	.5665
5	.2018	.2194	.2374	.2558	.2746	.2936	.3128	.3322	.3516	.3712
6	.0943	.1054	.1171	.1295	.1424	.1559	.1699	.1844	.1994	.2149
7	.0388	.0446	.0510	.0579	.0653	.0733	.0818	.0909	.1005	.1107
8	.0142	.0168	.0198	.0231	.0267	.0308	.0352	.0401	.0454	.0511
9	.0047	.0057	.0069	.0083	.0099	.0117	.0137	.0160	.0185	.0214
10	.0014	.0018	.0022	.0027	.0033	.0040	.0048	.0058	.0069	.0081
11	.0004	.0005	.0006	.0008	.0010	.0013	.0016	.0019	.0023	.0028
12	.0001	.0001	.0002	.0002	.0003	.0004	.0005	.0006	.0007	.0009
13	.0000	.0000	.0000	.0001	.0001	.0001	.0001	.0002	.0002	.0003
14	.0000	.0000	.0000	.0000	.0000	.0000	.0000	.0000	.0001	.0001

TABLE B.7   NORMAL DISTRIBUTION (RIGHT TAIL AREAS)

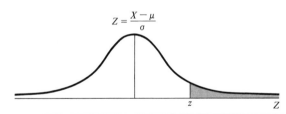

$$Z = \frac{X - \mu}{\sigma}$$

z	.00	.01	.02	.03	.04	.05	.06	.07	.08	.09
0.0	.5000	.4960	.4920	.4880	.4840	.4801	.4761	.4721	.4681	.4641
0.1	.4602	.4562	.4522	.4483	.4443	.4404	.4364	.4325	.4286	.4247
0.2	.4207	.4168	.4129	.4090	.4052	.4013	.3974	.3936	.3897	.3859
0.3	.3821	.3783	.3745	.3707	.3669	.3632	.3594	.3557	.3520	.3483
0.4	.3446	.3409	.3372	.3336	.3300	.3264	.3228	.3192	.3156	.3121
0.5	.3085	.3050	.3015	.2981	.2946	.2912	.2877	.2843	.2810	.2776
0.6	.2743	.2709	.2676	.2643	.2611	.2578	.2546	.2514	.2483	.2451
0.7	.2420	.2389	.2358	.2327	.2296	.2266	.2236	.2206	.2177	.2148
0.8	.2119	.2090	.2061	.2033	.2005	.1977	.1949	.1922	.1894	.1867
0.9	.1841	.1814	.1788	.1762	.1736	.1711	.1685	.1660	.1635	.1611
1.0	.1587	.1562	.1539	.1515	.1492	.1469	.1446	.1423	.1401	.1379
1.1	1357	.1335	.1314	.1292	.1271	.1251	.1230	.1210	.1190	.1170
1.2	.1151	.1131	.1112	.1093	.1075	.1056	.1038	.1020	.1003	.0985
1.3	.0968	.0951	.0934	.0918	.0901	.0885	.0869	.0853	.0838	.0823
1.4	.0808	.0793	.0778	.0764	.0749	.0735	.0721	.0708	.0694	.0681
1.5	.0668	.0655	.0643	.0630	.0618	.0606	.0594	.0582	.0571	.0559
1.6	.0548	.0537	.0526	.0516	.0505	.0495	.0485	.0475	.0465	.0455
1.7	.0446	.0436	.0427	.0418	.0409	.0401	.0392	.0384	.0375	.0367
1.8	.0359	.0351	.0344	.0336	.0329	.0322	.0314	.0307	.0301	.0294
1.9	.0287	.0281	.0274	.0268	.0262	.0256	.0250	.0244	.0239	.0233
2.0	.0228	.0222	.0217	.0212	.0207	.0202	.0197	.0192	.0188	.0183
2.1	.0179	.0174	.0170	.0166	.0162	.0158	.0154	.0150	.0146	.0143
2.2	.0139	.0136	.0132	.0129	.0125	.0122	.0119	.0116	.0113	.0110
2.3	.0107	.0104	.0102	.0099	.0096	.0094	.0091	.0089	.0087	.0084
2.4	.0082	.0080	.0078	.0075	.0073	.0071	.0069	.0068	.0066	.0064
2.5	.0062	.0060	.0059	.0057	.0055	.0054	.0052	.0051	.0049	.0048
2.6	.0047	.0045	.0044	.0043	.0041	.0040	.0039	.0038	.0037	.0036
2.7	.0035	.0034	.0033	.0032	.0031	.0030	.0029	.0028	.0027	.0026
2.8	.0026	.0025	.0024	.0023	.0023	.0022	.0021	.0021	.0020	.0019
2.9	.0019	.0018	.0018	.0017	.0016	.0016	.0015	.0015	.0014	.0014
3.0	.0013	.0013	.0013	.0012	.0012	.0011	.0011	.0011	.0010	.0010

TABLE B.8 EXPONENTIAL FUNCTIONS

x	e^x	$Log_{10}(e^x)$	e^{-x}	x	e^x	$Log_{10}(e^x)$	e^{-x}
0.00	1.0000	0.00000	1.000000	0.50	1.6487	0.21715	0.606531
0.01	1.0101	.00434	0.990050	0.51	1.6653	.22149	.600496
0.02	1.0202	.00869	.980199	0.52	1.6820	.22583	.594521
0.03	1.0305	.01303	.970446	0.53	1.6989	.23018	.588605
0.04	1.0408	.01737	.960789	0.54	1.7160	.23452	.582748
0.05	1.0513	0.02171	0.951229	0.55	1.7333	0.23886	0.576950
0.06	1.0618	.02606	.941765	0.56	1.7507	.24320	.571209
0.07	1.0725	.03040	.932394	0.57	1.7683	.24755	.565525
0.08	1.0833	.03474	.923116	0.58	1.7860	.25189	.559898
0.09	1.0942	.03909	.913931	0.59	1.8040	.25623	.554327
0.10	1.1052	0.04343	0.904837	0.60	1.8221	0.26058	0.548812
0.11	1.1163	.04777	.895834	0.61	1.8404	.26492	.543351
0.12	1.1275	.05212	.886920	0.62	1.8589	.26926	.537944
0.13	1.1388	.05646	.878095	0.63	1.8776	.27361	.532592
0.14	1.1503	.06080	.869358	0.64	1.8965	.27795	.527292
0.15	1.1618	0.06514	0.860708	0.65	1.9155	0.28229	0.522046
0.16	1.1735	.06949	.852144	0.66	1.9348	.28663	.516851
0.17	1.1853	.07383	.843665	0.67	1.9542	.29098	.511709
0.18	1.1972	.07817	.835270	0.68	1.9739	.29532	.506617
0.19	1.2092	.08252	.826959	0.69	1.9937	.29966	.501576
0.20	1.2214	0.08686	0.818731	0.70	2.0138	0.30401	0.496585
0.21	1.2337	.09120	.810584	0.71	2.0340	.30835	.491644
0.22	1.2461	.09554	.802519	0.72	2.0544	.21269	.486752
0.23	1.2586	.09989	.794534	0.73	2.0751	.31703	.481909
0.24	1.2712	.10423	.786628	0.74	2.0959	.32138	.477114
0.25	1.2840	0.10857	0.778801	0.75	2.1170	0.32572	0.472367
0.26	1.2969	.11292	.771052	0.76	2.1383	.33006	.467666
0.27	1.3100	.11726	.763379	0.77	2.1598	.33441	.463013
0.28	1.3231	.12160	.755784	0.78	2.1815	.33875	.458406
0.29	1.3364	.12595	.748264	0.79	2.2034	.34309	.453845
0.30	1.3499	0.13029	0.740818	0.80	2.2255	0.34744	0.449329
0.31	1.3634	.13463	.733447	0.81	2.2479	.35178	.444858
0.32	1.3771	.13897	.726149	0.82	2.2705	.35612	.440432
0.33	1.3910	.14332	.718924	0.83	2.2933	.36046	.436049
0.34	1.4049	.14766	.711770	0.84	2.3164	.36481	.431711
0.35	1.4191	0.15200	0.704688	0.85	2.3396	0.36915	0.427415
0.36	1.4333	.15635	.697676	0.86	2.3632	.37349	.423162
0.37	1.4477	.16069	.690734	0.87	2.3869	.37784	.418952
0.38	1.4623	.16503	.683861	0.88	2.4109	.38218	.414783
0.39	1.4770	.16937	.677057	0.89	2.4351	.38652	.410656
0.40	1.4918	0.17372	0.670320	0.90	2.4596	0.39087	0.406570
0.41	1.5068	.17806	.663650	0.91	2.4843	.39521	.402524
0.42	1.5220	.18240	.657047	0.92	2.5093	.39955	.398519
0.43	1.5373	.18675	.650509	0.93	2.5345	.40389	.394554
0.44	1.5527	.19109	.644036	0.94	2.5600	.40824	.390628
0.45	1.5683	0.19543	0.637628	0.95	2.5857	0.41258	0.386741
0.46	1.5841	.19978	.631284	0.96	2.6117	.41692	.382893
0.47	1.6000	.20412	.625002	0.97	2.6379	.42127	.379083
0.48	1.6161	.20846	.618783	0.98	2.6645	.42561	.375311
0.49	1.6323	.21280	.612626	0.99	2.6912	.42995	.371577
0.50	1.6487	0.21715	0.606531	1.00	2.7183	0.43429	0.367879

TABLE B.8 EXPONENTIAL FUNCTIONS (*continued*)

x	e^x	$Log_{10}(e^x)$	e^{-x}	x	e^x	$Log_{10}(e^x)$	e^{-x}
1.00	2.7183	0.43429	0.367879	1.50	4.4817	0.65144	0.223130
1.01	2.7456	.43864	.364219	1.51	4.5267	.65578	.220910
1.02	2.7732	.44298	.360595	1.52	4.5722	.66013	.218712
1.03	2.8011	.44732	.357007	1.53	4.6182	.66447	.216536
1.04	2.8292	.45167	.353455	1.54	4.6646	.66881	.214381
1.05	2.8577	0.45601	0.349938	1.55	4.7115	0.67316	0.212248
1.06	2.8864	.46035	.346456	1.56	4.7588	.67750	.210136
1.07	2.9154	.46470	.343009	1.57	4.8066	.68184	.208045
1.08	2.9447	.46904	.339596	1.58	4.8550	.68619	.205975
1.09	2.9743	.47338	.336216	1.59	4.9037	.69053	.203926
1.10	3.0042	0.47772	0.332871	1.60	4.9530	0.69487	0.201897
1.11	3.0344	.48207	.329559	1.61	5.0028	.69921	.199888
1.12	3.0649	.48641	.326280	1.62	5.0531	.70356	.197899
1.13	3.0957	.49075	.323033	1.63	5.1039	.70790	.195930
1.14	3.1268	.49510	.319819	1.64	5.1552	.71224	.193980
1.15	3.1582	0.49944	0.316637	1.65	5.2070	0.71659	0.192050
1.16	3.1899	.50378	.313486	1.66	5.2593	.72093	.190139
1.17	3.2220	.50812	.310367	1.67	5.3122	.72527	.188247
1.18	3.2544	.51247	.307279	1.68	5.3656	.72961	.186374
1.19	3.2871	.51681	.304221	1.69	5.4195	.73396	.184520
1.20	3.3201	0.52115	0.301194	1.70	5.4739	0.73830	0.182684
1.21	3.3535	.52550	.298197	1.71	5.5290	.74264	.180866
1.22	3.3872	.52984	.295230	1.72	5.5845	.74699	.179066
1.23	3.4212	.53418	.292293	1.73	5.6407	.75133	.177284
1.24	3.4556	.53853	.289384	1.74	5.6973	.75567	.175520
1.25	3.4903	0.54287	0.286505	1.75	5.7546	0.76002	0.173774
1.26	3.5254	.54721	.283654	1.76	5.8124	.76436	.172045
1.27	3.5609	.55155	.280832	1.77	5.8709	.76870	.170333
1.28	3.5966	.55590	.278037	1.78	5.9299	.77304	.168638
1.29	3.6328	.56024	.275271	1.79	5.9898	.77739	.166960
1.30	3.6693	0.56458	0.272532	1.80	6.0496	0.78173	0.165299
1.31	3.7062	.56893	.269820	1.81	6.1104	.78607	.163654
1.32	3.7434	.57327	.267135	1.82	6.1719	.79042	.162026
1.33	3.7810	.57761	.264477	1.83	6.2339	.79476	.160414
1.34	3.8190	.58195	.261846	1.84	6.2965	.79910	.158817
1.35	3.8574	0.58630	0.259240	1.85	6.3598	0.80344	0.157237
1.36	3.8962	.59064	.256661	1.86	6.4237	.80779	.155673
1.37	3.9354	.59498	.254107	1.87	6.4883	.81213	.154124
1.38	3.9749	.59933	.251579	1.88	6.5535	.81647	.152590
1.39	4.0149	.60367	.249075	1.89	6.6194	.82082	.151072
1.40	4.0552	0.60801	0.246597	1.90	6.6859	0.82516	0.149569
1.41	4.0960	.61236	.244143	1.91	6.7531	.82950	.148080
1.42	4.1371	.61670	.241714	1.92	6.8210	.83385	.146607
1.43	4.1787	.62104	.239309	1.93	6.8895	.83819	.145148
1.44	4.2207	.62538	.236928	1.94	6.9588	.84253	.143704
1.45	4.2631	0.62973	0.234570	1.95	7.0287	0.84687	0.142274
1.46	4.3060	.63407	.232236	1.96	7.0993	.85122	.140858
1.47	4.3492	.63841	.229925	1.97	7.1707	.85556	.139457
1.48	4.3929	.64276	.227638	1.98	7.2427	.85990	.138069
1.49	4.4371	.64710	.225373	1.99	7.3155	.86425	.136695
1.50	4.4817	0.65144	0.223130	2.00	7.3891	0.86859	0.135335

TABLE B.9   STUDENT'S t DISTRIBUTION

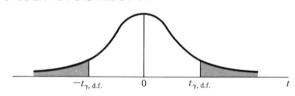

	PROBABILITY OF A VALUE GREATER IN ABSOLUTE VALUE THAN					
Degrees of Freedom	THE TABLE ENTRY					
	0.005	0.01	0.025	0.05	0.1	0.15
1	63.657	31.821	12.706	6.314	3.078	1.963
2	9.925	6.965	4.303	2.920	1.886	1.386
3	5.841	4.541	3.182	2.353	1.638	1.250
4	4.604	3.747	2.776	2.132	1.533	1.190
5	4.032	3.365	2.571	2.015	1.476	1.156
6	3.707	3.143	2.447	1.943	1.440	1.134
7	3.499	2.998	2.365	1.895	1.415	1.119
8	3.355	2.896	2.306	1.860	1.397	1.108
9	3.250	2.821	2.262	1.833	1.383	1.100
10	3.169	2.764	2.228	1.812	1.372	1.093
11	3.106	2.718	2.201	1.796	1.363	1.088
12	3.055	2.681	2.179	1.782	1.356	1.083
13	3.012	2.650	2.160	1.771	1.350	1.079
14	2.977	2.624	2.145	1.761	1.345	1.076
15	2.947	2.602	2.131	1.753	1.341	1.074
16	2.921	2.583	2.120	1.746	1.337	1.071
17	2.898	2.567	2.110	1.740	1.333	1.069
18	2.878	2.552	2.101	1.734	1.330	1.067
19	2.861	2.539	2.093	1.729	1.328	1.066
20	2.845	2.528	2.086	1.725	1.325	1.064
21	2.831	2.518	2.080	1.721	1.323	1.063
22	2.819	2.508	2.074	1.717	1.321	1.061
23	2.807	2.500	2.069	1.714	1.319	1.060
24	2.797	2.492	2.064	1.711	1.318	1.059
25	2.787	2.485	2.060	1.708	1.316	1.058
26	2.779	2.479	2.056	1.706	1.315	1.058
27	2.771	2.473	2.052	1.703	1.314	1.057
28	2.763	2.467	2.048	1.701	1.313	1.056
29	2.756	2.462	2.045	1.699	1.311	1.055
30	2.750	2.457	2.042	1.697	1.310	1.055
∞	2.576	2.326	1.960	1.645	1.282	1.036

Source: Reproduced by permission of the authors and publisher from Table III of Fisher and Yates: *Statistical Tables for Biological, Agricultural and Medical Research*, by Oliver & Boyd, Edinburgh.

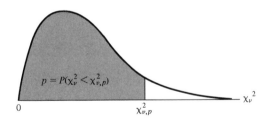

$$p = P(\chi_v^2 < \chi_{v,p}^2)$$

χ^2 DISTRIBUTION

% df	0.5	1	2.5	5	10	20	30	40	50	60	70	80	90	95	97.5	99	99.5	99.95
1	0.0001	0.0002	0.001	0.004	0.016	0.064	0.148	0.275	0.455	0.708	1.074	1.642	2.706	3.841	5.024	6.635	7.879	12.116
2	0.010	0.020	0.051	0.103	0.211	0.446	0.713	1.022	1.386	1.833	2.408	3.219	4.605	5.991	7.378	9.210	10.597	15.202
3	0.072	0.115	0.216	0.352	0.584	1.005	1.424	1.869	2.366	2.946	3.665	4.642	6.251	7.815	9.348	11.345	12.838	17.730
4	0.207	0.297	0.484	0.711	1.064	1.649	2.195	2.753	3.357	4.045	4.878	5.989	7.779	9.488	11.143	13.277	14.860	19.997
5	0.412	0.554	0.831	1.145	1.610	2.343	3.000	3.655	4.351	5.132	6.064	7.289	9.236	11.070	12.833	15.086	16.750	22.105
6	0.676	0.872	1.237	1.635	2.204	3.070	3.828	4.570	5.348	6.211	7.231	8.558	10.645	12.592	14.449	16.812	18.548	24.103
7	0.989	1.239	1.690	2.167	2.833	3.822	4.671	5.493	6.346	7.283	8.383	9.803	12.017	14.067	16.013	18.475	20.278	26.018
8	1.344	1.646	2.180	2.733	3.490	4.594	5.527	6.423	7.344	8.351	9.524	11.030	13.362	15.507	17.535	20.090	21.955	27.868
9	1.735	2.088	2.700	3.325	4.168	5.380	6.393	7.357	8.343	9.414	10.656	12.242	14.684	16.919	19.023	21.666	23.589	29.666
10	2.156	2.558	3.247	3.940	4.865	6.179	7.267	8.295	9.342	10.473	11.781	13.442	15.987	18.307	20.483	23.209	25.188	31.420
11	2.603	3.053	3.816	4.575	5.578	6.989	8.148	9.237	10.341	11.530	12.899	14.631	17.275	19.675	21.920	24.725	26.757	33.137
12	3.074	3.571	4.404	5.226	6.304	7.807	9.034	10.182	11.340	12.584	14.011	15.812	18.549	21.026	23.337	26.217	28.300	34.821
13	3.565	4.107	5.009	5.892	7.042	8.634	9.926	11.129	12.340	13.636	15.119	16.985	19.812	22.362	24.736	27.688	29.819	36.478
14	4.075	4.660	5.629	6.571	7.790	9.467	10.821	12.078	13.339	14.685	16.222	18.151	21.064	23.685	26.119	29.141	31.319	38.109
15	4.601	5.229	6.262	7.261	8.547	10.307	11.721	13.030	14.339	15.733	17.322	19.311	22.307	24.996	27.488	30.578	32.801	39.719
16	5.142	5.812	6.908	7.962	9.312	11.152	12.624	13.983	15.338	16.780	18.418	20.465	23.542	26.296	28.845	32.000	34.267	41.308
17	5.697	6.408	7.564	8.672	10.085	12.002	13.531	14.937	16.338	17.824	19.511	21.615	24.769	27.587	30.191	33.409	35.718	42.879
18	6.265	7.015	8.231	9.390	10.865	12.857	14.440	15.893	17.338	18.868	20.601	22.760	25.989	28.869	31.526	34.805	37.156	44.434
19	6.844	7.633	8.907	10.117	11.651	13.716	15.352	16.850	18.338	19.910	21.689	23.900	27.204	30.144	32.852	36.191	38.582	45.973
20	7.434	8.260	9.591	10.851	12.443	14.578	16.266	17.809	19.337	20.951	22.775	25.038	28.412	31.410	34.170	37.566	39.997	47.498
21	8.034	8.897	10.283	11.591	13.240	15.445	17.182	18.768	20.337	21.991	23.858	26.171	29.615	32.671	35.479	38.932	41.401	49.011
22	8.643	9.542	10.982	12.338	14.041	16.314	18.101	19.729	21.337	23.031	24.939	27.301	30.813	33.924	36.781	40.289	42.796	50.511
23	9.260	10.196	11.689	13.091	14.848	17.187	19.021	20.690	22.337	24.069	26.018	28.429	32.007	35.172	38.076	41.638	44.181	52.000
24	9.886	10.856	12.401	13.848	15.659	18.062	19.943	21.752	23.337	25.106	27.096	29.553	33.196	36.415	39.364	42.980	45.559	53.479
25	10.520	11.524	13.120	14.611	16.473	18.940	20.867	22.616	24.337	26.143	28.172	30.675	34.382	37.652	40.646	44.314	46.928	54.947
26	11.160	12.198	13.844	15.379	17.292	19.820	21.792	23.579	25.336	27.179	29.246	31.795	35.563	38.885	41.923	45.642	48.290	56.407
27	11.808	12.879	14.573	16.151	18.114	20.703	22.719	24.544	26.336	28.214	30.319	32.912	36.741	40.113	43.195	46.963	49.645	57.858
28	12.461	13.565	15.308	16.928	18.939	21.588	23.647	25.509	27.336	29.249	31.391	34.027	37.916	41.337	44.461	48.278	50.993	59.300
29	13.121	14.256	16.047	17.708	19.768	22.475	24.577	26.475	28.336	30.283	32.461	35.139	39.087	42.557	45.722	49.588	52.336	60.735
30	13.787	14.953	16.791	18.493	20.599	23.364	25.508	27.442	29.336	31.316	33.530	36.250	40.256	43.773	46.979	50.892	53.672	62.162
35	17.192	18.509	20.569	22.465	24.797	27.836	30.178	32.282	34.336	36.475	38.859	41.778	46.059	49.802	53.203	57.342	60.275	69.199
40	20.707	22.164	24.433	26.509	29.051	32.345	34.872	37.134	39.335	41.622	44.165	47.269	51.805	55.758	59.342	63.691	66.766	76.095
45	24.311	25.901	28.366	30.612	33.350	36.884	39.585	41.995	44.335	46.761	49.452	52.729	57.505	61.656	65.410	69.957	73.166	82.876
50	27.991	29.707	32.357	34.764	37.689	41.449	44.313	46.864	49.335	51.892	54.723	58.164	63.167	67.505	71.420	76.154	79.490	89.561
60	35.534	37.485	40.482	43.188	46.459	50.641	53.809	56.620	59.335	62.135	65.227	68.972	74.397	79.082	83.298	88.379	91.952	102.695
70	43.275	45.442	48.758	51.739	55.329	59.898	63.346	66.396	69.334	72.358	75.689	79.715	85.527	90.531	95.023	100.425	104.215	115.578
80	51.172	53.540	57.153	60.391	64.278	69.207	72.915	76.188	79.334	82.566	86.120	90.405	96.578	101.879	106.629	112.329	116.321	128.261
90	59.196	61.754	65.647	69.126	73.291	78.558	82.511	85.993	89.334	92.761	96.524	101.054	107.565	113.145	118.136	124.116	128.299	140.782
100	67.328	70.065	74.222	77.929	82.358	87.945	92.129	95.808	99.334	102.946	106.906	111.667	118.498	124.342	129.561	135.807	140.169	153.167
120	83.852	86.923	91.573	95.705	100.624	106.806	111.419	115.465	119.334	123.289	127.616	132.806	140.233	146.567	152.211	158.950	163.648	177.603
140	100.655	104.034	109.137	113.659	119.029	125.758	130.766	135.149	139.334	143.604	148.269	153.854	161.827	168.613	174.648	181.840	186.847	201.683
160	117.679	121.346	126.870	131.756	137.546	144.783	150.158	154.856	159.334	163.898	168.876	174.828	183.311	190.516	196.915	204.530	209.824	225.481
180	134.884	138.820	144.741	149.969	156.153	163.868	169.588	174.580	179.334	184.173	189.446	195.743	204.704	212.304	219.044	227.056	232.620	249.048
200	152.241	156.432	162.728	168.279	174.835	183.003	189.049	194.319	199.334	204.434	209.985	216.609	226.021	233.994	241.058	249.445	255.264	272.423

TABLE B.11 CRITICAL VALUES FOR THE F DISTRIBUTION

5% (ROMAN TYPE) AND 1% (BOLD FACE TYPE) POINTS FOR THE DISTRIBUTION OF F.

$f(F: n_1, n_2)$

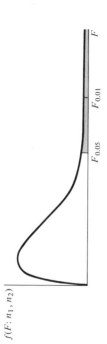

$F_{0.05}$ $F_{0.01}$ F

n_1 DEGREES OF FREEDOM (FOR GREATER MEAN SQUARE)

n_2	1	2	3	4	5	6	7	8	9	10	11	12	14	16	20	24	30	40	50	75	100	200	500	∞
1	161 **4,052**	200 **4,999**	216 **5,403**	225 **5,625**	230 **5,764**	234 **5,859**	237 **5,928**	239 **5,981**	241 **6,022**	242 **6,056**	243 **6,082**	244 **6,106**	245 **6,142**	246 **6,169**	248 **6,208**	249 **6,234**	250 **6,258**	251 **6,286**	252 **6,302**	253 **6,323**	253 **6,334**	254 **6,352**	254 **6,361**	254 **6,366**
2	18.51 **98.49**	19.00 **99.00**	19.16 **99.17**	19.25 **99.25**	19.30 **99.30**	19.33 **99.33**	19.36 **99.34**	19.37 **99.36**	19.38 **99.38**	19.39 **99.40**	19.40 **99.41**	19.41 **99.42**	19.42 **99.43**	19.43 **99.44**	19.44 **99.45**	19.45 **99.46**	19.46 **99.47**	19.47 **99.48**	19.47 **99.48**	19.48 **99.49**	19.49 **99.49**	19.49 **99.49**	19.50 **99.50**	19.50 **99.50**
3	10.13 **34.12**	9.55 **30.82**	9.28 **29.46**	9.12 **28.71**	9.01 **28.24**	8.94 **27.91**	8.88 **27.67**	8.84 **27.49**	8.81 **27.34**	8.78 **27.23**	8.76 **27.13**	8.74 **27.05**	8.71 **26.92**	8.69 **26.83**	8.66 **26.69**	8.64 **26.60**	8.62 **26.50**	8.60 **26.41**	8.58 **26.35**	8.57 **26.27**	8.56 **26.23**	8.54 **26.18**	8.54 **26.14**	8.53 **26.12**
4	7.71 **21.20**	6.94 **18.00**	6.59 **16.69**	6.39 **15.98**	6.26 **15.52**	6.16 **15.21**	6.09 **14.98**	6.04 **14.80**	6.00 **14.66**	5.96 **14.54**	5.93 **14.45**	5.91 **14.37**	5.87 **14.24**	5.84 **14.15**	5.80 **14.02**	5.77 **13.93**	5.74 **13.83**	5.71 **13.74**	5.70 **13.69**	5.68 **13.61**	5.66 **13.57**	5.65 **13.52**	5.64 **13.48**	5.63 **13.46**
5	6.61 **16.26**	5.79 **13.27**	5.41 **12.06**	5.19 **11.39**	5.05 **10.97**	4.95 **10.67**	4.88 **10.45**	4.82 **10.27**	4.78 **10.15**	4.74 **10.05**	4.70 **9.96**	4.68 **9.89**	4.64 **9.77**	4.60 **9.68**	4.56 **9.55**	4.53 **9.47**	4.50 **9.38**	4.46 **9.29**	4.44 **9.24**	4.42 **9.17**	4.40 **9.13**	4.38 **9.07**	4.37 **9.04**	4.36 **9.02**
6	5.99 **13.74**	5.14 **10.92**	4.76 **9.78**	4.53 **9.15**	4.39 **8.75**	4.28 **8.47**	4.21 **8.26**	4.15 **8.10**	4.10 **7.98**	4.06 **7.87**	4.03 **7.79**	4.00 **7.72**	3.96 **7.60**	3.92 **7.52**	3.87 **7.39**	3.84 **7.31**	3.81 **7.23**	3.77 **7.14**	3.75 **7.09**	3.72 **7.02**	3.71 **6.99**	3.69 **6.94**	3.68 **6.90**	3.67 **6.88**
7	5.59 **12.25**	4.74 **9.55**	4.35 **8.45**	4.12 **7.85**	3.97 **7.46**	3.87 **7.19**	3.79 **7.00**	3.73 **6.84**	3.68 **6.71**	3.63 **6.62**	3.60 **6.54**	3.57 **6.47**	3.52 **6.35**	3.49 **6.27**	3.44 **6.15**	3.41 **6.07**	3.38 **5.98**	3.34 **5.90**	3.32 **5.85**	3.29 **5.78**	3.28 **5.75**	3.25 **5.70**	3.24 **5.67**	3.23 **5.65**
8	5.32 **11.26**	4.46 **8.65**	4.07 **7.59**	3.84 **7.01**	3.69 **6.63**	3.58 **6.37**	3.50 **6.19**	3.44 **6.03**	3.39 **5.91**	3.34 **5.82**	3.31 **5.74**	3.28 **5.67**	3.23 **5.56**	3.20 **5.48**	3.15 **5.36**	3.12 **5.28**	3.08 **5.20**	3.05 **5.11**	3.03 **5.06**	3.00 **5.00**	2.98 **4.96**	2.96 **4.91**	2.94 **4.88**	2.93 **4.86**
9	5.12 **10.56**	4.26 **8.02**	3.86 **6.99**	3.63 **6.42**	3.48 **6.06**	3.37 **5.80**	3.29 **5.62**	3.23 **5.47**	3.18 **5.35**	3.13 **5.26**	3.10 **5.18**	3.07 **5.11**	3.02 **5.00**	2.98 **4.92**	2.93 **4.80**	2.90 **4.73**	2.86 **4.64**	2.82 **4.56**	2.80 **4.51**	2.77 **4.45**	2.76 **4.41**	2.73 **4.36**	2.72 **4.33**	2.71 **4.31**

df																								
10	2.54 **3.91**	2.55 **3.93**	2.56 **3.96**	2.59 **4.01**	2.61 **4.05**	2.64 **4.12**	2.67 **4.17**	2.70 **4.25**	2.74 **4.33**	2.77 **4.41**	2.82 **4.52**	2.86 **4.60**	2.91 **4.71**	2.94 **4.78**	2.97 **4.85**	3.02 **4.95**	3.07 **5.06**	3.14 **5.21**	3.22 **5.39**	3.33 **5.64**	3.48 **5.99**	3.71 **6.55**	4.10 **7.56**	4.96 **10.04**
11	2.40 **3.60**	2.41 **3.62**	2.42 **3.66**	2.45 **3.70**	2.47 **3.74**	2.50 **3.80**	2.53 **3.86**	2.57 **3.94**	2.61 **4.02**	2.65 **4.10**	2.70 **4.21**	2.74 **4.29**	2.79 **4.40**	2.82 **4.46**	2.86 **4.54**	2.90 **4.63**	2.95 **4.74**	3.01 **4.88**	3.09 **5.07**	3.20 **5.32**	3.36 **5.67**	3.59 **6.22**	3.98 **7.20**	4.84 **9.65**
12	2.30 **3.36**	2.31 **3.38**	2.32 **3.41**	2.35 **3.46**	2.36 **3.49**	2.40 **3.56**	2.42 **3.61**	2.46 **3.70**	2.50 **3.78**	2.54 **3.86**	2.60 **3.98**	2.64 **4.05**	2.69 **4.16**	2.72 **4.22**	2.76 **4.30**	2.80 **4.39**	2.85 **4.50**	2.92 **4.65**	3.00 **4.82**	3.11 **5.06**	3.26 **5.41**	3.49 **5.95**	3.88 **6.93**	4.75 **9.33**
13	2.21 **3.16**	2.22 **3.18**	2.24 **3.21**	2.26 **3.27**	2.28 **3.30**	2.32 **3.37**	2.34 **3.42**	2.38 **3.51**	2.42 **3.59**	2.46 **3.67**	2.51 **3.78**	2.55 **3.85**	2.60 **3.96**	2.63 **4.02**	2.67 **4.10**	2.72 **4.19**	2.77 **4.30**	2.84 **4.44**	2.92 **4.62**	3.02 **4.86**	3.18 **5.20**	3.41 **5.74**	3.80 **6.70**	4.67 **9.07**
14	2.13 **3.00**	2.14 **3.02**	2.16 **3.06**	2.19 **3.11**	2.21 **3.14**	2.24 **3.21**	2.27 **3.26**	2.31 **3.34**	2.35 **3.43**	2.39 **3.51**	2.44 **3.62**	2.48 **3.70**	2.53 **3.80**	2.56 **3.86**	2.60 **3.94**	2.65 **4.03**	2.70 **4.14**	2.77 **4.28**	2.85 **4.46**	2.96 **4.69**	3.11 **5.03**	3.34 **5.56**	3.74 **6.51**	4.60 **8.86**
15	2.07 **2.87**	2.08 **2.89**	2.10 **2.92**	2.12 **2.97**	2.15 **3.00**	2.18 **3.07**	2.21 **3.12**	2.25 **3.20**	2.29 **3.29**	2.33 **3.36**	2.39 **3.48**	2.43 **3.56**	2.48 **3.67**	2.51 **3.73**	2.55 **3.80**	2.59 **3.89**	2.64 **4.00**	2.70 **4.14**	2.79 **4.32**	2.90 **4.56**	3.06 **4.89**	3.29 **5.42**	3.68 **6.36**	4.54 **8.68**
16	2.01 **2.75**	2.02 **2.77**	2.04 **2.80**	2.07 **2.86**	2.09 **2.89**	2.13 **2.96**	2.16 **3.01**	2.20 **3.10**	2.24 **3.18**	2.28 **3.25**	2.33 **3.37**	2.37 **3.45**	2.42 **3.55**	2.45 **3.61**	2.49 **3.69**	2.54 **3.78**	2.59 **3.89**	2.66 **4.03**	2.74 **4.20**	2.85 **4.44**	3.01 **4.77**	3.24 **5.29**	3.63 **6.23**	4.49 **8.53**
17	1.96 **2.65**	1.97 **2.67**	1.99 **2.70**	2.02 **2.76**	2.04 **2.79**	2.08 **2.86**	2.11 **2.92**	2.15 **3.00**	2.19 **3.08**	2.23 **3.16**	2.29 **3.27**	2.33 **3.35**	2.38 **3.45**	2.41 **3.52**	2.45 **3.59**	2.50 **3.68**	2.55 **3.79**	2.62 **3.93**	2.70 **4.10**	2.81 **4.34**	2.96 **4.67**	3.20 **5.18**	3.59 **6.11**	4.45 **8.40**
18	1.92 **2.57**	1.93 **2.59**	1.95 **2.62**	1.98 **2.68**	2.00 **2.71**	2.04 **2.78**	2.07 **2.83**	2.11 **2.91**	2.15 **3.00**	2.19 **3.07**	2.25 **3.19**	2.29 **3.27**	2.34 **3.37**	2.37 **3.44**	2.41 **3.51**	2.46 **3.60**	2.51 **3.71**	2.58 **3.85**	2.66 **4.01**	2.77 **4.25**	2.93 **4.58**	3.16 **5.09**	3.55 **6.01**	4.41 **8.28**
19	1.88 **2.49**	1.90 **2.51**	1.91 **2.54**	1.94 **2.60**	1.96 **2.63**	2.00 **2.70**	2.02 **2.76**	2.07 **2.84**	2.11 **2.92**	2.15 **3.00**	2.21 **3.12**	2.26 **3.19**	2.31 **3.30**	2.34 **3.36**	2.38 **3.43**	2.43 **3.52**	2.48 **3.63**	2.55 **3.77**	2.63 **3.94**	2.74 **4.17**	2.90 **4.50**	3.13 **5.01**	3.52 **5.93**	4.38 **8.18**
20	1.84 **2.42**	1.85 **2.44**	1.87 **2.47**	1.90 **2.53**	1.92 **2.56**	1.96 **2.63**	1.99 **2.69**	2.04 **2.77**	2.08 **2.86**	2.12 **2.94**	2.18 **3.05**	2.23 **3.13**	2.28 **3.23**	2.31 **3.30**	2.35 **3.37**	2.40 **3.45**	2.45 **3.56**	2.52 **3.71**	2.60 **3.87**	2.71 **4.10**	2.87 **4.43**	3.10 **4.94**	3.49 **5.85**	4.35 **8.10**
21	1.81 **2.36**	1.82 **2.38**	1.84 **2.42**	1.87 **2.47**	1.89 **2.51**	1.93 **2.58**	1.96 **2.63**	2.00 **2.72**	2.05 **2.80**	2.09 **2.88**	2.15 **2.99**	2.20 **3.07**	2.25 **3.17**	2.28 **3.24**	2.32 **3.31**	2.37 **3.40**	2.42 **3.51**	2.49 **3.65**	2.57 **3.81**	2.68 **4.04**	2.84 **4.37**	3.07 **4.87**	3.47 **5.78**	4.32 **8.02**
22	1.78 **2.31**	1.80 **2.33**	1.81 **2.37**	1.84 **2.42**	1.87 **2.46**	1.91 **2.53**	1.93 **2.58**	1.98 **2.67**	2.03 **2.75**	2.07 **2.83**	2.13 **2.94**	2.18 **3.02**	2.23 **3.12**	2.26 **3.18**	2.30 **3.26**	2.35 **3.35**	2.40 **3.45**	2.47 **3.59**	2.55 **3.76**	2.66 **3.99**	2.82 **4.31**	3.05 **4.82**	3.44 **5.72**	4.30 **7.94**
23	1.76 **2.26**	1.77 **2.28**	1.79 **2.32**	1.82 **2.37**	1.84 **2.41**	1.88 **2.48**	1.91 **2.53**	1.96 **2.62**	2.00 **2.70**	2.04 **2.78**	2.10 **2.89**	2.14 **2.97**	2.20 **3.07**	2.24 **3.14**	2.28 **3.21**	2.32 **3.30**	2.38 **3.41**	2.45 **3.54**	2.53 **3.71**	2.64 **3.94**	2.80 **4.26**	3.03 **4.76**	3.42 **5.66**	4.28 **7.88**
24	1.73 **2.21**	1.74 **2.23**	1.76 **2.27**	1.80 **2.33**	1.82 **2.36**	1.86 **2.44**	1.89 **2.49**	1.94 **2.58**	1.98 **2.66**	2.02 **2.74**	2.09 **2.85**	2.13 **2.93**	2.18 **3.03**	2.22 **3.09**	2.26 **3.17**	2.30 **3.25**	2.36 **3.36**	2.43 **3.50**	2.51 **3.67**	2.62 **3.90**	2.78 **4.22**	3.01 **4.72**	3.40 **5.61**	4.26 **7.82**

Source: Reprinted by permission from *Statistical Methods*, 6th edition, by George W. Snedecor and William G. Cochran. © 1967 by The Iowa State University Press, Ames, Iowa.

TABLE B.11 CRITICAL VALUES FOR THE *F* DISTRIBUTION (*continued*)

n_1 DEGREES OF FREEDOM (FOR GREATER MEAN SQUARE)

n_2	1	2	3	4	5	6	7	8	9	10	11	12	14	16	20	24	30	40	50	75	100	200	500	∞	n_2
25	4.24 **7.77**	3.38 **5.57**	2.99 **4.68**	2.76 **4.18**	2.60 **3.86**	2.49 **3.63**	2.41 **3.46**	2.34 **3.32**	2.28 **3.21**	2.24 **3.13**	2.20 **3.05**	2.16 **2.99**	2.11 **2.89**	2.06 **2.81**	2.00 **2.70**	1.96 **2.62**	1.92 **2.54**	1.87 **2.45**	1.84 **2.40**	1.80 **2.32**	1.77 **2.29**	1.74 **2.23**	1.72 **2.19**	1.71 **2.17**	25
26	4.22 **7.72**	3.37 **5.53**	2.98 **4.64**	2.74 **4.14**	2.59 **3.82**	2.47 **3.59**	2.39 **3.42**	2.32 **3.29**	2.27 **3.17**	2.22 **3.09**	2.18 **3.02**	2.15 **2.96**	2.10 **2.86**	2.05 **2.77**	1.99 **2.66**	1.95 **2.58**	1.90 **2.50**	1.85 **2.41**	1.82 **2.36**	1.78 **2.28**	1.76 **2.25**	1.72 **2.19**	1.70 **2.15**	1.69 **2.13**	26
27	4.21 **7.68**	3.35 **5.49**	2.96 **4.60**	2.73 **4.11**	2.57 **3.79**	2.46 **3.56**	2.37 **3.39**	2.30 **3.26**	2.25 **3.14**	2.20 **3.06**	2.16 **2.98**	2.13 **2.93**	2.08 **2.83**	2.03 **2.74**	1.97 **2.63**	1.93 **2.55**	1.88 **2.47**	1.84 **2.38**	1.80 **2.33**	1.76 **2.25**	1.74 **2.21**	1.71 **2.16**	1.68 **2.12**	1.67 **2.10**	27
28	4.20 **7.64**	3.34 **5.45**	2.95 **4.57**	2.71 **4.07**	2.56 **3.76**	2.44 **3.53**	2.36 **3.36**	2.29 **3.23**	2.24 **3.11**	2.19 **3.03**	2.15 **2.95**	2.12 **2.90**	2.06 **2.80**	2.02 **2.71**	1.96 **2.60**	1.91 **2.52**	1.87 **2.44**	1.81 **2.35**	1.78 **2.30**	1.75 **2.22**	1.72 **2.18**	1.69 **2.13**	1.67 **2.09**	1.65 **2.06**	28
29	4.18 **7.60**	3.33 **5.42**	2.93 **4.54**	2.70 **4.04**	2.54 **3.73**	2.43 **3.50**	2.35 **3.33**	2.28 **3.20**	2.22 **3.08**	2.18 **3.00**	2.14 **2.92**	2.10 **2.87**	2.05 **2.77**	2.00 **2.68**	1.94 **2.57**	1.90 **2.49**	1.85 **2.41**	1.80 **2.32**	1.77 **2.27**	1.73 **2.19**	1.71 **2.15**	1.68 **2.10**	1.65 **2.06**	1.64 **2.03**	29
30	4.17 **7.56**	3.32 **5.39**	2.92 **4.51**	2.69 **4.02**	2.53 **3.70**	2.42 **3.47**	2.34 **3.30**	2.27 **3.17**	2.21 **3.06**	2.16 **2.98**	2.12 **2.90**	2.09 **2.84**	2.04 **2.74**	1.99 **2.66**	1.93 **2.55**	1.89 **2.47**	1.84 **2.38**	1.79 **2.29**	1.76 **2.24**	1.72 **2.16**	1.69 **2.13**	1.66 **3.07**	1.64 **2.03**	1.62 **2.01**	30
32	4.15 **7.50**	3.30 **5.34**	2.90 **4.46**	2.67 **3.97**	2.51 **3.66**	2.40 **3.42**	2.32 **3.25**	2.25 **3.12**	2.19 **3.01**	2.14 **2.94**	2.10 **2.86**	2.07 **2.80**	2.02 **2.70**	1.97 **2.62**	1.91 **2.51**	1.86 **2.42**	1.82 **2.34**	1.76 **2.25**	1.74 **2.20**	1.69 **2.12**	1.67 **2.08**	1.64 **2.02**	1.61 **1.98**	1.59 **1.96**	32
34	4.13 **7.44**	3.28 **5.29**	2.88 **4.42**	2.65 **3.93**	2.49 **3.61**	2.38 **3.38**	2.30 **3.21**	2.23 **3.08**	2.17 **2.97**	2.12 **2.89**	2.08 **2.82**	2.05 **2.76**	2.00 **2.66**	1.95 **2.58**	1.89 **2.47**	1.84 **2.38**	1.80 **2.30**	1.74 **2.21**	1.71 **2.15**	1.67 **2.08**	1.64 **2.04**	1.61 **1.98**	1.59 **1.94**	1.57 **1.91**	34
36	4.11 **7.39**	3.26 **5.25**	2.86 **4.38**	2.63 **3.89**	2.48 **3.58**	2.36 **3.35**	2.28 **3.18**	2.21 **3.04**	2.15 **2.94**	2.10 **2.86**	2.06 **2.78**	2.03 **2.72**	1.98 **2.62**	1.93 **2.54**	1.87 **2.43**	1.82 **2.35**	1.78 **2.26**	1.72 **2.17**	1.69 **2.12**	1.65 **2.04**	1.62 **2.00**	1.59 **1.94**	1.56 **1.90**	1.55 **1.87**	36
38	4.10 **7.35**	3.25 **5.21**	2.85 **4.34**	2.62 **3.86**	2.46 **3.54**	2.35 **3.32**	2.26 **3.15**	2.19 **3.02**	2.14 **2.91**	2.09 **2.82**	2.05 **2.75**	2.02 **2.69**	1.96 **2.59**	1.92 **2.51**	1.85 **2.40**	1.80 **2.32**	1.76 **2.22**	1.71 **2.14**	1.67 **2.08**	1.63 **2.00**	1.60 **1.97**	1.57 **1.90**	1.54 **1.86**	1.53 **1.84**	38
40	4.08 **7.31**	3.23 **5.18**	2.84 **4.31**	2.61 **3.83**	2.45 **3.51**	2.34 **3.29**	2.25 **3.12**	2.18 **2.99**	2.12 **2.88**	2.07 **2.80**	2.04 **2.73**	2.00 **2.66**	1.95 **2.56**	1.90 **2.49**	1.84 **2.37**	1.79 **2.29**	1.74 **2.20**	1.69 **2.11**	1.66 **2.05**	1.61 **1.97**	1.59 **1.94**	1.55 **1.88**	1.53 **1.84**	1.51 **1.81**	40
42	4.07 **7.27**	3.22 **5.15**	2.83 **4.29**	2.59 **3.80**	2.44 **3.49**	2.32 **3.26**	2.24 **3.10**	2.17 **2.96**	2.11 **2.86**	2.06 **2.77**	2.02 **2.70**	1.99 **2.64**	1.94 **2.54**	1.89 **2.46**	1.82 **2.35**	1.78 **2.26**	1.73 **2.17**	1.68 **2.08**	1.64 **2.02**	1.60 **1.94**	1.57 **1.91**	1.54 **1.85**	1.51 **1.80**	1.49 **1.78**	42
44	4.06 **7.24**	3.21 **5.12**	2.82 **4.26**	2.58 **3.78**	2.43 **3.46**	2.31 **3.24**	2.23 **3.07**	2.16 **2.94**	2.10 **2.84**	2.05 **2.75**	2.01 **2.68**	1.98 **2.62**	1.92 **2.52**	1.88 **2.44**	1.81 **2.32**	1.76 **2.24**	1.72 **2.15**	1.66 **2.06**	1.63 **2.00**	1.58 **1.92**	1.56 **1.88**	1.52 **1.82**	1.50 **1.78**	1.48 **1.75**	44

46	1.46 / 1.72	1.48 / 1.76	1.51 / 1.80	1.54 / 1.86	1.57 / 1.90	1.62 / 1.98	1.65 / 2.04	1.71 / 2.13	1.75 / 2.22	1.80 / 2.30	1.87 / 2.42	1.91 / 2.50	1.97 / 2.60	2.00 / 2.66	2.04 / 2.73	2.09 / 2.82	2.14 / 2.92	2.22 / 3.05	2.30 / 3.22	2.42 / 3.44	2.57 / 3.76	2.81 / 4.24	3.20 / 5.10	4.05 / 7.21	46
48	1.45 / 1.70	1.47 / 1.73	1.50 / 1.78	1.53 / 1.84	1.56 / 1.88	1.61 / 1.96	1.64 / 2.02	1.70 / 2.11	1.74 / 2.20	1.79 / 2.28	1.86 / 2.40	1.90 / 2.48	1.96 / 2.58	1.99 / 2.64	2.03 / 2.71	2.08 / 2.80	2.14 / 2.90	2.21 / 3.04	2.30 / 3.20	2.41 / 3.42	2.56 / 3.74	2.80 / 4.22	3.19 / 5.08	4.04 / 7.19	48
50	1.44 / 1.68	1.46 / 1.71	1.48 / 1.76	1.52 / 1.82	1.55 / 1.86	1.60 / 1.94	1.63 / 2.00	1.69 / 2.10	1.74 / 2.18	1.78 / 2.26	1.85 / 2.39	1.90 / 2.46	1.95 / 2.56	1.98 / 2.62	2.02 / 2.70	2.07 / 2.78	2.13 / 2.88	2.20 / 3.02	2.29 / 3.18	2.40 / 3.41	2.56 / 3.72	2.79 / 4.20	3.18 / 5.06	4.03 / 7.17	50
55	1.41 / 1.64	1.43 / 1.66	1.46 / 1.71	1.50 / 1.78	1.52 / 1.82	1.58 / 1.90	1.61 / 1.96	1.67 / 2.06	1.72 / 2.15	1.76 / 2.23	1.83 / 2.35	1.88 / 2.43	1.93 / 2.53	1.97 / 2.59	2.00 / 2.66	2.05 / 2.75	2.11 / 2.85	2.18 / 2.98	2.27 / 3.15	2.38 / 3.37	2.54 / 3.68	2.78 / 4.16	3.17 / 5.01	4.02 / 7.12	55
60	1.39 / 1.60	1.41 / 1.63	1.44 / 1.68	1.48 / 1.74	1.50 / 1.79	1.56 / 1.87	1.59 / 1.93	1.65 / 2.03	1.70 / 2.12	1.75 / 2.20	1.81 / 2.32	1.86 / 2.40	1.92 / 2.50	1.95 / 2.56	1.99 / 2.63	2.04 / 2.72	2.10 / 2.82	2.17 / 2.95	2.25 / 3.12	2.37 / 3.34	2.52 / 3.65	2.76 / 4.13	3.15 / 4.98	4.00 / 7.08	60
65	1.37 / 1.56	1.39 / 1.60	1.42 / 1.64	1.46 / 1.71	1.49 / 1.76	1.54 / 1.84	1.57 / 1.90	1.63 / 2.00	1.68 / 2.09	1.73 / 2.18	1.80 / 2.30	1.85 / 2.37	1.90 / 2.47	1.94 / 2.54	1.98 / 2.61	2.02 / 2.70	2.08 / 2.79	2.15 / 2.93	2.24 / 3.09	2.36 / 3.31	2.51 / 3.62	2.75 / 4.10	3.14 / 4.95	3.99 / 7.04	65
70	1.35 / 1.53	1.37 / 1.56	1.40 / 1.62	1.45 / 1.69	1.47 / 1.74	1.53 / 1.82	1.56 / 1.88	1.62 / 1.98	1.67 / 2.07	1.72 / 2.15	1.79 / 2.28	1.84 / 2.35	1.89 / 2.45	1.93 / 2.51	1.97 / 2.59	2.01 / 2.67	2.07 / 2.77	2.14 / 2.91	2.23 / 3.07	2.35 / 3.29	2.50 / 3.60	2.74 / 4.08	3.13 / 4.92	3.98 / 7.01	70
80	1.32 / 1.49	1.35 / 1.52	1.38 / 1.57	1.42 / 1.65	1.45 / 1.70	1.51 / 1.78	1.54 / 1.84	1.60 / 1.94	1.65 / 2.03	1.70 / 2.11	1.77 / 2.24	1.82 / 2.32	1.88 / 2.41	1.91 / 2.48	1.95 / 2.55	1.99 / 2.64	2.05 / 2.74	2.12 / 2.87	2.21 / 3.04	2.33 / 3.25	2.48 / 3.56	2.72 / 4.04	3.11 / 4.88	3.96 / 6.96	80
100	1.28 / 1.43	1.30 / 1.46	1.34 / 1.51	1.39 / 1.59	1.42 / 1.64	1.48 / 1.73	1.51 / 1.79	1.57 / 1.89	1.63 / 1.98	1.68 / 2.06	1.75 / 2.19	1.79 / 2.26	1.85 / 2.36	1.88 / 2.43	1.92 / 2.51	1.97 / 2.59	2.03 / 2.69	2.10 / 2.82	2.19 / 2.99	2.30 / 3.20	2.46 / 3.51	2.70 / 3.98	3.09 / 4.82	3.94 / 6.90	100
125	1.25 / 1.37	1.27 / 1.40	1.31 / 1.46	1.36 / 1.54	1.39 / 1.59	1.45 / 1.68	1.49 / 1.75	1.55 / 1.85	1.60 / 1.94	1.65 / 2.03	1.72 / 2.15	1.77 / 2.23	1.83 / 2.33	1.86 / 2.40	1.90 / 2.47	1.95 / 2.56	2.01 / 2.65	2.08 / 2.79	2.17 / 2.95	2.29 / 3.17	2.44 / 3.47	2.68 / 3.94	3.07 / 4.78	3.92 / 6.84	125
150	1.22 / 1.33	1.25 / 1.37	1.29 / 1.43	1.34 / 1.51	1.37 / 1.56	1.44 / 1.66	1.47 / 1.72	1.54 / 1.83	1.59 / 1.91	1.64 / 2.00	1.71 / 2.12	1.76 / 2.20	1.82 / 2.30	1.85 / 2.37	1.89 / 2.44	1.94 / 2.53	2.00 / 2.62	2.07 / 2.76	2.16 / 2.92	2.27 / 3.14	2.43 / 3.44	2.67 / 3.91	3.06 / 4.75	3.91 / 6.81	150
200	1.19 / 1.28	1.22 / 1.33	1.26 / 1.39	1.32 / 1.48	1.35 / 1.53	1.42 / 1.62	1.45 / 1.69	1.52 / 1.79	1.57 / 1.88	1.62 / 1.97	1.69 / 2.09	1.74 / 2.17	1.80 / 2.28	1.83 / 2.34	1.87 / 2.41	1.92 / 2.50	1.98 / 2.60	2.05 / 2.73	2.14 / 2.90	2.26 / 3.11	2.41 / 3.41	2.65 / 3.88	3.04 / 4.71	3.89 / 6.76	200
400	1.13 / 1.19	1.16 / 1.24	1.22 / 1.32	1.28 / 1.42	1.32 / 1.47	1.38 / 1.57	1.42 / 1.64	1.49 / 1.74	1.54 / 1.84	1.60 / 1.92	1.67 / 2.04	1.72 / 2.12	1.78 / 2.23	1.81 / 2.29	1.85 / 2.37	1.90 / 2.46	1.96 / 2.55	2.03 / 2.69	2.12 / 2.85	2.23 / 3.06	2.39 / 3.36	2.62 / 3.83	3.02 / 4.66	3.86 / 6.70	400
1000	1.08 / 1.11	1.13 / 1.19	1.19 / 1.28	1.26 / 1.38	1.30 / 1.44	1.36 / 1.54	1.41 / 1.61	1.47 / 1.71	1.53 / 1.81	1.58 / 1.89	1.65 / 2.01	1.70 / 2.09	1.76 / 2.20	1.80 / 2.26	1.84 / 2.34	1.89 / 2.43	1.95 / 2.53	2.02 / 2.66	2.10 / 2.82	2.22 / 3.04	2.38 / 3.34	2.61 / 3.80	3.00 / 4.62	3.85 / 6.66	1000
∞	1.00 / 1.00	1.11 / 1.15	1.17 / 1.25	1.24 / 1.36	1.28 / 1.41	1.35 / 1.52	1.40 / 1.59	1.46 / 1.69	1.52 / 1.79	1.57 / 1.87	1.64 / 1.99	1.69 / 2.07	1.75 / 2.18	1.79 / 2.24	1.83 / 2.32	1.88 / 2.41	1.94 / 2.51	2.01 / 2.64	2.09 / 2.80	2.21 / 3.02	2.37 / 3.32	2.60 / 3.78	2.99 / 4.60	3.84 / 6.64	∞

TABLE B.12 PERCENTAGE POINTS OF
THE LARGEST VARIANCE RATIO, s^2_{max}/s^2_0

UPPER 5% POINTS

$k*$ \ v	1	2	3	4	5	6	7	8	9	10
10	4.96	6.79	8.00	8.96	9.78	10.52	11.18	11.79	12.36	12.87
12	4.75	6.44	7.53	8.37	9.06	9.68	10.20	10.68	11.12	11.53
15	4.54	6.12	7.11	7.86	8.47	8.98	9.43	9.82	10.19	10.52
20	4.35	5.81	6.72	7.40	7.94	8.39	8.79	9.13	9.44	9.71
30	4.17	5.52	6.36	6.97	7.46	7.87	8.21	8.51	8.79	9.03
60	4.00	5.25	6.02	6.58	7.02	7.38	7.68	7.96	8.20	8.41
∞	3.84	5.00	5.70	6.21	6.60	6.92	7.20	7.44	7.65	7.84

UPPER 1% POINTS

k \ v	1	2	3	4	5	6	7	8	9	10
10	10.04	13.17	15.08	16.43	17.43	18.25	18.91	19.48	19.97	20.41
12	9.33	11.88	13.52	14.73	15.69	16.47	17.12	17.68	18.16	18.60
15	8.68	10.82	12.18	13.21	14.03	14.72	15.30	15.81	16.26	16.66
20	8.10	9.93	11.08	11.93	12.61	13.19	13.67	14.09	14.49	14.83
30	7.56	9.16	10.14	10.86	11.43	11.90	12.31	12.66	12.97	13.26
60	7.08	8.49	9.34	9.95	10.43	10.82	11.15	11.45	11.72	11.95
∞	6.63	7.88	8.61	9.15	9.54	9.87	10.16	10.41	10.62	10.82

* k is the number of independent variance estimates, each based on 1 degree of freedom, of which s^2_{max}. is the largest. v denotes the degrees of freedom of the independent 'error' mean square, s^2_0.

Source: Reprinted by permission from *Biometrika Tables for Statisticians*, edited by E. S. Pearson and H. O. Hartley, 4th ed., Cambridge University Press, 1965.

TABLE B.13   CONFIDENCE INTERVALS
FOR π (95 PERCENT CONFIDENCE LIMITS)

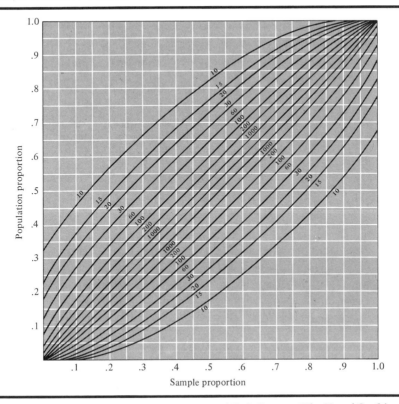

Source: Reprinted by permission from C. J. Clopper and E. S. Pearson, "The Use of Confidence or Fiducial Limits Illustrated in the Case of the Binomial," *Biometrika*, Vol. 26, p. 410.

TABLE B.14 RANDOM NUMBERS

00000	10097	32533	76520	13586	34673	54876	80959	09117	39292	74945
00001	37542	04805	64894	74296	24805	24037	20636	10402	00822	91665
00002	08422	68953	19645	09303	23209	02560	15953	34764	35080	33606
00003	99019	02529	09376	70715	38311	31165	88676	74397	04436	27659
00004	12807	99970	80157	36147	64032	36653	98951	16877	12171	76833
00005	66065	74717	34072	76850	36697	36170	65813	39885	11199	29170
00006	31060	10805	45571	82406	35303	42614	86799	07439	23403	09732
00007	85269	77602	02051	65692	68665	74818	73053	85247	18623	88579
00008	63573	32135	05325	47048	90553	57548	28468	28709	83491	25624
00009	73796	45753	03529	64778	35808	34282	60935	20344	35273	88435
00010	98520	17767	14905	68607	22109	40558	60970	93433	50500	73998
00011	11805	05431	39808	27732	50725	68248	29405	24201	52775	67851
00012	83452	99634	06288	98083	13746	70078	18475	40610	68711	77817
00013	88685	40200	86507	58401	36766	67951	90364	76493	29609	11062
00014	99594	67348	87517	64969	91826	08928	93785	61368	23478	34113
00015	65481	17674	17468	50950	58047	76974	73039	57186	40218	16544
00016	80124	35635	17727	08015	45318	22374	21115	78253	14385	53763
00017	74350	99817	77402	77214	43236	00210	45521	64237	96286	02655
00018	69916	26803	66252	29148	36936	87203	76621	13990	94400	56418
00019	09893	20505	14225	68514	46427	56788	96297	78822	54382	14598
00020	91499	14523	68479	27686	46162	83554	94750	89923	37089	20048
00021	80336	94598	26940	36858	70297	34135	53140	33340	42050	82341
00022	44104	81949	85157	47954	32979	26575	57600	40881	22222	06413
00023	12550	73742	11100	02040	12860	74697	96644	89439	28707	25815
00024	63606	49329	16505	34484	40219	52563	43651	77082	07207	31790
00025	61196	90446	26457	47774	51924	33729	65394	59593	42582	60527
00026	15474	45266	95270	79953	59367	83848	82396	10118	33211	59466
00027	94557	28573	67897	54387	54622	44431	91190	42592	92927	45973
00028	42481	16213	·97344	08721	16868	48767	03071	12059	25701	46670
00029	23523	78317	73208	89837	68935	91416	26252	29663	05522	82562
00030	04493	52494	75246	33824	45862	51025	61962	79335	65337	12472
00031	00549	97654	64051	88159	96119	63896	54692	82391	23287	29529
00032	35963	15307	26898	09354	33351	35462	77974	50024	90103	39333
00033	59808	08391	45427	26842	83609	49700	13021	24892	78565	20106
00034	46058	85236	01390	92286	77281	44077	93910	83647	70617	42941
00035	32179	00597	87379	25241	05567	07007	86743	17157	85394	11838
00036	69234	61406	20117	45204	15956	60000	18743	92423	97118	96338
00037	19565	41430	01758	75379	40419	21585	66674	36806	84962	85207
00038	45155	14938	19476	07246	43667	94543	59047	90033	20826	69541
00039	94864	31994	36168	10851	34888	81553	01540	35456	05014	51176
00040	98086	24826	45240	28404	44999	08896	39094	73407	35441	31880
00041	33185	16232	41941	50949	89435	48581	88695	41994	37548	73043
00042	80951	00406	96382	70774	20151	23387	25016	25298	94624	61171
00043	79752	49140	71961	28296	69861	02591	74852	20539	00387	59579
00044	18633	32537	98145	06571	31010	24674	05455	61427	77938	91936
00045	74029	43902	77557	32270	97790	17119	52527	58021	80814	51748
00046	54178	45611	80993	37143	05335	12969	56127	19255	36040	90324
00047	11664	49883	52079	84827	59381	71539	09973	33440	88461	23356
00048	48324	77928	31249	64710	02295	36870	32307	57546	15020	09994
00049	69074	94138	87637	91976	35584	04401	10518	21615	01848	76938

Source: Reprinted by permission from The Rand Corporation, *A Million Random Digits with 100,000 Normal Deviates*, The Free Press, 1955

TABLE B.14 RANDOM NUMBERS (*continued*)

00050	09188	20097	32825	39527	04220	86304	83389	87374	64278	58044
00051	90045	85497	51981	50654	94938	81997	91870	76150	68476	64659
00052	73189	50207	47677	26269	62290	64464	27124	67018	41361	82760
00053	75768	76490	20971	87749	90429	12272	95375	05871	93823	43178
00054	54016	44056	66281	31003	00682	27398	20714	53295	07706	17813
00055	08358	69910	78542	42785	13661	58873	04618	97553	31223	08420
00056	28306	03264	81333	10591	40510	07893	32604	60475	94119	01840
00057	53840	86233	81594	13628	51215	90290	28466	68795	77762	20791
00058	91757	53741	61613	62269	50263	90212	55781	76514	83483	47055
00059	89415	92694	00397	58391	12607	17646	48949	72306	94541	37408
00060	77513	03820	86864	29901	68414	82774	51908	13980	72893	55507
00061	19502	37174	69979	20288	55210	29773	74287	75251	65344	67415
00062	21818	59313	93278	81757	05686	73156	07082	85046	31853	38452
00063	51474	66499	68107	23621	94049	91345	42836	09191	08007	45449
00064	99559	68331	62535	24170	69777	12830	74819	78142	43860	72834
00065	33713	48007	93584	72869	51926	64721	58303	29822	93174	93972
00066	85274	86893	11303	22970	28834	34137	73515	90400	71148	43643
00067	84133	89640	44035	52166	73852	70091	61222	60561	62327	18423
00068	56732	16234	17395	96131	10123	91622	85496	57560	81604	18880
00069	65138	56806	87648	85261	34313	65861	45875	21069	85644	47277
00070	38001	02176	81719	11711	71602	92937	74219	64049	65584	49698
00071	37402	96397	01304	77586	56271	10086	47324	62605	40030	37438
00072	97125	40348	87083	31417	21815	39250	75237	62047	15501	29578
00073	21826	41134	47143	34072	64638	85902	49139	06441	03856	54552
00074	73135	42742	95719	09035	85794	74296	08789	88156	64691	19202
00075	07638	77929	03061	18072	96207	44156	23821	99538	04713	66994
00076	60528	83441	07954	19814	59175	20695	05533	52139	61212	06455
00077	83596	35655	06958	92983	05128	09719	77433	53783	92301	50498
00078	10850	62746	99599	10507	13499	06319	53075	71839	06410	19362
00079	39820	98952	43622	63147	64421	80814	43800	09351	31024	73167
00080	59580	06478	75569	78800	88835	54486	23768	06156	04111	08408
00081	38508	07341	23793	48763	90822	97022	17719	04207	95954	49953
00082	30692	70668	94688	16127	56196	80091	82067	63400	05462	69200
00083	65443	95659	18288	27437	49632	24041	08337	65676	96299	90836
00084	27267	50264	13192	72294	07477	44606	17985	48911	97341	30358
00085	91307	06991	19072	24210	36699	53728	28825	35793	28976	66252
00086	68434	94688	84473	13622	62126	98408	12843	82590	09815	93146
00087	48908	15877	54745	24591	35700	04754	83824	52692	54130	55160
00088	06913	45197	42672	78601	11883	09528	63011	98901	14974	40344
00089	10455	16019	14210	33712	91342	37821	88325	80851	43667	70883
00090	12883	97343	65027	61184	04285	01392	17974	15077	90712	26769
00091	21778	30976	38807	36961	31649	42096	63281	02023	08816	47449
00092	19523	59515	65122	59659	86283	68258	69572	13798	16435	91529
00093	67245	52670	35583	16563	79246	86686	76463	34222	26655	90802
00094	60584	47377	07500	37992	45134	26529	26760	83637	41326	44344
00095	53853	41377	36066	94850	58838	73859	49364	73331	96240	43642
00096	24637	38736	74384	89342	52623	07992	12369	18601	03742	83873
00097	83080	12451	38992	22815	07759	51777	97377	27585	51972	37867
00098	16444	24334	36151	99073	27493	70939	85130	32552	54846	54759
00099	60790	18157	57178	65762	11161	78576	45819	52979	65130	04860

TABLE B.15 DISTRIBUTION OF THE STUDENTIZED RANGE

Entry $= q_{p;v_1,v_2}$ where $\Pr[q_{v_1,v_2} < q_{p;v_1,v_2}] = p$ $v_1 =$ no. treats., $v_2 = d.f.$ associated with estimates of variance

$p = .95$

v_2 \ v_1	2	3	4	5	6	7	8	9	10	11	12	13	14	15	16	17	18	19	20
1	18.0	27.0	32.8	37.1	40.4	43.1	45.4	47.4	49.1	50.6	52.0	53.2	54.3	55.4	56.3	57.2	58.0	58.8	59.6
2	6.09	8.3	9.8	10.9	11.7	12.4	13.0	13.5	14.0	14.4	14.7	15.1	15.4	15.7	15.9	16.1	16.4	16.6	16.8
3	4.50	5.91	6.82	7.50	8.04	8.48	8.85	9.18	9.46	9.72	9.95	10.15	10.35	10.52	10.69	10.84	10.98	11.11	11.24
4	3.93	5.04	5.76	6.29	6.71	7.05	7.35	7.60	7.83	8.03	8.21	8.37	8.52	8.66	8.79	8.91	9.03	9.13	9.23
5	3.64	4.60	5.22	5.67	6.03	6.33	6.58	6.80	6.99	7.17	7.32	7.47	7.60	7.72	7.83	7.93	8.03	8.12	8.21
6	3.46	4.34	4.90	5.31	5.63	5.89	6.12	6.32	6.49	6.65	6.79	6.92	7.03	7.14	7.24	7.34	7.43	7.51	7.59
7	3.34	4.16	4.68	5.06	5.36	5.61	5.82	6.00	6.16	6.30	6.43	6.55	6.66	6.76	6.85	6.94	7.02	7.09	7.17
8	3.26	4.04	4.53	4.89	5.17	5.40	5.60	5.77	5.92	6.05	6.18	6.29	6.39	6.48	6.57	6.65	6.73	6.80	6.87
9	3.20	3.95	4.42	4.76	5.02	5.24	5.43	5.60	5.74	5.87	5.98	6.09	6.19	6.28	6.36	6.44	6.51	6.58	6.64
10	3.15	3.88	4.33	4.65	4.91	5.12	5.30	5.46	5.60	5.72	5.83	5.93	6.03	6.11	6.20	6.27	6.34	6.40	6.47
11	3.11	3.82	4.26	4.57	4.82	5.03	5.20	5.35	5.49	5.61	5.71	5.81	5.90	5.99	6.06	6.14	6.20	6.26	6.33
12	3.08	3.77	4.20	4.51	4.75	4.95	5.12	5.27	5.40	5.51	5.62	5.71	5.80	5.88	5.95	6.03	6.09	6.15	6.21
13	3.06	3.73	4.15	4.45	4.69	4.88	5.05	5.19	5.32	5.43	5.53	5.63	5.71	5.79	5.86	5.93	6.00	6.05	6.11
14	3.03	3.70	4.11	4.41	4.64	4.83	4.99	5.13	5.25	5.36	5.46	5.55	5.64	5.72	5.79	5.85	5.92	5.97	6.03
15	3.01	3.67	4.08	4.37	4.60	4.78	4.94	5.08	5.20	5.31	5.40	5.49	5.58	5.65	5.72	5.79	5.85	5.90	5.96
16	3.00	3.65	4.05	4.33	4.56	4.74	4.90	5.03	5.15	5.26	5.35	5.44	5.52	5.59	5.66	5.72	5.79	5.84	5.90
17	2.98	3.63	4.02	4.30	4.52	4.71	4.86	4.99	5.11	5.21	5.31	5.39	5.47	5.55	5.61	5.68	5.74	5.79	5.84
18	2.97	3.61	4.00	4.28	4.49	4.67	4.82	4.96	5.07	5.17	5.27	5.35	5.43	5.50	5.57	5.63	5.69	5.74	5.79
19	2.96	3.59	3.98	4.25	4.47	4.65	4.79	4.92	5.04	5.14	5.23	5.32	5.39	5.46	5.53	5.59	5.65	5.70	5.75
20	2.95	3.58	3.96	4.23	4.45	4.62	4.77	4.90	5.01	5.11	5.20	5.28	5.36	5.43	5.49	5.55	5.61	5.66	5.71
24	2.92	3.53	3.90	4.17	4.37	4.54	4.68	4.81	4.92	5.01	5.10	5.18	5.25	5.32	5.38	5.44	5.50	5.54	5.59
30	2.89	3.49	3.84	4.10	4.30	4.46	4.60	4.72	4.83	4.92	5.00	5.08	5.15	5.21	5.27	5.33	5.38	5.43	5.48
40	2.86	3.44	3.79	4.04	4.23	4.39	4.52	4.63	4.74	4.82	4.91	4.98	5.05	5.11	5.16	5.22	5.27	5.31	5.36
60	2.83	3.40	3.74	3.98	4.16	4.31	4.44	4.55	4.65	4.73	4.81	4.88	4.94	5.00	5.06	5.11	5.16	5.20	5.24
120	2.80	3.36	3.69	3.92	4.10	4.24	4.36	4.48	4.56	4.64	4.72	4.78	4.84	4.90	4.95	5.00	5.05	5.09	5.13
∞	2.77	3.31	3.63	3.86	4.03	4.17	4.29	4.39	4.47	4.55	4.62	4.68	4.74	4.80	4.85	4.89	4.93	4.97	5.01

$p = .99$

1	90.0	135	164	186	202	216	227	237	246	253	260	266	272	277	282	286	290	294	298
2	14.0	19.0	22.3	24.7	26.6	28.2	29.5	30.7	31.7	32.6	33.4	34.1	34.8	35.4	36.0	36.5	37.0	37.5	37.9
3	8.26	10.6	12.2	13.3	14.2	15.0	15.6	16.2	16.7	17.1	17.5	17.9	18.2	18.5	18.8	19.1	19.3	19.5	19.8
4	6.51	8.12	9.17	9.96	10.6	11.1	11.5	11.9	12.3	12.6	12.8	13.1	13.3	13.5	13.7	13.9	14.1	14.2	14.4
5	5.70	6.97	7.80	8.42	8.91	9.32	9.67	9.97	10.24	10.48	10.70	10.89	11.08	11.24	11.40	11.55	11.68	11.81	11.93
6	5.24	6.33	7.03	7.56	7.97	8.32	8.61	8.87	9.10	9.30	9.49	9.65	9.81	9.95	10.08	10.21	10.32	10.43	10.54
7	4.95	5.92	6.54	7.01	7.37	7.68	7.94	8.17	8.37	8.55	8.71	8.86	9.00	9.12	9.24	9.35	9.46	9.55	9.65
8	4.74	5.63	6.20	6.63	6.96	7.24	7.47	7.68	7.87	8.03	8.18	8.31	8.44	8.55	8.66	8.76	8.85	8.94	9.03
9	4.60	5.43	5.96	6.35	6.66	6.91	7.13	7.32	7.49	7.65	7.78	7.91	8.03	8.13	8.25	8.32	8.41	8.49	8.57
10	4.48	5.27	5.77	6.14	6.43	6.67	6.87	7.05	7.21	7.36	7.48	7.60	7.71	7.81	7.91	7.99	8.07	8.15	8.22
11	4.39	5.14	5.62	5.97	6.25	6.48	6.67	6.84	6.99	7.13	7.25	7.36	7.46	7.56	7.65	7.73	7.81	7.88	7.95
12	4.32	5.04	5.50	5.84	6.10	6.32	6.51	6.67	6.81	6.94	7.06	7.17	7.26	7.36	7.44	7.52	7.59	7.66	7.73
13	4.26	4.96	5.40	5.73	5.98	6.19	6.37	6.53	6.67	6.79	6.90	7.01	7.10	7.19	7.27	7.34	7.42	7.48	7.55
14	4.21	4.89	5.32	5.63	5.88	6.08	6.26	6.41	6.54	6.66	6.77	6.87	6.96	7.05	7.12	7.20	7.27	7.33	7.39
15	4.17	4.83	5.25	5.56	5.80	5.99	6.16	6.31	6.44	6.55	6.66	6.76	6.84	6.93	7.00	7.07	7.14	7.20	7.26
16	4.13	4.78	5.19	5.49	5.72	5.92	6.08	6.22	6.35	6.46	6.56	6.66	6.74	6.82	6.90	6.97	7.03	7.09	7.15
17	4.10	4.74	5.14	5.43	5.66	5.85	6.01	6.15	6.27	6.38	6.48	6.57	6.66	6.73	6.80	6.87	6.94	7.00	7.05
18	4.07	4.70	5.09	5.38	5.60	5.79	5.94	6.08	6.20	6.31	6.41	6.50	6.58	6.65	6.72	6.79	6.85	6.91	6.96
19	4.05	4.67	5.05	5.33	5.55	5.73	5.89	6.02	6.14	6.25	6.34	6.43	6.51	6.58	6.65	6.72	6.78	6.84	6.89
20	4.02	4.64	5.02	5.29	5.51	5.69	5.84	5.97	6.09	6.19	6.29	6.37	6.45	6.52	6.59	6.65	6.71	6.76	6.82
24	3.96	4.54	4.91	5.17	5.37	5.54	5.69	5.81	5.92	6.02	6.11	6.19	6.26	6.33	6.39	6.45	6.51	6.56	6.61
30	3.89	4.45	4.80	5.05	5.24	5.40	5.54	5.65	5.76	5.85	5.93	6.01	6.08	6.14	6.20	6.26	6.31	6.36	6.41
40	3.82	4.37	4.70	4.93	5.11	5.27	5.39	5.50	5.60	5.69	5.77	5.84	5.90	5.96	6.02	6.07	6.12	6.17	6.21
60	3.76	4.28	4.60	4.82	4.99	5.13	5.25	5.36	5.45	5.53	5.60	5.67	5.73	5.79	5.84	5.89	5.93	5.98	6.02
120	3.70	4.20	4.50	4.71	4.87	5.01	5.12	5.21	5.30	5.38	5.44	5.51	5.56	5.61	5.66	5.71	5.75	5.79	5.83
∞	3.64	4.11	4.40	4.60	4.76	4.88	4.99	5.08	5.16	5.23	5.29	5.35	5.40	5.45	5.49	5.54	5.57	5.61	5.65

Source: Reprinted with permission from E. S. Pearson and H. O. Hartley, *Biometrika Tables for Statisticians* (New York: Cambridge University Press, 1954).

TABLE B.16   FISHER z TRANSFORM (VALUES OF r FOR VALUES OF z)

z	.00	.01	.02	.03	.04	.05	.06	.07	.08	.09
.0	.0000	.0100	.0200	.0300	.0400	.0500	.0599	.0699	.0798	.0898
.1	.0997	.1096	.1194	.1293	.1391	.1489	.1586	.1684	.1781	.1877
.2	.1974	.2070	.2165	.2260	.2355	.2449	.2543	.2636	.2729	.2821
.3	.2913	.3004	.3095	.3185	.3275	.3364	.3452	.3540	.3627	.3714
.4	.3800	.3885	.3969	.4053	.4136	.4219	.4301	.4382	.4462	.4542
.5	.4621	.4699	.4777	.4854	.4930	.5005	.5080	.5154	.5227	.5299
.6	.5370	.5441	.5511	.5580	.5649	.5717	.5784	.5850	.5915	.5980
.7	.6044	.6107	.6169	.6231	.6291	.6351	.6411	.6469	.6527	.6584
.8	.6640	.6696	.6751	.6805	.6858	.6911	.6963	.7014	.7064	.7114
.9	.7163	.7211	.7259	.7306	.7352	.7398	.7443	.7487	.7531	.7574
1.0	.7616	.7658	.7699	.7739	.7779	.7818	.7857	.7895	.7932	.7969
1.1	.8005	.8041	.8076	.8110	.8144	.8178	.8210	.8243	.8275	.8306
1.2	.8337	.8367	.8397	.8426	.8455	.8483	.8511	.8538	.8565	.8591
1.3	.8617	.8643	.8668	.8692	.8717	.8741	.8764	.8787	.8810	.8832
1.4	.8854	.8875	.8896	.8917	.8937	.8957	.8977	.8996	.9015	.9033
1.5	.9051	.9069	.9087	.9104	.9121	.9138	.9154	.9170	.9186	.9201
1.6	.9217	.9232	.9246	.9261	.9275	.9289	.9302	.9316	.9329	.9341
1.7	.9354	.9366	.9379	.9391	.9402	.9414	.9425	.9436	.9447	.9458
1.8	.94681	.94783	.94884	.94983	.95080	.95175	.95268	.95359	.95449	.95537
1.9	.95624	.95709	.95792	.95873	.95953	.96032	.96109	.96185	.96259	.96331
2.0	.96403	.96473	.96541	.96609	.96675	.96739	.96803	.96865	.96926	.96986
2.1	.97045	.97103	.97159	.97215	.97269	.97323	.97375	.97426	.97477	.97526
2.2	.97574	.97622	.97668	.97714	.97759	.97803	.97846	.97888	.97929	.97970
2.3	.98010	.98049	.98087	.98124	.98161	.98197	.98233	.98267	.98301	.98335
2.4	.98367	.98399	.98431	.98462	.98492	.98522	.98551	.98579	.98607	.98635
2.5	.98661	.98688	.98714	.98739	.98764	.98788	.98812	.98835	.98858	.98881
2.6	.98903	.98924	.98945	.98966	.98987	.99007	.99026	.99045	.99064	.99083
2.7	.99101	.99118	.99136	.99153	.99170	.99186	.99202	.99218	.99233	.99248
2.8	.99263	.99278	.99292	.99306	.99320	.99333	.99346	.99359	.99372	.99384
2.9	.99396	.99408	.99420	.99431	.99443	.99454	.99464	.99475	.99485	.99495

Source: Reprinted by permission of the authors and publisher from Fisher and Yates: *Statistical Tables for Biological, Agricultural and Medical Research*, published by Oliver & Boyd, Edinburgh.

TABLE B.17   DURBIN–WATSON STATISTIC

SIGNIFICANCE POINTS OF d_L AND d_U: 1 %

n	$k' = 1$ d_L	d_U	$k' = 2$ d_L	d_U	$k' = 3$ d_L	d_U	$k' = 4$ d_L	d_U	$k' = 5$ d_L	d_U
15	0.81	1.07	0.70	1.25	0.59	1.46	0.49	1.70	0.39	1.96
16	0.84	1.09	0.74	1.25	0.63	1.44	0.53	1.66	0.44	1.90
17	0.87	1.10	0.77	1.25	0.67	1.43	0.57	1.63	0.48	1.85
18	0.90	1.12	0.80	1.26	0.71	1.42	0.61	1.60	0.52	1.80
19	0.93	1.13	0.83	1.26	0.74	1.41	0.65	1.58	0.56	1.77
20	0.95	1.15	0.86	1.27	0.77	1.41	0.68	1.57	0.60	1.74
21	0.97	1.16	0.89	1.27	0.80	1.41	0.72	1.55	0.63	1.71
22	1.00	1.17	0.91	1.28	0.83	1.40	0.75	1.54	0.66	1.69
23	1.02	1.19	0.94	1.29	0.86	1.40	0.77	1.53	0.70	1.67
24	1.04	1.20	0.96	1.30	0.88	1.41	0.80	1.53	0.72	1.66
25	1.05	1.21	0.98	1.30	0.90	1.41	0.83	1.52	0.75	1.65
26	1.07	1.22	1.00	1.31	0.93	1.41	0.85	1.52	0.78	1.64
27	1.09	1.23	1.02	1.32	0.95	1.41	0.88	1.51	0.81	1.63
28	1.10	1.24	1.04	1.32	0.97	1.41	0.90	1.51	0.83	1.62
29	1.12	1.25	1.05	1.33	0.99	1.42	0.92	1.51	0.85	1.61
30	1.13	1.26	1.07	1.34	1.01	1.42	0.94	1.51	0.88	1.61
31	1.15	1.27	1.08	1.34	1.02	1.42	0.96	1.51	0.90	1.60
32	1.16	1.28	1.10	1.35	1.04	1.43	0.98	1.51	0.92	1.60
33	1.17	1.29	1.11	1.36	1.05	1.43	1.00	1.51	0.94	1.59
34	1.18	1.30	1.13	1.36	1.07	1.43	1.01	1.51	0.95	1.59
35	1.19	1.31	1.14	1.37	1.08	1.44	1.03	1.51	0.97	1.59
36	1.21	1.32	1.15	1.38	1.10	1.44	1.04	1.51	0.99	1.59
37	1.22	1.32	1.16	1.38	1.11	1.45	1.06	1.51	1.00	1.59
38	1.23	1.33	1.18	1.39	1.12	1.45	1.07	1.52	1.02	1.58
39	1.24	1.34	1.19	1.39	1.14	1.45	1.09	1.52	1.03	1.58
40	1.25	1.34	1.20	1.40	1.15	1.46	1.10	1.52	1.05	1.58
45	1.29	1.38	1.24	1.42	1.20	1.48	1.16	1.53	1.11	1.58
50	1.32	1.40	1.28	1.45	1.24	1.49	1.20	1.54	1.16	1.59
55	1.36	1.43	1.32	1.47	1.28	1.51	1.25	1.55	1.21	1.59
60	1.38	1.45	1.35	1.48	1.32	1.52	1.28	1.56	1.25	1.60
65	1.41	1.47	1.38	1.50	1.35	1.53	1.31	1.57	1.28	1.61
70	1.43	1.49	1.40	1.52	1.37	1.55	1.34	1.58	1.31	1.61
75	1.45	1.50	1.42	1.53	1.39	1.56	1.37	1.59	1.34	1.62
80	1.47	1.52	1.44	1.54	1.42	1.57	1.39	1.60	1.36	1.62
85	1.48	1.53	1.46	1.55	1.43	1.58	1.41	1.60	1.39	1.63
90	1.50	1.54	1.47	1.56	1.45	1.59	1.43	1.61	1.41	1.64
95	1.51	1.55	1.49	1.57	1.47	1.60	1.45	1.62	1.42	1.64
100	1.52	1.56	1.50	1.58	1.48	1.60	1.46	1.63	1.44	1.65

Source: Reprinted by permission from J. Durbin and G. Watson, "Testing for Serial Correlation in Least Squares Regression," *Biometrika*, Vol. 38, 173 & 175.

TABLE B.17   DURBIN–WATSON STATISTIC (*continued*)

SIGNIFICANCE POINTS OF d_L AND d_U: 5%

n	$k' = 1$		$k' = 2$		$k' = 3$		$k' = 4$		$k' = 5$	
	d_L	d_U	d_L	d_U	d_L	d_U	d_L	d_U	d_L	d_U
15	1.08	1.36	0.95	1.54	0.82	1.75	0.69	1.97	0.56	2.21
16	1.10	1.37	0.98	1.54	0.86	1.73	0.74	1.93	0.62	2.15
17	1.13	1.38	1.02	1.54	0.90	1.71	0.78	1.90	0.67	2.10
18	1.16	1.39	1.05	1.53	0.93	1.69	0.82	1.87	0.71	2.06
19	1.18	1.40	1.08	1.53	0.97	1.68	0.86	1.85	0.75	2.02
20	1.20	1.41	1.10	1.54	1.00	1.68	0.90	1.83	0.79	1.99
21	1.22	1.42	1.13	1.54	1.03	1.67	0.93	1.81	0.83	1.96
22	1.24	1.43	1.15	1.54	1.05	1.66	0.96	1.80	0.86	1.94
23	1.26	1.44	1.17	1.54	1.08	1.66	0.99	1.79	0.90	1.92
24	1.27	1.45	1.19	1.55	1.10	1.66	1.01	1.78	0.93	1.90
25	1.29	1.45	1.21	1.55	1.12	1.66	1.04	1.77	0.95	1.89
26	1.30	1.46	1.22	1.55	1.14	1.65	1.06	1.76	0.98	1.88
27	1.32	1.47	1.24	1.56	1.16	1.65	1.08	1.76	1.01	1.86
28	1.33	1.48	1.26	1.56	1.18	1.65	1.10	1.75	1.03	1.85
29	1.34	1.48	1.27	1.56	1.20	1.65	1.12	1.74	1.05	1.84
30	1.35	1.49	1.28	1.57	1.21	1.65	1.14	1.74	1.07	1.83
31	1.36	1.50	1.30	1.57	1.23	1.65	1.16	1.74	1.09	1.83
32	1.37	1.50	1.31	1.57	1.24	1.65	1.18	1.73	1.11	1.82
33	1.38	1.51	1.32	1.58	1.26	1.65	1.19	1.73	1.13	1.81
34	1.39	1.51	1.33	1.58	1.27	1.65	1.21	1.73	1.15	1.81
35	1.40	1.52	1.34	1.58	1.28	1.65	1.22	1.73	1.16	1.80
36	1.41	1.52	1.35	1.59	1.29	1.65	1.24	1.73	1.18	1.80
37	1.42	1.53	1.36	1.59	1.31	1.66	1.25	1.72	1.19	1.80
38	1.43	1.54	1.37	1.59	1.32	1.66	1.26	1.72	1.21	1.79
39	1.43	1.54	1.38	1.60	1.33	1.66	1.27	1.72	1.22	1.79
40	1.44	1.54	1.39	1.60	1.34	1.66	1.29	1.72	1.23	1.79
45	1.48	1.57	1.43	1.62	1.38	1.67	1.34	1.72	1.29	1.78
50	1.50	1.59	1.46	1.63	1.42	1.67	1.38	1.72	1.34	1.77
55	1.53	1.60	1.49	1.64	1.45	1.68	1.41	1.72	1.38	1.77
60	1.55	1.62	1.51	1.65	1.48	1.69	1.44	1.73	1.41	1.77
65	1.57	1.63	1.54	1.66	1.50	1.70	1.47	1.73	1.44	1.77
70	1.58	1.64	1.55	1.67	1.52	1.70	1.49	1.74	1.46	1.77
75	1.60	1.65	1.57	1.68	1.54	1.71	1.51	1.74	1.49	1.77
80	1.61	1.66	1.59	1.69	1.56	1.72	1.53	1.74	1.51	1.77
85	1.62	1.67	1.60	1.70	1.57	1.72	1.55	1.75	1.52	1.77
90	1.63	1.68	1.61	1.70	1.59	1.73	1.57	1.75	1.54	1.78
95	1.64	1.69	1.62	1.71	1.60	1.73	1.58	1.75	1.56	1.78
100	1.65	1.69	1.63	1.72	1.61	1.74	1.59	1.76	1.57	1.78

TABLE B.18 LOGARITHMS (BASE 10 AND BASE *e*)

BASE 10

N	0	1	2	3	4	5	6	7	8	9	1	2	3	4	5	6	7	8	9
											*4	8	12	17	21	25	29	33	37
10	0000	0043	0086	0128	0170	0212	0253	0294	0334	0374									
11	0414	0453	0492	0531	0569	0607	0645	0682	0719	0755	4	8	11	15	19	23	26	30	34
12	0792	0828	0864	0899	0934	0969	1004	1038	1072	1106	3	7	10	14	17	21	24	28	31
13	1139	1173	1206	1239	1271	1303	1335	1367	1399	1430	3	6	10	13	16	19	23	26	29
14	1461	1492	1523	1553	1584	1614	1644	1673	1703	1732	3	6	9	12	15	18	21	24	27
15	1761	1790	1818	1817	1875	1903	1931	1959	1987	2014	*3	6	8	11	14	17	20	22	25
16	2041	2068	2095	2122	2148	2175	2201	2227	2253	2279	3	5	8	11	13	16	18	21	24
17	2304	2330	2355	2380	2405	2430	2455	2480	2504	2529	2	5	7	10	12	15	17	20	22
18	2553	2577	2601	2625	2648	2672	2695	2718	2742	2765	2	5	7	9	12	14	16	19	21
19	2788	2810	2833	2856	2878	2900	2923	2945	2967	2989	2	4	7	9	11	13	16	18	20
20	3010	3032	3054	3075	3096	3118	3139	3160	3181	3201	2	4	6	8	11	13	15	17	19
21	3222	3243	3263	3284	3304	3324	3345	3365	3385	3404	2	4	6	8	10	12	14	16	18
22	3424	3444	3464	3483	3502	3522	3541	3560	3579	3598	2	4	6	8	10	12	14	15	17
23	3617	3636	3655	3674	3692	3711	3729	3747	3766	3784	2	4	6	7	9	11	13	15	17
24	3802	3820	3838	3856	3874	3892	3909	3927	3945	3962	2	4	5	7	9	11	12	14	16
25	3979	3997	4014	4031	4048	4065	4082	4099	4116	4133	2	3	5	7	9	10	12	14	15
26	4150	4166	4183	4200	4216	4232	4249	4265	4281	4298	2	3	5	7	8	10	11	13	15
27	4314	4330	4346	4362	4378	4393	4409	4425	4440	4456	2	3	5	6	8	9	11	13	14
28	4472	4487	4502	4518	4533	4548	4564	4579	4594	4609	2	3	5	6	8	9	11	12	14
29	4624	4639	4654	4669	4683	4698	4713	4728	4742	4757	1	3	4	6	7	9	10	12	13
N	0	1	2	3	4	5	6	7	8	9	1	2	3	4	5	6	7	8	9

Proportional Parts

*Interpolation in this section of the table is inaccurate.

(continued)

TABLE B.18   LOGARITHMS (BASE 10 AND BASE e) (continued)

BASE 10

N	0	1	2	3	4	5	6	7	8	9	Proportional Parts 1	2	3	4	5	6	7	8	9
30	4771	4786	4800	4814	4829	4843	4857	4871	4886	4900	1	3	4	6	7	9	10	11	13
31	4914	4928	4942	4955	4960	4983	4997	5011	5024	5038	1	3	4	6	7	8	10	11	12
32	5051	5065	5079	5092	5105	5119	5132	5145	5159	5172	1	3	4	5	7	8	9	11	12
33	5185	5198	5211	5224	5237	5250	5263	5276	5289	5302	1	3	4	5	6	8	9	10	12
34	5315	5328	5340	5353	5366	5378	5391	5403	5416	5428	1	3	4	5	6	8	9	10	11
35	5441	5453	5465	5478	5490	5502	5514	5527	5539	5551	1	2	4	5	6	7	9	10	11
36	5563	5575	5587	5599	5611	5623	5635	5647	5658	5670	1	2	4	5	6	7	8	10	11
37	5682	5694	5705	5717	5729	5740	5752	5763	5775	5786	1	2	3	5	6	7	8	9	10
38	5798	5809	5821	5832	5843	5855	5866	5877	5888	5899	1	2	3	5	6	7	8	9	10
39	5911	5922	5933	5944	5955	5966	5977	5988	5999	6010	1	2	3	4	5	7	8	9	10
40	6021	6031	6042	6053	6064	6075	6085	6096	6107	6117	1	2	3	4	5	6	8	9	10
41	6128	6138	6149	6160	6170	6180	6191	6201	6212	6222	1	2	3	4	5	6	7	8	9
42	6232	6243	6253	6263	6274	6284	6294	6304	6314	6325	1	2	3	4	5	6	7	8	9
43	6335	6345	6355	6365	6375	6385	6395	6405	6415	6425	1	2	3	4	5	6	7	8	9
44	6435	6444	6454	6464	6474	6484	6493	6503	6513	6522	1	2	3	4	5	6	7	8	9
45	6532	6542	6551	6561	6571	6580	6590	6599	6609	6618	1	2	3	4	5	6	7	8	9
46	6628	6637	6646	6656	6665	6675	6684	6693	6702	6712	1	2	3	4	5	6	7	7	8
47	6721	6730	6739	6749	6758	6767	6776	6785	6794	6803	1	2	3	4	5	5	6	7	8
48	6812	6821	6830	6839	6848	6857	6866	6875	6884	6893	1	2	3	4	5	5	6	7	8
49	6902	6911	6920	6928	6937	6946	6955	6964	6972	6981	1	2	3	4	4	5	6	7	8
50	6990	6998	7007	7016	7024	7033	7042	7050	7059	7067	1	2	3	3	4	5	6	7	8
51	7076	7084	7093	7101	7110	7118	7126	7135	7143	7152	1	2	3	3	4	5	6	7	8
52	7160	7168	7177	7185	7193	7202	7210	7218	7226	7235	1	2	2	3	4	5	6	7	7
53	7243	7251	7259	7267	7275	7284	7292	7300	7308	7316	1	2	2	3	4	5	6	6	7
54	7324	7332	7340	7348	7356	7364	7372	7380	7388	7396	1	2	2	3	4	5	6	6	7

N	0	1	2	3	4	5	6	7	8	9	1	2	3	4	5	6	7	8	9
55	7404	7412	7419	7427	7435	7443	7451	7459	7466	7474	1	2	2	3	4	5	5	6	7
56	7482	7490	7497	7505	7513	7520	7528	7536	7543	7551	1	2	2	3	4	5	5	6	7
57	7559	7566	7574	7582	7589	7597	7604	7612	7619	7627	1	2	2	3	4	5	5	6	7
58	7634	7642	7649	7657	7664	7672	7679	7686	7694	7701	1	1	2	3	4	4	5	6	7
59	7709	7716	7723	7731	7738	7745	7752	7760	7767	7774	1	1	2	3	4	4	5	6	7
60	7782	7789	7796	7803	7810	7818	7825	7832	7839	7846	1	1	2	3	4	4	5	6	6
61	7853	7860	7868	7875	7882	7889	7896	7903	7910	7917	1	1	2	3	4	4	5	6	6
62	7924	7931	7938	7945	7952	7959	7966	7973	7980	7987	1	1	2	3	3	4	5	6	6
63	7993	8000	8007	8014	8021	8028	8035	8041	8048	8055	1	1	2	3	3	4	5	5	6
64	8062	8069	8075	8082	8089	8096	8102	8109	8116	8122	1	1	2	3	3	4	5	5	6
65	8129	8136	8142	8149	8156	8162	8169	8176	8182	8189	1	1	2	3	3	4	5	5	6
66	8195	8202	8209	8215	8222	8228	8235	8241	8248	8254	1	1	2	3	3	4	5	5	6
67	8261	8267	8274	8280	8287	8293	8299	8306	8312	8319	1	1	2	3	3	4	5	5	6
68	8325	8331	8338	8344	8351	8357	8363	8370	8376	8382	1	1	2	3	3	4	4	5	6
69	8388	8395	8401	8407	8414	8420	8426	8432	8439	8445	1	1	2	2	3	4	4	5	6
70	8451	8457	8463	8470	8476	8482	8488	8494	8500	8506	1	1	2	2	3	4	4	5	6
71	8513	8519	8525	8531	8537	8543	8549	8555	8561	8567	1	1	2	2	3	4	4	5	5
72	8573	8579	8585	8591	8597	8603	8609	8615	8621	8627	1	1	2	2	3	4	4	5	6
73	8633	8639	8645	8651	8657	8663	8669	8675	8681	8686	1	1	2	2	3	4	4	5	5
74	8692	8698	8704	8710	8716	8722	8727	8733	8739	8745	1	1	2	2	3	4	4	5	5
75	8751	8756	8762	8768	8774	8779	8785	8791	8797	8802	1	1	2	2	3	3	4	5	5
76	8808	8814	8820	8825	8831	8837	8842	8848	8854	8859	1	1	2	2	3	3	4	5	5
77	8865	8871	8876	8882	8887	8893	8899	8904	8910	8915	1	1	2	2	3	3	4	4	5
78	8921	8927	8932	8938	8943	8949	8954	8960	8965	8971	1	1	2	2	3	3	4	4	5
79	8976	8982	8987	8993	8998	9004	9009	9015	9020	9025	1	1	2	2	3	3	4	4	5

(continued)

TABLE B.18   LOGARITHMS (BASE 10 AND BASE e) *(continued)*

BASE 10

N	0	1	2	3	4	5	6	7	8	9		1	2	3	4	5	6	7	8	9
											Proportional Parts									
80	9031	9036	9042	9047	9053	9058	9063	9069	9074	9079		1	1	2	2	3	3	4	4	5
81	9085	9090	9096	9101	9106	9112	9117	9122	9128	9133		1	1	2	2	3	3	4	4	5
82	9138	9143	9149	9154	9159	9165	9170	9175	9180	9186		1	1	2	2	3	3	4	4	5
83	9191	9196	9201	9206	9212	9217	9222	9227	9232	9238		1	1	2	2	3	3	4	4	5
84	9243	9248	9253	9258	9263	9269	9274	9279	9284	9289		1	1	2	2	3	3	4	4	5
85	9294	9299	9304	9309	9315	9320	9325	9330	9335	9340		1	1	2	2	3	3	4	4	5
86	9345	9350	9355	9360	9365	9370	9375	9380	9385	9390		1	1	2	2	3	3	4	4	5
87	9395	9400	9405	9410	9415	9420	9425	9430	9435	9440		0	1	1	2	2	3	3	4	4
88	9445	9450	9455	9460	9465	9469	9474	9479	9484	9489		0	1	1	2	2	3	3	4	4
89	9494	9499	9504	9509	9513	9518	9523	9528	9533	9538		0	1	1	2	2	3	3	4	4
90	9542	9547	9552	9557	9562	9566	9571	9576	9581	9586		0	1	1	2	2	3	3	4	4
91	9590	9595	9600	9605	9609	9614	9619	9624	9628	9633		0	1	1	2	2	3	3	4	4
92	9638	9643	9647	9652	9657	9661	9666	9671	9675	9680		0	1	1	2	2	3	3	4	4
93	9685	9689	9694	9699	9703	9708	9713	9717	9722	9727		0	1	1	2	2	3	3	4	4
94	9731	9736	9741	9745	9750	9754	9759	9763	9768	9773		0	1	1	2	2	3	3	4	4
95	9777	9782	9786	9791	9795	9800	9805	9809	9814	9818		0	1	1	2	2	3	3	4	4
96	9823	9827	9832	9836	9841	9845	9850	9854	9859	9863		0	1	1	2	2	3	3	4	4
97	9868	9872	9877	9881	9886	9890	9894	9899	9903	9908		0	1	1	2	2	3	3	4	4
98	9912	9917	9921	9926	9930	9934	9939	9943	9948	9952		0	1	1	2	2	3	3	3	4
99	9956	9961	9965	9969	9974	9978	9983	9987	9991	9996		0	1	1	2	2	3	3	3	4
N	0	1	2	3	4	5	6	7	8	9		1	2	3	4	5	6	7	8	9

NATURAL OR NAPERIAN LOGARITHMS

To find the natural logarithm of a number which is $\frac{1}{10}$, $\frac{1}{100}$, $\frac{1}{1000}$, etc. of a number whose logarithm is given, subtract from the given logarithm $\log_e 10$, $2 \log_e 10$, $3 \log_e 10$, etc.

To find the natural logarithm of a number which is 10, 100, 1000, etc. times a number whose logarithm is given, add to the given logarithm $\log_e 10$, $2 \log_e 10$, $3 \log_e 10$, etc.

$\log_e 10 = 2.30258\ 50930$	$6 \log_e 10 = 13.81551\ 05580$
$2 \log_e 10 = 4.60517\ 01860$	$7 \log_e 10 = 16.11809\ 56510$
$3 \log_e 10 = 6.90775\ 52790$	$8 \log_e 10 = 18.42068\ 07440$
$4 \log_e 10 = 9.21034\ 03720$	$9 \log_e 10 = 20.72326\ 58369$
$5 \log_e 10 = 11.51292\ 54650$	$10 \log_e 10 = 23.02585\ 09299$

N	0	1	2	3	4	5	6	7	8	9
1.0	0.00000	.00995	.01980	.02958	.03922	.04879	.05827	.06766	.07696	.08618
.1	.09531	.10436	.11333	.12222	.13103	.13976	.14842	.15700	.16551	.17395
.2	.18232	.19062	.19885	.20701	.21511	.22314	.23111	.23902	.24686	.25464
.3	.26236	.27003	.27763	.28518	.29267	.30010	.30748	.31481	.32208	.32930
.4	.33647	.34359	.35066	.35767	.36464	.37156	.37844	.38526	.39204	.39878
.5	.40547	.41211	.41871	.42527	.43178	.43825	.44469	.45108	.45742	.46373
.6	.47000	.47623	.48243	.48858	.49470	.50078	.50682	.51282	.51879	.52473
.7	.53063	.53649	.54232	.54812	.55389	.55962	.56531	.57098	.57661	.58222
.8	.58779	.59333	.59884	.60432	.60977	.61519	.62058	.62594	.63127	.63658
.9	.64185	.64710	.65233	.65752	.66269	.66783	.67294	.67803	.68310	.68813
2.0	0.69315	.69813	.70310	.70804	.71295	.71784	.72271	.72755	.73237	.73716
.1	.74194	.74669	.75142	.75612	.76081	.76547	.77011	.77473	.77932	.78390
.2	.78846	.79299	.79751	.80200	.80648	.81093	.81536	.81978	.82418	.82855
.3	.83291	.83725	.84157	.84587	.85015	.85442	.85866	.86289	.86710	.87129
.4	.87547	.87963	.88377	.88789	.89200	.89609	.90016	.90422	.90826	.91228
.5	.91629	.92028	.92426	.92822	.93216	.93609	.94001	.94391	.94779	.95166
.6	.95551	.95935	.96317	.96698	.97078	.97456	.97833	.98208	.98582	.98954
.7	.99325	.99695	.00063	.00430	.00796	.01160	.01523	.01885	.02245	.02604
.8	1.02962	.03318	.03674	.04028	.04380	.04732	.05082	.05431	.05779	.06126
.9	.06471	.06815	.07158	.07500	.07841	.08181	.08519	.08856	.09192	.09527
3.0	1.09861	.10194	.10526	.10856	.11186	.11514	.11841	.12168	.12493	.12817
.1	.13140	.13462	.13783	.14103	.14422	.14740	.15057	.15373	.15688	.16002
.2	.16315	.16627	.16938	.17248	.17557	.17865	.18173	.18479	.18784	.19089
.3	.19392	.19695	.19996	.20297	.20597	.20896	.21194	.21491	.21788	.22083
.4	.22378	.22671	.22964	.23256	.23547	.23837	.24127	.24415	.24703	.24990
.5	.25276	.25562	.25846	.26130	.26413	.26695	.26976	.27257	.27536	.27815
.6	.28093	.28371	.28647	.28923	.29198	.29473	.29746	.30019	.30291	.30563
.7	.30833	.31103	.31372	.31641	.31909	.32176	.32442	.32708	.32972	.33237
.8	.33500	.33763	.34025	.34286	.34547	.34807	.35067	.35325	.35584	.35841
.9	.36098	.36354	.36609	.36864	.37118	.37372	.37624	.37877	.38128	.38379
4.0	1.38629	.38879	.39128	.39377	.39624	.39872	.40118	.40364	.40610	.40854
.1	.41099	.41342	.41585	.41828	.42070	.42311	.42552	.42792	.43031	.43270
.2	.43508	.43746	.43984	.44220	.44456	.44692	.44927	.45161	.45395	.45629
.3	.45862	.46094	.46326	.46557	.46787	.47018	.47247	.47476	.47705	.47933
.4	.48160	.48387	.48614	.48840	.49065	.49290	.49515	.49739	.49962	.50185

N	0	1	2	3	4	5	6	7	8	9
.5	.50408	.50630	.50851	.51072	.51293	.51513	.51732	.51951	.52170	.52388
.6	.52606	.52823	.53039	.53256	.53471	.53687	.53902	.54116	.54330	.54543
.7	.54756	.54969	.55181	.55393	.55604	.55814	.56025	.56235	.56444	.56653
.8	.56862	.57070	.57277	.57485	.57691	.57898	.58104	.58309	.58515	.58719
.9	.58924	.59127	.59331	.59534	.59737	.59939	.60141	.60342	.60543	.60744
5.0	1.60944	.61144	.61343	.61542	.61741	.61939	.62137	.62334	.62531	.62728
.1	.62924	.63120	.63315	.63511	.63705	.63900	.64094	.64287	.64481	.64673
.2	.64866	.65058	.65250	.65441	.65632	.65823	.66013	.66203	.66393	.66582
.3	.66771	.66959	.67147	.67335	.67523	.67710	.67896	.68083	.68269	.68455
.4	.68640	.68825	.69010	.69194	.69378	.69562	.69745	.69928	.70111	.70293
.5	.70475	.70656	.70838	.71019	.71199	.71380	.71560	.71740	.71919	.72098
.6	.72277	.72455	.72633	.72811	.72988	.73166	.73342	.73519	.73695	.73871
.7	.74047	.74222	.74397	.74572	.74746	.74920	.75094	.75267	.75440	.75613
.8	.75786	.75958	.76130	.76302	.76473	.76644	.76815	.76985	.77156	.77326
.9	.77495	.77665	.77834	.78002	.78171	.78339	.78507	.78675	.78842	.79009
6.0	1.79176	.79342	.79509	.79675	.79840	.80006	.80171	.80336	.80500	.80665
.1	.80829	.80993	.81156	.81319	.81482	.81645	.81808	.81970	.82132	.82294
.2	.82455	.82616	.82777	.82938	.83098	.83258	.83418	.83578	.83737	.83896
.3	.84055	.84214	.84372	.84530	.84688	.84845	.85003	.85160	.85317	.85473
.4	.85630	.85786	.85942	.86097	.86253	.86408	.86563	.86718	.86872	.87026
.5	.87180	.87334	.87487	.87641	.87794	.87947	.88099	.88251	.88403	.88555
.6	.88707	.88858	.89010	.89160	.89311	.89462	.89612	.89762	.89912	.90061
.7	.90211	.90360	.90509	.90658	.90806	.90954	.91102	.91250	.91398	.91545
.8	.91692	.91839	.91986	.92132	.92279	.92425	.92571	.92716	.92862	.93007
.9	.93152	.93297	.93442	.93586	.93730	.93874	.94018	.94162	.94305	.94448
7.0	1.94591	.94734	.94876	.95019	.95161	.95303	.95445	.95586	.95727	.95869
.1	.96009	.96150	.96291	.96431	.96571	.96711	.96851	.96991	.97130	.97269
.2	.97408	.97547	.97685	.97824	.97962	.98100	.98238	.98376	.98513	.98650
.3	.98787	.98924	.99061	.99198	.99334	.99470	.99606	.99742	.99877	.00013
.4	2.00148	.00283	.00418	.00553	.00687	.00821	.00956	.01089	.01223	.01357
.5	.01490	.01624	.01757	.01890	.02022	.02155	.02287	.02419	.02551	.02683
.6	.02815	.02946	.03078	.03209	.03340	.03471	.03601	.03732	.03862	.03992
.7	.04122	.04252	.04381	.04511	.04640	.04769	.04898	.05027	.05156	.05284
.8	.05412	.05540	.05668	.05796	.05924	.06051	.06179	.06306	.06433	.06560
.9	0.6686	.06813	.06939	.07065	.07191	.07317	.07443	.07568	.07694	.07819
8.0	2.07944	.08069	.08194	.08318	.08443	.08567	.08691	.08815	.08939	.09063
.1	.09186	.09310	.09433	.09556	.09679	.09802	.09924	.10047	.10169	.10291
.2	.10413	.10535	.10657	.10779	.10900	.11021	.11142	.11263	.11384	.11505
.3	.11626	.11746	.11866	.11986	.12106	.12226	.12346	.12465	.12585	.12704
.4	.12823	.12942	.13061	.13180	.13298	.13417	.13535	.13653	.13771	.13889
.5	.14007	.14124	.14242	.14359	.14476	.14593	.14710	.14827	.14943	.15060
.6	.15176	.15292	.15409	.15524	.15640	.15756	.15871	.15987	.16102	.16217
.7	.16332	.16447	.16562	.16677	.16791	.16905	.17020	.17134	.17248	.17361
.8	.17475	.17589	.17702	.17816	.17929	.18042	.18155	.18267	.18380	.18493
.9	.18605	.18717	.18830	.18942	.19054	.19165	.19277	.19389	.19500	.19611
9.0	2.19722	.19834	.19944	.20055	.20166	.20276	.20387	.20497	.20607	.20717
.1	.20827	.20937	.21047	.21157	.21266	.21375	.21485	.21594	.21703	.21812
.2	.21920	.22029	.22138	.22246	.22354	.22462	.22570	.22678	.22786	.22894
.3	.23001	.23109	.23216	.23324	.23431	.23538	.23645	.23751	.23858	.23965
.4	.24071	.24177	.24284	.24390	.24496	.24601	.24707	.24813	.24918	.25024
.5	.25129	.25234	.25339	.25444	.25549	.25654	.25759	.25863	.25968	.26072
.6	.26176	.26280	.26384	.26488	.26592	.26696	.26799	.26903	.27006	.27109
.7	.27213	.27316	.27419	.27521	.27624	.27727	.27829	.27932	.28034	.28136
.8	.28238	.28340	.28442	.28544	.28646	.28747	.28849	.28950	.29051	.29152
.9	.29253	.29354	.29455	.29556	.29657	.29757	.29858	.29958	.30058	.30158

TABLE B.19   CRITICAL AND QUASI-CRITICAL LOWER-TAIL VALUES OF W_+ (AND THEIR PROBABILITY LEVELS) FOR WILCOXON'S SIGNED-RANK TEST

W'_+ FOLLOWED BY $P(W_+ \leq W'_+)$

n	$\alpha = .05$		$\alpha = .025$		$\alpha = .01$		$\alpha = .005$	
5	0	.0313						
	1	.0625						
6	2	.0469	0	.0156				
	3	.0781	1	.0313				
7	3	.0391	2	.0234	0	.0078		
	4	.0547	3	.0391	1	.0156		
8	5	.0391	3	.0195	1	.0078	0	.0039
	6	.0547	4	.0273	2	.0117	1	.0078
9	8	.0488	5	.0195	3	.0098	1	.0039
	9	.0645	6	.0273	4	.0137	2	.0059
10	10	.0420	8	.0244	5	.0098	3	.0049
	11	.0527	9	.0322	6	.0137	4	.0068
11	13	.0415	10	.0210	7	.0093	5	.0049
	14	.0508	11	.0269	8	.0122	6	.0068
12	17	.0461	13	.0212	9	.0081	7	.0046
	18	.0549	14	.0261	10	.0105	8	.0061
13	21	.0471	17	.0239	12	.0085	9	.0040
	22	.0549	18	.0287	13	.0107	10	.0052
14	25	.0453	21	.0247	15	.0083	12	.0043
	26	.0520	22	.0290	16	.0101	13	.0054
15	30	.0473	25	.0240	19	.0090	15	.0042
	31	.0535	26	.0277	20	.0108	16	.0051
16	35	.0467	29	.0222	23	.0091	19	.0046
	36	.0523	30	.0253	24	.0107	20	.0055
17	41	.0492	34	.0224	27	.0087	23	.0047
	42	.0544	35	.0253	28	.0101	24	.0055
18	47	.0494	40	.0241	32	.0091	27	.0045
	48	.0542	41	.0269	33	.0104	28	.0052
19	53	.0478	46	.0247	37	.0090	32	.0047
	54	.0521	47	.0273	38	.0102	33	.0054
20	60	.0487	52	.0242	43	.0096	37	.0047
	61	.0527	53	.0266	44	.0107	38	.0053
21	67	.0479	58	.0230	49	.0097	42	.0045
	68	.0516	59	.0251	50	.0108	43	.0051
22	75	.0492	65	.0231	55	.0095	48	.0046
	76	.0527	66	.0250	56	.0104	49	.0052

Source: Body of table is reproduced, with changes only in notation, from Table II in F. Wilcoxon, S. K. Katti, and Roberta A. Wilcox, *Critical Values and Probability Levels for the Wilcoxon Rank Sum Test and the Wilcoxon Signed Rank Test*, American Cyanamid Company (Lederle Laboratories Division, Pearl River, N. Y.) and The Florida State University (Department of Statistics, Tallahassee, Fla.), August 1963, with permission of the authors and publishers.

TABLE B.19   CRITICAL AND QUASI-CRITICAL
LOWER-TAIL VALUES OF W_+ (AND THEIR PROBABILITY
LEVELS) FOR WILCOXON'S SIGNED-RANK TEST (*continued*)

$$W'_+ \text{ FOLLOWED BY } P(W_+ \leq W'_+)$$

n	$\alpha = .05$		$\alpha = .025$		$\alpha = .01$		$\alpha = .005$	
23	83	.0490	73	.0242	62	.0098	54	.0046
	84	.0523	74	.0261	63	.0107	55	.0051
24	91	.0475	81	.0245	69	.0097	61	.0048
	92	.0505	82	.0263	70	.0106	62	.0053
25	100	.0479	89	.0241	76	.0094	68	.0048
	101	.0507	90	.0258	77	.0101	69	.0053
26	110	.0497	98	.0247	84	.0095	75	.0047
	111	.0524	99	.0263	85	.0102	76	.0051
27	119	.0477	107	.0246	92	.0093	83	.0048
	120	.0502	108	.0260	93	.0100	84	.0052
28	130	.0496	116	.0239	101	.0096	91	.0048
	131	.0521	117	.0252	102	.0102	92	.0051
29	140	.0482	126	.0240	110	.0095	100	.0049
	141	.0504	127	.0253	111	.0101	101	.0053
30	151	.0481	137	.0249	120	.0098	109	.0050
	152	.0502	138	.0261	121	.0104	110	.0053
31	163	.0491	147	.0239	130	.0099	118	.0049
	164	.0512	148	.0251	131	.0105	119	.0052
32	175	.0492	159	.0249	140	.0097	128	.0050
	176	.0512	160	.0260	141	.0103	129	.0053
33	187	.0485	170	.0242	151	.0099	138	.0049
	188	.0503	171	.0253	152	.0104	139	.0052
34	200	.0488	182	.0242	162	.0098	148	.0048
	201	.0506	183	.0252	163	.0103	149	.0051
35	213	.0484	195	.0247	173	.0096	159	.0048
	214	.0501	196	.0257	174	.0100	160	.0051
36	227	.0489	208	.0248	185	.0096	171	.0050
	228	.0505	209	.0258	186	.0100	172	.0052
37	241	.0487	221	.0245	198	.0099	182	.0048
	242	.0503	222	.0254	199	.0103	183	.0050
38	256	.0493	235	.0247	211	.0099	194	.0048
	257	.0509	236	.0256	212	.0104	195	.0050
39	271	.0493	249	.0246	224	.0099	207	.0049
	272	.0507	250	.0254	225	.0103	208	.0051
40	286	.0486	264	.0249	238	.0100	220	.0049
	287	.0500	265	.0257	239	.0104	221	.0051
41	302	.0488	279	.0248	252	.0100	233	.0048
	303	.0501	280	.0256	253	.0103	234	.0050

TABLE B.19   CRITICAL AND QUASI-CRITICAL
LOWER-TAIL VALUES OF W_+ (AND THEIR PROBABILITY
LEVELS) FOR WILCOXON'S SIGNED-RANK TEST (*continued*)

	W'_+ FOLLOWED BY $P(W_+ \leq W'_+)$							
n	$\alpha = .05$		$\alpha = .025$		$\alpha = .01$		$\alpha = .005$	
42	319	.0496	294	.0245	266	.0098	247	.0049
	320	.0509	295	.0252	267	.0102	248	.0051
43	336	.0498	310	.0245	281	.0098	261	.0048
	337	.0511	311	.0252	282	.0102	262	.0050
44	353	.0495	327	.0250	296	.0097	276	.0049
	354	.0507	328	.0257	297	.0101	277	.0051
45	371	.0498	343	.0244	312	.0098	291	.0049
	372	.0510	344	.0251	313	.0101	292	.0051
46	389	.0497	361	.0249	328	.0098	307	.0050
	390	.0508	362	.0256	329	.0101	308	.0052
47	407	.0490	378	.0245	345	.0099	322	.0048
	408	.0501	379	.0251	346	.0102	323	.0050
48	426	.0490	396	.0244	362	.0099	339	.0050
	427	.0500	397	.0251	363	.0102	340	.0051
49	446	.0495	415	.0247	379	.0098	355	.0049
	447	.0505	416	.0253	380	.0100	356	.0050
50	466	.0495	434	.0247	397	.0098	373	.0050
	467	.0506	435	.0253	398	.0101	374	.0051

TABLE B.20 CRITICAL LOWER-TAIL VALUES
OF b FOR FISHER'S EXACT TEST FOR 2 × 2 TABLES,
FOLLOWED BY EXACT PROBABILITY LEVEL FOR TEST

Largest value of b, given *fixed* values of A, B, and a, for the ratio a/A to be just significantly larger than b/B in the following 2 × 2 table

a	$A - a$	A	
b	$B - b$	B	$A \geq B$
$a + b$	$(A + B) - (a + b)$	$A + B$	$a/A \geq b/B$

Thus largest value of b' for which

$$P(b \leq b' \,|\, A, B, a) = P\left\{ \left(\frac{a}{A} - \frac{b}{B} \right) \geq \left(\frac{a}{A} - \frac{b'}{B} \right) \right\} \leq \alpha$$

followed by $P(b \leq b' \,|\, A, B, a + b')$

			PROBABILITY			
		a	**0.05**	**0.025**	**0.01**	**0.005**
$A = 3$	$B = 3$	**3**	0.050	—	—	—
$A = 4$	$B = 4$	**4**	0.014	0.014	—	—
		3	0.029	—	—	—
$A = 5$	$B \times 5$	**5**	1.024	1.024	0.004	0.004
		4	0.024	0.024	—	—
		4	1.048	0.008	0.008	—
		5	0.040	—	—	—
		3	0.018	0.018	—	—
		5				
		2	0.048	—	—	—
		5				
$A = 6$	$B = 6$	**6**	2.030	1.008	1.008	0.001
		5	1.040	0.008	0.008	—
		4	0.030	—	—	—
		5	1.015$^+$	1.015$^+$	0.002	0.002
		6				
		5	0.013	0.013	—	—
		4	0.045$^+$	—	—	—
		4	1.033	0.005$^-$	0.005$^-$	0.005$^-$
		6				
		5	0.024	0.024	—	—
		3	0.012	0.012	—	—
		6				
		5	0.048	—	—	—
		2	0.036	—	—	—
		6				

Source: Body of table is an abridgement of D. J. Finney, R. Latscha, B. M. Bennett, and P. Hsu (Introduction by E. S. Pearson), *Tables for Testing Significance in a 2 × 2 Contingency Table*, London and New York: Cambridge University Press, 1963, with permission of the authors and publishers.

The table shows: (1) *in bold type*, for given A, B, and a, the value of b ($< a$) which is just significant at the probability level quoted (single-tail test): (2) *in small type*, for given A, B, and $a + b$, the exact probability (if there is independence) that b is equal to or less than the integer shown in bold type.

TABLE B.20   CRITICAL LOWER-TAIL VALUES
OF b FOR FISHER'S EXACT TEST FOR 2 × 2 TABLES,
FOLLOWED BY EXACT PROBABILITY LEVEL FOR TEST (*continued*)

		a	PROBABILITY			
			0.05	**0.025**	**0.01**	**0.005**
$A = 7$	$B = 7$	7	3.035^-	2.010^+	1.002	1.002
		6	1.015^-	1.015^-	0.002	0.002
		5	0.010^+	0.010^+	—	—
		4	0.035^-	—	—	—
	6	7	2.021	2.021	1.005^-	1.005^-
		6	1.025^+	0.004	0.004	0.004
		5	0.016	0.016	—	—
		4	0.049	—	—	—
	5	7	2.045^+	1.010^+	0.001	0.001
		6	1.045^+	0.008	0.008	—
		5	0.027	—	—	—
	4	7	1.024	1.024	0.003	0.003
		6	0.015^+	0.015^+	—	—
		5	0.045^+	—	—	—
	3	7	0.008	0.008	0.008	—
		6	0.033	—	—	—
	2	7	0.028	—	—	—
$A = 8$	$B = 8$	8	4.038	3.013	2.003	2.003
		7	2.020	2.020	1.005^+	0.001
		6	1.020	1.020	0.003	0.003
		5	0.013	0.013	—	—
		4	0.038	—	—	—
	7	8	3.026	2.007	2.007	1.001
		7	2.035^-	1.009	1.009	0.001
		6	1.032	0.006	0.006	—
		5	0.019	0.019	—	—
	6	8	2.015^-	2.015^-	1.003	1.003
		7	1.016	1.016	0.002	0.002
		6	0.009	0.009	0.009	—
		5	0.028	—	—	—
	5	8	2.035^-	1.007	1.007	0.001
		7	1.032	0.005^-	0.005^-	0.005^-
		6	0.016	0.016	—	—
		5	0.044	—	—	—
	4	8	1.018	1.018	0.002	0.002
		7	0.010^+	0.010^+	—	—
		6	0.030	—	—	—
	3	8	0.006	0.006	0.006	—
		7	0.024	0.024	—	—
	2	8	0.022	0.022	—	—

TABLE B.20   CRITICAL LOWER-TAIL VALUES
OF b FOR FISHER'S EXACT TEST FOR 2 × 2 TABLES,
FOLLOWED BY EXACT PROBABILITY LEVEL FOR TEST

		a	PROBABILITY			
			0.05	**0.025**	**0.01**	**0.005**
$A = 9$	$B = 9$	9	5.041	4.015$^-$	3.005$^-$	3.005$^-$
		8	3.025$^-$	3.025$^-$	2.008	1.002
		7	2.028	1.008	1.008	0.001
		6	1.025$^-$	1.025$^-$	0.005$^-$	0.005$^-$
		5	0.015$^-$	0.015$^-$	—	—
		4	0.041	—	—	—
	8	9	4.029	3.009	3.009	2.002
		8	3.043	2.013	1.003	1.003
		7	2.044	1.012	0.002	0.002
		6	1.036	0.007	0.007	—
		5	0.020	0.020	—	—
	7	9	3.019	3.019	2.005$^-$	2.005$^-$
		8	2.024	2.024	1.006	0.001
		7	1.020	1.020	0.003	0.003
		6	0.010$^+$	0.010$^+$	—	—
		5	0.029	—	—	—
	6	9	3.044	2.011	1.002	1.002
		8	2.047	1.011	0.001	0.001
		7	1.035$^-$	0.006	0.006	—
		6	0.017	0.017	—	—
		5	0.042	—	—	—
	5	9	2.027	1.005$^-$	1.005$^-$	1.005$^-$
		8	1.023	1.023	0.003	0.003
		7	0.010$^+$	0.010$^+$	—	—
		6	0.028	—	—	—
	4	9	1.014	1.014	0.001	0.001
		8	0.007	0.007	0.007	—
		7	0.021	0.021	—	—
		6	0.049	—	—	—
	3	9	1.045$^+$	0.005$^-$	0.005$^-$	0.005$^-$
		8	0.018	0.018	—	—
		7	0.045$^+$	—	—	—
	2	9	0.018	0.018	—	—
$A = 10$	$B = 10$	10	6.043	5.016	4.005$^+$	3.002
		9	4.029	3.010$^-$	3.010$^-$	2.003
		8	3.035$^-$	2.012	1.003	1.003
		7	2.035$^-$	1.010$^-$	1.010$^-$	0.002
		6	1.029	0.005$^+$	0.005$^+$	—
		5	0.016	0.016	—	—
		4	0.043	—	—	—
	9	10	5.033	4.011	3.003	3.003
		9	4.050$^-$	3.017	2.005$^-$	2.005$^-$
		8	2.019	2.019	1.004	1.004
		7	1.015$^-$	1.015$^-$	0.002	0.002
		6	1.040	0.008	0.008	—
		5	0.022	0.022	—	—

TABLE B.20 CRITICAL LOWER-TAIL VALUES
OF *b* FOR FISHER'S EXACT TEST FOR 2 × 2 TABLES,
FOLLOWED BY EXACT PROBABILITY LEVEL FOR TEST

		a	PROBABILITY 0.05	0.025	0.01	0.005
$A = 10$	$B = 8$	10	4.023	4.023	3.007	2.002
		9	3.032	2.009	2.009	1.002
		8	2.031	1.008	1.008	0.001
		7	1.023	1.023	0.004	0.004
		6	0.011	0.011	—	—
		5	0.029	—	—	—
	7	10	3.015$^-$	3.015$^-$	2.003	2.003
		9	2.018	2.018	1.004	1.004
		8	1.013	1.013	0.002	0.002
		7	1.036	0.006	0.006	—
		6	0.017	0.017	—	—
		5	0.041	—	—	—
	6	10	3.036	2.008	2.008	1.001
		9	2.036	1.008	1.008	0.001
		8	1.024	1.024	0.003	0.003
		7	0.010$^+$	0.010$^+$	—	—
		6	0.026	—	—	—
	5	10	2.022	2.022	1.004	1.004
		9	1.017	1.017	0.002	0.002
		8	1.047	0.007	0.007	—
		7	0.019	0.019	—	—
		6	0.042	—	—	—
	4	10	1.011	1.011	0.001	0.001
		9	1.041	0.005$^-$	0.005$^-$	0.005$^-$
		8	0.015$^-$	0.015$^-$	—	—
		7	0.035$^-$	—	—	—
	3	10	1.038	0.003	0.003	0.003
		9	0.014	0.014	—	—
		8	0.035$^-$	—	—	—
	2	10	0.015$^+$	0.015$^+$	—	—
		9	0.045$^+$	—	—	—

TABLE B.21 CRITICAL LOWER-TAIL
VALUES OF W_n FOR WILCOXON'S RANK-SUM TEST

Largest value of W'_n for which $P(W_n \leq W'_n) \leq \alpha$
where α is given in boldface

$n = 1$ 　　　　　　　　　　　　　　　　　　$n = 2$

m	.001	.005	.01	.025	.05	.10	$2\bar{W}$	.001	.005	.01	.025	.05	.10	$2\bar{W}$	m
2							4						—	10	2
3							5						3	12	3
4							6					—	3	14	4
5							7					3	4	16	5
6							8					3	4	18	6
7							9				—	3	4	20	7
8						—	10				3	4	5	22	8
9						1	11				3	4	5	24	9
10						1	12				3	4	6	26	10
11						1	13				3	4	6	28	11
12						1	14			—	4	5	7	30	12
13						1	15			3	4	5	7	32	13
14						1	16			3	4	6	8	34	14
15						1	17			3	4	6	8	36	15
16						1	18			3	4	6	8	38	16
17						1	19			3	5	6	9	40	17
18					—	1	20		—	4	5	7	9	42	18
19					1	2	21		3	4	5	7	10	44	19
20					1	2	22		3	4	5	7	10	46	20
21					1	2	23		3	4	6	8	11	48	21
22					1	2	24		3	4	6	8	11	50	22
23					1	2	25		3	4	6	8	12	52	23
24					1	2	26		3	4	6	9	12	54	24
25	—	—	—	—	1	2	27	—	3	4	6	9	12	56	25

$n = 3$ 　　　　　　　　　　　　　　　　　　$n = 4$

m	.001	.005	.01	.025	.05	.10	$2\bar{W}$	.001	.005	.01	.025	.05	.10	$2\bar{W}$	m
3					6	7	21								
4			—		6	7	24			—	10	11	13	36	4
5			6		7	8	27		—	10	11	12	14	40	5
6		—	7	8	9	30			10	11	12	13	15	44	6
7		6	7	8	10	33			10	11	13	14	16	48	7
8		—	6	8	9	11	36		11	12	14	15	17	52	8
9		6	7	8	10	11	39	—	11	13	14	16	19	56	9
10		6	7	9	10	12	42	10	12	13	15	17	20	60	10
11		6	7	9	11	13	45	10	12	14	16	18	21	64	11
12		7	8	10	11	14	48	10	13	15	17	19	22	68	12
13		7	8	10	12	15	51	11	13	15	18	20	23	72	13
14		7	8	11	13	16	54	11	14	16	19	21	25	76	14
15		8	9	11	13	16	57	11	15	17	20	22	26	80	15

Source: Body of table is reproduced, with changes only in notation, from Table 1 in L. R. Verdooren's "Extended Tables of Critical Values for Wilcoxon's Test Statistic," *Biometrika*, 50 (1963), 177–186, with permission of the author and editor.

TABLE B.21   CRITICAL LOWER-TAIL
VALUES OF W_n FOR WILCOXON'S RANK-SUM TEST (*continued*)

			$n = 3$								$n = 4$				
m	.001	.005	.01	.025	.05	.10	$2\bar{W}$	.001	.005	.01	.025	.05	.10	$2\bar{W}$	*m*
16	—	8	9	12	14	17	60	12	15	17	21	24	27	84	16
17	6	8	10	12	15	18	63	12	16	18	21	25	28	88	17
18	6	8	10	13	15	19	66	13	16	19	22	26	30	92	18
19	6	9	10	13	16	20	69	13	17	19	23	27	31	96	19
20	6	9	11	14	17	21	72	13	18	20	24	28	32	100	20
21	7	9	11	14	17	21	75	14	18	21	25	29	33	104	21
22	7	10	12	15	18	22	78	14	19	21	26	30	35	108	22
23	7	10	12	15	19	23	81	14	19	22	27	31	36	112	23
24	7	10	12	16	19	24	84	15	20	23	27	32	38	116	24
25	7	11	13	16	20	25	87	15	20	23	28	33	38	120	25

			$n = 5$								$n = 6$				
m	.001	.005	.01	.025	.05	.10	$2\bar{W}$	.001	.005	.01	.025	.05	.10	$2\bar{W}$	*m*
5		15	16	17	19	20	55								
6		16	17	18	20	22	60	—	23	24	26	28	30	78	6
7	—	16	18	20	21	23	65	21	24	25	27	29	32	84	7
8	15	17	19	21	23	25	70	22	25	27	29	31	34	90	8
9	16	18	20	22	24	27	75	23	26	28	31	33	36	96	9
10	16	19	21	23	26	28	80	24	27	29	32	35	38	102	10
11	17	20	22	24	27	30	85	25	28	30	34	37	40	108	11
12	17	21	23	26	28	32	90	25	30	32	35	38	42	114	12
13	18	22	24	27	30	33	95	26	31	33	37	40	44	120	13
14	18	22	25	28	31	35	100	27	32	34	38	42	46	126	14
15	19	23	26	29	33	37	105	28	33	36	40	44	48	132	15
16	20	24	27	30	34	38	110	29	34	37	42	46	50	138	16
17	20	25	28	32	35	40	115	30	36	39	43	47	52	144	17
18	21	26	29	33	37	42	120	31	37	40	45	49	55	150	18
19	22	27	30	34	38	43	125	32	38	41	46	51	57	156	19
20	22	28	31	35	40	45	130	33	39	43	48	53	59	162	20
21	23	29	32	37	41	47	135	33	40	44	50	55	61	168	21
22	23	29	33	38	43	48	140	34	42	45	51	57	63	174	22
23	24	30	34	39	44	50	145	35	43	47	53	58	65	180	23
24	25	31	35	40	45	51	150	36	44	48	54	60	67	186	24
25	25	32	36	42	47	53	155	37	45	50	56	62	69	192	25

			$n = 7$								$n = 8$				
m	.001	.005	.01	.025	.05	.10	$2\bar{W}$	.001	.005	.01	.025	.05	.10	$2\bar{W}$	*m*
7	29	32	34	36	39	41	105								
8	30	34	35	38	41	44	112	40	43	45	49	51	55	136	8
9	31	35	37	40	43	46	119	41	45	47	51	54	58	144	9
10	33	37	39	42	45	49	126	42	47	49	53	56	60	152	10
11	34	38	40	44	47	51	133	44	49	51	55	59	63	160	11
12	35	40	42	46	49	54	140	45	51	53	58	62	66	168	12
13	36	41	44	48	52	56	147	47	53	56	60	64	69	176	13
14	37	43	45	50	54	59	154	48	54	58	62	67	72	184	14
15	38	44	47	52	56	61	161	50	56	60	65	69	75	192	15

TABLE B.21   CRITICAL LOWER-TAIL
VALUES OF W_n FOR WILCOXON'S RANK-SUM TEST (*continued*)

			$n = 7$								$n = 8$				
m	.001	.005	.01	.025	.05	.10	$2\bar{W}$	.001	.005	.01	.025	.05	.10	$2\bar{W}$	m
16	39	46	49	54	58	64	168	51	58	62	67	72	78	200	16
17	41	47	51	56	61	66	175	53	60	64	70	75	81	208	17
18	42	49	52	58	63	69	182	54	62	66	72	77	84	216	18
19	43	50	54	60	65	71	189	56	64	68	74	80	87	224	19
20	44	52	56	62	67	74	196	57	66	70	77	83	90	232	20
21	46	53	58	64	69	76	203	59	68	72	79	85	92	240	21
22	47	55	59	66	72	79	210	60	70	74	81	88	95	248	22
23	48	57	61	68	74	81	217	62	71	76	84	90	98	256	23
24	49	58	63	70	76	84	224	64	73	78	86	93	101	264	24
25	50	60	64	72	78	86	231	65	75	81	89	96	104	272	25

			$n = 9$								$n = 10$				
m	.001	.005	.01	.025	.05	.10	$2\bar{W}$	.001	.005	.01	.025	.05	.10	$2\bar{W}$	m
9	52	56	59	62	66	70	171								
10	53	58	61	65	69	73	180	65	71	74	78	82	87	210	10
11	55	61	63	68	72	76	189	67	73	77	81	86	91	220	11
12	57	63	66	71	75	80	198	69	76	79	84	89	94	230	12
13	59	65	68	73	78	83	207	72	79	82	88	92	98	240	13
14	60	67	71	76	81	86	216	74	81	85	91	96	102	250	14
15	62	69	73	79	84	90	225	76	84	88	94	99	106	260	15
16	64	72	76	82	87	93	234	78	86	91	97	103	109	270	16
17	66	74	78	84	90	97	243	80	89	93	100	106	113	280	17
18	68	76	81	87	93	100	252	82	92	96	103	110	117	290	18
19	70	78	83	90	96	103	261	84	94	99	107	113	121	300	19
20	71	81	85	93	99	107	270	87	97	102	110	117	125	310	20
21	73	83	88	95	102	110	279	89	99	105	113	120	128	320	21
22	75	85	90	98	105	113	288	91	102	108	116	123	132	330	22
23	77	88	93	101	108	117	297	93	105	110	119	127	136	340	23
24	79	90	95	104	111	120	306	95	107	113	122	130	140	350	24
25	81	92	98	107	114	123	315	98	110	116	126	134	144	360	25

TABLE B.22   CRITICAL LOWER-TAIL VALUES OF *D*
(INDICATING POSITIVE CORRELATION) FOR HOTELLING
AND PABST'S SPEARMAN RANK-ORDER CORRELATION TEST

	Largest value of D' for which $P(D \leq D') \leq \alpha$ Significance level, α					
n	.001	.005	.010	.025	.050	.100
4	—	—	—	—	0	0
5	—	—	0	0	2	4
6	—	0	2	4	6	12
7	0	4	6	12	16	24
8	4	10	14	24	32	42
9	10	20	26	36	48	62
10	20	34	42	58	72	90
11	32	52	64	84	102	126
12	50	76	92	118	142	170
13	74	108	128	160	188	224
14	104	146	170	210	244	288
15	140	192	222	268	310	362
16	184	248	282	338	388	448
17	236	312	354	418	478	548
18	298	388	436	510	580	662
19	370	474	530	616	694	788
20	452	572	636	736	824	932
21	544	684	756	868	970	1090
22	650	808	890	1018	1132	1268
23	770	948	1040	1182	1310	1462
24	902	1102	1206	1364	1508	1676
25	1048	1272	1388	1564	1724	1910
26	1210	1460	1588	1784	1958	2166
27	1388	1664	1806	2022	2214	2442
28	1584	1888	2044	2282	2492	2742
29	1798	2132	2304	2562	2794	3066
30	2030	2396	2582	2866	3118	3414

Source: Adapted with permission of the authors and editor from Table 2 in G. J. Glasser and R. F. Winter's "Critical Values of the Coefficient of Rank Correlation for Testing the Hypothesis of Independence," *Biometrika*, 48 (1961), 444–448.

TABLE B.23   CRITICAL UPPER-TAIL VALUES OF S
FOR KENDALL'S RANK-ORDER CORRELATION TEST

Smallest value of S' for which $P(S \geq S') \leq \alpha$

n	$\alpha = .055$	$\alpha = .010$	$\alpha = .025$	$\alpha = .050$	$\alpha = .100$
4	8	8	8	6	6
5	12	10	10	8	8
6	15	13	13	11	9
7	19	17	15	13	11
8	22	20	18	16	12
9	26	24	20	18	14
10	29	27	23	21	17
11	33	31	27	23	19
12	38	36	30	26	20
13	44	40	34	28	24
14	47	43	37	33	25
15	53	49	41	35	29
16	58	52	46	38	30
17	64	58	50	42	34
18	69	63	53	45	37
19	75	67	57	49	39
20	80	72	62	52	42
21	86	78	66	56	44
22	91	83	71	61	47
23	99	89	75	65	51
24	104	94	80	68	54
25	110	100	86	72	58
26	117	107	91	77	61
27	125	113	95	81	63
28	130	118	100	86	68
29	138	126	106	90	70
30	145	131	111	95	75
31	151	137	117	99	77
32	160	144	122	104	82
33	166	152	128	108	86
34	175	157	133	113	89
35	181	165	139	117	93
36	190	172	146	122	96
37	198	178	152	128	100
38	205	185	157	133	105
39	213	193	163	139	109
40	222	200	170	144	112

Source: Body of table is reproduced from Table III in L. Kaarsemaker and A. van Wijngaarden's "Table for Use in Rank Correlation," *Statistica Neerlandica*, 7 (1953), 41–54 (reproduced as Report R73 of the Computation Department of the Mathematical Centre, Amsterdam), with permission of the authors, the Mathematical Centre, and the editor of *Statistica Neerlandica*.

TABLE B.24   CRITICAL VALUES OF *U* FOR TOTAL-NUMBER-OF-RUNS TEST

Largest value of U' for which $P(U \le U') \le \alpha$

$\alpha = .05$

n_2 \ n_1	2	3	4	5	6	7	8	9	10	11	12	13	14	15	16	17	18	19	20
4			2																
5		2	2	3															
6		2	3	3	3														
7		2	3	3	4	4													
8	2	2	3	3	4	4	5												
9	2	2	3	4	4	5	5	6											
10	2	3	3	4	5	5	6	6	6										
11	2	3	3	4	5	5	6	6	7	7									
12	2	3	4	4	5	6	6	7	7	8	8								
13	2	3	4	4	5	6	6	7	8	8	9	9							
14	2	3	4	5	5	6	7	7	8	8	9	9	10						
15	2	3	4	5	6	6	7	8	8	9	9	10	10	11					
16	2	3	4	5	5	6	7	8	8	9	10	10	11	11	11				
17	2	3	4	5	6	7	7	8	9	9	10	10	11	11	12	12			
18	2	3	4	5	6	7	8	8	9	10	10	11	11	12	12	13	13		
19	2	3	4	5	6	7	8	8	9	10	10	11	12	12	13	13	14	14	
20	2	3	4	5	6	7	8	9	9	10	11	11	12	12	13	13	14	14	15

$\alpha = .01$

n_2 \ n_1	2	3	4	5	6	7	8	9	10	11	12	13	14	15	16	17	18	19	20
5			2																
6			2	2	2														
7			2	2	3	3													
8			2	2	3	3	4												
9		2	2	3	3	4	4	4											
10		2	2	3	3	4	4	5	5										
11		2	2	3	4	4	5	5	5	6									
12		2	3	3	4	4	5	5	6	6	7								
13		2	3	3	4	5	5	6	6	6	7	7							
14		2	3	3	4	5	5	6	6	7	7	8	8						
15		2	3	4	4	5	5	6	7	7	8	8	8	9					
16		2	3	4	4	5	6	6	7	7	8	8	9	9	10				
17		2	3	4	5	5	6	7	7	8	8	9	9	10	10	10			
18		2	3	4	5	5	6	7	7	8	8	9	9	10	10	11	11		
19	2	2	3	4	5	6	6	7	8	8	9	9	10	10	11	11	12	12	
20	2	2	3	4	5	6	6	7	8	8	9	10	10	11	11	11	12	12	13

Source: Adapted with permission of the authors and editor from Table II in Frieda S. Swed and C. Eisenhart's "Tables for Testing Randomness of Grouping in a Sequence of Alternatives," *Annals of Mathematical Statistics*, 14 (1943), 66–87.

TABLE B.25   CRITICAL VALUES OF K^+ FOR KOLMOGOROV-
SMIRNOV MAXIMUM DEVIATION TESTS FOR GOODNESS OF FIT

Values of K' for which $P(K^+ \geq K') = \alpha$ [or for which $P(K \geq K') \cong 2\alpha$ when $\alpha \leq .05$]

n	$\alpha = .10$	$\alpha = .05$	$\alpha = .025$	$\alpha = .01$	$\alpha = .005$
1	.90000	.95000	97500	.99000	.99500
2	.68377	.77639	.84189	.90000	.92929
3	.56481	.63604	.70760	.78456	.82900
4	.49265	.56522	.62394	.68887	.73424
5	.44698	.50945	.56328	.62718	.66853
6	.41037	.46799	.51926	.57741	.61661
7	.38148	.43607	.48342	.53844	.57581
8	.35831	.40962	.45427	.50654	.54179
9	.33910	.38746	.43001	.47960	.51332
10	.32260	.36866	.40925	.45662	.48893
11	.30829	.35242	.39122	.43670	.46770
12	.29577	.33815	.37543	.41918	.44905
13	.28470	.32549	.36143	.40362	.43247
14	.27481	.31417	.34890	.38970	.41762
15	.26588	.30397	.33760	.37713	.40420
16	.25778	.29472	.32733	.36571	.39201
17	.25039	.28627	.31796	.35528	.38086
18	.24360	.27851	.30936	.34569	.37062
19	.23735	.27136	.30143	.33685	.36117
20	.23156	.26473	.29408	.32866	.35241
21	.22617	.25858	.28724	.32104	.34427
22	.22115	.25283	.28087	.31394	.33666
23	.21645	.24746	.27490	.30728	.32954
24	.21205	.24242	.26931	.30104	.32286
25	.20790	.23768	.26404	.29516	.31657
26	.20399	.23320	.25907	.28962	.31064
27	.20030	.22898	.25438	.28438	.30502
28	.19680	.22497	.24993	.27942	.29971
29	.19348	.22117	.24571	.27471	.29466
30	.19032	.21756	.24170	.27023	.28987
31	.18732	.21412	.23788	.26596	.28530
32	.18445	.21085	.23424	.26189	.28094
33	.18171	.20771	.23076	.25801	.27677
34	.17909	.20472	.22743	.25429	.27279
35	.17659	.20185	.22425	.25073	.26897
36	.17418	.19910	.22119	.24732	.26532
37	.17188	.19646	.21826	.24404	.26180
38	.16966	.19392	.21544	.24089	.25843
39	.16753	.19148	.21273	.23786	.25518
40	.16547	.18913	.21012	.23494	.25205

Source: Body of table is reproduced, with changes only in notation, from Table 1 in L. H. Miller's "Table of Percentage Points of Kolmogorov Statistics," *Journal of the American Statistical Association*, 51 (1956), 111–121, with permission of the author and editor.

TABLE B.26 CRITICAL VALUES OF k FOR SMIRNOV'S
MAXIMUM DEVIATION TESTS FOR IDENTICAL POPULATIONS

Smallest value of k for which $P(D^+ \geq k/n) \leq \alpha$ [or for which $P(D \geq k/n) \leq 2\alpha$], followed
followed by $P(D^+ \geq k/n)$ [which equals $\frac{1}{2}P(D \geq k/n)$]

	SIGNIFICANCE LEVEL, α					
$n = m$	.05	.025	.01	.005	.001	.0005
3	3 (.05000)	—	—	—	—	—
4	4 (.01429)	4 (.01429)	—	—	—	—
5	4 (.03968)	5 (.00397)	5 (.00397)	5 (.00397)		
6	5 (.01299)	5 (.01299)	6 (.00108)	6 (.00108)	—	—
7	5 (.02652)	6 (.00408)	6 (.00408)	6 (.00408)	7 (.00029)	7 (.00029)
8	5 (.04351)	6 (.00932)	6 (.00932)	7 (.00124)	8 (.00008)	8 (.00008)
9	6 (.01678)	6 (.01678)	7 (.00315)	7 (.00315)	8 (.00037)	8 (.00037)
10	6 (.02622)	7 (.00617)	7 (.00617)	8 (.00103)	9 (.00011)	9 (.00011)
11	6 (.03733)	7 (.01037)	8 (.00218)	8 (.00218)	9 (.00033)	9 (.00033)
12	6 (.04977)	7 (.01572)	8 (.00393)	8 (.00393)	9 (.00075)	10 (.00010)
13	7 (.02214)	7 (.02214)	8 (.00633)	9 (.00144)	10 (.00025)	10 (.00025)
14	7 (.02952)	8 (.00939)	8 (.00939)	9 (.00245)	10 (.00051)	11 (.00008)
15	7 (.03773)	8 (.01312)	9 (.00383)	9 (.00383)	10 (.00092)	11 (.00018)
16	7 (.04666)	8 (.01750)	9 (.00560)	10 (.00151)	11 (.00034)	11 (.00034)
17	8 (.02248)	8 (.02248)	9 (.00778)	10 (.00231)	11 (.00058)	12 (.00012)
18	8 (.02801)	9 (.01037)	10 (.00333)	10 (.00333)	11 (.00092)	12 (.00022)
19	8 (.03405)	9 (.01338)	10 (.00461)	10 (.00461)	12 (.00036)	12 (.00036)
20	8 (.04053)	9 (.01677)	10 (.00615)	11 (.00198)	12 (.00056)	13 (.00014)
21	8 (.04741)	9 (.02054)	10 (.00795)	11 (.00273)	12 (.00083)	13 (.00022)
22	9 (.02467)	9 (.02467)	11 (.00365)	11 (.00365)	13 (.00034)	13 (.00034)
23	9 (.02914)	10 (.01236)	11 (.00473)	11 (.00473)	13 (.00050)	13 (.00050)
24	9 (.03390)	10 (.01496)	11 (.00598)	12 (.00216)	13 (.00070)	14 (.00020)

Source: Adapted with permission of the authors and editor from Table 3 in Z. W. Birnbaum and
R. A. Hall's "Small Sample Distributions for Multi-Sample Statistics of the Smirnov Type," *Annals
of Mathematical Statistics*, 31 (1960), 710–720.

TABLE B.26   CRITICAL VALUES OF *k* FOR SMIRNOV'S
MAXIMUM DEVIATION TESTS FOR IDENTICAL POPULATIONS (*continued*)

	SIGNIFICANCE LEVEL, α					
$n = m$	.05	.025	.01	.005	.001	.0005
25	9	10	11	12	13	14
	(.03895)	(.01781)	(.00742)	(.00281)	(.00096)	(.00030)
26	9	10	11	12	14	14
	(.04425)	(.02090)	(.00904)	(.00357)	(.00042)	(.00042)
27	9	10	12	12	14	15
	(.04978)	(.02422)	(.00445)	(.00445)	(.00057)	(.00018)
28	10	11	12	13	14	15
	(.02776)	(.01281)	(.00545)	(.00213)	(.00076)	(.00025)
29	10	11	12	13	14	15
	(.03151)	(.01497)	(.00657)	(.00266)	(.00099)	(.00034)
30	10	11	12	13	15	15
	(.03544)	(.01729)	(.00782)	(.00327)	(.00045)	(.00045)
31	10	11	12	13	15	16
	(.03956)	(.01978)	(.00920)	(.00397)	(.00059)	(.00020)
32	10	11	13	13	15	16
	(.04384)	(.02243)	(.00476)	(.00476)	(.00075)	(.00027)
33	10	12	13	14	15	16
	(.04828)	(.01234)	(.00563)	(.00240)	(.00095)	(.00035)
34	11	12	13	14	16	16
	(.02819)	(.01409)	(.00660)	(.00289)	(.00045)	(.00045)
35	11	12	13	14	16	17
	(.03127)	(.01597)	(.00765)	(.00344)	(.00057)	(.00021)
36	11	12	13	14	16	17
	(.03450)	(.01797)	(.00880)	(.00405)	(.00071)	(.00027)
37	11	12	14	14	16	17
	(.03784)	(.02008)	(.00472)	(.00472)	(.00087)	(.00034)
38	11	12	14	15	17	17
	(.04130)	(.02230)	(.00547)	(.00248)	(.00042)	(.00042)
39	11	12	14	15	17	18
	(.04487)	(.02463)	(.00628)	(.00291)	(.00052)	(.00020)
40	11	13	14	15	17	18
	(.04854)	(.01430)	(.00715)	(.00338)	(.00064)	(.00025)

GLOSSARY

Addition Rule: If two outcomes are not mutually exclusive, the probability of either (or both) occurring is the sum of their separate probabilities of occurrence minus the probability of their joint occurrence.

Alternative Hypothesis: The hypothesis that is accepted if the null hypothesis is rejected.

Arithmetic Mean: The sum of the values of quantitative observations divided by the number of observations.

Average Deviation: The sum of the absolute values of the deviations from the mean divided by the number of observations.

Bayes Rule: The (posterior) probability of event B given that event A has occurred:

$$P(B|A) = \frac{P(B)P(A|B)}{P(A)}$$

Bernoulli Process: Two mutually exclusive and exhaustive event classes and their associated probabilities.

Beta Distribution: Probability distribution on the range 0 to 1 having parameters α and β:

$$f(x) = \begin{cases} \dfrac{\Gamma(\alpha + \beta)}{\Gamma(\alpha)\Gamma(\beta)} x^{\alpha-1}(1-x)^{\beta-1} & 0 < x < 1 \\ 0 & \text{elsewhere} \end{cases}$$

Binomial Distribution: Probability distribution of the number of successes in n trials or observations taken from a Bernoulli process:

$$P(x) = \frac{n!}{x!(n-x)!} \pi^x(1-\pi)^{n-x} \qquad x = 0, 1, \ldots, n$$

Bivariate Exponential Distribution: Joint distribution of independent random variables X_1 and X_2:

$$f(x_1, x_2) = \begin{cases} c_1 c_2 e^{-(c_1 x_1 + c_2 x_2)} & x_1, x_2 > 0 \\ 0 & \text{elsewhere} \end{cases}$$

where $c_1, c_2 > 0$.

Bivariate Normal Distribution: Joint distribution of independent random variables X_1 and X_2:

$$f(x_1, x_2) = \frac{1}{2\pi\beta_1\beta_2} e^{-(1/2)[(x_1 - \alpha_1)^2/\beta_1^2 + (x_2 - \alpha_2)^2/\beta_2^2]} \qquad \begin{matrix} -\infty < x_1 < \infty \\ -\infty < x_2 < \infty \end{matrix}$$

Categorical or Qualitative Data: Data that are grouped or classified according to some qualitative characteristic.

Central Limit Theorem: If X is a random variable having finite mean μ and finite variance σ^2, then the probability distribution of the random variable $(\bar{x} - \mu)/(\sigma/\sqrt{n})$ approaches the standard normal distribution as $n \to \infty$.

Chebyshev's Theorem: If X is a random variable whose distribution has mean μ and variance σ^2, then for any $k > 0$,

$$P(|x - \mu| < k\sigma) \geq 1 - \frac{1}{k^2}$$

or, equivalently,

$$P(|x - \mu| > k\sigma) < \frac{1}{k^2}$$

Chi-Square Distribution: Sampling distribution associated with variances of samples from normal distributions:

$$f(\chi^2) = \begin{cases} \dfrac{1}{2^{v/2}\Gamma\left(\dfrac{v}{2}\right)} (\chi^2)^{(v-2)/2} e^{-(1/2)\chi^2} & 0 < \chi^2 < \infty \\[2ex] 0 & \textit{elsewhere} \end{cases}$$

where v is the degrees of freedom.

Cluster Sampling: Partitioning of a population to be sampled on the basis of location or accessibility.

Coefficient of Determination: Squared value of the correlation coefficient, which gives the proportion of variance in the dependent variable that is accounted for by variance in the independent variable(s).

Coefficient of Variation: A relative measure of dispersion; the standard deviation divided by the mean.

Completely Randomized Design: Design in which experimental units are assigned to factor levels completely at random.

Conditional Density Function: Density function of a set of (continuous) random variables for fixed values of another set of (continuous) random variables, when the variables of both sets are jointly distributed.

Conditional Distribution: Probability distribution of a set of (discrete) random variables for fixed values of another set of (discrete) random variables, when the variables of both sets are jointly distributed.

Conditional Probability: Probability of the occurrence of an event given that another event has occurred.

Contingency Table: Two-way table that summarizes the cross-classification of sample elements according to two characteristics.

Continuous Random Variable: Random variable defined on a continuous sample space and assuming any of the (infinitely many) values within an interval or set of intervals.

Correlation Ratio: Proportion of the total variance in Y that is attributable to differences in the means of Y for different values of X:

$$E^2 = \frac{\sum\limits_{j=1}^{k} n_j(n_j - 1)s_{y_j}^2}{(n-1)s_y^2}$$

Cross-Section Data: Observations from a population defined at a specified time or for a specified (short) time period.

Cumulative Distribution Function: Function $F(a) = P(X \leq a)$ giving the probability that the random variable X assumes a value less than or equal to a.

Cumulative Frequency Distribution: Frequency distribution given in terms of the number of observations less than or equal to specified values.

Cycle: Component of a time series that represents some recurrent (not necessarily periodic) phenomenon.

Decision Theory: The study of quantitative techniques for formulating and analyzing decision problems.

Discrete Random Variable: Random variable defined on a discrete sample space and assuming a finite or countably infinite number of values.

Distribution-Free Test: Test that makes no assumption about the precise form of the sampled population.

Expectation: Measure of central tendency of the distribution of a random variable X, given by

$$E(X) = \sum_x xP(x)$$

for a discrete random variable and by

$$E(X) = \int_{-\infty}^{\infty} xf(x)\,dx$$

for a continuous random variable.

Experimental Unit: Amount of material to which a treatment is applied to obtain a single measurement of the response variable in analysis of variance.

Exponential Distribution: Probability distribution of the time until the first occurrence of an event and the time between occurrences of an event when the occurrence of the event has a Poisson distribution:

$$f(x) = \begin{cases} \lambda e^{-\lambda x} & x \geq 0 \\ 0 & x < 0 \end{cases}$$

Factor: Any induced or selected variation in experimental procedures or conditions whose effect is to be observed (also called a treatment).

Factorial Experiment: Experiment in which the effects of several factors are investigated simultaneously and the treatments consist of all possible combinations of levels of the factors.

F Distribution: Sampling distribution associated with ratios of variances of samples from normal distributions:

$$f(F) = \begin{cases} \dfrac{cF^{(v_1/2)-1}}{\left(1 + \dfrac{v_1}{v_2}F\right)^{(v_1+v_2)/2}} & x > 0 \\ 0 & \text{elsewhere} \end{cases}$$

where $c = \dfrac{\Gamma\left(\dfrac{v_1 + v_2}{2}\right)}{\Gamma\left(\dfrac{v_1}{2}\right)\Gamma\left(\dfrac{v_2}{2}\right)}\left(\dfrac{v_1}{v_2}\right)^{v_1/2}$ and v_1 and v_2 are degrees of freedom.

Finite Population Correction (f.p.c.): The factor $(N - n)/(N - 1)$ used in sampling theory to adjust formulas derived assuming infinite populations for use when populations are in fact finite.

Fixed-Effects Model: Analysis of variance model in which all the treatment levels are known constants (also called Model I).

Frequency Distribution: A table of classes or groups and the corresponding numbers of observations in each.

Frequency Polygon: Graphical representation of a frequency distribution obtained by plotting class frequencies and connecting successive points.

Game Theory: Analysis of decisions under conflict.

Gamma Distribution: Skewed probability distribution with parameters α and β:

$$f(x) = \begin{cases} \dfrac{1}{\beta^{\alpha}\Gamma(\alpha)} x^{\alpha-1} e^{-x/\beta} & x > 0 \\ 0 & \text{elsewhere} \end{cases}$$

Geometric Distribution: Special case of the negative binomial distribution when $k = 1$:

$$P(n) = \pi(1 - \pi)^{n-1} \qquad n = 1, 2, \ldots$$

(probability that the first success occurs on the nth trial).

Geometric Mean: Of a set of n numbers is the nth root of the product of the numbers.

Harmonic Mean: Of a set of n numbers is n divided by the sum of the reciprocals of the numbers.

Histogram: A diagram representing classes and frequencies by vertical bars.

Hypergeometric Distribution: Probability that a set of n elements randomly selected from a set of N Bernoulli elements contains x elements of the first type and $n - x$ elements of the second type:

$$P(x) = \dfrac{\dbinom{N\pi}{x}\dbinom{N - N\pi}{n - x}}{\dbinom{N}{n}} \qquad x = 0, 1, \ldots, n$$

Independent Random Variables: A set of random variables whose joint probability distribution (or density function) is equal to the product of their marginal probability distributions (or density functions).

Index Number: A descriptive statistic that is constructed to give an indication of the change in a variable over time.

Joint Density Function: Function $f(x, y)$ specified for all possible pairs of values (x, y) in the range of the random variables X and Y, such that $f(x, y) \geq 0$ and

$$\int_{\infty}^{\infty} \int_{-\infty}^{\infty} f(x, y)\, dx\, dy = 1.$$

Joint Probability Distribution: Probabilities $P(x, y)$ specified for all possible pairs of values (x, y) in the range of the random variables X and Y, such that $P(x, y) \geq 0$ and $\sum_{x,y} p(x, y) = 1$.

Kurtosis: The degree of peakedness of a distribution.

Laspeyres Index: A weighted aggregative price index with base period quantities as weights.

Level: General term referring to the characteristic or amount that defines or designates a particular category or classification of a factor in experimental design.

Level of Significance: The probability of committing a type I error in hypothesis testing.

Linear Combination of Population Means: Linear function of k population means given by

$$L = c_1\mu_1 + \cdots + c_k\mu_k = \sum_{j=1}^{k} c_j\mu_j$$

Linear Contrast of Population Means: Linear combination of k population means

$$L = \sum_{j=1}^{k} c_j \mu_j$$

for which $\sum_{j=1}^{k} c_j = 0$.

Log-Normal Distribution: If $\ln X$ is normally distributed with mean α and variance β^2, then X has the log-normal distribution

$$f(x) = \begin{cases} \dfrac{1}{x\beta\sqrt{2\pi}} e^{-(\ln x - \alpha)^2/2\beta^2} & x > 0, \beta > 0 \\ 0 & \text{elsewhere} \end{cases}$$

Marginal Density Function: Probability density function of one of the (continuous) random variables of a set of jointly distributed (continuous) random variables.

Marginal Distribution: Probability distribution of one of the (discrete) random variables of a set of jointly distributed (discrete) random variables.

Median: Measure of central tendency that divides a set of observations so that half are as large or larger than its value and half are smaller than its value.

Mode: Measure of central tendency that is the value, class, or category that occurs with the highest frequency.

Mixed Model: Analysis of variance model in which at least one treatment level is fixed and at least one treatment level is random.

Multicollinearity: Strong intercorrelation between or among the independent variables in a regression analysis.

Multinomial Distribution: Probability distribution of the occurrence of X_i observations of type A_i when $A_1, \ldots, A_k$ are mutually exclusive and exhaustive types of observations:

$$P(X_1, \ldots, X_k) = \frac{n!}{\prod\limits_{i=1}^{k} X_i!} \prod_{i=1}^{k} \pi_i^{X_i}$$

Multiplication Rule: The probability of a joint event or of a sequence of two events is equal to probability of the first event multiplied by the conditional probability of the second event given that the first event has occurred.

Mutually Exclusive: Events that cannot both occur in the same experiment or observation.

Negative Binomial Distribution: Probability of observing the kth success on the nth trial in sampling from a Bernoulli process:

$$P(n) = \binom{n-1}{k-1} \pi^k (1 - \pi)^{n-k} \qquad (0 < k \leq n)$$

(also called Pascal distribution).

Nonparametric Test: Test that does not state a hypothesis about the specific value of a parameter.

Normal Distribution: Symmetric bell-shaped probability distribution with parameters μ and σ^2:

$$f(x) = \frac{1}{\sigma\sqrt{2\pi}} e^{-1/2(x-\mu)^2/\sigma^2} \qquad -\infty < x < \infty$$

Null Hypothesis: A statement that is to be tested in a statistical context.

Numerical or Quantitative Data: Data that are grouped or classified according to a quantitative measurement.

Orthogonal Contrasts of Population Means: Linear Combinations of k population means

$$L_1 = \sum_{j=1}^{k} c_{j1}\mu_j \quad \text{and} \quad L_2 = \sum_{j=1}^{k} c_{j2}\mu_j$$

for which

$$\sum_{j=1}^{k} c_{j1}c_{j2} = 0$$

Orthogonal Variables: Variables among which there is no linear relationship.

Paasche Index: A weighted aggregative price index with given period quantities as weights.

Parameter: Characteristic or measure of a population that distinguishes one population from another similar population.

Pareto Distribution: Probability distribution having parameter α:

$$f(X) = \begin{cases} \alpha X_0{}^{\alpha}/X^{\alpha+1} & X > X_0, \text{ the minimum value of } X \\ 0 & \text{elsewhere} \end{cases}$$

where $1 < \alpha < 2$.

Partial Correlation Coefficient: Measure of the relationship between two variables after the effects of one or more other variables have been taken into account or "partialed out" (also called coefficient of partial correlation).

Payoff Matrix: A two-way table indicating the choices of strategies, the states of nature, and the payoff to the decision maker for the intersection of each choice and state of nature.

Pearsonian Coefficient of Skewness:

$$\text{SK} = \frac{3(\text{mean} - \text{median})}{\text{standard deviation}}$$

Percentage Frequency Distribution: A frequency distribution given in terms of percentages.

Poisson Distribution: Limiting form of the binomial distribution when $n \to \infty$ and $\pi \to 0$ with $n\pi$ constant:

$$P(x) = \frac{e^{-n\pi}(n\pi)^x}{x!} \qquad x = 0, 1, \ldots$$

or

$$P(x) = \frac{e^{-\lambda}\lambda^x}{x!} \qquad x = 0, 1, \ldots$$

Power Function: The function or curve that represents the probability of rejecting the null hypothesis when various alternatives are true.

Probability: A concept that expresses the degree of certainty with which an assertation is made.

Probability Density: Function $f(x)$ in terms of which the probability distribution of a continuous random variable X is defined.

Probability Distribution: Summary of all possible values of a discrete random variable and their corresponding probabilities.

Probability Function: Graph of the pairs (x_i, p_i), the possible values of X and their corresponding probabilities.

Probability Sampling: Any procedure for drawing a sample so that inferences can be stated unambiguously in terms of probability.

Random-Effects Model: Analysis of variance model in which all the treatment levels are randomly selected (also called Model II).

Randomization: Random assignment of experimental units to treatments in analysis of variance.

Randomization Distribution: Null distribution obtained by considering all possible assignments of observations to categories and the associated probabilities assuming equally likely occurrences.

Random Variable: Real-valued function defined on a sample space.

Range: Difference between the largest and the smallest observations.

Relative: Index number calculated as indicative of variation in relation to a base period.

Relative Efficiency: Relative efficiency of test A with respect to test B is b/a, where a is the number of observations required by test A to have the same power as test B has when based on b observations; both tests are for the same null hypothesis against the same alternative hypothesis at the same significance level.

Relative Frequency of an Event: Ratio of the number of ways an event can occur to the total number of possible occurrences.

Residual Variation: Extraneous variation that tends to obscure or distort the effects of factors of interest in analysis of variance (also called experimental error).

Robustness: A test is robust against violation of a particular assumption if its probabilities of type I and type II errors are not appreciably affected by the violation.

Saddle Point: Entry of a game matrix that is simultaneously a maximum of row minima and a minimum of column maxima.

Sampling Distribution: Distribution of a random variable corresponding to a statistic.

Scatter Plot: Plot of the points corresponding to pairs of observations (x_i, y_i) of the random variables X and Y, usually used for checking linearity and other assumptions in regression analysis (also called scatter diagram).

Seasonal Variation: Component of a time series that is the result of seasonality.

Set: A collection of elements.

Simple Linear Model: Relationship of the form

$$y_i = \beta_0 + \beta_1 x_i + \varepsilon_i \qquad i = 1, \dots, n$$

where Y is a random variable, X is a fixed variable, β_0 and β_1 are parameters to be estimated and the errors ε_i are independent normally distributed random variables with mean zero and variance σ^2 and are independent of the X_i.

Simple Random Sampling: Selection of a sample in such a way that every element in the population has an equal chance of being included in the sample, and every possible combination of elements in the population has an equal chance of being included in the sample.

Single-Factor Experiment: Experiment or investigation that concerns only one factor (also called one-way classification).

Stable Paretian Distribution: Probability distribution having four parameters and tails asymptotically distributed according to the Pareto distribution.

Standard Deviation: The positive square root of the variance.

Standard Error of Estimate: Standard deviation of the errors of estimation in regression analysis:

$$s = \sqrt{\frac{1}{n-2} \sum_{i=1}^{n} (y_i - \hat{y}_i)^2} = \sqrt{\frac{n-1}{n-2} (s_y{}^2 - \hat{\beta}_1 s_x{}^2)}$$

Standard Normal Distribution: Normal distribution having $\mu = 0$, $\sigma^2 = 1$:

$$f(x) = \frac{1}{\sqrt{2\pi}} e^{-1/2 x^2} \qquad -\infty < x < \infty$$

Statistic: Quantity that is determined or computed from sample values of a random variable.

Statistical Dependence of Events: Situation in which the occurrence of one event affects the probability of occurrence of the other event.

Statistical Independence of Events: Situation in which the occurrence of one event does not affect the probability of occurrence of the other event.

Stepwise Method: Version of forward selection of independent variables in regression that provides for deletion of a variable at a later step.

Stratified Sampling: Partitioning of a population to be sampled on the basis of some characteristic thought to be relevant.

Student-*t* Distribution: Sampling distribution associated with means of samples from normal distributions:

$$f(t) = \frac{\Gamma\left(\dfrac{v+1}{2}\right)}{\sqrt{\pi v}\, \Gamma\left(\dfrac{v}{2}\right)} \left(1 + \frac{t^2}{v}\right)^{-(v+1)/2} \qquad -\infty < t < \infty$$

where v is the degrees of freedom.

Time Series Data: Observations obtained over a period of time.

Trend: Component of a time series giving its general movement.

Type I Error: In hypothesis testing, the error of rejecting the null hypothesis when it is true.

Type II Error: In hypothesis testing, the error of accepting the null hypothesis when it is false.

Uniform Distribution (Continuous): Probability distribution of a homogeneous random variable:

$$f(x) = \begin{cases} \dfrac{1}{\beta - \alpha} & \alpha < x < \beta \\ 0 & \text{elsewhere} \end{cases}$$

Uniform Distribution (Discrete): Probability distribution of k equally likely outcomes:

$$P(x) = \frac{1}{k} \qquad x = x_1, x_2, \ldots, x_k$$

Universe: A set that contains all elements under consideration.

Utility: Quality that when possessed by something, causes it to render satisfaction.

Validity: A test is said to be valid at level α if the probability of type I error does not exceed α for any distribution.

Variance: Measure of dispersion of the distribution of a random variable x, given by

$$\sigma^2 = E(x - \mu)^2 = \sum_x (x - \mu)^2 P(x)$$

for a discrete random variable and by

$$\sigma^2 = E(x - \mu)^2 = \int_{-\infty}^{\infty} (x - \mu)^2 f(x)\, dx$$

for a continuous random variable.

ANSWERS TO ODD-NUMBERED PROBLEMS

CHAPTER 2

2.1 (a) $\sum\limits_{i=1}^{n} f_i x_i^2$ (d) $\sum\limits_{i=1,2,4,8}^{16} f(5i)$

(b) $\sum\limits_{i=1}^{50} x_i y_i$ (e) $\sum\limits_{i=1}^{4} \left[x^i + (x-1)^i \right]$

(c) $\sum\limits_{i=1}^{n} \left[i(i+1) \right]$ (f) $\sum\limits_{i=1}^{5} x^i y^i$

2.3 (a)
9.5–24.5	(b) 17 (c) 15
24.5–39.5	32
39.5–54.5	47
54.5–69.5	62
69.5–84.5	77
84.5–99.5	92

2.5 (a)
	(b)	
109.5–119.5	<120	2
119.5–129.5	<130	8
129.5–139.5	<140	19
139.5–149.5	<150	33
149.5–159.5	<160	42
159.5–169.5	<170	47
169.5–179.5	<180	50

(d) 140 lb

2.7 50 mph

2.9 $\bar{x} = 30$, $s = 6.30$

2.11 $Q_1 - Q_2 \neq Q_2 - Q_3$ for an asymmetric distribution

2.13 (a) 45

(b) Tarpon $0.20, Tampa $0.10, Tallahasse $0.033

(c) $.111

(d) 27

(e) arithmetic mean in (a), harmonic mean in (c)

2.15 (a) $184.50 (b) $175

(c) $150 and $200 (d) $79.36

(e) $62.35

2.17 Distribution whose mean and/or median cannot be obtained

2.19 (a) $\bar{x}_1 = 6$, $\bar{x}_2 = 6$

(b) $s_1 = 4$, $s_2 = 4$

(c) $sk_1 = -1.5$, $sk_2 = 0$

same mean and standard deviation, different shape; sample 1 negatively skewed, sample 2 symmetric

2.21 (a) $sk = .89$ (b) $\alpha_3 = 1.60$

(c) $\alpha_4 = 5.06$

CHAPTER 3

3.1 $y = 4719.75 + 169.73x$

3.3 Trend line: $\bar{y} = 1829.46 + 16.92t$

	1977 Forecast	Actual
Q_1	2287.09	2082
Q_2	2330.60	2247
Q_3	2134.70	2120
Q_4	1924.93	2168

3.5 (a) See table below.

(b) Bond prices moving down over time

3.5 (a)		1966	1967	1968	1969	1970	1971
	Govt.	4.75	5.01	6.14	6.53	8.21	7.02
	AAI	4.82	5.55	6.30	6.75	8.70	7.60
	T.C.	4.85	5.70	6.50	7.10	8.75	7.72
	A.U.	5.00	6.00	7.00	7.30	9.65	8.08
	Sum	19.42	22.26	25.94	27.68	35.31	30.42
		100.00	114.62	133.57	142.53	181.82	156.64

3.7 Trend line: $\bar{y} = 9774.54 + 41.160t$

3.9
	1972	1973	1974	1975	*Geom. Av.*	*Geom. Gr Rate*
Beef	100	94.2	103.3	115.0	37.1	$(8.6)i\%$
Veal	100	75.0	100.0	168.8	3.4	$(38.5)i\%$
L & M	100	93.3	86.7	77.1	9.3	$(8.1)i\%$
Pork	100	90.8	97.0	81.4	78.9	10%

3.11 Trend line: $\bar{y} = 156.615 - 5.33t$
Sales were erratic and declined from first half to second half; Trend line meaningless for forecasting

CHAPTER 4

4.1 $\frac{5}{18}$

4.3 (a) $\frac{1}{216}$ (b) $\frac{1}{36}$ (c) $\frac{5}{72}$
   (d) $\frac{5}{12}$ (e) $\frac{5}{9}$

4.5 (a) $\frac{8}{99}$ (b) $\frac{8}{99}$

4.7 .90

4.9 (a) $\frac{13}{20}$ (b) $\frac{5}{20}$
   (c) $\frac{3}{10}$ (d) $\frac{5}{8}$

4.11 $\frac{4}{9}$

4.13 $\frac{2}{3}$

4.15 $\frac{2}{5}$

4.17 (a) $\frac{3}{11}$ (b) $\frac{5}{16}$ (c) not independent

4.19 (a) $P(A) = \frac{1}{6}$, $P(B) = \frac{1}{30}$, $P(C) = \frac{1}{20}$
   (b) $\frac{1}{4}$

4.21 (a) .6976 (b) .488

4.23 $\frac{7}{24}$

4.25 $P(A) = \frac{3}{16}$, $P(B) = \frac{1}{16}$, $P(C) = \frac{1}{2}$

4.27 $\frac{1}{4}$

4.29 (a) 35 (b) 15

4.31 5250

4.33 13

CHAPTER 5

5.1 (a) .128 (b) .955

5.3 .855

5.5 (a) .15 (b) 1 hr 32 min

5.7 .187

5.9 .000236

5.11 .125, .0156

5.13 .338

5.15 (a) .393 (b) .135

5.17 (a) 5 min (b) .233

5.19 .340

5.21 (a) 17 (b) 17

5.23 .998

5.25 10

5.27 (a) .15 (b) .95

5.29 (a) .491 (b) .399 (c) .082

5.31 (a) .198 (b) 4.5

5.33 67

CHAPTER 6

6.1 (a) $\geq .96$ (b) 1

6.3 60

6.5 2.50

6.7 (a) 548 (b) 604.54

6.9 12

6.11 .001

6.13 (a) .1587 (b) 0

6.15 0

6.17 $< .01$

6.19 (a) .0668 (b) .8413

6.21 2.25

6.23 (a) .97 (b) .034

CHAPTER 7

7.5 (a) $25.01 \leq \mu \leq 26.19$
   (b) $2.63 \leq \sigma \leq 3.48$

7.7 $6.59 \leq \sigma^2 \leq 17.39$

7.9 $.021 \leq \pi \leq .079$

7.11 (a) 9604 (b) 1475

7.13 666

7.15 .97

7.17 $16.38 \leq \sigma \leq 27.76$

7.19 (a) 601 (b) 505

7.21 6

7.23 58

7.25 (a) 5.74 (b) 85

7.27 $50.97 \leq \mu \leq 54.45$; $6.84 \leq \sigma^2 \leq 27.08$

7.29 $.64 \leq \sigma^2 \leq 2.93$

CHAPTER 8

8.1 (a) .055 (b) 1

8.3 (a) .012 (b) .309 (c) .933

8.5 (a) $\bar{x} < 29.742$ and $\bar{x} > 30.258$
   (b) .0078
   (c) .00032

8.7 (a) .15 (b) .027 (c) 1

8.9 If 15 or more are defective in a sample of 100, stop the process.

8.11 (a) .095 (b) .652

8.13 (a) $H_0: \mu = 1.000$; $H_A: \mu = 0.920$
   (b) 56 (c) accept wire if $\bar{x} \geq 0.953$

CHAPTER 9

9.1 $z = -2.12$ so reject H_0

9.3 54

9.5 .62

9.7 (a) $t \doteq 4$ so reject H_0
   (b) $n_1 = 52, n_2 = 32$

9.9 (a) $z = 4$ so reject H_0

9.11 $t = 1.26$ so reject H_0

9.13 $t = 3.87$ so reject H_0

9.15 .994

9.17 $t = 1.4$ so accept H_0

9.19 $t = 2.22$ so reject H_0

9.21 $t = -2.09$ so reject H_0

9.23 $F = 3.25$ so reject H_0

9.25 $t = 1.44$ so accept H_0

9.27 $t = 1.67$ so accept H_0

9.29 $t = 0.57$ so accept H_0

9.31 $t = 13.40$ so reject H_0

9.33 $t = -1.732$ so accept H_0

9.35 (a) $t = 1$ so accept H_0
   (b) $t = 3.46$ so reject H_0

CHAPTER 10

10.1 (a) .046   (b) 0, .5, .5
    (c) .023; 0, 0.5, 1

10.3 (a) .046   (b) .5, 0, 0, .778

10.5 (a) $z = -1.75$ so reject H_0
    (b) reject H_0 if $p \le .753$ (c) .0005

10.7 (a) $H_0: \mu < .10; H_A: \mu \ge .10$ (b) 3

10.9 $\chi^2 = 62.68$ so reject H_0

10.11 $z = 1.54$ so reject H_0

10.13 $\chi^2 = 35.07$ so reject H_0

10.15 (a) $z = 6; \chi^2 = 36$ (b) reject H_0

10.17 (a) $\chi^2 = 4.375$ so accept H_0
     (b) $\chi^2 = 34.722$ so reject H_0

10.19 $\chi^2 = 4.2$ so accept H_0

10.21 $\chi^2 = 11$ so accept H_0

10.23 $\chi^2 = 9.73$ so reject H_0

10.25 (a) 21; 4
     (b) $\chi^2 = 3.13$ so accept H_0

10.27 $\chi^2 = 1.56$ so accept H_0

CHAPTER 11

11.3 246; $.10 \pm .03675$ or $.20 \pm .049$

11.5 $n_1 = 52, n_2 = 8$; $46,400,000

CHAPTER 12

12.1 $F = \dfrac{MS_{\text{trts}}}{MS_E} = 4$ so reject H_0

   T method: $\bar{y}_2$ vs. $\bar{y}_3$; S method: $\bar{y}_2$ vs. $\bar{y}_3$

12.3 $F = \dfrac{MS_{\text{trts}}}{MS_E} = 3.49$ so reject H_0

12.5 (a) $F = \dfrac{MS_{\text{trts}}}{MS_E} < 1$ so accept H_0

(b) $E(LS|CR) = .87$;
   $E(LS|RB \text{ rows}) = .875$;
   $E(LS|RB \text{ cols}) = .95$

12.7 $F = \dfrac{MS_{\text{trts}}}{MS_E} = 42.68$ so reject H_0

12.9 (a)

Source	df	ss	ms	F
Rows	2	6	3	1
Cols	2	6	3	1
Trts	2	150	75	25
Error	2	6	3	
Total	8	168		

(b) $F = \dfrac{MS_{\text{rows}}}{MS_E} = 1$ so accept H_0;

   $F = \dfrac{MS_{\text{cols}}}{MS_E} = 1$ so accept H_0;

   $F = \dfrac{MS_{\text{trts}}}{MS_E} = 25$ so reject H_0

(c) $F = 37.5$ so reject H_0; $F = 12.5$ so reject H_0

(d) $E(LS|CR) = 1$

12.11 (a) $F = \dfrac{MS_A}{MS_E} = 5.06$ so reject H_0;

    $F = \dfrac{MS_B}{MS_E} = 3.00$ so accept H_0;

    $F = \dfrac{MS_{AB}}{MS_E} = 28.28$ so reject H_0

(b) $F = 7.58$ so reject H_0; $F = 4.41$ so accept H_0

12.13 (a) $F = \dfrac{MS_{\text{tr}}}{MS_E} = 12.0$ so reject H_0;

    $F = \dfrac{MS_{\text{exp}}}{MS_E} = 35.0$ so reject H_0;

    $F = \dfrac{MS_{T \times E}}{MS_E} = 10.0$ so reject H_0

(b) as in (a) except $F = \dfrac{MS_{\text{tr}}}{MS_{T \times E}} = 1.2$ so accept H_0

12.15 (a) $F = \dfrac{MS_{\text{rows}}}{MS_E} = 5$ so reject H_0;

    $F = \dfrac{MS_{\text{cols}}}{MS_E} = 10$ so reject H_0;

    $F = \dfrac{MS_{R \times C}}{MS_E} = 5$ so reject H_0

(b) as in (a) except $F = \dfrac{MS_{\text{rows}}}{MS_{R \times C}} = 1$

so accept H_0

12.17 (a) 2 row means:

$\hat{L} - 0.74 \le L \le \hat{L} + 0.74$ (T-method)
$\hat{L} - 0.86 \le L \le \hat{L} + 0.86$ (S-method)

2 col means:

$\hat{L} - 0.65 \le L \le \hat{L} + 0.65$ (T-method)
$\hat{L} - 0.73 \le L \le \hat{L} + 0.73$ (S-method)

3 row means:

$\hat{L} - 1.48 \le L \le \hat{L} + 1.48$ (T-method)
$\hat{L} - 1.49 \le L \le \hat{L} + 1.49$ (S-method)

3 col means:

$\hat{L} - 1.30 \le L \le \hat{L} + 1.30$ (T-method)
$\hat{L} - 1.27 \le L \le \hat{L} + 1.27$ (S-method)

(b) 4 row means:

$\hat{L} - 1.22 \le L \le \hat{L} + 1.22$ (S-method)

4 col means:

$\hat{L} - 1.04 \le L \le \hat{L} + 1.04$ (S-method)

5 row means:

$\hat{L} - 3.33 \le L \le \hat{L} + 3.33$ (S-method)

5 col means:

$\hat{L} - 2.84 \le L \le \hat{L} + 2.84$ (S-method)

6 row means:

$\hat{L} - 1.49 \le L \le \hat{L} + 1.49$ (S-method)

(c) T shorter than S for 2 rows or columns, slightly shorter for 3 rows; intervals for columns shorter than for rows (fewer columns than rows)

CHAPTER 13

13.1 (a) $\tilde{y}_i = -85.67 + 1.53x_i$
(b) $r = .96$; $t = 9.70$ so reject H_0

13.3 $r = .99$; $t = 29.77$ so reject H_0

13.5 (a) $\tilde{y}_i = .5 + .5x_i$ (b) $\tilde{y}_i = 3$
(c) $r = .625$ (d) $s_{y \cdot x} = 1.27$

13.7 $69 \le \rho \le .87$

13.9 31

13.11 $z = -0.69$ so accept H_0

13.13 147

13.15 $F = 37.54$ so reject H_0

13.17 (a) $E^2 = .8$; $F = 36.36$ so reject H_0
(b) $F = 10$ so reject H_0

13.19 (a) $E^2 = .8$; $F = 100$ so reject H_0
(b) $F = 7.5$ so reject H_0

13.21 (a) $E^2 = .75$; $F = 38.33$ so reject H_0
(b) $F = 12.08$ so reject H_0

13.23 (a) $E^2 = .5$; $F = 11.43$ so reject H_0
(b) $F = 11.11$ so reject H_0

13.25 (a) $\tilde{y}_i = .4x_i$
(b) $s_{\beta_1} = .096$; $t = 4.17$ so reject H_0
(c) $r = .67$

13.27 $z = -1.80$ so accept H_0

13.29 (a) $\tilde{y}_i = 3.62 + 1.81x_i$
$\tilde{x}_i = -1.73 + 0.51y_i$
(b) $r = .961$ (c) $.92 \le \rho \le .98$

13.31 (a) $y = ce^{bx}$ (b) $y = bx^c + a$

(c) $y = \dfrac{x}{a + bx} + c$ (d) $y = ck^{ax}$

(e) $y = be^{ax} + k$ (f) $y = \dfrac{x}{c + kx}$

(g) $y = \dfrac{x}{c + kx}$ (h) $y = bx^c + k$

13.33 $\tilde{y}_i = 1.504x_i^{3.058}$

13.35 $\tilde{y}_i = 537.001^{-0.0297x_i}$

13.37 (a) errors heterogeneous
(b) errors autocorrelated
(c) errors randomly distributed
(d) errors heterogeneous and auto-correlated

13.39 $\varepsilon_t = \rho \varepsilon_{t-1} + u_t$
$H_0: \rho = 0$
$H_A: \rho \ne 0$
$d = 3.24$ so reject H_0

13.41 $\tilde{y}_i = \dfrac{x_i}{0.023 + 0.006x_i}$

13.43 $\tilde{y}_i = \dfrac{x_i}{0.662 + 0.006x_i} + 27$

CHAPTER 14

14.1 (a) $\tilde{y}_i = 6.686 - .125x_1 + 2.118x_2$
(b) $\tilde{y}_i = 6.374$
(c) $s_{y \cdot 12}^2 = 83.75$
(d) $R_{y \cdot 12}^2 = .1625$; $r_{y1 \cdot 2}^2 = .003$

14.3 $r_{12 \cdot 3} = .76$; $r_{13 \cdot 2} = -.74$

14.5 $r_{12 \cdot 3} = -.14$; $r_{13 \cdot 2} = .58$

14.7 $r_{12 \cdot 3} = .74$; $r_{13 \cdot 2} = .67$; $r_{23 \cdot 1} = -.25$

14.9 (a) $F = 3.5$ (b) $R^2 = .25$

14.11 (a) $R^2 = .25$ (b) $r_{12} = -.80$;
(c) $R^2 = .361$ $r_{y1 \cdot 2} = .042$

14.13 (a) 2, -2, 4, -3, 2, -2, 1, -3, 3, -1, 4, -4, 3, -1, 2
(b) $\varepsilon_t = \rho \varepsilon_{t-1} + u_t$; $H_0: \rho = 0$; $H_A: \rho \ne 0$; $d = 3.57$ so accept H_0

14.15 (a) -2, -1, 1, 1, 1, 2, -1, -1, 1, -2, 3, 1, 0, -1, -1, 1, 1, 1, -2, -2
(b) $\varepsilon_t = \rho \varepsilon_{t-1} + u_t$; $H_0: \rho = 0$; $H_A: \rho > 0$; $d = 1.71$ so accept H_0

14.17 $\tilde{y}_i = -2 + 3x_{1i} - x_{2i}$

CHAPTER 15

15.1 $0.417

15.3 $7.00

15.5 $p = \frac{1}{3}$

15.7 should serve pop; EOL = $175

15.9 should close temporarily; EOL = $1000

15.11 (a) $p_1 = 0, p_2 = 0, p_3 = \frac{2}{3}, p_4 = \frac{1}{3}$
$q_1 = \frac{3}{7}, q_2 = 0, q_3 = \frac{4}{7}, q_4 = 0$

(b) $p_1 = 1, p_2 = 0, p_3 = 0$
$q_1 = 1, q_2 = 0, q_3 = 0$

(c) $p_1 = 0, p_2 = 0, p_3 = 0, p_4 = \frac{5}{19}$,
$p_5 = \frac{14}{19}$
$q_1 = 0, q_2 = 0, q_3 = \frac{18}{19}, q_4 = \frac{1}{19}$

(d) $p_1 = \frac{8}{17}, p_2 = 0, p_3 = \frac{9}{17}, p_4 = 0$,
$p_5 = 0$
$q_1 = \frac{5}{17}, q_2 = 0, q_3 = \frac{12}{17}$

15.13 Process 3

15.15 $p_1 = 0, p_2 = \frac{15}{26}, p_3 = \frac{11}{26}$
$q_1 = \frac{15}{26}, q_2 = 0, q_3 = \frac{11}{26}$
Player 1 would win approximately $1.35 from Player 2

15.17 (a) $P(0) = \frac{1}{4}, P(1) = \frac{1}{2}, P(2) = \frac{1}{4}$

(b) should put money into management

(c) up to $300

15.19 (a) $p_1 = 0, p_2 = 0, p_3 = 0, p_4 = 0$,
$p_5 = 1, p_6 = 0$
$q_1 = 0, q_2 = 1, q_3 = 0, q_4 = 0$
$v = 8$

(b) $p_1 = 0, p_2 = \frac{1}{3}, p_3 = 0, p_4 = \frac{2}{3}$
$q_1 = 0, q_2 = 0, q_3 = \frac{1}{5}, q_4 = \frac{4}{5}$
$v = 2$

(c) $p_1 = 1, p_2 = 0$
$q_1 = q_2 = q_3 = q_4 = q_5 = q_6 = q_7 = 0, q_8 = 1$
$v = -8$

15.21 Choose bet 4; EOL = $11.25

15.23 (a) Multiple-choice test
(b) Essay test

15.25 (a) Bet Las with $p = \frac{3}{8}$, Vegas with $p = \frac{5}{8}$
(b) $p > \frac{5}{8}$

CHAPTER 16

16.1 $P(A|S) = .340; P(B|S) = .574;$
$P(C|S) = .085$

16.3 20.66

16.5 (a) .098, .472, .430
(b) .157, .636, .207

16.7 (a) Sell as $\frac{3}{16}$ (b) Sell as $\frac{1}{4}$

16.9 (a) Choose device A
(b) Choose device C

16.11 (a) Buy II (b) Buy I

CHAPTER 17

17.1 $K_{max} = .3830$ so accept H_0

17.3 $r_s = .771$ so reject H_0; $s = 9$ so accept H_0

17.5 $\chi^2 = 1.07$ so accept H_0; $z = -0.45$ so accept H_0

17.7 $r_s = .442$ so reject H_0 (did not decline)
$s = 46$ so reject H_0 (did not decline)
$a = 7 > b = 3$ so accept H_0 (did not decline)

17.9 $r_s = .943$ so reject H_0; $s = 87$ so reject H_0; $b = 0$ so reject H_0

INDEX